Engineering Geology for Society
and Territory – Volume 6

Giorgio Lollino · Daniele Giordan
Kurosch Thuro · Carlos Carranza-Torres
Faquan Wu · Paul Marinos
Carlos Delgado
Editors

Engineering Geology for Society and Territory – Volume 6

Applied Geology for Major Engineering Projects

Part I

Springer

Editors

Giorgio Lollino
Daniele Giordan
Institute for Geo-Hydrological Protection
National Research Council (CNR)
Turin
Italy

Kurosch Thuro
Department of Engineering Geology
Technical University of Munich
Munich
Germany

Carlos Carranza-Torres
Department of Civil Engineering
University of Minnesota Duluth
Duluth, MN
USA

Faquan Wu
Institute of Geology and Geophysics
Chinese Academy of Sciences
Beijing
China

Paul Marinos
Geotechnical Engineering Department
National Technical University of Athens
Athens
Greece

Carlos Delgado
Escuela Universitaria de Ingeniería Técnica
   de Obras Públicas
Universidad Politécnica de Madrid
Madrid
Spain

ISBN 978-3-319-09059-7   ISBN 978-3-319-09060-3   (eBook)
DOI 10.1007/978-3-319-09060-3
Springer Cham Heidelberg New York Dordrecht London

Library of Congress Control Number: 2014946956

Cover Illustration: Pont Ventoux, Val di Susa, north western Italy. Tunnel Boring Machine(TBM) used during the construction of the gallery used as deviation channel for a hydroelectric power plant. The TBM was used to drill a gallery of 4.3 km long and with circular diameter of 4.05 meters. *Photo*: Giorgio Lollino.

Printed on acid-free paper

Springer is part of Springer Science+Business Media (www.springer.com)

Foreword

It is our pleasure to present this volume as part of the book series on the Proceedings of the XII International IAEG Congress, Torino 2014.

For the 50th anniversary, the Congress collected contributions relevant to all themes where the IAEG members were involved, both in the research field and in professional activities.

Each volume is related to a specific topic, including:

1. Climate Change and Engineering Geology;
2. Landslide Processes;
3. River Basins, Reservoir Sedimentation and Water Resources;
4. Marine and Coastal Processes;
5. Urban Geology, Sustainable Planning and Landscape Exploitation;
6. Applied Geology for Major Engineering Projects;
7. Education, Professional Ethics and Public Recognition of Engineering Geology;
8. Preservation of Cultural Heritage.

The book series aims at constituting a milestone for our association, and a bridge for the development and challenges of Engineering Geology towards the future.

This ambition stimulated numerous conveners, who committed themselves to collect a large number of contributions from all parts of the world, and to select the best papers through two review stages. To highlight the work done by the conveners, the table of contents of the volumes maintains the structure of the sessions of the Congress.

The lectures delivered by prominent scientists, as well as the contributions of authors, have explored several questions ranging from scientific to economic aspects, from professional applications to ethical issues, which all have a possible impact on society and territory.

This volume testifies the evolution of engineering geology during the last 50 years, and summarizes the recent results. We hope that you will be able to find stimulating contributions, which will support your research or professional activities.

Giorgio Lollino   Carlos Delgado

Preface

Engineering geology, a relatively young field, emerged through recognition of the need for geologic input into engineering projects. Today, this primary field has expanded as the statutes of its learned society, the IAEG, define: "Engineering geology is the science devoted to the investigation, study and solution of the engineering and environmental problems which may arise as the result of the interaction between geology and the works and activities of man as well as to the development of measures for prevention or remediation of geological hazards."

The role of engineering geology for major engineering projects and infrastructure construction is well represented in the papers included in this volume of the proceeding of the 12th IAEG congress, devoted to major engineering projects. The geologic input is not only confined to the initial stage of such projects but the contribution of engineering geology includes all stages for their completion, reflecting the present standing of engineering geology in geotechnical engineering.

A retrospective review of the development of engineering geology shows that in the early days, up to the 1950s or even the 1960s, what was understood as engineering geology was restricted to assessments, with general and qualitative engineering descriptions. Then this is followed by a second period of development until about the 1980s. The demands of the development of society required more knowledge for the behaviour of the ground. Now meaningful geological models could be provided. However, the quantitative component was weak, and contributions to the design of structures were limited. Although improved, the understanding of geology in the engineering milieu is not satisfactory. A third period starts from the 1980s but mainly from the 1990s. Engineering geology, keeping the core values so far developed, is now evolving towards geoengineering.

Indeed, today engineering geology not only offers services but is also a substantial and an integral component of geotechnical engineering in construction. It is present in all phases of investigation, design and construction:

1. Engineering geology defines the geological conditions, provides the geological model (formations, tectonics and structure), and translates it into engineering terms, providing suitable ground profiles at the appropriate scale. Its role is decisive for detecting the presence of geological hazards, in the selection of the site or the alignment of the engineering structure and for the basic principles of the construction method. It makes no sense to proceed without a sound knowledge of the geological model. Let us be a little dogmatic here: in the absence or misinterpretation of the geological model the construction or operation will almost certainly be associated with problems either small or large, as accidents, delays, cost over-runs or even failures may occur. On the contrary, if this model is known from the very beginning of the design, half the game has already been won ... *if at the very start the geological structure of the site is misinterpreted, then any subsequent ... calculation may be so much labour in vain.* (Glossop 1968, 8th Rankine Lecture). Therefore: start from the forest and then look at the trees.

2. After having understood the behaviour of the ground, engineering geology contributes to the definition of the properties of the geometrical, the selection of suitable design parametres and of the appropriate criteria. This a stage with a close synergy with engineering. An understanding of in situ stresses and groundwater conditions complete this stage.

3. Engineering geology is and should also be present at the design phase to ensure that calculations and simulations do not misinterpret the geological reality. John Knill in his first Hans Cloos lecture, in 2002, expressed strong concern that the *effectiveness of the integration of engineering geology within the geotechnical engineering remains to be improved*. This integration is a field of development in today's engineering geology, and papers in this volume contribute towards such advance.

4. Engineering geology is involved in construction in order to validate the assumptions of the design, to contribute in the application of measures in unforeseen or unforeseeable circumstances and to secure the implementation of the contract.

And, undoubtedly, geological and engineering judgement should never be neglected in this whole process of creating an engineering project. Next to knowledge, experience is needed for this judgment. Mark Twain said *Good judgment comes from experience. But where does experience come from? Experience comes from bad judgment.* However, the correct application of geological and engineering principles means that experience can also come from good judgement.

It is very satisfactory that this volume of proceedings of the 12th congress of IAEG embraces all the above mentioned, and a large variety of cases of engineering works is presented. Dams and tunnels are the majority of these cases but also foundations, offshore structures, roads, railroads, slope design, construction material, tailings, repositories are dealt with. Papers on engineering properties and geotechnical classifications, site investigation issues and influence of groundwater are present together with contributions on the behaviour of soft rocks and weak rock masses. Active tectonics also attract special attention.

The volume is expected to constitute a valuable and lasting source of reference in the field of engineering geology, in particular, and in geotechnical engineering, in general.

Contents

Part VII   Engineering Geological Problems in Deep Seated Tunnels

Part XVIII   Uncertainty and Risk in Engineering Geology

Consiglio Nazionale delle Ricerche
Istituto di Ricerca per la
Protezione Idrogeologica

The Istituto di Ricerca per la Protezione Idrogeologica (IRPI), of the Italian Consiglio Nazionale delle Ricerche (CNR), designs and executes research, technical and development activities in the vast and variegated field of natural hazards, vulnerability assessment and geo-risk mitigation.

We study all geo-hydrological hazards, including floods, landslides, erosion processes, subsidence, droughts, and hazards in coastal and mountain areas. We investigate the availability and quality of water, the exploitation of geo-resources, and the disposal of wastes. We research the expected impact of climatic and environmental changes on geo-hazards and geo-resources, and we contribute to the design of sustainable adaptation strategies. Our outreach activities contribute to educate and inform on geo-hazards and their consequences in Italy.

We conduct our research and technical activities at various geographical and temporal scales, and in different physiographic and climatic regions, in Italy, in Europe, and in the World. Our scientific objective is the production of new knowledge about potentially dangerous natural phenomena, and their interactions with the natural and the human environment. We develop products, services, technologies and tools for the advanced, timely and accurate

detection and monitoring of geo-hazards, for the assessment of geo-risks, and for the design and the implementation of sustainable strategies for risk reduction and adaptation. We are 100 dedicated scientists, technicians and administrative staff operating in five centres located in Perugia (headquarter), Bari, Cosenza, Padova and Torino. Our network of labs and expertizes is a recognized Centre of Competence on geo-hydrological hazards and risks for the Italian Civil Protection Department, an Office of the Prime Minister.

Problems in Buildings and Public Works Derived from Soils with Unesteable Structure and Soils with Large Volume Instability

Carlos Delgado Alonso-Martirena

1.1 Introduction

It is customary to analyze the behavior of foundations under the implicit assumption that the ground will suffer deformations and settle as a result of the increasing load transmitted by a construction in progress. Usually an estimation has to be made of the final movements to be expected in the construction (whether a building or a public work) in the short or in the long term.

Unfortunately, certain soils may exhibit large volume changes upon variations in moisture content, even while under constant external load. Since this affects completed structures under service (closed structures) and the resulting foundation movements progress rapidly, noticeable cracking occurs. If no remedial action is adopted, the pathology is prone to increase because the changes in moisture content occur seasonally.

The initial purpose of this communication is to identify two main groups of soils largely affected by moisture contents:

- Soils with metastable structure, more usually known as "collapsible" or "collapsing" soils.
- Soils with large volume instability. They are usually referred to as "swelling soils". However, this terminology may lead to error. In fact, swelling may be reduced to unharmful values by adequate contact pressures in foundations, that is to say, by adequately dimensioning the foundations. In arid regions, though during large periods of drought, the moisture content of high plasticity clays decrease to depths, below the round surface, that may reach 6 m and even more. The resulting "shrinkage" settlements are large and of rapid progress, as pointed out for collapsing soils, although their source is different. Collapse occurs though moisture increase, shrinkage is produced by moisture decrease. Engineering geology

must be well aware of those problems with moisture contact changes, and has to evaluate them properly through laboratory testing in order to come to adequate recommendations in the project geotechnical report, as it will be synthetically referred to subsequently. A final purpose of the communication will be to draw attention to the possibilities offered by the technique of reinforced grouting/through hydraulic fracturing of ground, using stable mixes of cement-bentonite) to stabilize those problematic soils, under foundations defectively designed and constructed.

1.2 Laboratory Identification of "Collapsing" and "Swelling" Soils

Figure 1.1, reproduced from Peck et al. (1974), shows the behaviors, in a double oedometer (or paired confined compression) test, characterizing those two types of peculiar soils.

As it is usual, the oedometric curves are represented in a semi logarithmic plot, where pressures are figured in log scale, as abscissae, and void ratios in arithmetic scale, as ordinates.

Curve **a** corresponds to a test started at the natural water content, and to which no water is allowed to access.

Curves **b** and **c** correspond to tests on samples to which water is added from the start, as it is normal practice in oedometer tests.

If curves **a** and **b** represent the behavior of the same soil (curve **b** lies entirely below curve **a**) the soil is said to have collapsed. Under field conditions, at effective presume p, the soil would exhibit void ratio e_0 being unsaturated (positive suction). The increase of water content until saturation would cause de void ratio, at pressure p1, to decrease to e1, with corresponding unit settlement of —— occurring suddenly, under the form of a collapse.

On the other hand, if the soil shows a behavior represented by curves **a** and **c** lying entirely above curve **a**) the soil is said to swell. The addition of water, under pressure p_1

C. Delgado Alonso-Martirena (✉)
Civil Engineering School, Technical University of Madrid
(U.P.M.), Alfonso XII 3 y 5, 28014, Madrid, Spain
e-mail: carlos.delgado@upm.es

G. Lollino et al. (eds.), *Engineering Geology for Society and Territory – Volume 6,*
DOI: 10.1007/978-3-319-09060-3_1, © Springer International Publishing Switzerland 2015

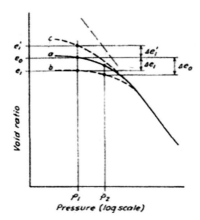

Fig. 1.1 Diagrams obtained in double oedometer tests

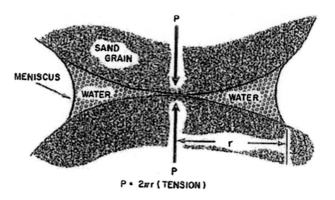

Fig. 1.2 Capillary produced contact pressure

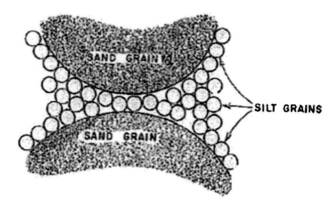

Fig. 1.3 Schematic arrangement of silt grains between two sand grains

would increase void ratio e_0 to e'_1, corresponding to a unit swelling of ——.

1.3 Essential Features and Locations of Collapsing Soils

Dudley (1970) published a valuable "Review of Collapsing Soils", which includes an extensive bibliography.

For collapse to occur the soil must have a structure that leads itself to this phenomenon. All cases reported by Dudley have a honeycomb structure of bulky shaped grains, with the grains held in place by some material or force.

In many cases the temporary strength is due to capillary tension. As the soil dries below the shrinkage limit, the water withdraws into narrow spaces close to the junction of the soil grains as shown in Fig. 1.2. In the expression for effective stress $\sigma = \sigma - u$, the excess pore water pressure **u** become negative. The actual effective stress becomes longer than the total stress applied by the load. For unsaturated silts, like the constituting loess materials (particles ranging in diameter from 0.02 to 0.002 mm), the effective stresses may be in the range of 0.35–3.5 kg/cm². For fine sands (beach sands) the maximum effective intergranular stress due to moisture films is in the order of 0.14 kg/cm².

The collapsing material may consist of sand with some silt binder. In this case the capillary forces apply around the silt to silt contacts and the silt to sand contacts as shown in Fig. 1.3.

When the bulky grains are bound by clay, the history of the soil formation is important, since a variety of arrangements are possible. The clay may be either formed in place or transported.

Clays may form in place by water acting on feldspars. One of the arrangements that would be produced is shown in Fig. 1.4. The figure shows a close-packed parallel arrangement of clay particles. Under desiccated conditions the

strength may be considerable. The addition of water would cause the clay particles to separate to some extent, thereby producing a loss of strength.

Fig. 1.4 Schematic arrangement of sand grains, with aggregated clay particles

In areas of high rainfall much of the clay formed in place could be leached out, but when the rainfall is small, if the clay particles were dispersed in the pore fluid, the situation shown in Fig. 1.5 could develop. The resulting buttresses support and hold together the bulky grains large capillary tensions can also be present. When water is added the capillary tensions would be relieved and the ion concentration in the fluid would be reduced. This would increase the repulsive force existing between particles, as shown in Fig. 1.6.

However Warkentin and Yong (1900) found that, al constant void ratios, both kaolinite and montmorillonite had higher strength at lesser salt concentrations. It may occur that the void ratios and the temperature change when the salt concentration changes in situ the resistance to consolidation caused by the presence of the clay buttresses is a function of salt concentration, void ratio within the clay structure, and probably temperature. Many clays, a a matter of fact, expand as they cool in the vicinity of room temperature.

In the case of mudflows where the initial water content is not much more than required to attain a fluid condition, the ion concentration is probably high, and even the constant shearing action while in movement cannot maintain a dispersed arrangement. The clay particles would tend to cluster around the bulky grains in a flocculated structure. As drying progresses, some of the clay may be caught between the bulky grains, and other portions of the clay could be drawn into the narrow wedges adjacent to the bulky grains, as shown in Fig. 1.7.

Due to the variety of soil arrangement prone to collapse shown in Figs. 1.2, 1.3, 1.4, 1.5 and 1.7, those soils are present in extensive areas of the world. They have been associated with regions of moisture deficiency in any continent. Soils exhibiting this behavior must have an open structure, like those of aeolian origin, but as it has been shown, their origin may be also alluvial or even residual. Brink and Kantey (1961) described residual decomposed

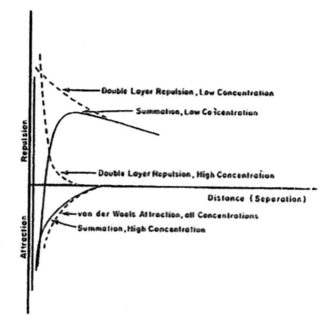

Fig. 1.6 Repulsive and attractive energy al high and low ion concentrations

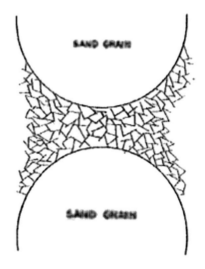

Fig. 1.7 Mud flow type of flocculated clay between two sand grains

granites near Cape Town, in Swaziland, in Northern Rhodesia and in Northern Transvaal that were found to collapse.

1.4 Essential Features and Locations of Swelling Soils

The potential capacity of a clayey soil to swell, as a result of water content increase, under moderate and constant unit load, is usually related to its plasticity index.

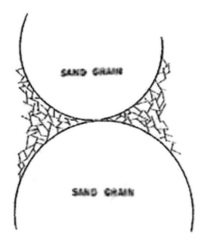

Fig. 1.5 Schematic arrangement of two sand grains with ring aggregation of clay

Peck et al. (1974) propose the following table:

Swelling potential	Plasticity index
Low	0–15
Medium	10–35
High	20–55
Very high	>35

However, the possibility, for a soil with a high swelling potential, to develop effectively this potential depends upon several conditions:

- The main condition is the difference existing between the soil moisture at the beginning of the construction, and the equilibrium moisture content of the soil, after the construction has been completed. Should this difference be high, that is to say, if the water content of the soil increases noticeably after construction, an important expansion of the unloaded or lightly loaded soil will occur. On the other hand, if the expansion is completely restricted, high and destructive swelling pressures would develop.
  Conversely, if the equilibrium water content were less than the initial water content, the soil would shrink instead of expanding.

- A second condition is the degree of compaction of the soil. The presence of a soil is placed at a high degree of compaction, or on a natural soil highly over consolidated, favor the expansion of a potentially high swelling soil if its water content is increased.

- A third condition is the unit load transmitted by the foundation once the construction has been completed. The higher the unit load, the lesser the swelling.

According to Peck et al. (1974) useful quantitative information may be obtained from the variation of swelling pressure obtained in an oedometer test in which example, with the water content corresponding to the start of the construction process, is subjected to a selected vertical pressure, and previously to let free access to water, is allowed to reach the equilibrium under that vertical pressure.

The selected vertical pressure is taking to be approximately twice the effective stress acting on the sample in the ground, before extraction. If a high swelling potential is expected, the selected vertical pressure may be taken even higher. Once the oedometer has been saturated the swelling is measured until equilibrium is reached. The vertical pressure is then reduced to half its initial value and, again, the equilibrium deformation is obtained. A new reduction to half is introduced in the vertical pressure, and the equilibrium deformation is registered. The process ends for a vertical pressure reduced to zero.

The results obtained are represented by a graph similar to that shown in Fig. 1.8.

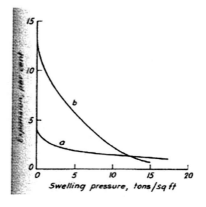

Fig. 1.8 Typical results obtained in swelling tests

- Curve **a** shows the behavior of a soil exhibiting a high swelling pressure for a reduced volume increase, but developing only a reduced swelling pressure after a moderate expansion is allowed. This type of behavior is typical of the swelling tertiary clays of the southern of Madrid (Spain).

- Curve **b** shows a more dangerous swelling behavior. Although the swelling pressure is moderate after a moderate expansion is allowed to occur, after the pressure is reduced to zero, the swelling is high.

A swelling test of this type just described is useful not only to determine the pressure necessary to avoid swelling of soil at certain depth, or to limit this swelling to an allowable value, but also to estimate the expected movements at the bottom of a basement excavation.

It is necessary to remark that, due to the disturbance of soil samples in the field and in laboratory activities, the test results are to be considered with care.

Swelling pathologies develop especially in light constructions with shallow foundations, as shown in Fig. 1.9 taken from Parcher and Means (1968).

Peck et al. (1974), in Fig. 1.10 show schematically the pathologic trends associated with a pier-supported structure in which grade beams would be in contact with a highly swelling soil.

Fig. 1.9 Pathology of brick wall with shallow foundation in swelling clay

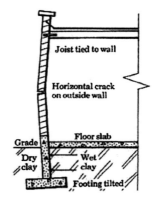

Fig. 1.10 Pathology of grade beam resting on swelling soil

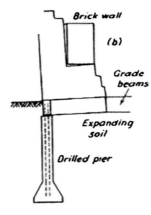

When the zone of seasonal variations in moisture extends to a depth greater than that below which drilled piers have been designed, attempts to form the bells of such piers may be unsuccessful because the clay in the zone of moisture changes likely to be highly slicken sided and successive caving of the blocks between slickensides might occur.

Even if the grade beams and floors of a structure, on piers of adequate depth, are not subjected to uplift forces (a clearance has been designed between the grade beams and the soil surface), the swelling soil tends to grip the shafts of the piers and lift them.

On the other hand, if the construction is located on a loping ground, it has to be taken into account that drilled piers have only a reduced resistance against lateral displacement and should not be expected to restrain down brill movements associated with creep.

Swelling soils are widespread all over the world. In fact they are montmorillonitic clays that might develop from the decomposition of volcanic glasses.

However although the expansion of those soils may be remediated in almost any case of construction (building or public work) by proper design measures (adequate contact pressure and depth of foundations, or soil substitution in the zone of more critical moisture changes in subsoil), in arid climates successive and important drought periods may cause the reduction of soil moisture and consequent shrinkage of soil down to important depths (6–7 in Sevilla, Spain).

An important parameter to ascertain the problem of shrinkage of soils is the shrinkage limit, very low (10–11 %) in montmorillonitic clays. If the equilibrium moisture content below a construction is considerably higher than the final moisture content after "a chain of brought periods" considerable settlements may affect the construction independently of the load it transmits to the ground.

On the other hand, the settlement distribution may not be uniform since the water content reduction may be different at the outside limits of the foundation than under the central part of the building plan extension.

As it will be established in the next sections, a very effective means of solving the problem in all types of existing constructions (new constructions with structures and foundations made of reinforced concrete or old palaces and monuments of historic or architectonic high value) is the application of the technique of soil improvement by reinforced grouting.

1.5 Grouting by Hydraulic Fracture Through Sleeve Pipes

Although grouting has been, for a long time, considered as a procedure of ground impregnation (filling the voids of the soil with grout mixes, without disturbing the preexisting particle arrangement), very soon after the technique of sleeve pipes (Fig. 1.11) was introduced, the interest of hydraulically fracturing the ground was recognized in applications for alluvial soils sealing off. The initial permeability of the ground was lowered by creating inclusions of cement-bentonite that, after setting, would cause water to divert from straight paths, with a rapid increase of head loss.

If this hydraulic fracturing is systematically introduced in the ground, through successive injection steps (Fig. 1.12), the inclusions of hardened mix produced in the ground by neighboring sleeve-pipes may interfere with one another o, creating a "skeleton" within the ground mass to be improved (Fig. 1.13).

This will promote the improvement of the ground mechanical characteristics, namely increasing strength and stiffness. Figure 1.14, initially taken from Morgenstern and Vaughan (1963), would show maximum injection pressure

Fig. 1.11 Sleeve pipes grouting

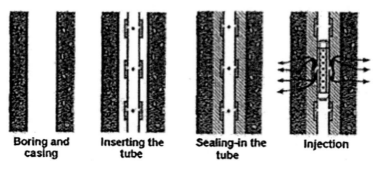

Boring and casing   Inserting the tube   Sealing-in the tube   Injection

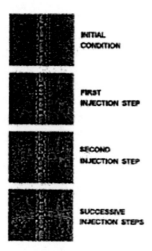

Fig. 1.12 Hydraulic fracture injection (successive steps)

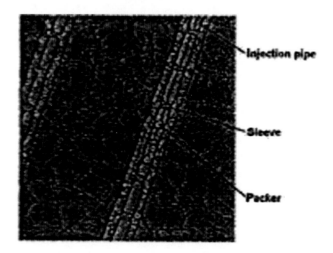

Fig. 1.13 Skeleton of hardened inclusions

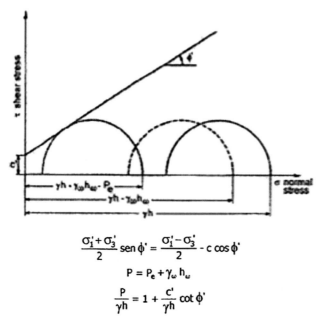

$$\frac{\sigma_1' + \sigma_3'}{2} \, sen \, \phi' = \frac{\sigma_1' - \sigma_3'}{2} - c \cos \phi'$$

$$P = P_e + \gamma_\omega h_\omega$$

$$\frac{P}{\gamma h} = 1 + \frac{c'}{\gamma h} \cot \phi'$$

Fig. 1.14 Assumed mechanism for hydraulic fracture

P, for a saturated ground, to be reached without shear fracture, c and $\varnothing'$ representing the effective shear strength parameters of the ground. However, the ground, under hydraulic pressure, cracks (rupture in tension) before reaching plastic shear, as Vaughan himself established later for cores in the earth dams.

Santos et al. (2000) and Santos and Cuellar (2000) established that in case where P would represent the final "shut-in pressure" which would be sustained in the vicinity of an open "manchette" without any further mix take variation, the final relationship shown in Fig. 1.14 would link P with the c' and $\varnothing$ parameters of the treated ground. *Therefore, a ground which could not be permeated by cement—bentonite mix, could be improved, to a predesired condition, by hydraulic fracture grouting with prefixed shut-in-pressures.*

1.6 Use of Hydraulic Fracture Grouting to Solve Foundation Problems in Collapsing and Swelling Soils

1.6.1 Negative Effects of Collapse, Swelling and Shrinking of Soils on Shallow and Deep Foundations

In relation to foundations on collapsible soils, Peck et al. (1974) established that if the possibility of wetting could not be ruled out and if the ensuing settlements were excessive, the foundation had to be established below the soil of potential collapse. Although they mentioned the possibility of inducing the collapse before building the structure, the real fact is that this procedure has been used successfully only to treat the foundations of earth dams or dikes that load completely the soil during construction and can usually tolerate settlements of more than 1 m. In connection with foundations for buildings or bridges it has not been successful.

In the case of piles or piers established in the collapsible soils, they need to be embedded adequate lengths below the bottom of the unstable foundation, and it has to be additionally taken into account that the subsequent wetting and collapse of the unstable soil are likely to induce negative skin friction on the foundation units, not to speak of the possibility of parasitic flexures on the shafts to occur, if the bottom of the collapsible foundation is sloping.

As it has been established in foregoing paragraphs, the term swelling soils imply not only the tendency to increase in volume when water is available, but also to decrease in volume, or shrink, if water is removed.

Although adequately determined permanent contact pressures on swelling soils and the design of the floor structures with proper clearance from the surface of the expansive soil may cope with volume increase, only the location of the foundation below the critical depth of soil seasonal shrinkage could face the potential volume decrease. In the arid regions, in the presence of highly plastic clays or marls (as it occurs, for instance in southern Spain) this depth may be in excess of 6–7 m.

Deep foundations, in those highly swelling soils, are subject to parasitic tensions and flexures, and their construction is complicated because those materials, of high resistance for water contents below the plastic limit, may dramatically lose strength by water addition, because their extremely high suction potential (values of nearly 100 kg/cm^2 at 98 % degree of saturation have been measured at the Centro de Estudios y Experimentacion de Obras Publicas, CEDEX, in marls of southeastern Spain with 250 kg/cm^2 unconfined compressive strength).

1.6.2 Stabilization of Collapsing and Swelling Soils and Soft Rocks Means of Hydraulic Fracture Grouting and Bolting Through Sleeve Pipes

Santos et al. (2000) presented, in the book "Geotecnia en el año 2000" (Geothechnics in the year 2000), an in-depth analysis of the basic principles, execution and control of the hydraulic fracture grouting through a sleeve pipes, along with a number of outstanding applications of this technique. Their chapter in the book was entitled "Procedure for a predetermined ground improvement compatible with milimetric movements of the surroundings".

In fact, three main aspects of the technique of fracture grouting need to be pointed out:
- It is possible to pre-establish the volume of ground to be improved the cometrical "solid" to be treated) along with its final mechanical properties, insuring, during the process, only mill metric movements (2–3 mm) of structures above or in the vicinity of the treated ground.
- Both the control of movements of the structures and the resulting mechanical conditions of the improved ground can readily and quickly be controlled.
As matter of fact, as it was verified by Santos and Cuellar (2000), testing of the ground by cross-hole geophysical technique, or by pressure meter tests show perfectly correspondent results. The comparison between the mechanical conditions of argillaceous marl, before and after the grouting could be achieved by the authors through load testing of two shallow footings (2 × 2 m and

1.50 m deep) on treated and untreated ground. The combined effects of grouting and bolting through sleeve-pipes resulted in multiplying by a factor of 10 the initial, already high, modulus of deformation of the marl.

- As long as the sleeve-pipes are properly distributed within the geometrical volume to be treated, their orientation may vary, that is to say, it is possible to achieve the drilling and grouting from any working area around the zone to be improved. This allows, for instance, in combination with the much reduced movements generated during the treatment, to reinforce the foundation of a construction without disturbing its normal operation.

Successful stabilization of collapsing soils has been achieved below and behind existing constructions in the eastern zone of Spain (collapsible silts in the provinces of Alicante and Castellon). A special case of treating very loose salty sands was affected (and controlled by CEDEX) behind so-called "muro de cipres" (cypress wall) at the Generalife and below several foundations of the Alhambra de Granada (Spain).

In the southern Spain, light constructions (some of them family houses and schools) have been protected, from the pathologies created by high plasticity clayey subsoil in arid climate, by means of grouting of the soil down to a safe depth (around 7 m.) without disturbing the normal life in those constructions.

The southern wing of the Parador in Carmona (Sevilla, Spain) was about to be torn down because of its continuing movements, even after having been under primed with micro piles, and anchored, because having been built in a cliff supported by both collapsible formation (Albero) and swelling marls (Mayos del Guadalquivir) underneath. In 2005 the problems was completely solved by grouting through sleeve- pipes of both formation, and the Parador remains in perfect housing condition.

Delgado (2011) has shown that this technique is implacable to allow buildings construction on slopes in swelling soils, subject to seasonal creep. This is due to the fact that the treated ground may be dimensioned as high inertia counterforts that allow to stabilize the slope without any need for anchors or ties.

It has to be mentioned that the so-called MPSP technique (multiple packers sleeve pipes) allows the use of grouting through sleeve pipes even in fractured or fissured rock materials.

1.7 Conclusions

Many books in Soil Mechanics, an even a number of standards in Geotechnics, fail to point out that the foundations may suffer deformations and settle after the construction operations have ended of the subsoil, below the contact level of those foundations, suffer changes in water content.

This is especially important in soils with metastable structure, that may collapse under soaking, and are therefore referred to as "collapsible soils", and in soils prone to severe changes of volume with changes in water content, termed "swelling soils" although their shrinkage when water content diminishes is even more serious.

The paper presents a summary of conditions leading to collapse or swelling/shrinkage, and shows that those problematic foundations are present in many countries of the world, although generally associated to arid climates.

Finally, reference is made to the technique of hydraulic fracture grouting that may solve the problems derived from those soils under existing constructions, and even in the present of slopes or cliffs, because the geometrical volume, to be achieved, and their final mechanical conditions may be designed in advance, executed with only mill metric movements of the ground below existing foundations, and the working surfaces may be selected with large flexibility. It is possible to check the differences between the ground conditions, before and after the treatment by the geophysical "cross-hole" technique, very easy to effect.

References

Brink ABA, Kantey BA (1961) Collapsible grain structure in residual granite soils in southern Africa. In: Proceedings fifth international conference on soil mechanics and foundation engineering, pp 611–614

Delgado C (2011)

Dudley JH (1970) Review of collapsing soils. J Soil Mech Found Div. In: Proceedings of American Society of Civil Engineers, SM3, May, pp 925–947

Parcher JV, Means RE (1968) Soil mechanics and foundations. Charles E. Merrill Publishing Co, Columbus

Peck RB, Hanson WE, Thornburn TH (1974) Foundation engineering, 2nd edn. Wiley, New York

Santos A, Martinez JM, Garcia JL, Garrido C (2000) Sistema de mejora prefijada del terreno compatible con movimientos milimetricos del entorno, Libro homenaje a Jose Antonio Jimenez Salas, Geotecnia en el ano 2000, Sociedad Espanola de Mecanica del Suelo y CEDEX, Madrid, pp 217–223

Santos A, Cuellar V (2000) Mechanical improvement of argillaceous marl through cement—based reinforced grouting. In: Rathmayer EH (ed) Proceedings of the 4th international conference on grouting, soil improvement, geosystems including reinforcement, Finnish Geotechnical Society, Building Information Ltd, Helsinki

Warkentin BP, Yong RN (1900) Shear strength of montmorillonite and kaolinite related to interparticle forces. Clay sand clay minerals, (NAS-NCR), Macmillan Co., London, pp 210–218 (vol 11 of Earth Science, C.R)

Translating Geotechnical Risk in Financial Terms

Alessandro Palmieri

Abstract

Practitioners know that geotechnical uncertainty never ends until a tunnel is completed. In some cases, uncertainty extends into operation. The present note summarizes relevant project financing elements such as viability, risk allocation, and bankability. Main financial instruments for different project structures are outlined, highlighting their likely ranges of application. Two key instruments for managing project risks, the Geotechnical Baseline Report, and the Project Risk Register, are presented and their joint use illustrated. The importance of carrying over uncertainty along the entire project cycle (planning, construction, and operation) is elaborated by using a concept borrowed from the hydropower sector.

2.1 Project Sustainability

Achieving project sustainability is a pre-requisite for financing, together with project's technical and economical viability. A recurrent message is that "the project cannot be implemented because of lack of financing". While that is true in several cases, it is equally true that, in many instances, financing could be available with good project preparation and a robust financial architecture.

So what does it take to prepare a "good project"? Over the years, the threshold of environmental and social acceptability for large projects has significantly raised, and it would be very unwise to get financially involved in any operation where these aspects have not been fully addressed.

A group of international financing institutions have set out minimum requirements for a project to be financed. These principles, referred to as the "Equator Principles," were first designed in 2003 in conjunction with the International Finance Corporation (IFC—the private sector arm of the World Bank); the most recent version is dated June 2013 (for details see www.equator-principles.com).

2.2 Financial Viability

Government decision-making is based on the economic value of a project to a nation, but the financing of that project depends on its financial viability. Financial viability is the measure of the commercial strength of a project, generally assessed over a period of 15–20 years. It determines whether the project is robust enough to repay loans at commercial rates of interest even under a downside scenario, and whether it is likely to provide a sufficiently high return on equity to attract private investors.

Water infrastructure projects often fall in the gap between economic and financial viability. A project can be economically attractive and represent the preferred option when seen from a long-term national perspective, but when considered as a commercial investment it may be unable to generate adequate financial returns. Xiaolangdi Multipurpose Project represents a relevant example (Table 2.1).

The content of this paper reflects the experience of the author and, as such, does not necessarily represent policies or practices of the Salini-Impregilo Group.

A. Palmieri (✉)
Salini-Impregilo SpA, Milan, Italy

Table 2.1 Xiaolangdi multipurpose project on the yellow river (China)

Total costs US$3.5 billion, US$1 billion for resettlement. Completed 1 year ahead of schedule; cst savings 300 MUS$
Multipurpose reservoir: flood control, sedimentation management, maintaining adequate in stream flows, water supply, irrigation, hydropower replacing old, coal fuelled power plants
Economic rate of return unchanged from project appraisal (17.5–17.9 %,) but financial rate of return unsatisfactory because only energy sales accounted for. All other benefits accounted as public goods and not reflected in the financial analysis

Closing the gap between economic and financial viability requires consideration of project financing partnership.

2.3 Risk

A risk is anything that can have a negative effect on the project outcome. All risks ultimately translate into financial terms, and an investor will tend to judge her risk exposure by the amount she could lose compared with the amount she expects to gain at any particular stage in the lifecycle of the project. There are three main types of risks; mitigation measures are different for each of them.

Project specific risks, as related to contracting risks (delay and cost overruns), are very difficult to insure. Physical catastrophes like collapses and fires may, in some cases, be insurable. Insurance will generally only cover single events rather than systemic problems. Even then it may not cover the full losses; for example it might cover the cost of reinstatement but not necessarily the consequential losses resulting from the delay—and the latter can often be the larger element (Table 2.2).

In all cases, risk has a cost and risk cost depends on how risk is allocated among stakeholders. Combination of financial viability and risk assessment results in project

bankability, i.e. how attractive the project is for financial institutions.

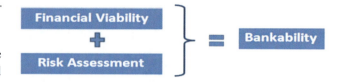

The most attractive projects rate AAA (triple A). BBB + is generally considered the minimum level for a project to generate investment interest. Bankability determines the interest rate and the tenure (duration) of the loan.

2.4 Main Financial Instruments and Project Structure

Several financial instruments, from private equity to concessionary finance, have been used for financing infrastructure projects; their choice is strongly dependent upon project structure and ownership. There is a wide spectrum of financing instruments, ranging from publicly sourced grants and soft loans through to financing on strictly commercial terms. With some generalisation it is possible to group these

Table 2.2 Project specific risks

Type of risk	Examples	Mitigation
Political (country)	Risk of nationalization	Guarantees
	Changes in law affecting status or financial position of project company	
Commercial	Market	Partially insurable
	Risk to revenue such as change in regulation or difficulties in enforcing payment	
	Defaulting off-taker	
Project	Site-specific risks such as cost and time overruns during construction	Usually not insurable
	Difficulties in obtaining necessary environmental permits and clearances	
	Uncertainty of addressing social issues which may arise	
	Hydrological risk	
	Transmission interconnection	

Table 2.3 Main financing instruments

Financing instrument	Source
Concessionary finance	Grants or soft loans (low interest or long tenure), usually from bilateral or multilateral aid agencies
Public equity	Public investment with the support of the government, often indirectly funded from bilateral and multilateral development banks (MDB) sources
Public debt	Project-specific loans from the government or from bilateral and multilateral development banks
Export credit agencies and guarantees	Finance direct from the export credit agencies, or from private commercial banks using guarantees from public MDBs
Private "commercial" debt	Loans from private banks, and from the commercial arms of the public MDBs. Also occasionally bond issues
Private equity	Direct investments made by private sponsors and other private investors, and by the public MDBs

disparate sources of finance into six broad categories of instruments (Table 2.3).

In general, the wider the gap between economic and financial viability in a project, the more that project will require concessionary and/or public finance. A financially strong project can be fully sustained by private financing.

The financial architecture of a project will depend on its viability and on the extent to which project risks can be mitigated. Project will not be attractive to the private sector if risks are not likely to be substantially mitigated. In that case, if economic value is large and the project is a national priority, financing will have to be public. A project can still attract private sector participation if one or more financially viable components can be "sliced" from the project, e.g. public financing for the dam and private for the powerhouse. The following diagram (Head 2005) exemplifies the decision-making process for assessing the appropriate project structure.

2.5 Geotechnical Risk and Project Risk Management

Geotechnical risks, in the form of unexpected geological conditions, are a serious factor in cost and schedule control on all major civil engineering projects. The amounts of money, involved in claims arising from geotechnical problems, are enormous and are taken very seriously by financing agencies. In spite of numerous attempts to deal with these situations by the incorporation of various clauses in contract documents, the problems persist. The best course of action is to define the geological conditions as early and as accurately as possible so that surprises are minimised. Unfortunately, that is not possible, or not considered possible, in most of the cases. At the same time, sharing risks associated with unpredictable events can substantially improve the success of a contract both in terms of cost and schedule control.

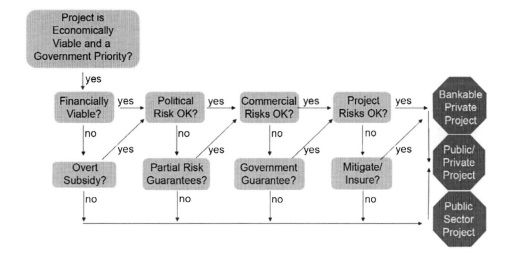

Table 2.4 Rock mass classification system

Rock class	Rock mass rating (RMR)	Percentage of excavation volume
I and II	61–80	80
III	41–60	10
IV	21–40	10

Table 2.5 Baseline statements regarding groundwater-associated trouble areas

Geotechnical feature	Baseline conditions
Peak groundwater inflows	Peak of 500 l/s with sustained inflows up to 125 l/s over a 100 m length of tunnel
Steady state groundwater inflows	100 l/s over 1,000 m
Hot water springs	At three locations during underground excavation with temperatures up to 70 °C and flow rates of 20 l/s
Groundwater pressure	About 250 m head of water at localized areas such as *creek X* and *creek Y* where there are perennial streams

Where the overall financial and contractual arrangements permit, it may be possible for the parties to agree on some form of risk sharing package; two key tools for that purpose are:

• Geotechnical Baseline Report (GBR), and
• Project Risk Register (PRR)

The GBR (ASCE 2007) aims to establish a contractual understanding of the site conditions, referred to as the geotechnical/geological baseline. Risks associated with conditions consistent with or less adverse than the baseline are allocated to the contractor and the owner accepts conditions significantly more adverse than the baseline. The more clearly defined the anticipated conditions, the more easily the encountered conditions can be evaluated. Therefore, the baseline statements shall be described using quantitative terms that can be measured and verified during construction. How the baseline has been set determines risk allocation and has a great influence on risk acceptance, bid prices, quantity of change orders and the final cost of the project.

Typical baseline conditions are those pertaining to distribution of rock types along tunnel route; they are generally expressed in terms of a rock mass classification system which has to be clearly defined in the tender documents, e.g. (Table 2.4).

The following are examples of baseline statements regarding groundwater-associated trouble areas, which are expected during construction of a tunnel (Table 2.5).

The Geotechnical Baseline Report (GBR) is a key element for the preparation of the Project Risk Register (PRR). The latter covers also risk elements such as design, technical/technological, labor, health and safety, etc. Several risk scenarios are identified in each category. For each scenario, the following elements are assessed:

• Frequency or probability of occurrence (as appropriate),
• Preventive measures,
• Potential consequences, before remedial measures,

• Remedial measures along with associated resources and costs,
• Schedule and cost consequences after remedial measures. Jointly, GBR and PRR allow to:
• inform decision making on the most appropriate project technology and procurement strategy,
• inform contract documents preparation, and allocate contingency funds,
• prepare Health and Safety Management Plans to be implemented during construction,
• manage design variations and associated claims during construction.

Not all project developers/owners have the same attitude towards such a transparent approach. Many still believe in the possibility of loading all the risks on the contractor, possibly with a turn-key, fixed cost, contractual arrangement. Experience has repeatedly proven that such expectation is, at best, very optimistic and, in reality, almost impossible to achieve. Unreasonable risk allocation strategies will keep good bidders away and attract entities who are ready to take advantage of the situations with pre-defined claims at the bid preparation stage. A review (The World Bank 1996) of water infrastructure projects, featuring important underground works, revealed significant schedule and cost overruns. It is the author belief that a large part of those overruns can be attributed to the contractual practices in use at the time of those projects (late 70' and 80'); recent practices, increasingly incorporating GBR and PRR elements, have proved to be conducive to better results.

2.6   Uncertainty Management

Large civil engineering projects like dams, hydropower schemes, tunnels, underground caverns, etc. inevitably involve significant uncertainties that translate into financial and other types of risks. As much as risks cannot be totally removed, so uncertainties cannot be cancelled regardless of the amount of studies, investigations, contractual arrangements, financial engineering, etc.

The best way to manage uncertainties is to the carry them over along the planning process by periodically re-assessing the relative implications on safety, engineering, and financial

aspects. At the pre-feasibility level of a project, uncertainties should be used to carefully plan studies and investigations for feasibility purposes.

Once feasibility is confirmed, residual uncertainties, including those that, meanwhile, have added to the list, should guide definition of contingency measures, including financial ones, for tender design purposes. Construction contracts, whatever their form (traditional, turn key, concessions, etc.), should incorporate measures to address residual uncertainties. The remaining ones, after construction and commissioning, should guide the preparation of operation and maintenance plans.

In a paper on hydro plant rehabilitation, Gummer and Obermoser (2008), refer to the concept of "unknown unknowns" (uK-uKs), which the US politician Donald Rumsfeld used in one summary of progress in Iraq (2002). They argue that the "uK-uKs" concept makes a lot of sense in apportioning contractual risk in hydro plant rehabilitation works. The concept is equally suitable in tunneling projects. The following plate exemplifies the "uK-uKs" concept in a tunneling context (Table 2.6).

"Known knowns" should be dealt with by a good design based on an adequate site investigation (Hoek and Palmieri 1998).

"Known unknowns" should be mitigated by appropriate contractual architecture; to that end it is advantageous to build sufficient flexibility into the contract so that design can be adapted during construction according to rock mass properties actually encountered. Such refinements can be based on back-analysis of measurements of excavation deformation and observations of excavation behavior. The following plate outlines an example of such approach, referring to tunneling in squeezing rock.

Convergence-based rock mass reinforcement

- GBR will specify expected baseline deformations δ values for different rock mass conditions.
- If δ values, as measured 2D away from the face, exceed the baseline value, additional support, pre-established in GBR, is installed.

Table 2.6 "uK-uKs" concept in a tunneling context

"uK-uKs"	Tunneling context	What to do
"Known knowns"	General geology, overburden, expected rock types, groundwater, etc	The problems lie in the detail, i.e. adequate site investigations at the planning stage
"Known unknowns"	Actual distribution of rock types along tunnel alignment, extent of fault areas, sudden water inflows, etc	Make adequate resources available, and provide for contractual flexibility
"Unknown unknowns"	Un-anticipated extensive fault area, large karst cavity with water and debris filling, mud-like soil within hard rock, etc	Make provision for investigations during construction (probe drilling, gas detection, etc.)

- Should excessive deformation be attributable to excavation (e.g. poor blasting), or excessive time lag in installing supports by the Contractor, the latter will bear the cost of additional support.
- In areas where baseline δ exceeds 1/3 of final lining's thickness, excavation diameter will be increased by δ.

A contract that imposes rigid designs and inflexible construction methods will almost certainly result in an inefficient and costly tunneling project.

"Unknown unknowns" can be minimised if investigation is embedded in the construction stage. A very important element in this respect is the stipulation, in contract documents, of mandatory probe drilling ahead of the tunnel face, at least in the stretches where the most problematic conditions are expected to occur. Comprehensive plotting of forecasting data and preparation of performance and geological forecasting report are also recommended.

Finally, residual uncertainties, after construction completion, should be incorporated in the Operation and Maintenance Plan of the Project.

References

ASCE (2007) Geotechnical baseline report for construction—suggested guidelines. American Society of Civil Engineers, Reston, VA

Head C (2005) The financing of water infrastructure—a review of case studies. World Bank, Washington DC

Hoek E, Palmieri A (1998) Geotechnical risks on large civil engineering projects IAEG congress. Vancouver, Canada

Gummer JH, Obermoser H (2008) A new approach to defining risk in rehabilitation works. Int J Hydropower Dams (5)

The World Bank (1996) Estimating construction costs and schedules—experience with power generation projects in developing countries. World Bank technical paper number 325

Large Deformation of Tunnel in Slate-Schistose Rock

A Case Study on Muzhailing Tunnel, Lan-Yu Railway, China

Faquan Wu, Jinli Miao, Han Bao, and Jie Wu

3

Abstract

Large deformation of soft rock tunnel has been one of significant problems in the railway construction in Northwestern China. The main causes led to the problem are large depth with high crustal stresses, soft rock with strong anisotropy and quick excavation. The paper is to introduce the large deformation of the slate-schistose rock tunnel in Lan-Yu railway construction, the essential factors affecting the deformation of carbonaceous slate tunnel including geological stress, special properties of the rock and its soften mechanism caused by excavation, rock pressure and its strengthening effect; some lessons from the geological investigation and design of tunnels.

Keywords

Carbonaceous slate tunnel • Large deformation • Softening of carbonaceous slate • Strengthening of rock pressure

3.1 Background

Large deformation of soft rock tunnels has been a difficult problem in the railway construction in mountain area of Northwestern China. Lan (Lanzhou)-Yu (Chongqing) railway is one of the typical cases. Large amount of tunnels of the railway line has to slow down or stop the excavation because of the lining systems has been broken by strong deformation. A big proportion of designed supporting system of tunnels has been changed because the class of rock quality has to be reduced from the original classification. And consequently a remarkable amount of investment has to be input for the change.

F. Wu · H. Bao
Institute of Geology and Geophysics, Chinese Academy
of Sciences, Beijing 100029, China

J. Miao
China Railway Sixth Group Co. Ltd, Beijing 100036, China

J. Wu (✉)
Universidad Politecnica de Madrid, 28014 Madrid, Spain
e-mail: wj86716@hotmail.com

As well known, pararock formed from the clastic rocks of Paleozoic and Mesozoic eras are widely distributed in Northwestern China, particularly in the Qilian-Qinling area as shown in Fig. 3.1. Most of the rock shows laminar and schistose feature with significant anisotropy because slight or strong metamorphism.

On the other hand, the depth of railway tunnels going across Qilian-Qinling mountain area usually reaches several hundreds of meters. And meanwhile there is generally a certain amount of crustal stress because it is in a tectonic active belt. Geological stress along the railway line have been measure by hydraulic fracture method as Fig. 3.2 which indicates that the maximal horizontal stress varies among 10.47–33.82 MPa. Strong rock pressure is one of the driving forces for the large deformation.

The common procedure for tunnel design in China is in the following steps: engineering geological survey, classification of rock and modification based on the condition of geo-stress and groundwater in different sections of the tunnel; according to the current technical standards, design of the supporting system based on class of rock in each section. However, for lots of tunnels, it may be difficult to accurately classify the rock quality in the stage of geological survey because of the insufficient data. This will actually leads to

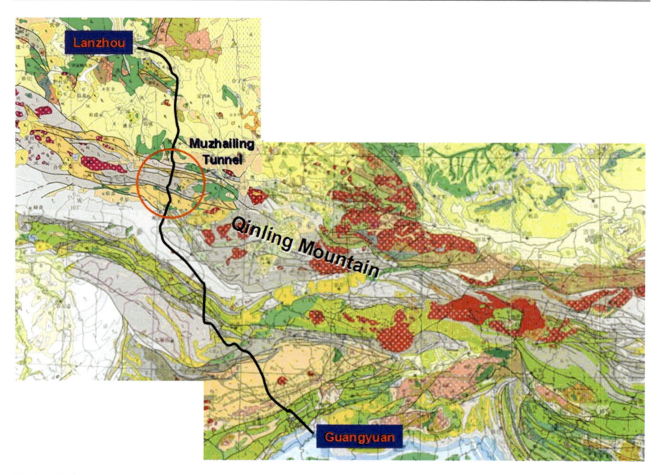

Fig. 3.1 Distribution of pararock along Lan-Yu railway line

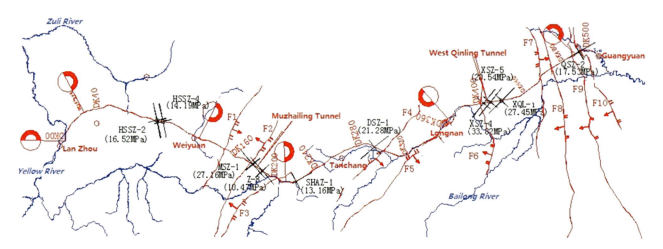

Fig. 3.2 Geological stresses measured along the railway line from China Railway First Survey and Design Inst

the inappropriate design for the lining system, for the case of Lan-Yu railway it may be weaker than required for the problematic sections of the tunnels.

Tunnels are mainly distributed in the northern part of Lan-Yu line from Lanzhou to Guangyuan as shown in Fig. 3.2. Sixty-six tunnels with length of 343 km take 70 % of the total length of the section, i.e. 493 km. Nine extremely long tunnels are the most difficult ones in the construction.

As some examples, Table 3.1 lists the data of 6 double-line tunnels in the Triassic slate which have changed the designed lining due to the reduction of rock class during the construction. Around 72.0 % of the excavated length of tunnels has strengthened their supporting system.

A series of research work has been conducted to find the causes of large deformation and explanation for the change of design and rock classes. The paper is to introduce the large deformation of slate tunnels, taking Muzhailing tunnel as an example, the phenomena, causes of deformation and lessons learnt from it.

3.2 A Case of Large Deformation at Muzhailing Tunnel

Muzhailing tunnel is located at the section from Zhang County to Min County, which is designed mainly as two single line tunnels, but partly merged as a double lines tunnel. The length of left tunnel is 19,020 m and right 19,080 m respectively. The largest depth of tunnel reaches 715 m. For speeding the excavation process, 8 inclined shafts have been opened to increase work faces. However, the excavation of the tunnel still not finished after 5 years since started in March of 2009.

The strata in the area are mainly carbonaceous slate and sand stone from Devonian to Triassic systems. The thickness of a single layer of the rock varies around 2–10 cm and connected with carbonaceous films. This makes the rock very deformable while excavated (Fig. 3.3). The quality of the rock has been classified into III to IV grade based on the

Fig. 3.3 Structure of the carbonaceous slate at Muzhailing tunnel

drilling and geophysical exploration at the geological survey stage.

Geological stress has been measured as shown in Fig. 3.2 near the tunnel and the maximal horizontal stress $\sigma_H = 27.16$ MPa at the direction of NE29 ~ 68°, a small inter-angle with the axis of the tunnel. The value of the measured stress indicates that it is in a horizontal stress state, i.e. $\sigma_1 = \sigma_H$.

The lining system of tunnel has been designed in a safer consideration comparing to the current technical standard. Table 3.2 lists the design parameters for the double-line

Table 3.1 Part of tunnels with lining changed[a]

Name of tunnel	Excavated (m)	Length of lining changed according to modification of rock class (m)							
		III to III+	III to IV	III to V	IV to IV+	IV to V	V to V+	Sum	Proportion (%)
Majiashan	3,910	40	3,049	25	558	60	0	3,732	95.4
Tongzhai	3,618	85	1,743	15	330	0	20	2,193	60.6
Qinggang	3,050	49	1,628	0	0	0	0	1,677	55.0
Xinchengzi	2,823	0	943	0	180	20	36	1,179	41.8
Maoyushan	2,863	0	2,193	50	447	10	8	2,708	94.6
Tianchiping	5,091	10	3,162	71	322	321	0	3,876	76.1
Sum	21,355	184	12,718	161	1,837	411	64	15,365	72.0

[a] From China Railway First Survey and Design Inst

Table 3.2 Parameters for the lining of double-line tunnel in IV–V class rock[a]

Class of rock	Reserved deformation space (cm)	Primary supporting system						Second lining, C35 reinforced concrete (cm)
		C30 Jet concrete (cm)	φ 22 Bolts		φ 8 Steel mesh (cm)	H150 Steel frame, spacing (m)		
			Length (m)	Spacing (m)				
IV	30	25	6.0	1.0 × 0.8	20 × 20	1.0		55 ~ 65
V	35	30	6.0–8.0	1.0 × 0.8	20 × 20	0.5		60 ~ 70

[a] From China Railway First Survey and Design Inst

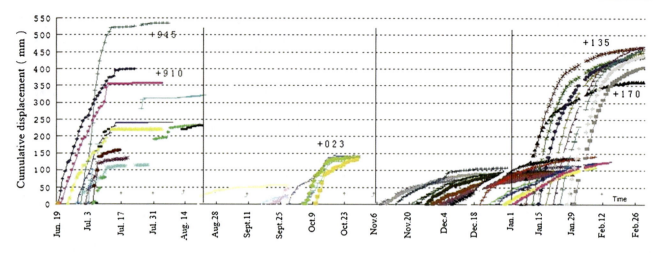

Fig. 3.4 Convergence deformation of the tunnel near Dazhangou shaft from China Railway First Survey and Design Inst

Table 3.3 Change of supporting grade[a]

	Excavated	III to III+	III to IV	IV to IV+	IV to V	V to V+	Sum	Proportion (%)
Left line (m)	7,687	125	855	3,275	5	480	4,739	61.6
Right line (m)	8,999	192	573	2,987	8	540	4,350	48.3

[a] From China Railway First Survey and Design Inst

tunnel in the IV and V class rock with high geo-stress condition. However, the deformation of tunnel still could not be controlled. Figure 3.4 shows the convergent deformation along with the excavation of the tunnel. The largest deformation could reach 530 mm in around 10 days.

Some experiences from the references (Mao and Yang 2005; Du 2011) indicate that to increase the stiffness of the lining system maybe one of the effective way for the control of the deformation of the tunnel. Four stages of tests have been conducted for different supporting measures. According to the results of the tests, 48.3–61.6 % in length of the tunnels have changed their design of linings (Table 3.3), mainly changed from grade IV to IV+ for strengthen their support systems.

3.3 Analysis on the Causes of the Deformation

Besides lots of field monitoring and tests, theoretical analysis has been done to find the reasons of the deformation of the tunnels. Two factors have been recognized as the main causes for the abnormal deformation, i.e., softening and anisotropisation of the carbonaceous slate by failure of inter-slice connection; increase effect of the rock pressure caused by softening of the rock. These two factors have not been taken into account in any of the current technical standards for tunneling, though there has been some modification for the classification of rock quality.

1. Softening and anisotropisation of carbonaceous slate

The slate is a kind of pararock composed of slices of sandstone cohered by carbonaceous films. The feature of the rock may be remarkably weakened due to the excavation of the tunnel. This is one of the reasons led to the softening of deformability and strength of the rock (Gao et al. 2011; Zhang 2010; Zhao et al. 2011; Zhao 2011).

The mechanism of the process could be explained by the diagrammatic sketch as Fig. 3.5. The slate rock is made of a series of hard slices and cemented by flaky minerals. It is hard

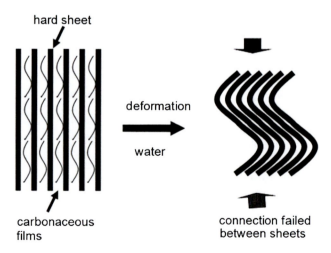

Fig. 3.5 Sketch showing the mechanism of softening and anisotropisation of slate

(a)   **(b)**

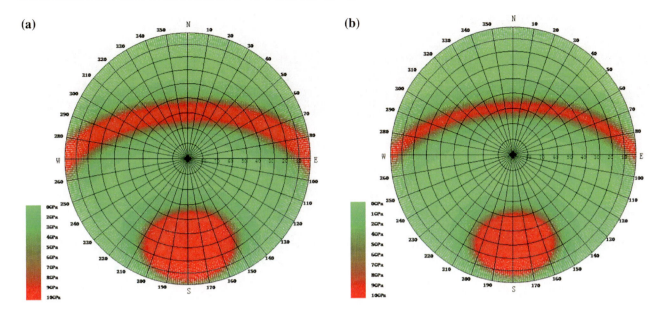

Fig. 3.6 Variation of modulus between original and disturbed state of the slate. **a** $E = 10$ GPa, $v = 0.4$, $a = 1$ m, $\varphi = 20°$, $c = 1$ MPa; **b** $E = 10$ GPa, $v = 0.4$, $a = 1$ m, $\varphi = 20°$, $c = 0$ MPa

originally like a brick, however, after excavation of tunnel, the deformation will easily break the connection between the slices. Thus the "brick" will be changed into "a stack of paper" with weaker mechanical property and notable anisotropy.

This can also be illustrated by the calculation of Yang's modulus of rock. The modulus of rock, E_m, can be calculated by the following formula Wu (1993):

$$E_m = \frac{E}{1 + \frac{32(1-v^2)}{\pi(2-v)} \lambda a h^2 \sin^2 \delta}$$

where E, v are the Yang's modulus and Poison's ratio of rock block, λ, a and δ are the density, average radius of a set of discontinuities and the inter-angle between the normal of the planes and acting stress, $h = \frac{\tau-(c+\sigma \tan \phi)}{\tau}$ is the ratio of residual shear stress on a discontinuity.

Figure 3.6 shows the calculated modulus of slate for the original state (relatively high pressure and well cohered) and after disturbed (pressure unloaded and cohesion between slices lost), where taking the parameters of rock block as $E = 10$ GPa, $v = 0.4$, the average radius and the cohesion of the discontinuities as $a = 1$ m and $c = 1$ MPa. The frictional angle of the planes has been measured in the field as $\varphi = 20°$, and after failure of the planes, the cohesion will be totally lost, i.e. $c = 0$ MPa.

The calculation has shown that the average modulus of rock before failure is 3.32 GPa, and it is reduced to 2.52 GPa after excavation, a 24.1–31.7 % reduction comparing with the average values.

2. Increase of the rock pressure

It is a common knowledge that the rock pressure to the lining system will increase while the mechanical property was weakened. It could be affirmed from the classic Kastner formula (Kastner 1951; Cai 2002)

$$p = (p_0 + c \, \cot \varphi)(1 - \sin \varphi)\left(\frac{a}{R_p}\right)^{\frac{2 \sin \varphi}{1 - \sin \varphi}} - c \, \cot \varphi$$

which infers that the rock pressure will definitely increase with the reduction of the strength of rock, i.e., c and φ. Where p_0, a and R_p are remote stress, radii of tunnel and plastic region of the surrounding rock respectively.

Numerical calculation has also shown that the rock pressure acting on the rigid lining of tunnel will significantly increase while the rock weakened. Taking a tunnel at a depth of 500 m, supported with 30 cm thick reinforced concrete lining, and the mechanical parameters are considered as $E_m = 20$ GPa, $v = 0.25$, $\varphi = 40°$, c = 12 MPa, $\sigma_t = 6$ MPa for hard rock and $E_m = 5$ GPa, $v = 0.4$, $\varphi = 30°$, c = 1 MPa, $\sigma_t = 0.5$ MPa for soft rock, Fig. 3.7 indicates that, the rock pressure for soft rock will be about two times of the hard rock.

On the other hand, the rock pressure will not only increase in quantity but also show unsymmetrical distribution due to the anisotropisation of the rock. A calculation by 3DEC has shown that the surrounding rock will move and bend inwards to the tunnel in the normal direction of beddings (Fig. 3.8a), which has been widely illustrated in practical phenomena.

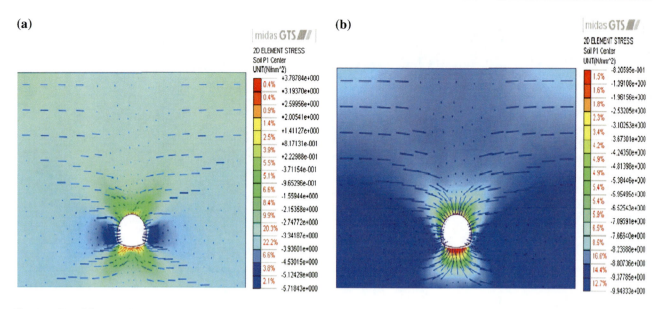

Fig. 3.7 The differences in rock pressure between hard rock and soft rock. **a** E_m = 20 GPa, v = 0.25, φ = 40°, c = 12 MPa, σ_t = 6 MPa; **b** E_m = 5 GPa, v = 0.4, φ = 30°, c = 1 MPa, σ_t = 0.5 MPa

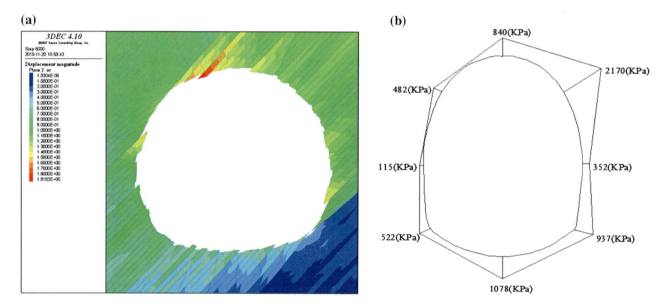

Fig. 3.8 Deformation Rock pressure of tunnel in layered rock. **a** Deformation of layered rock tunnel simulated by 3DEC; **b** Rock pressure measured at a tunnel of Lan-Yu railway

Rationally, there will be stronger rock pressure to the lining of tunnel as the deformation is restricted. A monitoring of rock pressure in Lan-Yu railway tunnel has affirmed that the contacting pressure has reached 2.17 MPa (Fig. 3.8b) which is much higher than our experiences, and the point with highest pressure located just at the place we discussed above.

No doubt, the higher the rock pressure, the stronger deformation at the case without rigid lining.

3.4 Lessons Learnt from the Case

Some lessons have to be learnt from the practice of tunneling in Lan-Yu railway construction, mainly regarding to the current technical standards.

1. The special property of slate and schistose rock with weak cement should be taken into account of the classification of rock.

The current classification takes the rock as isotropic and with invariable mechanical property without considering any change by engineering disturbance. However, the schistose rock with weak cement by flaky minerals like carbonaceous films is much easy to be broken into anisotropic and soft medium when excavated. There should be some rules to reflect the softening and anisotropisation of rock due to the change of condition.

2. The deformation pressure should be considered in the rock pressure calculation for deep buried and soft rock tunnel.

The current standards for tunnel design mainly considers the broken-rock pressure, q, by the following formula

$$q \approx \gamma h, \quad h = 0.41 \times 1.79^s$$

where γ, h, and s are the unit weight, height of broken rock and the grade of rock. Because of the worst grade of rock is $s = 6$ in China, thus $h = 0.739 \sim 13.5$ m, and $q = 18.5 \sim 337.5$ kPa.

However, we have already known that the rock pressure for deep buried and soft rock tunnel will be much higher than the upper limit mentioned above. This indicates that a big element from deformation has not been involved into the rock pressure calculation. And it is just the crucial reason why we could not predict the large deformation and failure of surrounding rock of tunnels.

3. Asymmetric supporting system should be carried out in design of tunnel lining system

Obviously, evenly distributed bolts and symmetrical supporting system are not appropriate to the deformation control for schistose rock tunnel not only because they could not match the distribution of deformation, but also they could not work efficiently. For instance, a bolt parallel to the schistose plane could never strengthen the connection of the slices, but when goes across the planes will be much more effective. This is definitely a matter need to be taken into account in the current standards.

References

Cai M (2002) Rock mechanics and engineering, Science Press, 2002 (in Chinese)

Du Y (2011) Study on large deformation mechanics and controlling technique of soft-slate rock tunnel. Central South University

Gao C, Xu J, Li Z et al (2011) Experimental study of anisotropically mechanical characteristics of sandy slate in Xuefeng mountain tunnel. Rock Soil Mech 32(5):1360–1364

Mao H, Yang C (2005) Study on effects of discontinuities on mechanical characters of slate. Chin J Rock Mech Eng 24 (20):3651–3656 (in Chinese)

Wu F (1993) Principles of statistical mechanics of rock mass. Press of China University of Geosciences, 1993 (in Chinese)

Zhang X (2010) Study on large deformation cause analysis and engineering practice of carbonaceous slate tunnel. Chang'an University, Xi'an

Zhao Z (2011) Large excavation deformation analysis and controlling technical of carbonaceous slate in Tongzhai tunnel. West-China Explor Eng 12:184–186

Zhao S, Lin A, Yan X (2011) Large deformation analysis and controlling construction technical of vertical slate rock in Hadapu tunnel. Mod Tunn Technol 48(2):46–49

Part II

Addressing Geological Uncertainties in Major Engineering Projects

Convener Dr. Clark Fenton—*Co-convener* Pedro Refinetti Martins

Large civil engineering projects, including dams, lifelines and offshore energy developments are often situated in regions of spatially complex geology. Regardless of how detailed the site investigation is, such projects will always face a degree of uncertainty as to the exact nature of the ground conditions. Engineering geology has a central role in addressing these uncertainties. Using a thorough understanding of the geological history of a site, the materials present and of the current geological processes, in addition to those which operated in the recent past and those that may affect the site during the project lifetime and even during decommissioning, Engineering Geology provides the tools to identify, quantify and manage uncertainties in ground behaviour, engineering performance and environmental impacts. This session will highlight both current and emerging approaches for addressing geological uncertainties, including conventional deterministic and more novel probabilistic methods.

Effect of Petrogenesis on the Suitability of Some Pelitic Rocks as Construction Aggregates in the Tropics

4

Tochukwu A.S. Ugwoke and Celestine O. Okogbue

Abstract

Ten rock samples collected from five rock quarry units of Albian Asu-River Group (southeastern Nigeria) were studied megascopically and subjected to XRD to assess their petrography. The samples were subjected to degradability test to simulate their resistance to repeated wetting and drying common in tropical regions while nine of them, pelitic in composition, were further subjected to abrasion test to determine their abrasion value (LAAV) and impact test to determine their impact value (AIV). The field and petrographic studies showed that rocks of varying petrogenetic origins notably; hydrothermally altered pelitic rocks and volcanic bombs, pyroclastic rock, pelitic argillites and hornfels occur in the quarry units. XRD revealed that none of the rock types has significant amount of siliceous minerals implying that none is susceptible to alkali-aggregate reaction. Results of degradability test showed that the two pelitic argillites and one of the four hydrothermally altered pelitic rocks, having percentage mass loss ranging from 1.28 to 26.76 %, showed significant deterioration implying that the three rocks are not suitable for construction of structures like embankment and unpaved roads in tropical regions because of their petrogenesis and mineralogy. Results of the LAAV and AIV tests ranged from 9.40 to 14.00 and 19.00 to 24.00 respectively indicating that all the rocks are suitable for construction of all pavements sections. In general, all these results show that mechanical degradation of rocks is not only dependent on petrography but also on petrogenesis.

Keywords

Petrogenesis • Pelitic rocks • Aggregates • Degradability test • Tropics

4.1 Introduction

Aggregates are non-renewable solid geologic materials used for construction purposes. Aggregates are either loose materials (e.g. sand and clay) or rock (e.g. igneous and metamorphic rocks). Works by Hudec (1980), George et al. (1990) and Bell (2007) had shown that suitability of rock as construction aggregate has most often being assessed based on the physical and mechanical properties of the rock with little or no attention paid on the possible influence of the rock's petrogenesis.

This work assesses the influence of petrogenesis on the suitability of some pelitic rocks as aggregate in tropical regions.

4.2 Regional Geology

The pelitic rocks studied in this work belong to the Albian Asu-River Group (southeastern Nigeria), which is the oldest lithostratigraphic unit of southern Benue Trough. Grant (1971) and Burke et al. (1971) reconstructed that the Benue Trough evolved as the third failed arm of a triple rift system

T.A.S. Ugwoke (✉) · C.O. Okogbue
Department of Geology, University of Nigeria, Nsukka, Nigeria
e-mail: tcugwoke@yahoo.com

C.O. Okogbue
e-mail: celeokogbue@yahoo.com

G. Lollino et al. (eds.), *Engineering Geology for Society and Territory – Volume 6*,
DOI: 10.1007/978-3-319-09060-3_4, © Springer International Publishing Switzerland 2015

due to separation of South American and African plate, which was associated with faulting and subsidence of the major crustal blocks. According to Olade (1975), Ofoegbu (1983) and Ojoh (1990), the southern part of Benue Trough experienced three tectonic upheavals, which were characterized by volcanic eruptions, in Aptian/Pre-Albian, Turonian and Santonian Stages. The eruptions intruded the Asu-River Group that is mostly composed of low-grade regionally metamorphosed calcareous/silty shales (Obiora and Umeji 2004; Obiora and Charan 2010).

4.3   Field Studies and Laboratory Analyses

Ten rock samples collected from five mapped rock quarry units were studied megascopically and subjected to X-ray diffraction and degradability tests. The samples were code-named following the locations from they were collected.

About 3 g each of dry pulverized sample passing through sieve 150 µm was analyzed from $0° - 70°$ 2θ scan range using Shimadzu X-ray diffractometer (XRD-6000) to generate diffractogram. Prominent peaks of the diffractogram were matched and labeled by the mineral data card software of the diffractometer from which the dominant minerals contained in the sample were identified. Degradability test involved soaking dry lumps of rock samples each weighing between 120 and 180 g (W_{dry}), in potable water of about 24 °C contained in non-corrodible can for 48 h. Each sample was removed from water, washed with fresh water and finger pressure to detect particle(s) that might have lost cohesion. Particle(s) that got detached during soaking or washing was (were) carefully picked and air-dried to constant weight (W_p) to achieve complete dryness while the intact rock was air-dried for 24 h to achieve partial drying and thereafter subjected to another cycle. The experiment was repeated for 15 cycles and the cumulative percentage of mass lost (M_{lost}) at the 15th cycle was calculated using the equation:

$$M_{lost} = \left[\frac{\sum_{n=15}^{i} W_p}{W_{dry}} \right] \times 100\%$$

Nine samples that were pelitic in composition were subjected to Los Angeles Abrasion (LAAV) and Impact value (AIV) tests following the grade A of IS: 383 and IS: 2386 (1963) standards respectively.

4.4   Results and Discussion

Table 4.1 shows the field observations, megascopic description, mineralogy, M_{lost}, LAAV and AIV of the analyzed samples. Although AGU1 and AGU2 occur at the

same location (Agu-Akpu) they are different rocks implying that there is an unconformity at that location, which was correlated with one observed at Okposi (OKP1, OKP2). From field and megascopic studies, AGU1, OKP1, ENY1 and ENY2 are the same rock type while AGU2 and EZZ are the same rock type. Hypabyssal features and co-occurrence of ONY1, ONY2 and ONY3 at Onyikwa rock units reveal that they are of same volcanic/igneous origin but their actual field occurrence had been distorted by a post Santonian tectonic event. The reconstructed petrogenesis, from field observations and petrography, is that ONY1 is a porphyroblastic hornfel that occurred between the pyroclastic rock (ONY2) and baked margin (ONY3). The implication of the above observations, when regional geology of the area is taken into consideration, is that AGU1, OKP1, ENY1 and ENY2 are hydrothermally altered pelitic rocks; AGU2, and EZZ are low-grade regionally metamorphosed pelites (argillite); OKP2 is hydrothermally altered volcanic bomb; ONY1 and ONY3 are pelitic hornfels while ONY2 is pyroclastic rock.

Table 4.1 reveals that the pelitic argillites (EZZ and AGU2), which show highest degree of deterioration (highest M_{lost}), are neither richer in water absorbent/soluble minerals nor more porous than other samples while the igneous rocks (ONY2 and OKP2) that show lowest degree of deterioration are neither more deficient in water absorbent/soluble minerals nor least in effective porosity. The pelitic hornfels (ONY1 and ONY3) contain more water absorbent/soluble minerals than the pelitic argillites (EZZ and AGU2) but do not show significant ($\geq 1\%$) deterioration.

Also, out of the four hydrothermally altered pelitic rocks (AGU1, OKP1, ENY1 and ENY2), OKP1 that is rich in water-absorbent/soluble minerals shows significant deterioration. It follows that deterioration (durability) of rocks due to repeated wetting and drying, characteristic of tropical regions, is dependent not only on the petrography but also on the petrogenesis of the rock. The implication of the degradability test results is that the three rocks, EZZ, AGU2 and OKP1, that show significant deterioration due to their petrogenesis and/or petrography, will obviously not resist repeated wetting and drying that occurs in tropical regions and are therefore not suitable aggregates for construction of structures like unpaved roads, embankment and facade in tropical regions.

Based on IRC (1970) standard, the LAAV and AIV indicate that all the rocks analyzed are suitable for constructing all sections of concrete and bituminous pavements. Contrary to general belief that all coarse-grained rocks are more prone to mechanical degradation than fine-grained ones, ONY1, which is coarse-grained, does not have higher LAAV and AIV relative to other samples that are much finer in grain size. This characteristic is attributed to the fact that ONY1, which is a pelitic hornfel, has attained significant

Table 4.1 Field occurrence, megascopic descriptions, mineralogy, cumulative percentage of mass lost, LAAV and AIV of the analyzed samples

S/n	Location/ rock unit	Sample	Field occurrence	Megascopic description	Dominant minerals and group	M_{lost} (%)	LAAV (%)	AIV (%)
1.	Enyigba	ENY1	Massive and non-fractured	Grey coloured, fine-grained and smooth surface. Spits into flaky portions when hammered	Paragonite (mica), Halloysite (kaolinite-serpentine), Pyrophyllite (talc), Chlorite-vermiculite-montmorillonite (mixed clay)	0.21	14.00	23.00
2.	Enyigba	ENY2	Massive and non-fractured	Very similar to ENY1 but less flaky	Antigorite (serpentine), Tremolite (amphibole), Lizardite (serpentine), Illite (clay), Osumillite (milarite), Hopeite (Phosphate)	0.12	13.60	24.00
3.	Okposi	OKP1	Unconform-able surface, hypabyssal	Grey coloured and micaceous, silty and sub-rough surfaced.	Truscottite, Faujasite (zeolite), paragonite (mica), Sepiolite, Illite (clay)	1.28	9.80	19.00
4.	Okposi	OKP2	Randomly enveloped in OKP1	Sub-spherical in shape, grey coloured, silty sand-grained and micaceous. Resistant to hammering.	Lizzardite (serpentine), Chrysotile (serpentine), Ferropargasite (amphibole), Muscovite (mica)	0.00	9.40	20.00
5.	Agu-Akpu	AGU1	Massive	Similar to OKP1 but not micaceous	Truscottite, Faujasite (zeolite), Talc	0.12	9.70	22.95
6.	Agu-Akpu	AGU2	No visible bedding surface	Grayish ash coloured and fine-grained (silty)	Parahopeite (oxide), Muscovite (mica), Tremolite (amphibole), Riebeckite (amphibole), Osumillite (milarite), Chrysotile (serpentine)	11.63	12.00	22.95
7.	Onyikwa	ONY1	Fractured and hypabyssal features	Porphyroblastic/ poikiloblastic and very rough surfaced	Muscovite (mica), Riebeckite (amphibole), Kaolinite (clay), Anthophyllite (amphibole), Phlogopite (mica), Dickite (kaolinite-serpentine), Illite (clay), Ferropargasite (amphibole)	0.59	11.30	21.21
8.	Onyikwa	ONY2	The same as ONY1	Pyroclastic texture and rough surfaced.	Grunerite (amphibole), Osumillite (milarite), Montmorillonite (clay), Tremolite (amphibole), Ferropargasite (amphibole), Antigorite (serpentine), Riebeckite (amphibole)	0.03	NA	NA
9.	Onyikwa	ONY3	The same as ONY1	Ash-coloured, massive, smooth surfaced and angular edges.	Osumillite (milarite), Muscovite (mica), Tremolite (amphibole), Ferropargasite (amphibole), Phlogopite (mica)	0.06	11.00	22.00
10.	Ezzamgbo	EZZ	Tilt bedded	Grayish ash coloured and fine-grained (silty)	Anthophyllite (amphibole), Tremolite (amphibole), Sepiolite, Antigorite (Serpentine), Grunerite (amphibole)	26.76	10.00	20.99

NA Not analyzed

hardening due to thermal baking and is therefore more resistant to mechanical degradation than expected. The finding implies that resistance of rock to abrasion and impact may not be solely controlled by petrography (texture) but also by the petrogenesis.

All the rocks are deficient of siliceous mineral(s) and so none of them is susceptible to alkali-silicate reaction. However, two of the four hydrothermally altered pelitic rocks (ENY1 and ENY2) will be unsuitable for structural concrete in the tropics as they are likely to be prone to stripping/popout due to their smooth surface and flaky nature. In general, only the hydrotheramally altered volcanic bombs (OKP2), pelitic hornfels (ONY1 and ONY3) and one of the hydrothermally altered pelites (AGU1) can be said to be suitable for construction of all types of civil engineering structures in the tropics.

4.5 Conclusion

This work has shown that suitability of rock as construction aggregate is not only dependent on its physical and mechanical properties but also on the petrogenesis of the rock particularly in the tropics where rocks are exposed to repeated rainfall and high temperature.

References

Bell FG (2007) Engineering geology (2nd edn). Butterworth-Heinemann, Elsevier. Linacre House, Jordan Hill, Oxford. 581 p

Burke K, Dessauvagie TFJ, Whiteman AJ (1971) Opening of the gulf of Guinea and geological history of the Benue depression and Niger Delta. Nature (Physical Science) 233:51–55

George SA, James MB, Edward WS (1990) Building with stone in Northern New Mexico. New Mexico geological survey guidebook. In: 41st field conference, Southern Sangre de-Cristo mountains, New Mexico, pp 405–416

Grant NK (1971) The South Atlantic, Benue trough and gulf of Guinea cretaceous triple junction. Bull Geol Soc Am 82:2295–2298

Hudec PP (1980) Durability of carbonate rocks as function of their thermal expansion, water sorption and mineralogy. ASTM Tech Pub 691:497–508

IRC:15 (1970) Indian Roads Congress. Tentative specification (for various types of construction methods)

IS: 2386 part IV (1963) Indian standard institution. Indian standard methods of test for aggregate for concrete

Obiora SC, Charan SN (2010) Geochemical constraints on the origin of some intrusive igneous rocks from the Lower Benue rift, Southeastern Nigeria. J Afr Earth Sci 58:197–210

Obiora SC, Umeji AC (2004) Petrographic evidence for regional burial metamorphism of sedimentary rocks in the Lower Benue rift. J Afr Earth Sci 38:269–277

Ofoegbu CO (1983) A model for the tectonic evolution of Benue Trough of Nigeria. Geol Rundschau 73:1007–1018

Ojoh K (1990) Cretaceous geodynamics evolution of the southern part of the Benue Trough (Nigeria) in the equatorial domain of the South Atlantic: stratigraphy, basin analysis and paleogeography. Bull Centers Resh Explor—Prod EIF—Aquitaine 14:419–442

Olade MA (1975) Evolution of Nigeria's Benue trough (Aulacogen). A Tectonic Model. Geol Mag 112:575–583

IS: 383, Indian standards Institution. Indian standard specification for coarse and fine aggregates from natural sources

Geological Society of London Engineering Group Working Party on Periglacial and Glacial Engineering Geology

David Giles, Martin Culshaw, Laurance Donnelly, David Evans, Mike de Freitas, James Griffiths, Sven Lukas, Christopher Martin, Anna Morley, Julian Murton, David Norbury, and Mike Winter

Abstract

In 2012 the Engineering Group of the Geological Society of London established a Working Party to undertake a state-of-the-art review on the ground conditions associated with former Quaternary periglacial and glacial environments and their materials, from an engineering geological viewpoint. The final report was not intended to define the geographic extent of former periglacial and glacial environments around the world but to concentrate on ground models that would be applicable to support the engineering geological practitioner. Key aspects of ground condition uncertainty would be addressed and developed within these ground models. The Working Party considered the following topics with respect to engineering geology: Quaternary Setting, Geomorphological Framework, Glacial Conceptual Ground Models, Periglacial Conceptual Ground Models, Engineering Materials and Hazards, Engineering Investigation and Assessment along with Design and Construction Considerations.

Keywords

Glacial • Periglacial • Ground model • Quaternary

5.1 Introduction

In 2012 the Engineering Group of the Geological Society of London established a Working Party to undertake a state-of-the-art review of ground conditions associated with former Quaternary periglacial and glacial environments and their materials, from an engineering geological viewpoint. The final report will concentrate on the development of new ground models that would be applicable to support the engineering geological practitioner, enhancing current knowledge, whilst focusing on their applicability to the engineering geologist. The ground models will be developed to communicate the complex and variable ground conditions that could be expected in these former periglaciated and glaciated terrains. The Working Party considered the following topics with respect to engineering geology: Quaternary Setting, Geomorphological Framework, Glacial Conceptual Ground Models, Periglacial Conceptual Ground Models, Engineering Materials and Hazards, Engineering Investigation and Assessment along with Design and Construction Considerations. Former glacial and periglacial settings present the engineering geologist with a complexity of vertically and laterally varying ground conditions with a high degree of uncertainty which require the use of conceptual ground models to fully understand and interpret, for example the complexities of the ice-marginal environment as conceptualized in Fig. 5.1. Such complexity can also be seen in Fig. 5.2, an example of the varying ground conditions associated with superficial valley disturbances in a former clay pit in Devon, UK.

D. Giles (✉) · M. Culshaw · L. Donnelly · D. Evans · M. de Freitas · J. Griffiths · S. Lukas · C. Martin · A. Morley · J. Murton · D. Norbury · M. Winter
Geological Society of London, Burlington House, Piccadilly, London, W1J 0BG, UK
e-mail: dave.giles@port.ac.uk

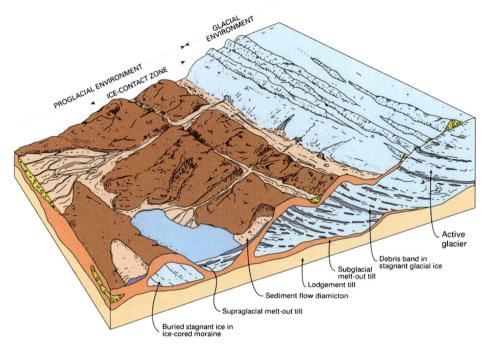

Fig. 5.1 Current supraglacial and ice-contact ground model (McMillan and Powell 1999)

Fig. 5.2 Periglacial environment: Superficial valley disturbances, Newbridge Ball Clay Quarry, Devon, UK. (Dump trucks in *bottom right* corner for scale)

5.2   Terms of Reference

The Periglacial & Glacial Engineering Geology Working Party (PGEGWP) has been established by the Engineering Group of the Geological Society and comprises officers and specialist participating members who will act as lead authors. The PGEGWP will produce a report, in book format, to complement the previous report on Tropical Residual Soils produced by an earlier Working Party of the Engineering Group, first published in 1990 and republished in book format in 1997 (Fookes 1997). A similar format was adopted by the Hot Deserts Working Party, which published their final report in 2012 (Walker 2012). It is intended that the report will be a state-of-the-art review on the ground conditions associated with former Quaternary periglacial and glacial environments and their materials, from an engineering geological viewpoint. There necessarily will be appropriate coverage of the modern processes and environments that formed these materials. A key aspect of the report will be to integrate soil description methodologies utilized by Quaternary scientists, engineering geologists and geotechnical engineers. Field workshops have been organized (Figs. 5.3 and 5.4) to consider various glaciogenic classification schemes specifically with their regard to their applicability to engineering geology.

It is not intended to define the geographic extent of former periglacial and glacial environments around the world, but to concentrate on ground models that would be applicable to support the engineering geological practitioner. The aim of the PGEGWP is to produce a report that will act as an essential reference handbook as well as a valuable textbook

Table 5.1 Example of a Terrain Unit definition table from the geomorphological setting chapter

Terrain unit	Relict frost mounds /relict ramparted ground-ice depressions: pingos
Image	
	Small pingo remnant (approx. 30 m diameter) near to Thompson, Norfolk
Form/topography	Pingos are ice-cored mounds or hills developed in permafrost. Relict pingos and other ground ice mounds formed during Quaternary cold stages may be indicated by circular or ovate depressions, often surrounded by raised rampart-like rims with a peat or soft ground core. Two forms are identified, closed system (or hydrostatic) pingos and open system (or hydraulic) pingos. The former occur in lowland settings within the continuous permafrost zone, and the latter are more common in valley bottom and footslope localities in both discontinuous and continuous permafrost. Pingos can reach up to 70 m in height and up to 600 m in diameter
Landsystem	Lowland Periglacial Terrain
Process of formation	Formed by injection of water into near surface permafrost to form an ice core. Water under sufficient pressure to overcome overburden stress. Pressure can develop in two ways; Closed System where water is expelled from saturated coarse grained sediments during the refreezing of a talik (a zone of unfrozen sediment within a continuous permafrost) or Open System where artesian water pressures within a sub permafrost aquifer cause upward injection and freezing of water
Modern analogue	
	Active Pingo, Innerhytte, Svalbard
Associated features	Related smaller ground ice phenomena associated with permafrost regions are lithalsas, mineral palsas, and seasonal ground ice mounds
Engineering significance	Compressible soils; differential settlement
References	Harris and Ross (2007), Hutchinson (1980, 1991, 1992)

for practicing professionals and students. The style will be concise and digestible by the non-specialist, yet be authoritative, up-to-date and extensively supported by data and collations of technical information. The use of jargon will be minimized and necessary specialist terms will be defined in an extensive glossary. There will be copious illustrations, many of which will be original, and many good quality photographs. The content of the report will embrace a full

Fig. 5.3 Working Party field discussion of glaciogenic soil classification and description methodologies at Barmston, East Yorkshire, UK. Section shows Skipsea Till overlain by subglacial canal fills

Fig. 5.4 Working Party field visit to North East England to discuss potential glacial sediment nomenclatures and ground models to be included in the final publication. Field description and assessment methodologies have been a core discussion point of the Working Party (Table 5.1)

range of topics, from the latest research findings to practical applications of existing information. Likely directions of research and predictions of future developments will be highlighted where appropriate. The report will be based on world-wide experience in periglacial and glacial terrain and will draw upon the experience of its members and publications on periglacial and glacial conditions.

The Working Party members will be collectively responsible for the whole report. Although each participating member will be the named author or co-author of one or more chapters, all members will be expected to review and

contribute to the chapters drafted by other members and would be acknowledged as such. Individual book chapters will be included in the Thomson Book Citation Index.

5.3 Chapter Listing with Lead Authors

The Working Party is chaired by Chris Martin, (BP) who will also draw together the diverse inputs that will be required for the introductory first chapter. Anna Morley

(Arup) is the Secretary of the Working Party while Professor Jim Griffiths (University of Plymouth) is the current Editor in Chief for the Engineering Geology Special Publications series. The chapter titles and lead authors of the remaining chapters are as follows:

- Quaternary Setting
  - Dr Sven Lukas, Queen Mary University of London
- Geomorphological Framework
  - David Giles, University of Portsmouth
- Glacial Conceptual Ground Models
  - Professor David Evans, Durham University
- Periglacial Conceptual Ground Models
  - Professor Julian Murton, University of Sussex
- Engineering Behaviour & Properties
  - Professor Martin Culshaw, University of Birmingham
- Engineering Investigation and Assessment
  - Dr Mike de Freitas, Imperial College
- Design & Construction Considerations
  - Dr Mike Winter, Transport Research Laboratory
- Geohazards and Problematic Ground
  - Dr Laurance Donnelly, Wardell Armstrong

References

Fookes PG (ed) (1997) Tropical Residual Soils: A Geological Society Engineering Group Working Party Revised Report. Geological Society, London

Harris C, Ross N (2007) Pingos and pingo scars. In: Elias SA (ed) Encyclopedia of Quaternary Science, Vol 2. Elsevier, Amsterdam, pp 2200–2207

Hutchinson JN (1980) Possible late Quaternary pingo remnants in central London. Nature 284:253–255. doi:10.1038/284253a0

Hutchinson JN (1991) Periglacial and slope processes. In: Forster A, Culshaw MG, Cripps JC, Moon CF (Eds) Quaternary Engineering Geology, Geological Society Engineering Geology Special Publication No. 7, 283–331. doi:10.1144/GSL.ENG.1991.007.01.27

Hutchinson JN (1992) Engineering in relict periglacial and extraglacial areas in Britain. In: Gray JM (Ed) Applications of Quaternary Research, Quaternary Proceedings No. 2. Quaternary Research Association, Cambridge, 49–65

McMillan A, Powell J (1999) BGS rock classification scheme. Volume 4, classification of artificial (man-made) ground and natural superficial deposits: applications to geological maps and datasets in the UK. British Geological Survey Research Report, RR 99–04, Keyworth

Walker MJ (ed) (2012) Hot Deserts: Engineering, Geology and Geomorphology: Engineering Group Working Party Report (No. 25). Geological Society, London

New Methods of Determining Rock Properties for Geothermal Reservoir Characterization

6

Mathias Nehler, Philipp Mielke, Greg Bignall, and Ingo Sass

Abstract

The influence of hydrothermal alteration on permeability, thermal conductivity and thermal diffusivity was investigated for more than 300 drill cores from the wells THM18, TH18 and THM19 of the Tauhara Geothermal Field (Wairakei, New Zealand). The measurements were performed with newly developed, portable laboratory devices. The anisotropic, intrinsic permeability was measured with a gas pressure Columnar-Permeameter, while the thermal conductivity and thermal diffusivity were measured with a device based on the optical scanning method. The hydrothermal alteration rank (argillic or propylitic) was determined semi-quantitative by methylene blue dye adsorption tests in combination with thin section analyses. Samples from the Huka Fall Formation and the Waiora Formation, composed of layered mud-, silt- and sandstones as well as pumice-rich tuffs deposited in a limnic environment as well as associated rhyolitic and andesitic intrusive rocks were examined. A prograde alteration with depth is indicated by an increasing amount of illite and the corresponding decrease of smectite. Generally lithologies of higher primary permeabilities are more affected by hydrothermal alteration. With an increase of secondary clay minerals the permeability decreases.

Keywords

Hydrothermal alteration • Thermal conductivity • Permeability

6.1 Introduction

The Tauhara Geothermal Field is part of the Taupo Volcanic Zone (TVZ). This zone is an active volcanic arc and back arc basin on the central North Island of New Zealand of Pliocene to Quaternary age (Wilson et al. 1995). Numerous wells were already drilled and field conditions are geologically as well as chemically and physically well defined. The epithermal system is characterized by 300 °C hot water, which rises from a depth of 5–8 km. Samples have been taken from well TH18, located at the resistivity boundary outside of the active geothermal area and from wells THM18 and THM19, which are situated within the center of the Tauhara Field, close to subsidence bowls. The samples were taken at intervals between 5 and 15 m depending on the heterogeneity of the respective stratigraphy.

Petrophysical measurements and petrographic characterizations of the samples were used to evaluate the effects of lithology and hydrothermal alteration on permeability, thermal conductivity and thermal diffusivity. In order to explain the measurement results more accurately, the samples were classified by lithological criteria. Nine distinct lithologies can be identified: silty mudstones, silty sandstones, pumice-rich crystal tuffs, sedimentary breccias, andesitic lavas, hydrothermal andesite breccias, rhyolitic lavas, rhyolitic breccias and igneous breccias.

M. Nehler (✉) · P. Mielke · I. Sass
Chair of Geothermal Science and Technology, Technische Universität Darmstadt, Schnittspahnstraße 9, 64287 Darmstadt, Germany
e-mail: M_Nehler@gmx.de

G. Bignall
GNS Science, 114 Karetoto Road, 3377 Wairakei, New Zealand

G. Lollino et al. (eds.), *Engineering Geology for Society and Territory – Volume 6*,
DOI: 10.1007/978-3-319-09060-3_6, © Springer International Publishing Switzerland 2015

Fig. 6.1 Schematic sketch of the preliminary sample preparation

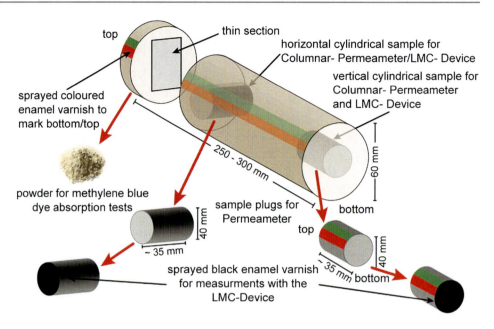

6.2 Sample Preparation

Oriented plugs (vertical and horizontal) were drilled out of existing drill cores with a diameter of 60 mm (Fig. 6.1). The cylindrical plugs must have a diameter of 40 mm and lengths between 30 and 45 mm for the measurements. Oriented thin sections were also prepared. Approximately 4 g of powder was pulverized for the semi-quantitative methylene blue dye adsorption tests (MEB) to identify the clay minerals. Plugs were dried at 40 °C for 48 h. After measuring the samples with the Columnar-Permeameter, they were prepared for the measurements with the Lambda Measuring Center (LMC). Therefore, the planar surface of the samples was sprayed with acrylic matt black enamel to achieve identical initial conditions.

6.3 Permeability Measurements

The intrinsic permeability (k_i) was determined according to Klinkenberg (1941) with a pressurized air driven, portable Columnar-Permeameter, which was invented by Hornung and Aigner (2002) and is described in detail by Arndt and Bär (2011). The specified measuring range is between 0.001 and 1,000 mD.

The gas permeability is measured at five different pressure stages (1,050, 1,250, 1,500, 2,000, 3,000 and 5,000 mbar) for each sample and is extrapolated to calculate the effective gas permeability for air under infinitely high air pressure. The pressure difference for all five stages remains identical and lies between 50 mbar for porous and up to a maximum of 1,000 mbar for slightly porous samples. The apparent permeability k_a [m^2] is calculated for every single measurement by the means of Darcy´s law for compressible fluids and is plotted against the reciprocal mean pressure [1/p*] in the corresponding pressure stage (Klinkenberg plot). The apparent permeability k_a is associated with the intrinsic permeability k_i by the means of the Klinkenberg-factor b. The approach is used for permeabilities $<5 \times 10^{-14}$ m^2 and Klinkenberg-factors > 0.24 bar.

6.4 Thermal Measurements

The portable Lambda Measuring Center (LMC) was used to determine thermal conductivity (λ) and thermal diffusivity (a). It is a contactless method for measuring solid materials based on the optical scanning method with a fixed point heat source (Popov et al. 1999). The measuring range is between 0.5 W/(m K) and 5.0 W/(m K). To calculate λ of a sample it is necessary to use a standard of known thermal properties. Sample and standard are linked by Eq. (1):

$$\lambda_{sample} = \lambda_{standard} \cdot \frac{\Delta T_{standard}}{\Delta T_{sample}} \cdot K \qquad (6.1)$$

If steady state conditions are confirmed the sample surface is heated up for 2 s with 15 % power of the 150 W$_e$ Osram lamp. The temperature of the sample will then decrease at a rate depending on λ. The temperature difference ΔT is measured and the software determines λ by Eq. 6.1. To calculate a the sample is heated up again and the temperature is measured at a distance x to the heating point. The temperature maximum at a time t is determined by the software. With t and x the thermal diffusivity a can be calculated by

Fig. 6.2 Stratigraphic logs show the measurement results of the wells THM18, TH18 and THM19

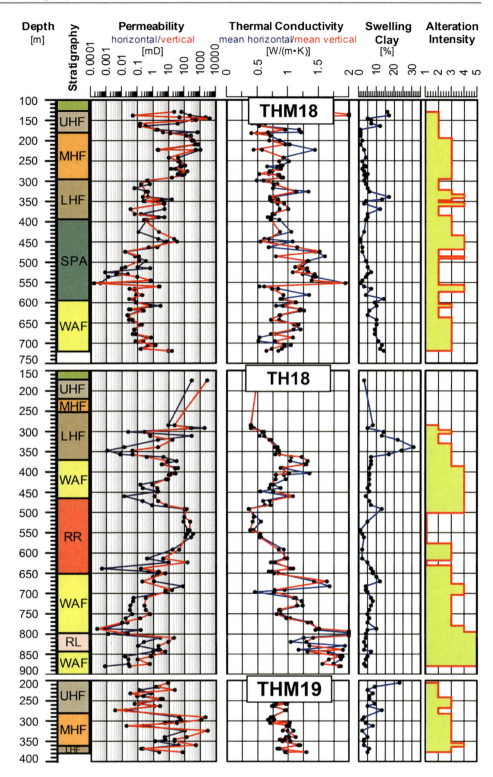

Eq. (6.2) (Hamm and Theusner 2010). The thermal diffusivity can be calculated with an analogues relation to the thermal conductivity including the time t instead of ΔT.

$$t = \frac{x^2}{a \cdot 6} \qquad (6.2)$$

6.5 Methylene Blue Dye Adsorption Test (MEB)

The MEB is a semi-quantitative method to determine the amount of swelling clay (mainly smectite) in rocks and soil materials (Gunderson et al. 2000). The test is a common method for determining the swelling clay content in water based drilling fluids, but can also be used to estimate the smectite/smectite-illite clay content in hydrothermal systems. These clays represent the predominant rock alteration products in the 50–200 °C zone above many high temperature, pH neutral, geothermal systems (Browne 1978). The methylene blue is an organic dye that shows a high selectivity for adsorption by reactive clay minerals as smectite, but is unaffected for adsorption by common clay minerals. This standardized (API 1988) technique uses a concentration of 3.74 g/l of methylene blue for the testing procedure. At this concentration the addition of 1 ml methylene blue dye solution is equivalent to an exchange capacity of 1 milliequivalent (1 meq) per 100 g cation. Based on the fact that smectites have an average cation exchange capacity of 100 meq/100 g, 1 ml of methylene blue is equivalent to 1 % of swelling clay content. The method is called semi-quantitative because of the actual cation exchange capacity of the swelling clays, which may vary from 80 to 150 meq/100 g.

6.6   Results

The results are characterized by a great variability depending on the source rocks (educts), the rank and intensity of the hydrothermal alteration, depth, primary permeability etc. (compare to Fig. 6.2).

The permeability is highly variable, but generally decreases with depth, while the thermal conductivity increases with depth. The pumice-rich crystal tuffs of the Middle Huka Fall Formation (MHF) and Waiora Formation (WAF) are characterized by relatively constant values of around 10 mD. Lithologies of greater permeabilities such as the tuffs are more affected by hydrothermal alteration and therefore show significant changes in their properties. With an increase of the fine fraction the permeability decreases.

The thermal conductivity varies between 0.35 and 2.50 W/(mK) showing a negative correlation with the permeability. Mud- and siltstones are characterized by a small range of measured values between 0.5 and 1.4 W/(m K). Silt- and sandstones show increased thermal conductivities with increasing alteration intensities. Pumice-rich crystal tuffs also have very high thermal conductivities when intensively altered. Compact lava units and the igneous breccia's show the highest thermal conductivities with maxima greater than 2.0 W/(m K). As might be expected, with increasing rock strength the thermal conductivity also increases.

A decreased amount of swelling clay (smectite) indicates a higher alteration rank. Peak concentrations occur at the UHF and LHF. A high amount of clay in highly altered sedimentary lithologies leads to low permeabilities, typically for caprocks. Smaller maxima occur within the WAF, indicating a general decrease of smectite with depth due to its natural stability range (70–160 °C).

With increasing depth the temperature, intensity and rank of the hydrothermal alteration also increases. The alteration type changes from argillic in the shallower parts to propylitic at greater depths (below 600 m). The primary mineral assemblage is predominantly replaced by clay minerals, calcite and secondary quartz. Therefore, hydrothermal alteration is generally prograde. Intensively altered rocks occur only in the deeper parts of the system, influenced by the propylitic alteration including silification processes. Greater permeabilities, such as fractures seem to facilitate this process. However, the hydrothermal alteration also depends on many other factors like mineralogy, texture, primary permeability and fluids.

References

API (1988) Recommended practice standard procedure for field testing drilling fluids. 12th edition 54 S., Washington, DC (American Petroleum Institute)

Arndt D, Bär K (2011) Forschungs- und Entwicklungsprojekt. 3D-Modell der geothermischen Tiefenpotenziale von Hessen. Abschlussbericht.—218 S., Darmstadt

Browne P (1978) Hydrothermal alteration in active geothermal fields-Annu. Rev Earth Planet Sci 6(1):229–250

Gunderson R, Cumming W, Astra D, Harvey C (2000) Analysis of smectite clays in geothermal drill cuttings by the methylene blue method for well site geothermometry and resistivity sounding correlation: WGC. Kyushu—Tohoku, Japan

Hamm K, Theusner M (2010) Lambda-Mess-Center LMC1. Installation und Bedienung. Manual—HTM Hamm & Theusner GbR. Erzhausen, Germany

Hornung J, Aigner T (2002) Reservoir architecture in a terminal alluvial plain: An outcrop analogue study (Upper Triassic, Southern Germany) part 1: Sedimentology and petrophysics. J Petroleum Geol 25(1):3–30

Klinkenberg LJ (1941) The permeability of porous media to liquids and gases. Drilling and production practice - Shell Development Co., pp 200–213 (American Petroleum Institute)

Popov YA, Pribnow DFC, Sass JH, Williams CF, Burkhardt H (1999) Characterization of rock thermal conductivity by high-resolution optical scanning. Geothermics 28(2):253–276

Rosenberg MD, Bignall G, Rae AJ (2009) The geological framework of the Wairakei–Tauhara Geothermal System, New Zealand. Special issue on the Wairakei Geothermal Field, New Zealand. 50 Years Generating Electricity 38(1):72–84

Wilson CJN, Houghton BF, McWilliams MO, Lanphere MA, Weaver SD, Briggs RM (1995) Volcanic and structural evolution of Taupo Volcanic Zone, New Zealand: a review. Taupo Volcanic Zone, New Zealand 68(1–3):1–28

Application of Reliability Methods to Tunnel Lining Design in Weak Heterogeneous Rockmasses

7

John C. Langford, N. Vlachopoulos, M.S. Diederichs, and D.J. Hutchinson

Abstract

Tunnel design in weak, heterogeneous materials such as flysch poses a variety of engineering challenges. The complex depositional and tectonic history of these materials leads to significant in situ variability in rockmass behaviour. Additionally, the alterations of sandstone and pelitic layers make rockmass characterization using traditional methods difficult. As a result, significant uncertainty exists in the ground response for a tunnel through such materials. Reliability-based methods can be used to better understand the impact this uncertainty has on convergence and tunnel lining performance. By assessing the impact of input uncertainty on ground response, the probability of failure can be evaluated for a given limit state. A quantitative risk approach can then be used to select the optimum design option on the basis of both safety and cost. This paper explores this issue further and presents a reliability-based, quantitative risk approach for the design of the Driskos tunnel along the Egnatia Odos highway in northern Greece.

Keywords

Reliability methods • Weak rock tunnelling • Support design • Squeezing

7.1 Introduction

Weak, heterogeneous rockmasses such as flysch pose a serious design challenge for geological engineers. Due to the complex depositional environment and tectonic history, such materials exhibit generally low rockmass strength and a high degree of variability. As a result, a range of possible squeezing conditions can be encountered when excavating within a single unit. Given the safety and cost implications associated with squeezing in flysch, much work has been done to properly characterize these materials. To obtain reliable estimates of rockmass strength, a firm understanding of the relative presence of competent and incompetent layers is required. As these percentages will vary over the tunnel alignment, a conservative estimate is typically used to ensure a robust lining design is selected that is capable of withstanding the "worst" anticipated loading conditions. Such an approach leads to over-conservatism, which can have a substantial negative impact on both the project schedule and cost.

Reliability-based design (RBD) methods, when used in conjunction with more traditional design methods, can provide a more rational approach to quantify design risk in such highly variable rockmasses. By assessing the impact of input uncertainty on ground response, the probability of failure can be assessed for a given failure mechanism. This allows for a greater understanding of support performance and the application of a quantitative risk approach for design.

J.C. Langford
Hatch Mott MacDonald, Vancouver, Canada
e-mail: connor.langford@gmail.com

N. Vlachopoulos (✉) · M.S. Diederichs · D.J. Hutchinson
GeoEngineering Centre, Queen's University-Royal Military College, Kingston, Canada
e-mail: vlachopoulos-n@rmc.ca

M.S. Diederichs
e-mail: diederim@queensu.ca

D.J. Hutchinson
e-mail: hutchinj@queensu.ca

G. Lollino et al. (eds.), *Engineering Geology for Society and Territory – Volume 6*,
DOI: 10.1007/978-3-319-09060-3_7, © Springer International Publishing Switzerland 2015

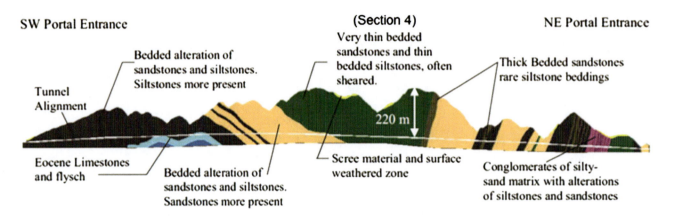

Fig. 7.1 Longitudinal topographic profile and idealized cross section for the Driskos tunnel (modified after Egnatia Odos S.A. 2003, Hoek and Marinos 2000)

This paper applies the reliability approach outlined in Langford et al. (2013) to perform a quantitative risk assessment for a number of support options at the Driskos twin tunnel along the Egantia Odos highway in northern Greece. This tunnel was excavated in a complex sequence of sandstone and siltstone layers (flysch) that experienced compressional deformations. Given the challenging geological conditions, excessive deformations and overstressing of the temporary support systems were experienced during excavation. The difficulties experienced and the scale of the project make the Driskos tunnel an excellent case study to illustrate the validity and benefits of a comprehensive risk-based design approach.

7.2 Case Study Area: Driskos Twin Tunnel, Egnatia Odos Highway

The Driskos twin tunnel is located within the Epirus region in the northwest corner of Greece. It was constructed as part of the Egantia Odos highway, which is a 670 km long construction project that consists of 76 twin tunnels raging in length from 800 m to 4,600 m and over 1,600 bridges. Each tunnel is horseshoe shaped with an internal diameter of 11 m internal diameter and a separation distance of approximately 13 m. The tunnels are approximately 4.6 km long and cross the NE Greek Pindos mountain chain under a maximum overburden of 220 m. The tunnel was constructed using a conventional drill and blast sequential excavation based on an observational design approach. A series of support categories were used based on the rockmass quality encountered.

7.2.1 Local Geology

The Driskos tunnel is situated in a series of varying lithological features of the Ionian tectonic unit adjacent to the

Pindos isopic unit. The material is less tectonically disturbed than the Pindos Flysch meaning there is an absence of extensive chaotic zones within the Ionian Flysch. Based on the site investigation, a longitudinal section was prepared that details the major rock units and topography along the tunnel length (Fig. 7.1). The alignment was subsequently divided into 14 sections in Vlachopoulos et al. (2013) on the basis of geology, rockmass quality and in situ stress conditions. Of specific interest to this analysis is the area identified as Sect. 4, which extends from chainage 8 + 385 to 9 + 035, as significant squeezing issues were encountered.

7.2.2 Rockmass Characterization

In order to predict tunnelling problems in flysch, reliable estimates of the rockmass strength and stiffness must be obtained. Unfortunately, the heterogeneity and variability within flysch makes the determination of intact parameters extremely challenging. The classification system developed by Marinos and Hoek (2001) addresses this concern and allows an appropriate flysch category to be selected based on a Geological Strength Index (GSI). The GSI value considers the structure present and the relative composition of the rockmass with respect to sandstone and siltstone layers. After determining the flysch category, a weighted average approach can be used to estimate the uniaxial compressive strength (UCS), Hoek-Brown material constant (m_i) and the Modulus Ratio (MR) based on intact strength parameters for sandstone and siltstone.

The advantage of this approach is that it allows standardized inputs for the generalized Hoek-Brown method to be obtained. As such, an appropriate failure criterion can be developed for the flysch rockmass based on its intact strength and rockmass quality (GSI) over the tensile and compressive regions. The estimate of rockmass strength is based on the assumption that the rock behaves in an isotropic fashion at the

Fig. 7.2 Detailed geological section showing division of Sect. 4 into Sects. 4.1–4.5 (modified after Egnatia Odos S.A. 1998)

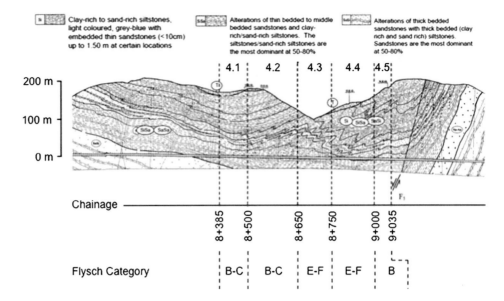

scale of the excavation due to the presence of several closely spaced discontinuities, which is appropriate in this case.

7.2.3 Support Categories and Excavation Approach

A series of five (I to V) support categories were developed on the basis of the expected rockmass quality conditions and initial estimates of support requirements were determined during the design phase. Categories III (15 cm unreinforced shotcrete), IV (20 cm shotcrete, HEB 140 steel sets with 2 m centres) and V (25 cm shotcrete, HEB 160 steel sets with 2 m centres) are of particular interest for this analysis. Support was typically installed 2 m back from the face and a sequential, heading and bench excavation was used.

7.2.4 Uncertainty in Ground Conditions

Uncertainty in geological systems is typically divided into two categories: variability caused by random processes (aleatory) and knowledge-based uncertainty (epistemic). The natural variability in rockmass and in situ stress parameters is typically considered to be aleatory as the process of formation results in a real variation in properties from one spatial location to another. As this variability is inherent in the material, continued testing will not eliminate the uncertainty, but will provide a more complete understanding of it. Conversely, epistemic uncertainty exists as a consequence of limited information as well as measurement, statistical estimation, transformation and modelling uncertainty. As these components are a result of imperfect techniques, they should be reduced as much as possible.

In order to quantify uncertainty in the ground conditions, a series of homogeneous domains were established along Sect. 4 (Fig. 7.2). These Sects. (4.1–4.5) were developed on the basis of lithology, rockmass quality, presence of rockmass alteration and in situ stress conditions. The classification system by Marinos and Hoek (2001) was used to provide an indirect means of quantifying uncertainty for the intact strength and stiffness parameters for each of these domains. Based on the GSI values obtained for each domain, an appropriate flysch category was assigned and the corresponding weighted average was selected. A mean and standard deviation were then calculated for the UCS, m_i and MR for each domain based on the design values as well as acceptable ranges for sandstone and siltstone parameters. For this analysis, the in situ stress conditions in each domain were considered deterministically and calculated based on the overburden depth for each section. Hydrostatic stress conditions were assumed.

7.2.5 Analysis Method

For this study a two-dimensional, plane strain model was developed in the finite element modelling program 'Phase 2' by Rocscience Incorporated (http://www.rocscience.com). The full-face excavation of a single tunnel was considered for simplicity. Three-dimensional advance of the tunnel was simulated in a multi-staged two-dimensional model. The convergence-confinement method was used to describe the reduction in radial resistance at a particular point along a tunnel as the face advances. The approach by Vlachopoulos and Diederichs (2009) was also used to determine the timing for stiff support installation based on the longitudinal displacement profile (LDP) for the unsupported tunnel.

(a)

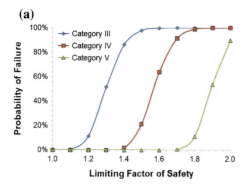

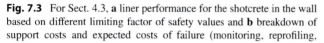

(b)

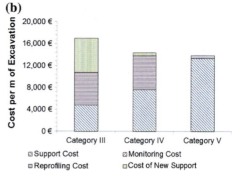

Fig. 7.3 For Sect. 4.3, **a** liner performance for the shotcrete in the wall based on different limiting factor of safety values and **b** breakdown of support costs and expected costs of failure (monitoring, reprofiling, installation of new support) for each liner category (costs are per metre of tunnel)

The structural stability of the lining systems was calculated using bending moment and thrust-shear force support capacity diagrams based on limiting factor of safety (FS) values developed according to Carranza-Torres and Diederichs (2009). The envelope of failure shown in these capacity diagrams is a graphical representation of the critical failure surface, separating the combinations of loads that are acceptable and those that exceed allowable limits. For this study, a limiting FS value of 1.3 was selected to ensure an appropriate level of safety for the tunnel.

7.2.6 Reliability Analysis

To assess the performance of each support category, reliability methods were used to determine the variability in ground response and liner loads. Unlike deterministic analyses, reliability methods directly incorporate the natural variability of the inputs into the design process. From this, a probability failure (p_f) can be established with respect to a specific failure mode, with "failure" defined as either the complete collapse of the structure (ultimate limit state, or ULS) or a loss of functionality (serviceability limit state, or SLS).

For this study, the modified point estimate method (PEM) proposed by Langford and Diederichs (2013) was used to determine the variability in liner loads and the probability of failure with respect to a limiting capacity curve (FS = 1.3). A quantitative risk assessment was used that considered both the probability and consequence of lining failure. With respect to the consequence of failure, two conditions were considered: (a) moderate damage, which would require monitoring and assessment by an engineer (failure of shotcrete in Category IV or V), and (b) complete failure of the lining, which would require re-excavation and the application of a higher class of support (failure of shotcrete in Category III or steel reinforcement in Category IV or V). For this analysis, support and failure costs were developed based on records from the Driskos tunnel project.

7.2.7 Results and Analysis

Each support category was modeled and uncertainty in thrust, bending moment and shear forces were calculated at each liner node based on the modified PEM approach. Based on the calculated liner load distributions, probabilities of failure were calculated based on different limiting capacity curves. In the interests of space, only the results from the shotcrete analysis in Sect. 4.3 are presented in Fig. 7.3a. The results illustrate the expected trend; a more robust support category will be able to sustain a greater rock load and therefore has a lower p_f.

To select the most appropriate support class for this section, a risk assessment was performed using the limiting p_f value (Fig. 7.3b). In this case, while Category III has the lowest support cost, the high probability of shotcrete failure leads to the highest reprofiling and risk costs. When Category IV and V are compared, it is clear that the expected reprofiling cost is significantly greater for the HEB140 steel sets than the HEB160. As such, Category V is considered to be the optimum support for Sect. 4.3 on the basis of economic risk.

As can be seen, this approach allows the improvement in safety for a given support category to be quantified, providing additional information to the Contractor and Owner with which to make design decisions.

References

Carranza-Torres C, Diederichs MS (2009) Mechanical analysis of circular liners with particular reference to composite supports. Tunn Undergr Space Technol 24(5):506–532

Langford JC, Diederichs MS (2013) Reliability-based approach to tunnel lining design using a modified point estimate method. Int J Rock Mech Min Sci 60:263–276

Marinos P, Hoek E (2001) Estimating the geotechnical properties of heterogeneous rock masses such as flysch. Bull Eng Geol Environ 60(2):85–92

Vlachopoulos N, Diederichs MS (2009) Improved longitudinal displacement profiles for convergence confinement analysis of deep tunnels. Rock Mech Rock Eng 42(2):131–146

Vlachopoulos N, Diederichs MS, Marinos V, Marinos P (2013) Tunnel behaviour associated with the weak Alpine rock masses of the Driskos Twin Tunnel, Egnatia Odos highway. Can Geotech J 50(1):91–120

Geological and Geotechnical Difference on Both Sides of the Same Tunnel

8

Pedro Olivença and Vítor Santos

Abstract

The excavation of Marão tunnel, located in the Northeast of Portugal, started simultaneously from both portals (East and West). The same method of excavation (Drill and Blasting—D and B), similar procedures for blasting and the primary support applied was the same and equivalent equipment used on both sides. Considering the same period of excavation, on the East side were excavated 2,300 m, while on the West side the excavation just 1,350 m were excavated. This differential on excavation between both sides of the tunnel, make it necessary to evaluate the causes that could justify this abnormal difference in excavation rates. To determine the cause of this difference in productivity, it was analyzed the geological features, the result of rock mass classifications and the geotechnical characteristics of the rock masses, which could influence the behavior of the excavation. Using descriptive statistics and multivariate analysis of data, applied to rock mass characteristics in each side of the tunnel, was possible to verify the existing differences, as well as the characteristics of the rock mass with greater relevance in the description of the geotechnical zoning for each side of the tunnel.

Keywords

Tunnel • Heterogeneity • Rock mass classification • Productivity

8.1 Introduction

The Marão tunnel excavation was advanced from both ends with the Drill and Blast (D and B) method. However, after the same time period, the east side advanced 2,300 m, while

P. Olivença (✉) · V. Santos
Consultores Para Estudos de Geologia E Engenharia, CEGE, Algés, Portugal
e-mail: pedro.olivenca@cege.pt

V. Santos
CICEGe—Centro de Investigação Em Ciência E Engenharia Geológica, Faculdade de Ciências E Tecnologia Da Universidade Nova de Lisboa, Lisbon, Portugal
e-mail: vitor.santos@cege.com.pt

P. Olivença
GeoFCUL—Departamento de Geologia Da Faculdade de Ciências Da, Universidade de Lisboa, Lisbon, Portugal

the west side only 1,350 m. At this point, the excavation was stopped due to contractual litigation between the Portuguese government and the concession company.

Marão tunnel is part of the A4 motorway that will link the cities of Amarante and Vila Real, in the Northeast of Portugal. This future infrastructure will have two parallel tunnels, each with a length of 5,600 m, a horseshoe cross-section with an invert arch and approximately 100 m^2 of section.

This tunnel will cross Marão mountain which reaches at its highest point around 1,200 m, in the alignment of the tunnel, with the maximum overburden of approximately 500 m.

Although the construction methodologies used on both ends of the tunnel were identical, two different construction teams were involved in the excavation, each one working on a different side of the tunnel, the productivity was quite different. It was important to find causes for this differences in productivity.

Table 8.1 Rock mass zoning (CJC 2009)

Zone	RMR	Support	Excavation	Progress
ZG4	<20	Steel ribs and 30 cm of shotcrete	Partial excavation—top heading and invert excavation. Temporary invert on the top heading	0.6 m a 1.0 m
ZG3	20–35	15 cm of shotcrete reinforced with metalic fibres and 5 m Swellex rockbolts	Partial excavation—top heading and invert excavation	1.4 m a 2.0 m
ZG2	35–50	10 cm of shotcrete reinforced with metalic fibres and 5 m Swellex rockbolts	Full-face excavation	1.8 m a 3.0 m
ZG1	>50	5–10 cm of shotcrete reinforced with metalic fibres and 5 m Swellex rockbolts	Full-face excavation	2.6 m a 4.0 m

Although causes could be various, this article addresses the differences in geology and geotechnical parameters observed on each side of the tunnel, highlighting the regional geological context and geotechnical characteristics of the rock mass.

The project includes four geotechnical zones, with the characteristics shown in Table 8.1.

8.2 Methodology

Given these objectives, the geotechnical conditions encountered along the tunnel alignment were examined as one of the main factors influencing the productivity of any tunnel excavation (Costa-Pereira 1985).

RMR (Bieniawski 1989) was calculated along the tunnel alignment with the objective to determine the quality of the rock mass for tunneling and classify each excavation cycle in the respective geotechnical zone, as defined in the design.

By applying descriptive statistics to RMR values, it is possible to determine the quality of the rock mass occurring at each end of the tunnel.

The geotechnical zoning of the tunnel results from the RMR value, and was calculated systematically along the excavation, so for each side of the tunnel, it is important to calculate the percentage of occurrence of each zone.

As shown in Table 8.1, the better the quality of the rock mass, the less support needed to be applied, to ensure its stability and bigger lengths of excavation are possible.

As defined by Bieniawski, RMR results from the arithmetic sum of weights, assigned to a set of parameters. The multi correspondences analysis (MCA), widely applied for dimensionality reduction of variables (Davis 2002; Hill and Lewicki 2007) aims to determine the characteristics of the rock mass that have the greatest influence on geotechnical zoning.

8.3 Geology

The Marão mountain is composed of autochthonous formations of Cambrian to Lower Devonian age (Pereira 1987).

The major geological structure is an anticline formed during the Variscan Orogeny. Geological mapping of the region is shown in Fig. 8.1. The following units are present (Sá et al. 2005):

- Desejosa formation (Cambrian): interbedded shales and metasiltstone, present in all 2,300 m excavated from the east portal.
- Vale de Bojas formation (middle and lower Floian): characterized by polymictic conglomerates and thinner layers of psamitic metatuff. The tunnel excavations have not encountered this formation.
- Marão formation (middle Floian): consists of quartzites alternating with phyllites or psamitic rocks. The upper member of this formation has been encountered in west side of tunnel excavation.
- Moncorvo formation: a monotonous sequence of gray shale and is present in the west portal of the tunnel.

A Hercynian granitic intrusion is located at the western area of the tunnel. This caused regional metamorphism of the Moncorvo formation (west side), giving it greater resistance, which led to a brittle deformation behavior, developing open discontinuities, with blocks of substantial size (Coke and Santos 2012).

The Cambrian shales (east side), away from the granitic intrusion, were not affected by contact metamorphism.

8.4 Characterization of Rock Mass

The distribution of RMR values calculated systematically during excavation allowed us to evaluate the rock mass quality at each end of the tunnel. RMR values are higher on

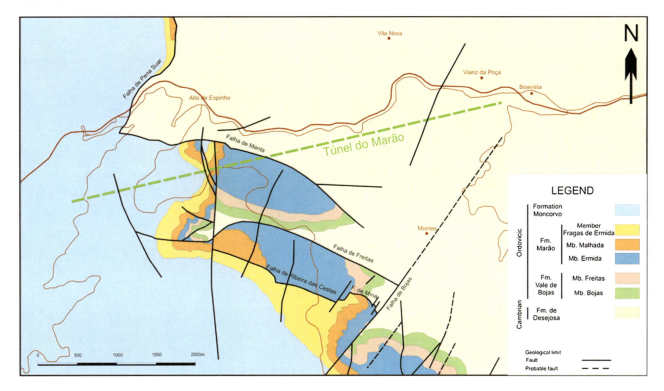

Fig. 8.1 Geological map of the tunnel area (Coke 2000)

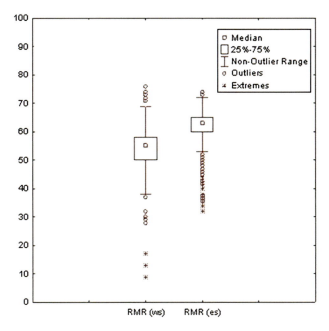

Fig. 8.2 Statistics RMR observed for the west side and east side of the tunnel

the east side, but show minor dispersion at the west end, which indicates greater heterogeneity of the rock mass in west side (Fig. 8.2).

RMR was used to divide the rock mass into various geotechnical zones, as defined in the design. The relative length of witch zone is present in Fig. 8.3.

Applying the MCA is possible to see that in the west side of the tunnel (Fig. 8.4a left), the ZG1(better geotechnical zone) is defined "equally" by all parameters of Bieniawski classification. The major factors in the definition of ZG2 were the discontinuities spacing—F2 and RQD values in the interval of 50–75 %. The ZG3 is defined by UCS results between 25 and 50 MPa and ZG4 (weaker geotechnical zone) is characterized by UCS results of 5–25 MPa and RQD values <25 %.

At the east side (Fig. 8.4b), the MCA shows that ZG1 all parameters of Bieniawski classification have influenced the RMR values obtained. The ZG2 is mainly conditioned by the characteristics of the weak discontinuities and ZG3 is defined by characteristics of the very weak discontinuities. The ZG4 (weaker geotechnical zone) was not found in the east side of the tunnel.

As known, different parameters are involved in RMR value, but some are more important to define each geotechnical zone and they are different for each side of the tunnel. While at the west end the characteristics associated with weaker geotechnical zones (ZG2, ZG3 and ZG4) are the UCS and RQD, at the east end the characteristics related with the discontinuities prevail (east side has no ZG4). On

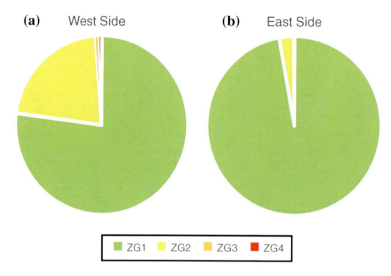

Fig. 8.3 Frequency of geotechnical zone

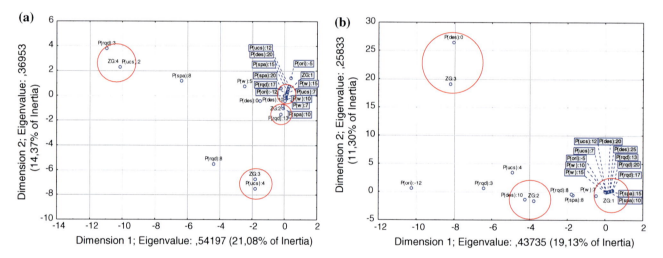

Fig. 8.4 MCA of rock mass characteristics. West side (*left*). East side (*right*)

the better geotechnical zone (ZG1) all the parameters are important and present more or less in the same way on both sides of the tunnel.

8.5   Conclusions

Along Marão tunnel, the regional geological environment plays an important role in the characteristics of the rock mass encountered. The presence of contact metamorphism at the west end of the tunnel, affected the behavior of the rock mass. The effects of contact metamorphism, are not present at the east end of the tunnel.

The rock mass on the west side is more heterogeneous when compared with the east side, which resulted in a different distribution of the values of RMR observed, with generating lower values on the west side.

Among the parameters used to calculate RMR, some are particularly associated with a geotechnical zone and are different on each end of the tunnel. With identical lithologies at both ends of the tunnel, this fact reinforces that the importance of regional geological environment, is controlling the geotechnical characteristics present in the rock mass.

The lower quality of the rock mass observed on the west side of the tunnel, implies a distribution of rock mass classification in all geotechnical zones defined on design, the presence of ZG4 and higher percentages of occurrence of ZG2 and ZG3, when compared with the east side of the tunnel.

The presence of more weaker zones in the west side of the tunnel, involved the need to apply larger quantities of support, lower excavation lengths and sometimes half-section excavation. All these factors significantly influence the productivity of excavation.

Acknowledgments The authors acknowledge the Infratúnel authorization to use of data collected during the technical assistance of tunnel excavation, without which it would be impossible to develop this article.

Bibliography

Bieniawski Z (1989) Rock mass classifications. Wiley, New York, p 251

CJC (2009) Memória descritiva e justificativa do projecto de execução do Túnel do Marão (p 154)

Coke C (2000) Evolução geodinâmica do ramo sul da Serra do Marão um caso de deformação progressiva em orógenos transpressivos. Universidade de Trás-os-Montes e Alto Douro (in Portuguese)

Coke C, Santos V (2012) Geologia estrutural na caracterização do comportamento geotécnico da escavação do Túnel do Marão. *Congresso Nacional de Geotecnia* (p. CD–ROM). Lisboa: Sociedade Portuguesa de Geotecnia (in Portuguese)

Costa-Pereira (1985) A geologia de engenharia no planeamento e projecto de túneis em maciços rochosos. Faculdade de Ciências e Tecnologia da Universidade Nova de Lisboa (in Portuguese)

Davis JC (2002) Statistics and data analysis in geology, 3rd edn. Wiley, New York, p 638

Hill T, Lewicki P (2007) Statistics: methods and applications. StatSoft, Tulsa, p 800

Pereira E (1987) Estudo geológico estrutural da região de Celorico de Bastos e sua interpretação geodinâmica. Faculdade de Ciencias da Universidade de Lisboa (*in Portuguese*)

Sá A, Meireles C, Coke C, Gutiérrez-Marco J (2005) Unidades litoestratigráficas do Ordovícico da região de Trás-os-Montes (Zona Centro-Ibérica, Portugal) (in Portuguese)

Development of Probabilistic Geotechnical Ground Models for Offshore Engineering

Konstantinos Symeonidis and Clark Fenton

Abstract

Traditional offshore site investigation (SI) practice focuses on characterising ground conditions around a single asset and its spatially-limited foundations. Applying a conventional approach to both widely distributed and deep water sites often limits the scope of geotechnical data gathering to principally remote geo-physical sensing combined with sparse sampling of questionable representative-ness. SI design can be improved to cope with distributed assets and multiple geo-hazards, while better SI sequencing and recent advances in geophysical techniques have improved the SI process considerably. However, the time and cost implications of applying these advances are potentially unacceptable when dealing with multiple facility footprints distributed over broad areas of seabed with complex, heterogeneous ground conditions, *e.g.*, wind farm developments on the UK continental shelf. A cost-effective alternative that integrates the inter-disciplinary SI functions better and embraces probabilistic ground models is required. Applying techniques developed for seismic hazard assessment with limited data sets, probability distribution functions can be derived allowing rational, fact-based 'forecasts'. This approach permits limited datasets to be evaluated for both epistemic uncertainty (data paucity) and aleatory (natural) variability, allowing the selection of representative geotechnical parameters. Probabilistic methods and spatial analysis techniques are applied to synthetic models of the seabed for the purpose of testing the effect of sampling, size and pattern, in accurately determining soil parameters, such as the undrained shear strength and friction angle or engineering parameters like pile penetration depth. A number of different sampling patterns are examined. The results suggest that there is a relation between pattern efficiency in describing the uncertainty and the existence of spatial trends in soil parameters or the existence of features like buried channels. These approaches have the potential to increase the efficiency of offshore SI, leading to more cost effective foundation design.

Keywords

Site investigation • Probabilistic analysis • Spatial distribution

9.1 Introduction

The increasing development of multi-asset offshore projects has resulted into the need for better understanding of the complexities involved in their foundation design. It is important to understand how representative are the collected and tested samples in terms of geotechnical strength parameters and seafloor conditions. It is important to

K. Symeonidis · C. Fenton (✉)
Department of Civil and Environmental Engineering,
Imperial College, London, UK
e-mail: c.fenton@imperial.ac.uk

K. Symeonidis
Mott MacDonald, Glasgow, UK

understand how the chosen sampling patterns may affect the efficiency of the investigation. Natural phenomena commonly exhibit variability in their characteristics, which means that they cannot be predicted with absolute certainty. Deterministic geotechnical analysis can be more easily applied when parameter uncertainties are low and materials and their geometries are known with a degree of accuracy. However, in geotechnics most parameters used in analysis are uncertain, often because of a limited sampling programme. Engineers typically deal with this uncertainty by choosing conservative values for these parameters (Nadim 2007). In the offshore environment uncertainties include design loads and the structure's resistance (Baecher and Christian 2003).

Probabilistic techniques can complement traditional deterministic analysis, by quantifying the degree of uncertainty, evaluating the data acquisition strategies and assessing hazards (Fenton 1997).

9.2   Sampling Strategies

Sampling strategies are considered during the site investigation phase of a project and decisions should be made on sample size, pattern and density. Typical sampling schemes include: random; gridded; uniform; clustered; and traversed. The selection of sampling pattern depends on the geological setting and the expected variations within the sampled population. The sampling scheme must avoid under-sampling or even over-sampling a sub-population, for example stratification, thereby introducing data bias. Spatial

functions can be used to describe the variation of geological and geotechnical parameters. Spatial functions are continuous and typically observations closely spaced are auto-correlated. Dealing with spatial functions using classical statistics may not be adequate, thus regionalised variables may be required (Symeonidis 2012). This considers the properties of the spatial function and disregards the nature of the physical phenomenon (Olea 1984). Using probabilistic modelling and spatial analysis it is possible to evaluate the effect of the sampling parameters and to measure the influence in site investigation design.

9.3   Model Development and Statistical Evaluation

Symeonidis (2012) developed an approach to evaluate the efficiency of differing sampling strategies in obtaining representative geotechnical data. Synthetic data are created for two ground models Scenario 1 and Scenario 2 (Fig. 9.1). Each model describes a relatively simple setting consisting of an upper clay layer and a lower sand layer. For each model a number of parameters are provided along with their coordinates. These parameters are later considered as random variables in our analysis. These model parameters are:

- Mudline undrained shear strength (kPa)
- Variation in undrained shear strength for clay with depth (kPa/m)
- Angle of friction (°) for the lower sand layer
- Clay layer thickness (m)
- Pile penetration depth (m) calculated using API-RP2A.

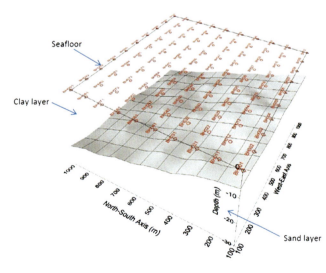
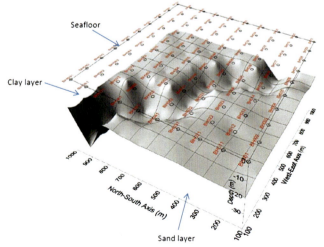

Fig. 9.1 *Scenario 1* a seabed defined by a clay layer and an underlying sand layer. The surface between the two layers is inclined towards the North. For this scenario 100 boreholes are given arranged in an equally spaced grid with separation distance 100 m, aligned to the North–South and East–West directions. *Scenario 2* a seabed defined by a clay layer

and an underlying sand layer. The sand layer forms a trough that it is roughly directed from NW to SE. Also, for this scenario 100 boreholes are given arranged in an equally spaced grid with separation distance 100 m, aligned to the North–South and East–West directions. The *grey surface* marks the boundary between the two layers

The data are provided in the form of a 10 by 10 rectangular grid of synthetic boreholes (BH) at 100 m intervals. Each BH contains the geotechnical parameters described above. The area of data coverage for each model is 1 km^2.

Different sampling patterns are the defined (Fig. 9.2). Then the completeness of these samples is statistically compared to the global population (herein called the Representative Sample [RS]) for each model. The analytical procedure involves the following approaches for each model

parameter: A: Descriptive statistics (histograms and frequency unit area diagrams). B: Inferential statistics (normal and lognormal distributions). C: Spatial analysis (contour maps with kriging and trend analysis with plane surface fit using polynomial regression). D: Sampling comparison. For each sampling pattern and sampling effort the sample is compared to the corresponding RS. The measures used for the comparisons are: (1) percentage difference between the mean values of the samples and the RS and (2) the maximum

Fig. 9.2 Typical sampling patters. **a** Random, **b** grid, **c** traverse 1, **d** traverse 2, **e** regular clustered, and **f** intersecting traverse

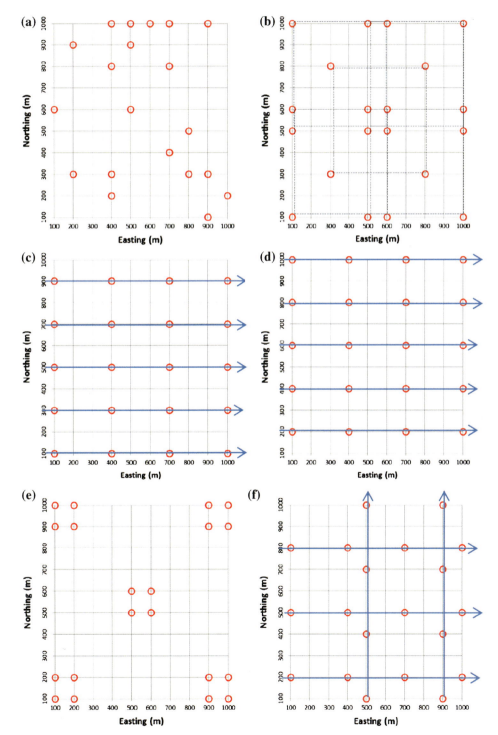

absolute difference of the theoretical distributions (normal and lognormal) defined by the sample and the RS, based on the K–S test formulation. From these comparisons the sampling patterns are ranked according to their efficiency. E: Kriging comparison: using the depth to the layer interface (derived from the model parameter z). Contour surfaces are created for each sampling pattern at the 3rd sampling effort (12 BHs) for Scenario 1 and for 3rd (12 BHs) and 5th (20 BHs) sampling efforts for Scenario 2. These contour surfaces are compared to the contour surfaces of the representative samples.

Sampling patterns (Fig. 9.2) are defined and tested on each models. The same maximum number 20 out 100 of sampling points for each pattern is applied. For each pattern, five (5) increments of four (4) new sampling points are added at each test iteration. Each of these increments is referred to as a sampling effort. For each pattern a custom Visual Basic (VBA) routine is created that applies the sampling procedure consisting of five sampling efforts (Fig. 9.2).

The first step in setting up the probabilistic modeling analysis is defining the procedures applied. For each parameter the workflow is applied for its Representative Sample (RS) and then for each sampling effort of the different sampling patterns. The formulation of the probabilistic model for each model parameter is performed in two parts (Fig. 9.3). Initially calculation of the descriptors of randomness is performed. Secondly the statistical distribution that best fits the empirical distribution derived in the initial stage is defined. Following selection of the distribution describing data variability, parameters that uniquely define this distribution are derived, along with tests that quantify the degree of fitness. Furthermore, based on the above analysis probabilities can be calculated from the proposed model distributions.

The spatial variability of the model parameters is displayed using contour maps that utilize point kriging. Trend analysis is then applied for each model parameter. This is modelled by a linear equation in the more simplistic approach along certain traverses at each model, using the least squares method. Trend analysis using the polynomial regression is also used to define large-scale trends and patterns in the data. Finally residual analysis is conducted for traverses along the model where the autocorrelation function is modelled for each RS for all model parameters.

Comparisons are performed in order to evaluate the efficiency of each sampling pattern in representing as accurate as possible each model parameter. Also, the performance of each sampling pattern is evaluated as the sampling efforts increase the sample size. The measures used are:

- Percentage difference between the mean values of the RS and each sample.
- Maximum absolute difference of the theoretical distributions (normal and lognormal) defined from the RS and each sample.
- The kriging technique for the parameter.

Contour surfaces are created for each sampling pattern. The difference between each sampling effort and the RS are mapped using kriging technique in order to evaluate the degree of difference. Sampling patterns with the smallest differences spatially are more efficient.

9.4 Results

Based on the statistical differences between the mean values for the parameters in the sampling efforts and the RS for each scenario the following observations are made:

Fig. 9.3 Schematic of the probabilistic modelling workflow

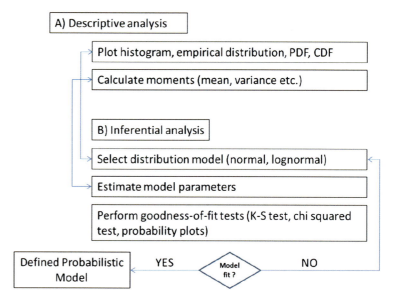

- Grid type sampling patterns (grid, traverse, and intersecting traverses) perform better when there is a distinct spatial trend in the parameter variability
- When the RS shows no spatial trend or when the range of variability is limited grid sampling shows limited advantage over random or clustered sampling patterns
- Sampling perpendicular to spatial trends has a poorer than expected efficiency. Intersecting traverse lines are more efficient than parallel traverses in this case.

These results are similar to those obtained by Olea (1984) using a universal kriging technique. In terms of efficiency of capturing the variability of the RS, regular sampling followed by stratified, then random and finally clustered sampling patterns, are most efficient.

When investigating depth parameters (the location of the clay-sand interface in the scenario models) inclined planar surfaces are best approximated using regular grid sampling. Elongated topography (the buried channel in Scenario 2) is best approximated using traverse sampling perpendicular to the trend of the feature. With increasing sampling points a uniform grid pattern becomes more efficient.

References

Baecher GB, Christian JT (2003) Reliability and statistics in geotechnical engineering. Wiley, Hoboken

Fenton GA (1997) Probabilistic methods in geotechnical engineering. Workshop presented at ASCE GeoLogan

Nadim, F. (2007) Tools and strategies for dealing with uncertainty in geotechnics. In: Probabilistic methods in geotechnical engineering, pp 71–95

Olea RA (1984) Sampling design optimization for spatial functions. Math Geol 16(4):369–392

Symeonidis K (2012) Development of probabilistic geotechnical ground models for offshore engineering, Unpublished M.Sc. thesis, Imperial College London, p 115

Baixo Sabor (Portugal) Upstream Dam Foundation: From Design Geological Predictions to Construction Geological Facts and Geotechnical Solutions

Jorge Neves, Celso Lima, Fernando Ferreira, and João Machado

Abstract

The upstream scheme of Baixo Sabor Hydroelectric Development includes a 123 m high dam with a crest length of 505 m. Two important geological faults that could affect the foundation of the dam were identified and characterized during the exploration works executed in the design phases. The foundation mapping performed during the construction phase allowed a detailed knowledge of these geological structures and complemented the existing data used by EDP for the detailed design of the engineering foundation treatment solutions.

Keywords

Sabor • Dam • Foundation • Fault • Treatment

10.1 Introduction

The Baixo Sabor Hydroelectric Development that EDP—Energias de Portugal, S.A. has under construction at the north of Portugal, at the Sabor River, a right bank tributary of the Douro River, includes two schemes located 12.6 and 3 km upstream of Sabor river mouth.

The upstream scheme (Fig. 10.1) is the largest and includes a concrete arch dam 123 m high with a crest length of 505 m and a total concrete volume of 670,000 m³. A controlled surface spillway is located in the central part of the dam, having four 16 m wide spans controlled by radial gates with a discharge capacity of 5,000 m³/s into a downstream plunge pool basin. At the right bank, the powerhouse is equipped with two reversible units, located in independent 79 m high and 11.5 m diameter shafts, topped by an unloading and erection building. Two independent and approximately parallel 5.7 m diameter headrace tunnels intersect the rock mass under the dam, connecting the reservoir with the powerhouse, with a 94 m head. The tailrace includes 2 tunnels and an outlet structure to operate the sluice gate.

The upstream dam is placed in a 1 km long NE–SW valley segment, with a deep, narrow and slightly asymmetrical transversal profile, 25 m wide at the base and 440 m at the crest level. The dam foundation consists in a medium to coarse grained, two mica, porphyritic granite, frequently showing mica orientation.

During the design phases, from mid-1990s to 2006, the performed geological and geotechnical investigations and characterization studies included detailed mapping of 13 trenches (2,300 m) and 6 galleries (180 m), 15 seismic refraction profiles and 3 electrical apparent resistivity profiles, 38 diamond drill holes (2,186 m), 388 Lugeon tests, 14 seismic refraction cross-hole sections, 41 borehole dilatometer (BHD) and 4 large flat jacks (LFJ) tests for in situ rock mass deformability evaluation, 3 stress tensor tube tests (STT) for in situ stress assessment, 31 joint shear tests and 18 sound velocity and 18 unconfined compressive strength tests.

J. Neves (✉) · C. Lima
EDP—Energias de Portugal, SA, Porto, Portugal
e-mail: jorgepacheco.neves@edp.pt

C. Lima
e-mail: celso.lima@edp.pt

F. Ferreira · J. Machado
Geoárea—Consultores de Geotecnia e Ambiente, Alfragide, Amadora, Portugal
e-mail: fernando.ferreira@geoarea.pt

J. Machado
e-mail: joao.machado@geoarea.pt

These exploration works and geotechnical characterization tests allowed the foundation to be subdivided into 3 zones. From the shallowest to the deepest, these zones are characterized, in average, by the following parameters:

- ZG3: W3–W5, F3–F5, RQD < 50 %, RMR < 40, $E_m \leq 5$ GPa, $\sigma_c < 35$ MPa
- ZG2: W2–W3, F2–F4, 50 % < RQD < 90 %, 37 < RMR < 64, $E_m = 7.5$ GPa, $\sigma_c = 53$ MPa
- ZG1: W1–W2, F1–F3, RQD > 90 %, 54 < RMR < 72, $E_m = 17.5$ GPa, $\sigma_c = 126$ MPa.

10.2 Design Geological Predictions

The geological investigations performed at the upstream Baixo Sabor dam site, particularly the trenches and galleries mapping complemented by drill hole data, allowed the identification and characterization of two important bedrock faults that influenced the foundation design and the slope excavations of adjacent slopes.

The first (fault n° 24) was identified at the left bank trenches, striking N10°E and dipping 65°E, filled with a 10–15 m thick quartz vein at the hanging wall and 3–4 m thick of sheared kaolinitized granitic mylonite (Fig. 10.2) at the footwall, predictably affecting the upstream slope excavation and the dam left abutment foundation. This clayey mylonitic zone was of particular concern to EDP designers due to its significant thickness, high deformability and unfavorable location that could cause slope stability and foundation deformability and permeability difficulties.

The second important geological structure was fault n° 11, a 10–15 m thick fault zone at the bottom of the valley, inferred by geomorphological interpretation and detected by rotary core drilling (core recovery losses, gouge fragments, intense rock mass fracturing and weathering) and later on, by seismic refraction cross hole sections performed inside exploration drill holes.

This geological structure was interpreted as a sub vertical fault zone that consisted in a few approximately parallel, N30°E, sub vertical minor faults (Fig. 10.3) with decimetre- to metre-scale thick gouge filling, separated by intensely sheared ZG3 granite with low deformability modulus. The fault zone low to moderate permeability (2–8 Lugeon units) was attributed to fault gouge washing and fracture filling.

10.3 Construction Geological Facts

The rock mass geotechnical zoning analysis confirmed that the dam foundation was mostly composed of fresh to moderately weathered (W1–W3) granite with moderate to completely weathered (W3–W5) zones associated with faults (Fig. 10.4). Joints, wide to closely spaced (F2–F4), were assigned to 6 major sets. Tectonic structures such as quartz and pegmatite veins less than 0.5 m thick occur throughout the surface. Faults, assigned to 3 sets, are generally filled with clayey gouge and rock particles up to 0.5 m thick, bordered by irregular zones of highly weathered (W4–W5) rock mass with closely spaced discontinuities.

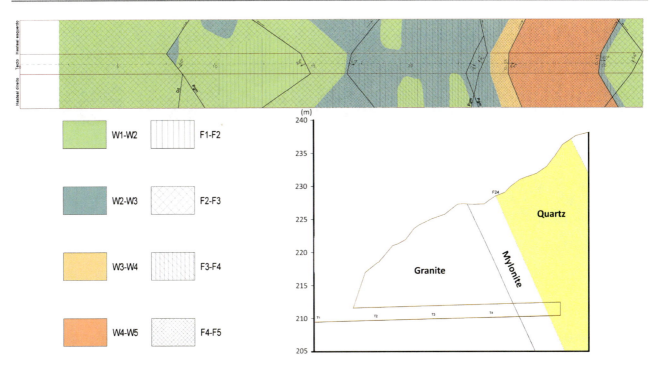

Fig. 10.2 Ceiling and wall mapping of GE1 gallery (walls folded to horizontal). W4–W5 zone corresponds to fault's mylonitic footwall, below quartz vein

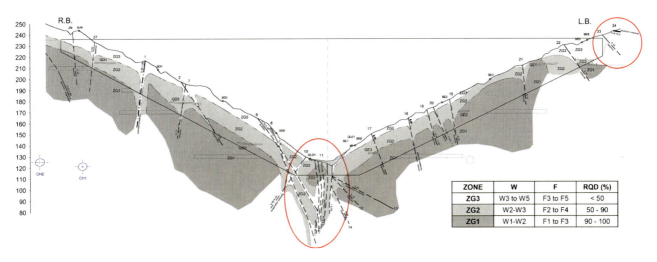

ZONE	W	F	RQD (%)
ZG3	W3 to W5	F3 to F5	< 50
ZG2	W2-W3	F2 to F4	50 - 90
ZG1	W1-W2	F1 to F3	90 - 100

Fig. 10.3 EDP's design phase geological and geotechnical zoning—section through dam reference surface. Fault zones referred in text are inside *red ellipses*

Fault nº 24, at the left abutment, N8ºW-0º-26ºE, dipping 50º–70ºESE, has a 2–5 m thick footwall mylonitic zone with sandy clay fill, gouge fragments and intensely weathered and fractured granite and quartz. At the hanging wall occurs a quartz vein of 8.5–14 m thick mixed with hydrothermally altered granite (Figs. 10.4 and 10.5).

A large fault zone occurs at the bottom of the valley, bordered by 2 sub parallel major alignments (A and C), N10º-40ºE, dipping 60ºNW-90º (Figs. 10.4 and 10.6). They are filled with 0.05–0.5 m of clayey gouge at the hanging wall and have a 2–3 m thick weathered and highly fractured

rock mass zone at the footwall. These faults are linked by a dip-slip fault (B), N55ºE, dipping 70ºSE-90º-80ºNW with a 0.1–0.2 m thick mylonitic sandy clay fill and a 1.5 m thick associated weathered and highly fractured rock mass zone. A and B faults were included in fault zone nº 11, identified during the design phase and fault C corresponds to fault nº 10.

Fault E (Fig. 10.4) is a N20ºE sub vertical fault that occurs at the right bank of the Sabor river. Its fill includes a1–2 m thick quartz mass and a clayey mylonite with rock particles. Several other discontinuities occur throughout the

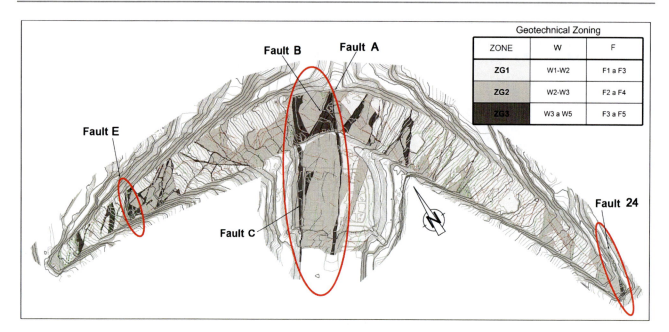

Fig. 10.4 Geoárea's simplified geological map and dam foundation geotechnical zoning. Fault zones referred in text are inside *red ellipses*

Fig. 10.5 a Fault nº 24. **b** Support wall and pre-stressed anchor beam

dam foundation but they weren't considered a geotechnical problem due to the reduced thickness of their fills and associated alteration zones.

10.4 Geotechnical Solutions

Dam foundation excavation depths were designed by EDP in order to guarantee an adequate embedding, mostly on good quality granitic rock (ZG1-dark grey zone in Fig. 10.3). The total excavation volume reached 560,000 m³ and some foundation reconstitutions had to be made in lower quality

fault related sectors, such as nº 10/11 (A, B, C) and nº 24, but also in smaller ones like fault E.

The treatment solution executed on fault nº 24 near the left abutment, consisted of a support wall 8.5 m high and 30 m long, concreted against the upstream slope of the dam foundation excavation, below an anchored beam built as shown in Fig. 10.5, with the main objective of stabilizing that slope and the rock mass above it.

At the river bed, fault material was replaced by reinforced concrete several meters deep, on faults A, B and C (Figs. 10.4 and 10.6). Between the contraction joints 19 and 20, an additional gallery parallel to fault C was left inside the

Fig. 10.6 a River bed faults excavations (*facing downstream*). **b** Placed reinforced concrete and PVC molds for posterior cement grout injection (*facing upstream*)

Fig. 10.7 Plan showing grout treatment holes configuration for faults A, B and C

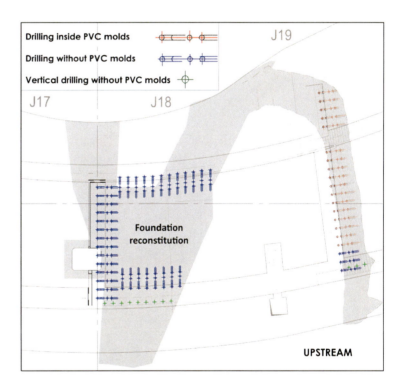

dam, in order to enable an adequate foundation treatment that consisted in fault gouge substitution by cement grout injection along these faults dip direction (Figs. 10.6 and 10.7). When dam concrete reached enough height, these localized treatments were followed by generalized cement grout injection (still ongoing) for foundation rock mass consolidation and waterproofing.

10.5   Final Remarks

The design phase predictions concerning geotechnical zoning and the main tectonic accidents intersecting Baixo Sabor upstream dam foundation were generally confirmed by the detailed geological mapping performed during the construction phase. These data and those gathered during the design phases allowed the design of dam foundation treatment by generalized cement grout injection and fault zones treatment design solutions, to further improve foundation geotechnical characteristics, i.e. deformability, shear strength and permeability.

The Foundations of Constructions in Dobrogea—Romania, on Water Sensitive Soils, Loess

11

Gabriela Brîndusa Cazacu, Nicolae Botu, and Daniela Grigore

Abstract

This article presents the geotechnical characteristics of loess, wetting sensitive soil in Dobrogea. These lands are of Quaternary age, are found just below the topsoil and most buildings are founded on it. Problems can arise when the foundation on these lands is due to any softening of foundation soil with water from different sources, permanent or casual. It will present the parameters of geotechnical solutions for improvement when appropriate.

Keywords

Loess • Foundation • Dobrogea • Sensitive soil

11.1 Introduction

The loess is a category featured among continental, Quaternary, sedimentary formations.

The name loess was introduced in 1834 by C. Lyell, coming from the German *lose* or *loss,* used in Rhineland, with the meaning of loose, porous, brittle. Loess lands occupy about 10 % of the entire surface of the continents, a spread of loess in the world is shown in Fig. 11.1a, b, North America and China, and in Fig. 11.2 in Europe.

Loess deposits in Romania occupies an area representing 17 % of the entire country. In Dobrogea there are areas where loess thickness is up to 60 m.

11.2 Properties of Loess

Most buildings in Dobrogea are founded on loess and loess soils, hence the need to understand the behavior of these soils and changes in terms of land and loads on which the construction transmits it.

The minimum and maximum values of geotechnical parameters of loess, in the natural state, in the Dobrogea area are listed in Table 11.1.

11.3 The Collapse Risk of the Loess

Problems can arise when the foundation on these regions is due to any softening of foundation soil with water from different sources, permanent or casual.

Because of the extra moisture, the loess can become collapsible.

Depending on the behavior of the loess, increasing of moisture content it has been classified in two categories:

(A) lands which not settles under the geological load, but of deformation under the influence of the loads transmitted by the construction, are not in risk of collapsing.

(B) lands which settles in geological load, they may be in risk of collapsing.

G.B. Cazacu (✉)
S.C. Geotech Dobrogea SRL, 900532, Constanta, Romania
e-mail: brandusacazacu@yahoo.com

N. Botu · D. Grigore
Faculty of Civil Engineering and Building Services, Technical University "Gheorghe Asachi" Iasi, 700050, Iasi, Romania

G. Lollino et al. (eds.), *Engineering Geology for Society and Territory – Volume 6,*
DOI: 10.1007/978-3-319-09060-3_11, © Springer International Publishing Switzerland 2015

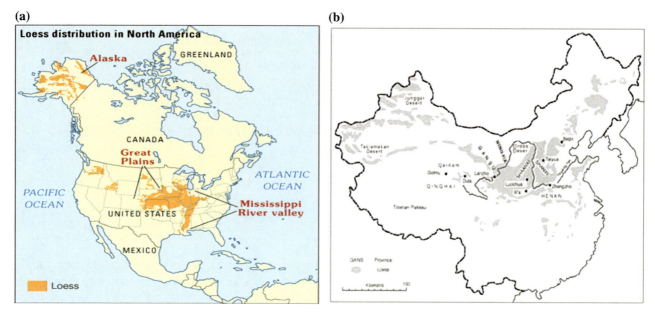

Fig. 11.1 Loess distribution. **a** In America de Nord. **b** China (*source* http://gec.cr.usgs.gov)

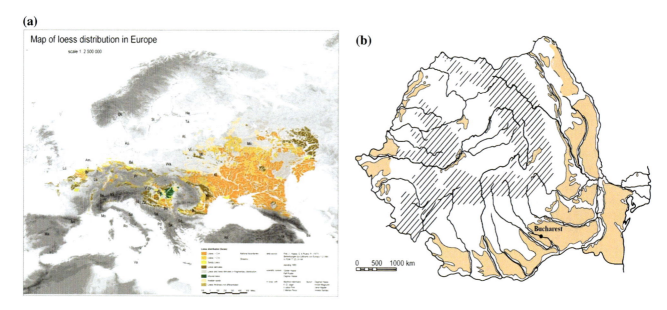

Fig. 11.2 Loess distribution (*source* Dagmar Haase/UFZ). **a** Europe. **b** Romania

Table 11.1 The minimum and maximum values of geotechnical parameters

Param	Clay (%)	Silt (%)	Sand (%)	w_L (%)	w_P (%)	w (%)	n (%)	Sr (%)	M2-3 (daN/cmp)	im3 (cm/m)	φ (grade)	c (kPa)
Min	14	50	3	32	12	7.8	46	0.4	18.7	0.6	5	5
Max	29	80	18	40	17	28.5	54	1	107	15	30	48

In order to make a correct classification of a soil, a comprehensive analysis of all the following parameters are required:

- parameters defining its composition and physical properties (granulometry, porosity);

- mechanical parameters: values of the index of the specific subsidence by wetting below the mark of 300 kPa and the structural strength obtained from endometrium testing;
- the thickness of loess, found in the history of the works from the area.

Fig. 11.3 Diagrams of compression subsidence

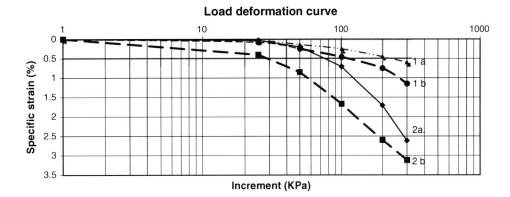

Load deformation curve

In the Dobrogea region loess of both categories occur, and their structural strength is based on the category they fall into: for 25–60 kPa is loess of category B and for 80–100 kPa is loess of category A.

In Fig. 11.3 are presented the diagrams of compression subsidence for the loesses with a low porosity, soil denoted by 1 and loesses with a high porosity, denoted by 2; (a) samples with natural moisture and (b) samples that were previously flooded.

11.4 Foundation Solutions

To ensure the normal behavior of buildings founded on land susceptible to wetting, there is a need for rational foundation systems and appropriate measures to avoid flooding of the foundation soil, both during and after construction. The humidity action on soils susceptible to wetting must be viewed from two perspectives, namely: the reduction in the bearing capacity and the growth of settlements under construction load transmitted (additional subsidence due to moisture). The foundation solutions are adopted according to the nature of the soil foundation, the hydrostatic level and construction characteristics:

- direct foundation on loess;
- foundation reinforced trough layers of cohesive soils (loess, loess lined with various waterproofing solutions, loess mixed with cement or lime);
- loess consolidation by intensive compaction (hard mallet and super hard) or with different injection solutions, heat treatment;
- adopting a foundation system that exceed sensitive soil layer wetting (deep foundations, piles, columns, etc.) and that are embedded in the insensitive wetting layer.

11.5 Conclusions

The infrastructure and construction on land susceptible to collapse under wetting have created problems. Their stability is not problematic under natural field conditions, but problems can arise if there are additional moistening that can cause deformations of buildings located on these moisted lands.

The special character of these lands is reflected in the fact that for them has been developed a normative—NP 125:2010 "Normative for foundation of construction on land susceptible to wetting". This legislation provides elements for identification, classification, conditions that have to be taken into account in designing, the recommended constructive and operation measures, maintenance and monitoring of buildings and facilities located on such lands.

References

Andrei S et al. La systematisation, le stockage et la reutilisation de informations geotechniques. Principe d' organisation d'une banque des donnees geotechniq

Andrei S et al (2006) The systematization and storing methods of information concerning the geotechnical parameters. XIII Danube European conference on geotechnical engineering, Ljubljana

Cazacu GB Contributions concerning the methods of establishing of geotechnical parameters of soils moisture sensitivity, Ph.D. thesis

Dianu VD (1982) Loess deposits as founding soils. Editura Tehnica

Ferreira RC, Monteiro LB (1985) Identification and evaluation of collapsibility of colluvial soils that occur in the Sao Paulo State. First international conference on geomechanics in tropical lateritic and saprolitic soils, vol I, Brazilia, pp 269–280

Muhs DR (2006) Late quaternary loess in northeastern Colorado: part I-age and paleoclimatic significance. Geol Soc Am Bull 111:1861–1875

Andrei S, Cazacu GB, Zarojanu D The systematization and storing methods of information concerning the geotechnical parameters. XIII Danube European conference on geotechnical engineering, Ljubljana

Influence of Micro-texture on the Geo-engineering Properties of Low Porosity Volcanic Rocks

12

Ündül Ömer and Amann Florian

Abstract

The geo-engineering properties of rocks often depend on their petrographic, mineralogical and micro-structural features, and the interaction of micro-texture and physico-mechanical properties is often relevant. A series of petro-physical and mechanical tests on low porosity volcanic rocks suggest that small changes in porosity or unit weight can cause strength variations. Petro-physical and quantitative mineralogical analysis were utilized to understand these variations. In addition, quantitative petrographic studies focusing on distribution of minerals and mineral dimensions were conducted. Microstructural studies were carried out on thin sections before and after mechanical loading to analyse the distribution of micro- and macro-cracks which formed during unconfined compression tests. The results of petro-physical, petrographic, micro-structural, and mineralogical analysis suggest, that both, the peak strength and crack initiation threshold are strongly influenced by the distribution of phenocrysts (e.g. biotite, plagioclase) and the ratio between the total content of phenocrysts to the fine-grained groundmass. On the other hand it was found that variations in petro-physical properties (e.g. unit weight) and Young's Modulus are associated with the mass fraction of minerals.

Keywords

Crack initiation • Elastic properties • Unconfined compressive strength • Quantitative petrography • Volcanic rocks

12.1 Introduction

Variations in geo-engineering properties associated with textural variations have been examined by many researchers (Prikryl 2006). Most of these studies investigated changes in petro-physical and mechanical properties based on qualitative and/or semi quantitative petrographical and mineralogical analysis. More recent studies were focused on variations in crack initiation (CI), peak strength (i.e. UCS), and crack propagation associated with micro-textural properties (Eberhardt et al. 1999; Nicksiar and Martin 2013). In this study the effect of micro-texture and mineral constituents on petro-physical (i.e. ultrasonic p-wave velocity V_p) and mechanical properties [i.e. UCS, CI, Young's modulus (E)] were investigated utilizing quantitative mineralogical and petrographic studies.

12.2 Methods

Mineralogy of the samples was determined with X-ray powder diffraction analysis (XRD). The quantitative mineral composition of the samples was determined with the Rietveld program AutoQuan (GE SEIFERT). Petrographic studies were used to trace the boundaries of individual minerals (e.g. plagioclase, amphibole, biotite) to quantify

Ü. Ömer (✉)
Geological Engineering Department, Istanbul University,
34320 Istanbul, Turkey
e-mail: oundul@istanbul.edu.tr; omer.undul@erdw.ethz.ch

Ü. Ömer · A. Florian
Engineering Geology, Swiss Federal Institute of Technology,
8092 Zurich, Switzerland

G. Lollino et al. (eds.), *Engineering Geology for Society and Territory – Volume 6*,
DOI: 10.1007/978-3-319-09060-3_12, © Springer International Publishing Switzerland 2015

their area and distribution. Each mineral grain with a major axis >200 μm was traced. Mineral grains with a major axis <200 μm were considered as groundmass.

Unconfined compression tests were accomplished on a servo-hydraulic rock testing device with a digital feedback control. Axial load was applied in such a way as to maintain a constant radial displacement rate of 0.03 mm/min). Two strain based methods were used to determine the stress at CI: (1) according to Brace et al. (1966) and (2) according to Lajtai (1974).

12.3 Results and Interpretation

12.3.1 Mineralogical, Petrographic and Microstructural Properties

Macroscopically, the samples are composed of varying amounts of plagioclase, amphibole and biotite as phenocrysts and fine grained groundmass (Table 12.1). The areas of phenocrystals in respect to the investigated areas are given in Table 12.2. It should be noted that quartz and orthoclase were very rarely observed as phenocrystals, but the mass proportion of these minerals were considerably high (Table 12.1). These findings suggest that almost all of the orthoclase and most of the quartz exist within the groundmass. Thin sections taken after mechanical loading revealed that cracks initiate along boundaries between different minerals, and along the mineral—groundmass boundary. Cracks penetrating groundmass and Biotite phenocrystal are interpreted to form at higher axial stress (Ündül et al. 2013).

12.3.2 Physical and Mechanical Properties

With increasing orthoclase content the unit weight tend to decrease. Increasing content of quartz and plagioclase increase the unit weight (Fig. 12.1). This is most probably related to the higher density of quartz (2.63–2.65) and plagioclase (2.62–2.76) relative to orthoclase (2.55–2.63). An increasing mass fraction in plagioclase was also associated with an increase in V_p. On the other hand, the normalized area of basic microstructural components (Tables 12.2 and 12.3) did not provide any relations between unit weight and V_p.

Table 12.2 Normalized area of basic microstructural components obtained from thin section studies

SG	Groundmass (%)	Amphibole (%)	Biotite (%)	Feldspars (%)
1	51–59	2–16	2–9	13–34
2	54–56	5–21	4–13	18–24
3	47–63	11–26	0–13	12–34
4	53–61	4–20	1–9	16–40

SG sample group, *Feldspars* plagioclase felds + alkali felds

The study also suggests that with increasing mass fractions of quartz and plagioclase the Young's modulus increases (Fig. 12.2). An opposing effect on E was found for an increasing mass fraction of orthoclase. These tendencies might be associated with the different Young's modulus of these minerals [e.g. Quartz: 91–105 GPa; Orthoclase: 73–74 GPa; Plagioclase: 81–106 GPa, (Data provided from Prodaivoda et al. 2004)] and their net contribution to the bulk Young's modulus. Changes in Poisson's ratio were only dependent on changes in the normalized area of the ground mass. The Poisson's ratio decreases as the normalized area of the groundmass increases.

Unconfined compressive strength tests revealed UCS values ranging from 103–289 MPa. Only some specimen showed macroscopic cracks at failure. The crack initiation level was found to be 0.45–0.46 × UCS (Fig. 12.3).

It was observed that with an increasing normalized area of plagioclase and amphibole phenocrysts (i.e. normalized area of the groundmass decreases), CI and UCS increase (Fig. 12.4). Furthermore the data suggest that the UCS decreases with an increase in the normalized area of Biotite phenocrystal.

Ündül et al. (2013) showed on UCS test on Andesite specimens that cracks emanate from the boundaries of larger grains and penetrate the groundmass which primarily consists of orthoclase and quartz (Table 12.1). For larger normalized area of the groundmass, cracks, which formed under unconfined compression, were composed of cracks parallel to the maximum applied load and oblique cracks. With increasing phenocrystal area axial cracks dominate. This change in failure mechanism was associated with an increase in the radial strain in respect to the axial strain for axial stresses in excess of CI.

Table 12.1 Variations in mass fractions of dominant mineralogical components obtained with XRD studies

SG	Amph (%)	Biotite (%)	CM (%)	Orth. (%)	Plag. (%)	Quartz (%)
1	3.1–4.9	2.1–2.5	1.1–2.0	22.8–23.5	45.6–47.8	13.0–13.5
2	17.6–19.4	0.5–1.8	1.2–1.6	24.0–25.1	25.0–28.9	10.2–12.1
3	15.2–17.1	0.5–1.3	1.2–1.5	21.2–21.9	35.9–39.4	14.0–14.6
4	5.7–6.9	1.6–1.9	3.6–5.4	21.0–22.1	40.9–42.7	11.8–12.4

SG sample group, *Amph* amphibole, *CM* clay minerals, *Orth* orthoclase, *Plag* plagioclase

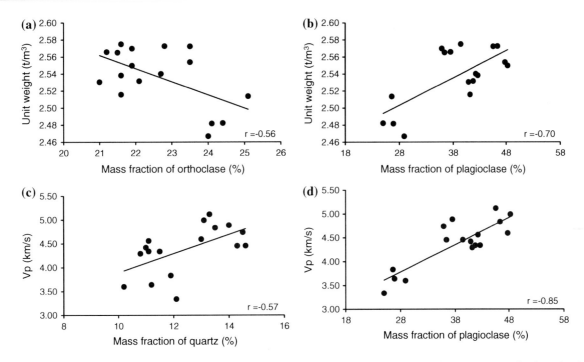

Fig. 12.1 The change of basic physical properties with mass fractions of different minerals. **a** Unit weight versus mass fraction of orthoclase. **b** Unit weight versus mass fraction of plagioclase. **c** V_p versus mass fraction of quartz. **d** V_p versus mass fraction of plagioclase

Table 12.3 Minimum and maximum values of some mechanical properties

SG	UCS (MPa)	σ_{ci} (MPa)[a]	σ_{ci} (MPa)[b]	Young's modulus (GPa)
1	108–271	107–126	101–127	55.3
2	103–199	47–78	46–79	30.2
3	148–289	62–130	63–129	52.4
4	144–163	69–74	65–73	39.3

SG sample group
[a] According to Brace et al. (1966)
[b] According to Lajtai (1974)

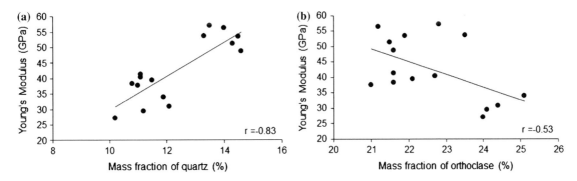

Fig. 12.2 The relation of Young's modulus with (**a**) the mass fraction quartz (**b**) the mass fraction orthoclase

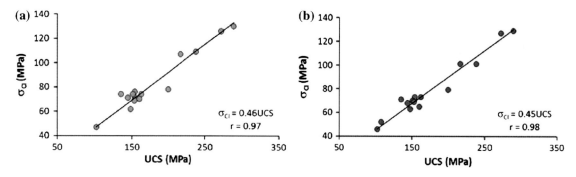

Fig. 12.3 The relation of crack initiation with UCS (**a**) crack initiation values obtained according to Brace et al. (1966) (**b**) crack initiation values obtained according to Lajtai (1974)

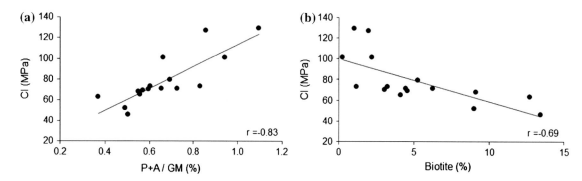

Fig. 12.4 The influence of the ratio of selected phenocrystals to groundmass on CI. P + A, represents the sum of the normalized areas of plagioclase and amphibole. GM represents the groundmass area (Ündül et al. 2013)

12.4 Conclusion

In this study quantitative petrographic and mineralogical data were utilized to quantify variations in micro-textural characteristics of andesitic rocks and their effect on crack initiation, UCS, Young's modulus and petro-physical properties. The synthesis of results showed that:

- The unit weight of the utilized rocks increases with increasing mass fractions of quartz and plagioclase, but decreases with increasing orthoclase content.
- No relation between the spatial distribution of minerals and petro-physical properties can be found. Petro-physical properties are dependent on the mass fractions of orthoclase, plagioclase and quartz.
- The crack initiation threshold was 0.45–0.46 × UCS.
- With increasing normalized area of plagioclase and amphibole phenocrystals the UCS increased. An increase in the normalized area of biotite phenocrysts caused a decrease in CI and UCS.
- With increasing mass fraction of orthoclase the Young's modulus decreased.
- Poisson ratio increases with decreasing normalized area of the groundmass.

- With increasing phenocrystal area the failure mode changes and was associated with an increased number of axial cracks.

References

Brace WF, Paulding BR, Scholz C (1966) Dilatancy in fracture of crystalline rocks. J Geophys Res 71(16):3939–3953

Eberhardt E, Stimpson B, Stead D (1999) Effect of grain size on the initiation and propagation threshold of stress-induced brittle fracture. Rock Mech Rock Eng 32(2):81–99

Lajtai EZ (1974) Brittle fracture in compression. Int J Fract 10(4):525–536

Nicksiar M, Martin CD (2013) Crack initiation stress in low porosity crystalline and sedimentary rocks. Eng Geol 154:64–76

Prikryl R (2006) Assessment of rock geomechanical quality by quantitative rock fabric coefficients: limitations and possible source of misinterpretations. Eng Geol 87:149–162

Prodaivoda GT, Maslov BP, Prodaivoda TG (2004) Estimation of thermoelastic properties of rock-forming minerals. Russ Geol Geophys 45(3):389–404

Ündül Ö, Amann F, Aysal N, Plötze M (2013) Micro-textural controlled variations of the geomechanical properties of andesites. In: Kwasniewski M, Lydzba D (eds) Eurock 2013 rock mechanics for resources, energy and environment. CRC Press, Boca Raton, pp 357–361

Conceptual Geological Models, Its Importance in Interpreting Vadose Zone Hydrology and the Implications of Being Excluded

Matthys A. Dippenaar and J. Louis van Rooy

Abstract

Vadose zone conditions are becoming increasingly important in site investigation, including, for instance, (1) the protection of the phreatic zone from surficial contaminants (aquifer vulnerability), (2) surface water–groundwater interaction and biodiversity, (3) water influencing infrastructure development and (4) problem soil behaviour resulting in surface expressions of subsurface volume change. This multidisciplinary paradigm involves a wide range of specialists, but at the root of issues pertaining to the vadose zone is the geological regime in which it occurs. A conceptual model should include all fundamental branches of geology, viz. stratigraphy, mineralogy and petrology, structural geology and physical geology. The importance of a proper geological model is addressed at the hand of selected case studies in urban and peri-urban South Africa. In all instances, the sites were developed with subsequent issues arising due to inadequate vadose zone investigation. To evaluate, geological models were compiled based on available geological and geomorphological information. Physical properties such as detailed soil profiles and grading analyses were inferred to address vertical and spatial material variability. Mineralogy of the various soil horizons was used in conjunction with bulk dry densities to determine porosity and to address pedogenetic and eluviation processes. Hydrological data include percolation tests and Atterberg limits and were inferred onto the model to clarify anticipated hydrological behaviour. The additional geological data improve the understanding of the case studies incorporating ephemeral hillslope wetlands, constructed fill and water addition through leaking pipelines and irrigation, through the addition of knowledge overlooked by hydraulic testing exclusively and accentuate the importance of proper geological understanding prior to hydrological interpretation. The major issues arising, apart from damage to infrastructure and contamination of water resources, are excessive rehabilitation costs, decrease in aesthetic value and general discontent of land owners and proximate residents. The geological model is imperative and should not be excluded or overlooked due to increasingly popular alternative methods of investigation.

Keywords

Geological model • Vadose zone • Hillslope wetland • Percolation • Granite gneiss

M.A. Dippenaar (✉) · J.L. van Rooy
Engineering Geology and Hydrogeology, Department of Geology, University of Pretoria, Private Bag X20, Hatfield, Pretoria, 0028, South Africa
e-mail: madip@up.ac.za; matthys.dippenaar@up.ac.za; madippenaar@gmail.com
URL: www.up.ac.za/geology

13.1 Introduction

Urban development induces changes to the vadose zone, which is readily overlooked in compilation of a conceptual model prior to development. Made ground is of different grading and compaction than in situ materials, saturation is increased by irrigation or leaking pipelines, and surfaces are

Fig. 13.1 Perspectives on the vertical succession of earth materials [adapted from (**a**) Foster (1984, 2012), (**b**) Koita et al. (2013), (**c**) Hillel (2003), (**d**) Weinert (1980), Brink (1985); combined from Dippenaar and Van Rooy (2014)]

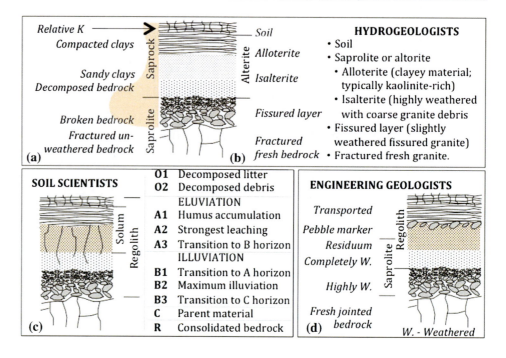

sealed as land is being developed. These are just come factors contributing to the already complex vadose zone.

The importance of proper vadose zone conceptualization, notably in urban settings, is explained at the hand of five case studies underlain by Lanseria tonalite gneiss of the Johannesburg Dome. Granites in South Africa are—given its age—highly variable due to an intricate tectonic and geomorphological history. These case studies illustrate the importance of conceptual vadose zone models, as well as the implications of exclusion.

13.2 Conceptual Models

Perspectives regarding the vertical succession of earth materials and the classification of soils vary between disciplines. Albeit based on different intentions, an all-encompassing multi-disciplinary approach may significantly improve information gained from soil profile descriptions (e.g. Dippenaar 2012). Various approaches to the classification of weathered earth materials are shown in Fig. 13.1.

Additional to the vertical variation, the spatial variation and subsurface hydrology can best be simplified by the catena concept, which has been well-documented (e.g. Schaetzl and Anderson 2005).

The importance of conceptual models is also well documented, e.g.:

- The importance thereof in groundwater models (Izady et al. 2013)
- The lack of proper understanding in conceptual hydrological models on land (Lahoz and de Lannoy 2013)

- The special circumstances in karst settings (Bakalowicz 2005)
- Incorporation of multidisciplinary data and different scales of observation (Dewandel et al. 2005)
- Considerations of different potential conceptual models in assessment of aquifer vulnerability (Seifert et al. 2007)
- Inclusion of subsurface flow in shallow saprolite and deep bedrock (Banks et al. 2009)
- Challenges and trends in geological visualisation and modelling (Turner 2006)
- The inclusion of historical data to potentially replace the conceptual model (Royse et al. 2009)
- The influence of alternative conceptual models on predicting beyond calibration base of the flow model (Troldborg et al. 2007)
- The concept of hydrostratigraphy contributing to proper conceptual modelling (Allen et al. 2007; Angelone et al. 2009; Heinz and Aigner 2003)
- The incorporation of geology, engineering geology, hydrogeology and geomorphology in advancing conceptual model quality Dippenaar and Van Rooy (2014).

Essentially the principles of the triangle of geomechanics and the triangle of engineering geology apply as shown in Fig. 13.2 (Bock 2006) after which actual data can be inferred to model scenarios.

The hydrological cycle is an intricate interaction between the lithosphere, atmosphere and biosphere and represents the movement of water in different phases. With respect to the vadose zone, more attention is generally given to the soil or plant root zone and the processes of surface water–groundwater interaction and evapotranspiration, often with the

Fig. 13.2 Triangle of geomechanics (*left*) and of engineering geology (*right*) (Bock 2006)

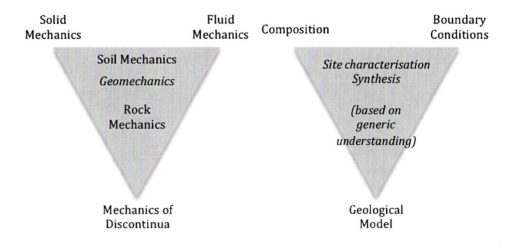

Fig. 13.3 Movement of water as depicted through the hydrological cycle (adapted from Todd and Mays 2005)

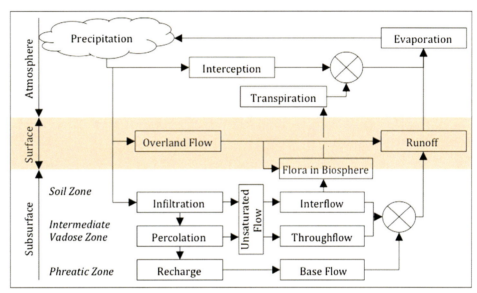

distinct exclusion of the deeper intermediate vadose zone (Fig. 13.3). Numerous aspects have to be considered when compiling a comprehensive vadose zone model. Some of these are illustrated in Fig. 13.4 and indicate the vast range of input in such conceptual models (Dippenaar et al. 2014):

1. **Atmosphere–Biosphere**
   1.1. Climate–precipitation, evaporation
   1.2. Plant water availability and transpiration
   1.3. Surface water–groundwater interaction
   1.4. Sensitive ecosystems
   1.5. Land use and land cover
2. **Soil Zone**
   2.1. Infiltration
   2.2. Perched water tables
   2.3. Interflow, throughflow
   2.4. Translocation and pedogenesis
3. **Intermediate Vadose Zone**
   3.1. Percolation to eventual recharge
   3.2. Soil vadose zone

3.3. Fractured rock vadose zone
3.4. Variable saturation.

13.3 Case Studies

For the purposes of explanation, geology and climate are kept as constant parameters and all case studies presented as situated on tonalite gneiss (Lanseria Gneiss, Johannesburg) within the Midrand and Johannesburg areas of South Africa.

13.3.1 Pedogenesis and Ephemeral Hillslope Wetlands

Randjesfontein, situated on tonalite gneiss of the Archaean basement complex in Midrand (RSA), was investigated during two subsequent stages. During the first, vegetation was absent due to veldt fire in the dry winter months. During

Fig. 13.4 Considerations in the compilation of a vadose zone model

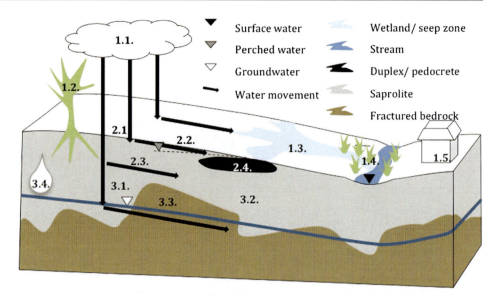

Fig. 13.5 Shallow interflow through glaubular to honeycomb ferricrete resulting in the formation of an ephemeral hillslope wetland on Archaean tonalite gneiss (Midrand, RSA) (adapted from Dippenaar et al. 2014)

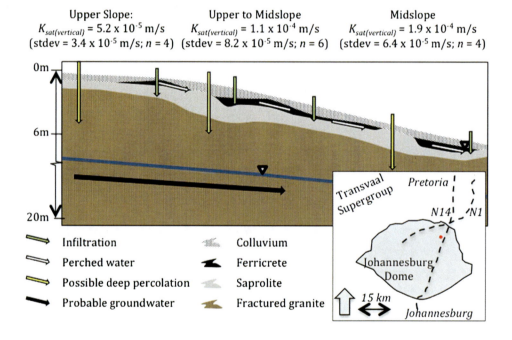

the second, the geotechnical investigation noted marshy areas at the site and deduced that the ferricrete in the profiles represents a periodically perched water table system. These markers were overlooked and, using only the absence of wetland vegetation indicators as proof, the site was excavated for construction. With changing seasons, the wetland wetted up again and water influx into the excavation resulted in cessation of construction as water collected at the excavation floor and wetland conditions regenerated on barren bedrock.

Of interest in the conceptual understanding of this study area are the following (Fig. 13.5):

- Ferricrete appears much more heterogeneously and anisotropically as anticipated. Given the 150 × 50 × 10 m excavation, the varying thickness, as well as the localized absence, of ferricrete can clearly be seen. Important for the conceptual model is to properly infer the vertical and spatial extent of the pedocretes.
- In the instance of this study site, the ferricrete changed its role in the subsurface hydrology. With initial perching probably on low permeability bedrock, the process of pedogenesis progressively resulted in varying stages of ferruginization in residuum and the pebble marker. In the stages before nodular/glaebular and in the hardpan stage,

the ferricrete is of low permeability and porosity with water perched above. However, in the honeycomb phase, the porosity is low due to the cementation, but the calculated porosity of 0.15 as opposed to the overlying 0.21 exists as large connected pore spaces where parent soil has been washed out. The perching, therefore, occurs localized within the ferricrete, and not above.

13.3.2 Made Ground and Variably Compacted Fill

Urban hydrogeology has an increased variable: that of anthropogenic materials. These include a wide variety of materials of variable compaction and grading (e.g. uncompacted imported fill), as well as variable composition (e.g. contaminated land). In two separate studies in the suburbs of Alexandra and Linbro Park respectively, the following was found:

- Cut-and-fill operations have to be included as significant impacts on shallow interflow. This is notable in all three studies and the excavation through illuviation zones result in water being subjected to other lateral pathways and not necessarily deeper percolation. Compaction of imported materials may have higher porosity and/or permeability, fill for drainage systems may interrupt flow paths in unsaturated materials and induce imbibition rather than drainage, and retaining wall systems may become unstable and exhibit seepage through faces.
- Non-uniform building rubble and fill material may influence downslope properties and may create periodical waterlogged conditions in poorly compacted fill materials. This is notable in the central urban portions where old buildings are demolished for the construction of high rises, requiring deeper foundations and basements.

13.3.3 Increased Infiltration and Interflow

Increased water addition can be explained at a case study in Fourways (residential golf course). Significant additional volumes of water was released, resulting in the following:

- The natural systems were in hydrological equilibrium where water infiltrated and sporadic small wetlands occurred in Fourways. These areas were developed as the golf course and the residential areas on the perimeters and downslope.
- Increased golf course irrigation resulted in the formation of a perched water system where the high saturation of surface soils induced interflow on the bedrock interface.
- The main damage is with water imbibing into porous plaster and mortar, resulting in extensive damage to buildings.

13.4 Discussion

The geological model should include the vadose zone and all variability, whether natural or induced, in material properties and moisture contents. Urban development notably exacerbates the uncertainty in the vadose zone due to, for instance, surface sealing, disruption of natural structure and compaction, inducing increased flow, and diverting flow paths.

The variation in properties between the soil zone and intermediate vadose zone, and between the soils, saprolite and bedrock should be understood properly to ensure optimal construction. Damage to infrastructure on-site and off-site, drainage of wetlands, generation of waterlogged conditions, and weakening of retaining structures are just some of the significant results of proper understanding of the vadose zone.

The importance of the vadose zone exceeds vulnerability assessments and optimizing irrigation practices. It should be seen as a fundamental component of all hydrological and geotechnical assessments and should be included as a detailed component of any conceptual model.

Acknowledgments The authors wish to acknowledge the South African Water Research Commission (www.wrc.org.za) for funding of project K5/2052 on Multidisciplinary Vadose Zone Hydrology (Dippenaar et al. 2014).

References

Allen DM, Schuurman N, Deshpande A, Scibek J (2007) Data integration and standardization in cross-border hydrogeological studies: a novel approach to hydrostratigraphic model development. Environ Geol 53:1441–1453

Angelone M, Cremisini C, Piscopo V, Proposito M, Spaziani F (2009) Influence of hydrostratigraphy and structural setting on the arsenic occurrence in groundwater of the Cimino-Vico volcanic area (central Italy). Hydrogeol J 17:901–914

Bakalowicz M (2005) Karst groundwater: a challenge for new resources. Hydrogeol J 13:148–160

Banks EQ, Simmons CT, Love AJ, Cranswick R, Werner AD, Bestland EA, Wood M, Wilson T (2009) Fractured bedrock and saprolite hydrogeologic controls on groundwater/surface–water interaction: a conceptual model (Australia). Hydrogeol J 17:1969–1989

Bock H (2006) Common ground in engineering geology, soil mechanics and rock mechanics: past, present and future. Bull Eng Geol Environ 65:209–216

Brink ABA (1985) Engineering geology of southern Africa, vol 4. Post-Gondwana deposits. Building Publications, Pretoria

Dewandel B, Lachassagne P, Boudier F, Al-Hattali S, Ladouche B, Pinault J-L, Al-Suleimani Z (2005) A conceptual hydrogeological model of ophiolite hard-rock aquifers in Oman based on a multiscale and a multidisciplinary approach. Hydrogeol J 13:708–726

Dippenaar MA (2012) How we lose ground when earth scientists become territorial: defining "soil". Nat Resour Res 21(1):137–142

Dippenaar MA, Van Rooy JL (2014) Review of engineering, hydrogeological and vadose zone hydrological aspects of the Lanseria Gneiss, Goudplaats-Hout River Gneiss and Nelspruit Suite Granite (South Africa). J Afr Earth Sc 91:12–31

Dippenaar MA, Van Rooy JL, Breedt N, Huisamen A, Muravha SE, Mahlangu S, Mulders JA (2014) Vadose zone hydrology: concepts and techniques. WRC report TT 584/13. Water Research Commission, Pretoria

Foster S (1984) African groundwater development: the challenges for hydrological science. IAHS. Wallingford UK. 3–12

Foster S (2012) Hard-rock aquifers in tropical regions: using science to inform development and management policy. Hydrogeol J 20:659–672

Heinz J, Aigner T (2003) Hierarchical dynamic stratigraphy in various quaternary gravel deposits, Rhine glacier area (SW Germany): implications for hydrostratigraphy. Int J Earth Sci (Geol Rundsch) 92:923–938

Hillel D (2003) Introduction to environmental soil physics. Academic Press, Waltham, p 494 (eBook)

Izady A, Davary K, Alizadeh A, Ziaei AN, Alipoor A, Joodavi A, Brusseau ML (2013) A framework toward developing a groundwater conceptual model. Arab J Geosci. doi:10.1.1007/s12517-013-0971-9

Koita M, Jourde H, Koffi KJP, Da Silveira KS, Biaou A (2013) Characterization of weathering profile in granite and volcanosedimentary rocks in West Africa under humid tropical climate conditions. Case of the Dimbokro Catchment (Ivory Coast). J Earth Syst Sci 122(3):841–854

Lahoz WA, De Lannoy GJM (2013) Closing the gaps in our knowledge of the hydrological cycle over land; conceptual problems. Surv Geophys. doi:10.1007/s10712-013-9221-7

Royse KR, Rutter HK, Entwisle DC (2009) Property attribution of 3D geological models in the Thames Gateway, London: new ways of visualising geoscientific information. Bull Eng Geol Environ 68:1–16

Schaetzl RJ, Anderson S (2005) Soils: genesis and geomorphology. Cambridge University Press, New York

Seifert D, Sonnenborg TO, Scharling P, Hinsby K (2007) Use of alternative conceptual models to assess the impact of a buried valley on groundwater vulnerability. Hydrogeol J 16:659–674

Todd DK, Mays LW (2005) Groundwater hydrology, 3rd edn. Wiley, New Jersey

Troldborg L, Refsgaard JC, Jensen KH, Engesgaard P (2007) The importance of alternative conceptual models for simulation of concentrations in a multi-aquifer system. Hydrogeol J 15:843–860

Turner KA (2006) Challenges and trends for geological modelling and visualisation. Bull Eng Geol Environ 65:109–127

Weinert HH (1980) The natural road construction materials of South Africa. Academica, Cape Town

Treatment of Fossil Valley in Dam Area: A Case Study

14

A.K. Singh and Bhatnagar Sharad

Abstract

Himalaya occupies a unique position internationally providing ample opportunities for hydro power development through its fast flowing rivers. The total potential of three major river system of Himalaya i.e. Indus, Ganga & Brahmaputra is estimated to be 65,623 MW (at 60 % Load factor). However, only about 25 % of this potential has been tapped so far. In recent times, when ideal locations for dam in Himalayan region have almost exhausted, many dams of variable dimensions are being constructed in adverse geological settings, taking into consideration various ground improvement techniques. These dam structures are often faced with the problems of sheared and fractured rock at the foundation grade, deep overburden, slope instability of the abutments, presence of solution cavities etc. Occasionally, in glaciated valleys, fossil valleys are formed due to change in river course by sudden blockage of river in geological past. These geomorphic features are easily picked up on the ground but to establish their extent at depth requires detailed subsurface investigations. Suitable remedial measures are formulated if these features are located in close vicinity of any engineering structures. The present paper deals with the fossil valley treatment carried out on the right abutment of Parbati Dam, Stage-II under construction by NHPC Ltd in Himachal Pradesh where extensive grouting using TAM technology was carried out in phased manner utilizing various grout mix, ultrafine cement, admixtures etc. Grouting pattern was evolved so that the whole fluvioglacial material lying in the vicinity of dam area could be grouted. Post grouting check holes and permeability tests showed marked reduction in permeability values before & after the grouting, ascertaining successful treatment of the fossil valley.

Keywords

Fossil valley • Tube a manchhettes • Permeation grouting • Admixtures

14.1 Introduction

Parbati H.E. Project-St-II (800 MW) is one of the largest hydroelectric projects in Himachal Pradesh, under construction by NHPC Ltd, a premier organization in India for hydropower development. The project is located on river Parbati, a tributary of Beas river in Distt Kullu of Himachal Pradesh. The project envisages construction of an 85 m high concrete gravity dam near village Barsheni/Pulga to divert River Parbati into the main water conductor system. The project is an μinter basin transfer scheme. The proposed 31.5 km long, 6 m finished dia HRT shall conduct water from Pulga via Garsa/Sheelagarh valley which shall ultimately be discharged into river Sainj, a downstream tributary of river Beas at Suind village where a surface power house is under construction for generating 800 MW of power. In order to augment the power generation it is planned to divert the discharge of Jigrai nala located downstream of Pulga

A.K. Singh · B. Sharad (✉)
Parbati H.E.Project, Stage-II, NHPC Ltd, Faridabad, Himachal Pradesh, India
e-mail: bhatnagarsharad2008@gmail.com

A.K. Singh
e-mail: aksighnhpc@gmail.com

Dam into the reservoir. The discharge of other intermediate nalas i.e. Hurla nala and its tributaries (Manihar and Pancha nala) and flow of Jiwa nala shall also be diverted to HRT through small diversion structures i.e. trench weirs, feeder tunnels and drop shafts. The layout of the project is shown as Fig. 14.1.

14.2 Brief Regional Geology and Geo-Morphology of Project Area

The project is located in lesser Himalaya. The site is characterized by elevation ranging between El ± 1,200 m to El ± 2,200 m above MSL. The great difference in elevation is suggestive of a very young and immature topography conforming to the late orogenic uplift of the Himalayas marked by high mountains, narrow & glaciated valley, fossil valleys with steep slopes and escarpments.

In project area a variety of rock formations are exposed which have undergone extensive structural deformations due to tectonic activity associated with Himalayan orogeny. The deformations are exhibited in the form of folds, faults and thrusts. The main rock types in this region belong to Jutogh Banjar and Larji formations ranging in age from Precambrian to Permian and are separated by thrusted margins.

14.3 Geo-Technical Assessment of Dam Area

The dam site is located on Parbati river, near Pulga/Barseni village, in a narrow 'U' shape gorge, about 150 m downstream of its confluence with Tosh nala. At dam site the left right abutment rises steeply from the river bed with an average slope of 80°. Schistose quartzites with thin intercalated bands of contorted mica schist bands belonging to Kullu member of Jutogh Formation are exposed on both the abutments.

The rockmass is dissected by 3 sets of joints with foliation joints (040-060°/35–45°) favourably dipping in upstream direction. The other two sets having attitudes 180–210°/45°—and 250–310°/50–60° are also distinct.

The right bank is extensively covered with thick fluvioglacial material up to El ± 2,270 m (Ref: Photo 14.1) whereas on the left bank rock is exposed up to El ± 2,230 m with thin cover of slope wash material, however, beyond El ± 2,240 m the entire slope on left bank is also covered with thick fluvioglacial material.

The dam foundation has been laid in sound rock comprising of schistose quartzite rock, after removing 9–18 m of river borne material. Presently the intake structures & spillway blocks are under construction.

14.3.1 Fossil Valley

In close vicinity of dam axis, on right bank, the signature of the old course of Parbati river exists in the form of fossil valley. The depth of this valley is about 101 m and is filled with glacial moraines comprising of large boulders, pebbles & cobbles in silty matrix. This has also been established by the geological investigations carried out by Geological Survey of India & NHPC. Large accumulation of fluvioglacial material observed on the right bank of Parbati river is indicative of concealment of its original course which got buried under a thick pile of moraine deposits. It is believed that after the recession of glaciers in Pleistocene period, river

Fig. 14.1 Layout of Parbati stage-II project

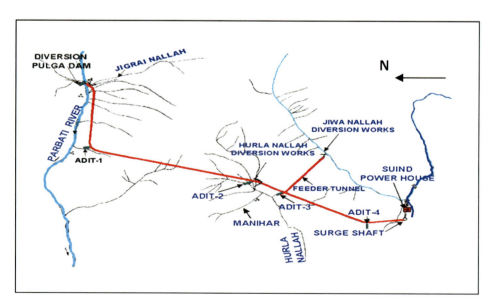

Photo 1 Exposures of fluvioglacial material on right bank

Parbati would have started flowing through the valley earlier occupied by glaciers. The valley seems to be occupied by moraines deposited by receding glaciers, leaving very limited space for river to flow, thereby reducing its transportation and eroding capacity which resulted in impoundment of a lake. The river then over toppled the barrier and followed a new course .The old river channel got filled with moraines and resulted in 'fossil valley'.

14.3.1.1 Geotechnical Investigations for Fossil Valley

Besides detailed geological mapping on 1:500 scale, the dam area in general and the fossil valley in particular was extensively investigated to establish the shape and extent of fossil valley. Following subsurface explorations were carried out:

14.3.1.2 Geophysical Surveys

Four seismic profiles namely P11-P14, aggregating to 800 m length was laid on the right bank in the fossil valley area. These profiles have established overburden depth of 80–90 m in the fossil valley area. A seismic velocity of 2,200 m/s was recorded indicating loose to semi compacted river-borne/fluvioglacial material.

14.3.1.3 Exploratory Drilling

The fossil valley area was explored by eight drill holes aggregating to 473 m depth. The deepest hole was 124 m. Permeability tests were also carried out in these holes. A brief account of drill holes is as under: (Table 14.1).

In general these holes established that fossil valley is mainly composed of four distinct type of material:

0-25 m-Yellowish brown to grey sand with boulders of granitic gneiss, quartzite & mica schist. Percentage of fine is more in comparison to coarse material.

25–30 m-Fine silt without pebble or boulders

30–60 m-Angular to sub angular boulders & pebbles of quartzite and schist

60–100 m-Silty sand with large boulders of gneiss, granitic gneiss.

14.3.1.4 Permeability and Groutability Tests

Variable overburden permeability ranging from 1×10^{-3} to 6×10^{-5} was noticed in fossil valley. Similarly the grout intake pattern indicated irregular & anisotropic condition of overburden. No effect of grouting (with normal cement) on reduction in permeability values were observed down to 22.6 m depth beyond which marginal decrease in permeability value were observed.

14.3.1.5 Exploratory Drift

A 220 m long drift was excavated on the right bank at El ± 2,166 m to explore the extent of the fossil valley. The drift started in rock comprising of schistose quartzites with band of biotite schist up to 130 m length. Between RD 130–210 m fluvioglacial material in silty matrix was observed. Beyond RD 210–220 m again schistose quartzite with thin bands of biotite schist were observed (Ref. Fig. 14.2). The above explorations established an 80 m wide and about 100 m deep fossil valley extending from u/s of dam to d/s of dam, near the diversion tunnel outlet portal.

14.4 Treatment of Fossil Valley

Leakage is apprehended through fossil valley, at the time of filling of the reservoir. Hence, it has been planned to create a cut off grout curtain by permeation grouting using TAM (*Tube a Manchhettes*) technique. This cut-off is proposed to

Table 14.1 Exploratory drill holes in fossil valley

Drill Hole no.	Location	Collar El (M)	Total depth (m)	Bedrock depth	Overburden/Bedrock characteristics
DH-2	Centre of fossil valley	2,204	116 m	101 m(El 2,103)	Quartz mica gneiss with schist bands
DH-3	u/s of DH-2	2,158	64	53 m (El 2,105)	Gneiss
DT-3	40 m u/s of dam axis extreme end of fossil valley)	2,175	45	21 m(El 2,154)	Quartzite with schistose quartzite bands
DT-5	D/s of dam axis(extreme end of fossil valley)	2,270	56	Bedrock not encountered	Riverborne/fluvioglacial material up to excavated depth
DDH/II/01	Fossil valley	2,201	60	Bedrock not encountered	Riverborne/fluvioglacial material up to excavated depth
DDH/II/03	Fossil valley	2,201	60	Bedrock not encountered	Riverborne/fluvioglacial material up to excavated depth
DDH/II/04	Fossil valley	2,197	61.5	Bedrock not encountered	Riverborne/fluvioglacial material up to excavated depth
DDH/II/05	Centre of fossil valley	2,200	124	101(EL2099)	Quartzite with mica schist bands

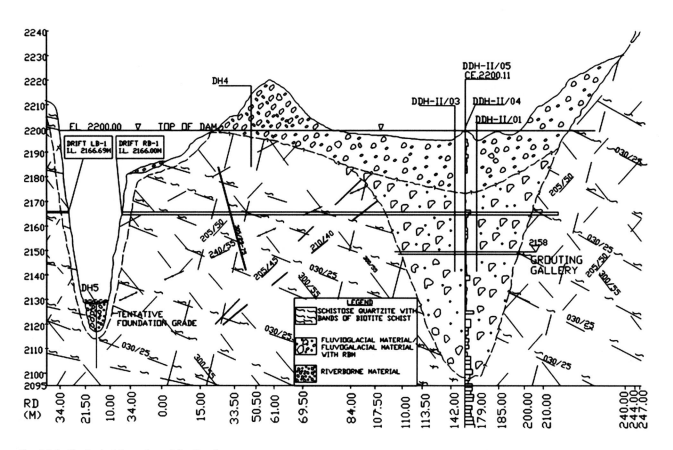

Fig. 14.2 Geological L-section of fossil valley

extend from the surface to alluvial/bedrock contact and up to 10 m into the bedrock. A 3.5 m × 3.8 m grouting gallery with portal at El ± 2,148 m was constructed to facilitate grouting. The grouting was carried out in two stages: In the first stage grouting was performed from surface at El ± 2,200 m to grouting gallery. This was designated as upper grout curtain. The length of upper grout curtain was 215 m. Drill hole ranging in depth from few meters to 50–55 m were drilled from the surface up to the level of grouting gallery at El ± 2,150 m. In the second stage the media between the grouting gallery & bedrock was grouted. This was designated as lower grout curtain. The length of the

Parbati Fossil Valley
Sketch of grout curtain pattern arrangement

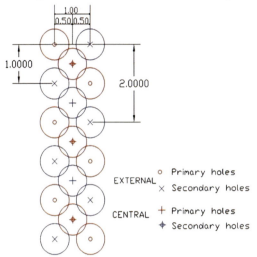

Fig. 14.3 Grouting pattern

lower grout curtain was 65–70 m. Overlapping between the upper grout curtain and lower grout curtain was ensured to have continuity.

14.4.1 Drilling/Grouting Equipment

Drilling was performed using three drilling rigs (Rodio SR52). One of the rig was electrically operated to be accommodated in grouting gallery. The holes were provided with a sleeve pipe (*tube a manchettes*) after retrieving the drill rods and before withdrawal of the casing. The borehole trajectory was measured utilizing the surveying instrument 'BORETRACK MKII'. The grouting was carried out by double acting piston pump type CIRO 10, operating at a maximum discharge pressure of 7.5 MPa. The plant had a computerized control & record unit to record the flow rate, grouting pressure and total grouted volume.

14.4.2 Grouting Pattern

Preliminary trials for selection of grouting components and composition were carried out in laboratory at Mumbai. Grouting was carried out in three rows i.e. upstream, downstream and central row. The distance between u/s & d/s hole was kept as 1 m from centre to centre. Holes of each row were also separated by a distance of 1 m. (Ref. Fig. 14.3). Grouting was achieved in two phases:

In the first phase u/s and d/s holes were grouted using fine cement (Blaine-4,500),while the second phase of grouting was carried out in the central hole using ultra fine cement (Blaine 10,000),after completing grouting of the external rows. Grouting was controlled by Pmax/Vmax (Pressure/volume Ratio). P max varied from 5bars to a max. 35 bars, whereas V max for external holes was set at 180 liters while for central it was set at 350 liters.

14.4.3 Grouting Materials

Following grouting materials were used for treatment:
1. **Cement**: Cement of blaine >4,500 cm^2/g was used for permeation grouting.
2. **Ultrafine cement**: Ultrafine cement of 8,000 cm^2/g blaine was used to fill the voids between very fine particles.
3. **Admixtures**: Bentonite, silica gel (Sodium silicate), admixtures were also used.

For first phase of grouting following grout mixes were used:

Cement/water ratio = 0.5, Bentonite/water ratio = 2 %, Additive/water ratio = 0.5 %

For second phase of grouting carried out in central hole for upper curtain:

Cement/water ratio = 0.225, Bentonite/water ratio = 2.5 % & Additive/water ratio = 0.3 % were used.

Similar ratio was used for carrying out grouting in the lower curtain. In order to seal the hollow space between the grout hole and Tam sleeve pipe, a cement/water ratio of 0.5 and bentonite/water ratio of 5 % were used (Fig. 14.4).

In all 536 nos of boreholes of variable depth were done in a stretch of 175 m in the Upper Curtain including 179 holes in upstream, 181 in downstream and 176 in central row, whereas in lower curtain 232 nos of holes were done including 77 holes each in upstream and downstream row respectively and 78 holes in centre row. Six no of check holes were also done to a certain the efficacy of the grout curtain. Permeability test carried out in check holes have given a value of $1.54–5.3 \times 10^{-5}$ against the targeted value of 1×10^{-4}. Similarly, for lower grout curtain 3 check holes have indicated permeability value between $2.6 \times 10^{-5}–2.12 \times 10^{-6}$ against the targeted 10^{-4} value. In addition, peizometers have been installed downstream of the curtain to monitor the variation in water level before and after impounding, to verify the grout efficacy. Overall, the test results showed marked improvement in ground condition after grouting.

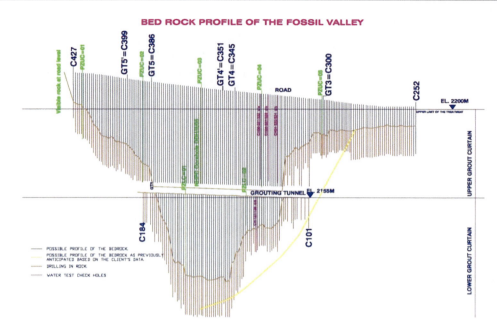

Fig. 14.4 Bed rock profile in fossil valley

14.5   Conclusion

The presence of an old fossil valley on the right bank of Parbati river was the main constraint in dam construction in Parbati H.E.Project (800 MW).Due to topographical condition it was not possible to shift the dam location elsewhere. Effective grouting techniques using TAM technology were carried out to consolidate the huge pile of fluvioglacial material. The grouting carried out in phased manner have shown considerable reduction in the permeability of the glacial material, however, it will be interesting to observe the efficacy of grouting during impoundment of the reservoir.

References

A report on reconnaissance geological investigations of Parbati Hydroelectric scheme, Kulu Distt, H.P, G.Pant & S.E.Hasan, Unpublished field season report by, GSI (March, 1970)

Cutoff/Grout curtain for fossil valley treatment, Parbati Hydroelectric Project-Stage-II, 800(4 × 200 MW)-Himachal Pradesh, India by RODIO, Trevi (March, 2005)

Detailed Project Report (Revised) on Parbati H.E. Project, Vol-IV Site Investigations & Geology, NHPC Ltd, Faridabad (September, 2000)

Part III

Applied and Active Tectonics

Convener Prof. Deffontaines Benoit—*Co-convener* Gerardo Fortunato

Tectonics (active or not) play a major role in engineering geology all over the world not only in crustal plate boundaries but also in intraplate areas. Urban geology, buildings and all anthropic constructions face major problems that we need to better take into consideration from the scientific as well as engineering point of view. Coseismic deformations due to earthquakes, as well as intersismic creep have to be better characterized and constrained from both new efficient methods and from all the past experiences we have had on active tectonic areas. This "Applied and Active Tectonics" session will focus on better localized, characterized, quantified, and modelized tectonic processes and their associated phenomena enriched by the numerous past experiences and by proposing new efficient methodologies and technologies. We hope also in this session to reinforce the links between scientists and engineering geologists in order to face seismic hazards and their disastrous implications.

A Case Study of Three-dimensional Determination of Stress Orientation to Crystalline Rock Samples in Wenchuan Earthquake Fault Scientific Drilling Project Hole-2

15

Weiren Lin, Lianjie Wang, Junwen Cui, Dongsheng Sun, and Manabu Takahashi

Abstract

Stress state around an earthquake fault is a key parameter to understand mechanisms of the fault rupturing. We tried to determine three-dimensional in-site stress orientations by anelastic strain recovery (ASR) measurements. As a case study, we applied this core based ASR method to two drill core samples retrieved from Wenchuan Earthquake Fault Scientific Drilling Project Hole-2. The core samples used for ASR experiments are chloritized diorite classified as a crystalline rock and retrieved from depths of 1,444 and 1,469 m in the hole-2, respectively. Anelastic strains of a core sample in nine directions, including six independent directions, were measured after its in situ stress was released by drilling. We obtained anelastic strain variation with time due to relaxation after its stress release. Then, the three-dimensional principal orientations of the in situ stress tensor at the two depths were successfully determined by determining the three dimensional principal orientations of the anelastic strain tensor. Our preliminary results showed both the stress regimes at the two depths are nearly strike slip faulting stress regime, and the maximum horizontal stress orientations are northwest–southeast or north-northwest–south-southeast. In addition, the results also suggested that the ASR method is applicable and useful for such in situ stress measurements in deep drilling projects.

Keywords

In situ stress • Stress orientation • Anelastic strain • Crystalline rock • Scientific drilling

15.1 Introduction

Determination of in situ stress state is very important both in geoscience and geoengineering fields. A great and destructive earthquake (Ms 8.0; Mw 7.9), the Wunchuan earthquake struck on the Longmenshan foreland trust zone in Sichuan province, China on 12 May 2008. Two almost parallel surface ruptures were observed, the main coseismic rupture developed along the Yingxiu-Beichuan-Qinchuan fault over a length of 275 km; and the other rupture was along Guanxian-Anxian fault over a length of 100 km approximately. In November 2008 about a half of year after the main shock of the Wunchuan earthquake, a very rapid scientific deep drilling project (Wunchuan Earthquake Fault Scientific Drilling, WFSD) started at Hongkou, Dujianyan,

W. Lin (✉)
Japan Agency for Marine-Earth Science and Technology (JAMSTEC), Nankoku, 783-8502, Japan
e-mail: lin@jamstec.go.jp

L. Wang · D. Sun
Institute of Geomechanics, Chinese Academy of Geological Sciences, Beijing, 100081, China

J. Cui
Institute of Geology, Chinese Academy of Geological Sciences, Beijing, 100037, China

M. Takahashi
National Institute of Advanced Industrial Science and Technology (AIST), Tsukuba, 305-8567, Japan

Sichuan to investigate physical and chemical properties of the active faults that slipped during the Wenchuan earthquake (Li et al. 2013). This drilling project consists of total five drilling holes which were or will be drilled to maximum depths of 0.5–3 km to penetrate in different faults at different sites. Following the first shallower pilot hole WFSD-1, the deeper main hole WFSD-2 was drilled to a total depth (TD) 2,284 m. WFSD-2 which completed in 2012 penetrated through the main fault zone which ruptured during the 2008 Mw7.9 Wenchuan earthquake.

To determine the in situ stress state after the Wenchuan earthquake in WFSD holes, we applied a core-based method called anelastic strain recovery (ASR) method into this drilling project. Following the first results carried out in WFSD-1 (Cui et al. 2013), we report our preliminary results on stress orientations determined from two chloritized diorite core samples of WFSD-2.

15.2 Methods

The principle idea behind the ASR method is that stress-induced elastic strain is released first instantaneously (i.e., as time-independent elastic strain), followed by a more gradual or time-dependent recovery of anelastic strain. The ASR method takes advantage of the time-dependent strain. Voight (1968) first proposed that anelastic strain could provide constraints on in situ stress; and then Teufel (1983) applied this in petroleum industry as a two-dimensional method. Matsuki (1991) showed that the method could be extended to three-dimensional stress and that it could constrain stress magnitudes. Recently, this three-dimensional method were successfully applied in various deep drilling projects (Lin et al. 2006 etc.). In principle, the anelastic strain is induced by stress release of the core sample accompanying drilling. Therefore, the stress constrains obtained by ASR measurement are of the present-day stress state.

Matsuki (1991) showed that the orientations of the three principal in situ stresses coincide with the orientations of the three principal anelastic strains for isotropic viscoelastic materials. Thus, the orientations of the principal in situ stresses can be determined by calculating the orientations of principal strains based on anelastic strain data measured in at least six independent directions. In this study, we conducted the ASR experiments based on the basic principle suggested by Matsuki (1991) and employed the same test procedures and apparatuses as Cui et al. (2013). We carried the ASR measurements out at onsite laboratory of WFSD.

A local coordinate system in which the Z axis was parallel to the borehole axis and the X and Y axes were properly defined by referring to core surface situation (Fig. 15.1a). It means to allow selection of homogenous, crack-free and smooth locations for gluing strain gauges on the cylindrical surface of ASR sample. Six cross-type wire strain gauges and six single strain gauges were mounted on each ASR sample for measuring anelastic normal strain. Thus, the nine directions of the strain measurements were X (direction cosines: 1, 0, 0), Y (0, 1, 0), Z (0, 0, 1), XY (0.707, 0.707, 0), $-XY$ (−0.707, 0.707, 0), YZ (0, 0.707, 0.707), $-YZ$ (0, −0.707, 0.707), ZX (0.707, 0, 0.707), and $-ZX$ (0.707, 0, −0.707). In addition, two strain gauges were used for each of the nine directions (Fig. 15.1a, b).

The ASR measurement system used in this study consists of a data logger with a scanning box for recording strain and temperature data, a water bath (Constant Temperature Chamber) and a circulator for keeping the water temperature

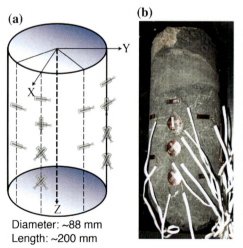
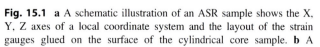
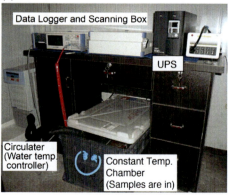

Fig. 15.1 **a** A schematic illustration of an ASR sample shows the X, Y, Z axes of a local coordinate system and the layout of the strain gauges glued on the surface of the cylindrical core sample. **b** A photograph of the ASR core sample taken from 1,469 m in WFSD-2. **c** A photograph of ASR measurement system set in the WFSD drilling onsite laboratory. *UPS* uninterruptible power supply

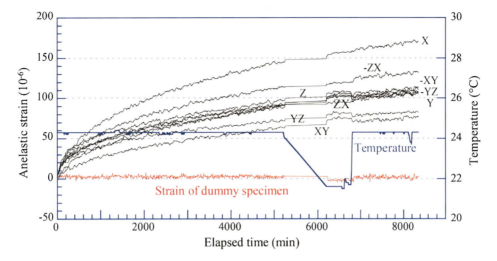

Fig. 15.2 Anelastic strain curves (labels of X, Y, Z etc. showing its measurement direction) were measured in nine directions during approximate 1 week of a core sample taken from 1,469 m in WFSD-2 as an example of anelastic normal strain recovery raw data

constant, and an uninterruptible power supply (UPS) to prevent measurement problems arising from electric power failures (Fig. 15.1c). Measurements of the ASR core sample (s) and a dummy rock sample, which did not undergo any deformation except thermal expansion, were acquired simultaneously. Purpose of the measurement of the dummy sample is to monitor the drift of the system and to correct the measured strain data if necessary. The strain gauges and two high-resolution thermistor thermometers were connected directly to the data logger, and the digital data were recorded every 10 min.

15.3 Preliminary Results

As an example of anelastic strain curves obtained from the WFSD-2 hole, raw data of anelastic strain in nine directions of a core sample taken from 1,469 m is shown in Fig. 15.2. The duration of the measurement period was approximate 6 days. During the experiment, a trouble made the electric power supply to shut down for about 16 h (from 5,200 to 6,200 min in Fig. 15.2) which is too long for capability of the UPS, thus the data was lack and the temperature in the chamber decreased. Except the no power supply duration, the constant temperature controller and chamber worked correctly, so the temperature change was less than ± 0.1 °C. As a result, the anelastic strains in all directions were extensional; all of the curves varied smoothly and similarly with increasing time. The values of the strain in the various directions, continuously measured for about 6 days depended on the orientation, the largest one (X direction) reached

more than 150 microstrains (0.015 %). The dummy sample showed that the drift of the measurement system was very small relative to the anelastic strain of the ASR sample. It indicates the strain of ASR samples were anelastic strain induced by the stress release accompanying drilling. Thus, these data could be used for the three-dimensional analysis to determine the orientations of principal strains.

From the measured anelastic normal strains in the nine directions, which included six independent directions, the anelastic strain tensor was calculated by least-squares analysis. By using a data set of the anelastic strain tensor for an arbitrary elapsed time, a data set of orientations of the three principal strains corresponding to that time can be determined. The determined orientations of and their variations as elapsed time increases are depicted as the curves from the beginning (open diamond symbols) to the end (solid diamond symbols) in Fig. 15.3 for core samples from 1,444 m and 1,469 m in WFSD-2, respectively. Then, the average (solid circles) of each principal orientation can be calculated by using the data from the beginning to the end. The orientations of the three principal anelastic strains must be the same as the orientations of the three principal in situ stresses. The three-dimensional stress orientations show a very good consistency with each other between the two samples. Our preliminary results show that both the stress regimes at the two depths are nearly strike slip faulting stress regime, and the maximum horizontal stress orientations are northwest–southeast or north-northwest–south-southeast which is coherence with the present known regional stress pattern. In general, the results can be considered reasonable and may suggest that the ASR method is well suited for the applications in directly

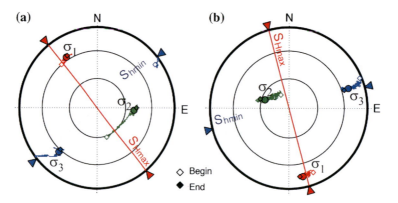

Fig. 15.3 Stereo projections (lower hemisphere) of orientations of three-dimensional principal stresses which are same as those of the principal anelastic strains from ASR measurements with respect to the true north coordinate system. The stress orientations were calculated from anelastic strain values at a certain range of elapsed times, thus the orientations vary with time increases. Open diamond symbol shows the beginning point and solid symbol shows the final point respectively; and the larger solid circles shows the average of the stress orientation from the beginning to the final point. **a** WFSD-2, 1,444 m, **b** WFSD-2, 1,469 m

determining the directions of principal in situ stresses in three dimensions in scientific deep drillings.

15.4 Summary

To determine three-dimensional principal stress orientations, we applied ASR (anelastic strain recovery) measurements using drill core samples taken from Wenchuan Earthquake Fault Scientific Drilling Project Hole-2. Here, we reported preliminary stress orientation results of the two chloritized diorite samples. We glued strain gauges on their cylindrical surface, and successfully obtained high quality anelastic strain data in at least six directions. And then, we determined the three-dimensional stress orientations by the strain–time curves. At the depths around 1,444 m and 1,469 m, the in situ stress regime about 3 years after the earthquake were nearly strike slip faulting stress regime, and the maximum horizontal stress orientations are northwest–southeast or north-northwest–south-southeast which is approximately perpendicular to the coseismic surface rupture.

Acknowledgements The drill core samples used were provided by Wenchuan Earthquake Fault Scientific Drilling Project. We gratefully acknowledge constructive scientific discussions by Haibing Li, Zhiqin Xu and the other WFSD colleagues. W. Lin was financially supported by Grant-in-Aid for Scientific Research No. 25287134 (JSPS), Japan.

References

Cui J et al (2013) Determination of three-dimensional in situ stress by anelastic strain recovery in Wenchuan earthquake fault Scientific Drilling Project Hole-1 (WFSD-1). Tectonophys, in-press, doi:10.1016/j.tecto.2013.09.013

Li H et al (2013) Characteristics of the fault-related rocks, fault zones and the principal slip zone in the Wenchuan earthquake fault Scientific Drilling Project Hole-1 (WFSD-1). Tectonophysics 584:23–42. doi:10.1016/j.tecto.2012.08.021

Lin W, Kwasniewski M, Imamura T, Matsuki K (2006) Determination of three-dimensional in situ stresses from anelastic strain recovery measurement of cores at great depth. Tectonophysics 426:221–238

Matsuki K (1991) Three-dimensional in situ stress measurement with anelastic strain recovery of a rock core. In: Wittke W (ed) Proceedings of 7th international congress rock mech., Aachen, 1, 557–560

Teufel LW (1983) Determination of in situ stress from anelastic strain recovery measurements of oriented core. In: SPE paper 11649, SPE/DOE symposium on low permeability. Denver, CO, 421–430

Voight B (1968) Determination of the virgin state of stress in the vicinity of a borehole from measurements of a partial anelastic strain tensor in drill cores. Felsmech Ingenieurgeol 6:201–215

Mathematical-Numerical Modeling of Tectonic Fault Zone (Tadzhikistan)

16

Ernest V. Kalinin, Olga S. Barykina, and Leili L. Panasyan

Abstract

The following paper formulates the criteria of fractured zone size assessment. This article is based on the thesis that the fault zone has to be regarded as a special engineering geologic massif, which is characterized by the stretched form, zonal structure and unfavorable engineering geological conditions. The authors suggest to pay principle attention to the detailed study of the fault structure, composition, sizes and other fault peculiarities. The site of Rogun Hydroelectric Station was reiterated as the object of study. It is situated in the highly complicated rock massif. The site region is characterized by the high activity of modern tectonic movements, and it provides for high strain state of the massif. With the help of boundary element method the questions of the conformity of observable changes in massif—to real width of faults' dynamic zone, its configuration, the location of stress concentration areas, free surface moving and variation of this parameters during the changes of different conditions were considered. A simulation shows that the zone of stress-strain state changing, received in calculation, is twice a large than it is fixed in nature.

Keywords

Fractured zone • Mathematical-numerical modeling • Stress-strain state • Surface movements

16.1 Introduction

The study of tectonic faults is one of the most complex and actual problems of engineering geologic theory and practice. Fault tectonics plays a special role among the factors influencing the engineering geologic conditions. At the same time, there are a limited number of papers which consider the main elements of fault zones, their sizes, composition, properties and their influence on the hydroelectric facilities as well. The estimation of fractured zones influence is rather difficult and insufficiently explored problem in engineering geological investigations. Faults' dynamic areas (*fdz*) as fractured zones, determined by S.I.Sherman, are not distinctly fixed in nature during investigations (Sherman et al. 1983). The zone which is determining by the formation character and its subsequent life is understood as this area. Within this zone the rock massif is undergoes by mechanical, structural and petrographic changes. So such zones play the negative role during the construction and operation of the facilities. During the construction the adverse phenomena are possible in this zone (such as: busting, inrushes, displacements of rocks and others), which are accompanied by the changes of stress-strain intensity near the faults.

The following paper formulates the criteria of fractured zone sizes assessment. With the help of boundary element method the questions of the conformity of observable changes in massif—to real width of faults' dynamic zone, its configuration, the location of stress concentration areas, free surface moving and variation of this parameters during the changes of different conditions.

E.V. Kalinin · O.S. Barykina (✉) · L.L. Panasyan
Moscow Lomonosov State University, Leninskie Gory 1, GSP-1,
119991 Moscow, Russia
e-mail: barykina@geol.mail.ru

G. Lollino et al. (eds.), *Engineering Geology for Society and Territory – Volume 6*,
DOI: 10.1007/978-3-319-09060-3_16, © Springer International Publishing Switzerland 2015

16.2 Methodology of Mathematical-Numerical Modeling

Calculations were done by boundary element method. This method is not deserved rarely used for the solving those engineering geological problems, when the need of strain-stress state research of rock massif and its changing under natural and man-triggered factors is arisen. With the help of the boundary element method (*BEM*) the significant semi-infinite areas, which are often the subject of study in geology and engineering geology, are simply explored. *BEM* lets to solve complicated contact problems, associated with registration not only elastic interaction on the contiguous boundaries, but with irreversible deforming on the contacts, that is important in behavior of creviced environment studying (Crouch and Starfield 1987). Moreover, due to reduction on the geometric unit dimension, the problems of *BEM* use reduce the spending of the information preparation, memory and time of computation. So during the solution of number of engineering geological problems the *BEM* using is highly progressive.

16.3 The Geological Description of the Rogun Hydroelectric Station Site

The site of Rogun Hydroelectric Station, broken by numerous faults of different width, was selected as the investigation object. The Rogun Hydroelectric Station, at the Vakhsh River (Tadzhikistan), is situated in the complex geological media massif of cretaceous sand-stones and silt-stones. The site region is characterized by the high activity

of the present tectonic movements, and it provides for high strain state of the massif. All the complex of station constructions is situated within the limits of an invisible tectonic block, limited by sub parallel Ionakhsh and Gulizindan Faults. The tunnel crossing the fault zone through has "flexible" construction. Complex tilt metric and deformational investigations (Starkov 1987) registered vertical movement in 1–3 mm/year of Ionakhsh Fault sides.

16.4 Schematization of Geological Conditions

The aim of mathematical-numerical modeling was the size estimation of near-fault changes (and so width *fdz* definition), the detection of stress spreading features within this zone and the move of free surface in main units. It was studied for this: firstly: stress-strain state of fault zone under gravitation; secondly: under gravitation and tectonic force; and thirdly: the direction and the value of the free surface move under gravitation and tectonic force. The rock massif considered as isotropic linear-elastic medium. The influence of the following factors was taken into account: topographic inequality; fracturing dislocation by the fault; force of gravity; horizontal contractive force, simulated tectonic force.

The width of Ionakhsh Fault together with crushing zone was taken as 20 m at the surface. The rocks are characterized by closed values of density and lateral deformation rate, so in modeling scheme it was taken: module of deformation—30 MPa; Poisson's ratio—0,3; density—2,6 g/sm3, horizontal contractive force—12 MPa.

Fig. 16.1 The scheme of mathematical-numerical modeling of tectonic fault zone for stress and move studying (*1* sand-stones, *2* aleurolites, *3* argillites, *4* fault, *5* depth, where stress is determined, *6* points of surface and their move direction; *7* the scale, *8* isolines of horizontal stress field, MPa)

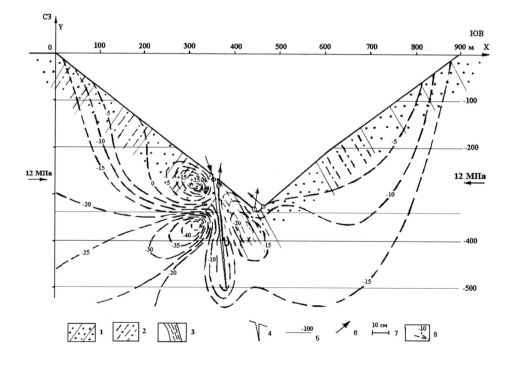

16.5 Analysis of the Mathematical-Numerical Modeling Results

The modeling scheme and the results of stress-strain and movement studying are presented on the Fig. 16.1. The calculations show that the zone of faults' dynamic influence is limited 200 m at the lying side and 70 m at hanging side; and stress concentration zones are at the fault creek and at average part of the fault at the lying side and between hanging side of the fault and valley floor. Analysis of free surface movement shows that its direction is observed in massif and has uplift direction.

According to engineering geological investigations the width of Ionakhsh Fault is 80 m and with the crushing zone it is about 120 m. But according to the mathematical-numerical modeling its summary width is about 270 m. So a simulation shows that the zone of stress-strain state changing, received in calculation, is double the number of field investigations.

16.6 Conclusion

The mathematical-numerical modeling of fractured zones allows to predict currently unknown regularities, to work out the rational methods of the study of tectonic faults being in various engineering geological conditions, especially in the active tectonic areas, taking into consideration the construction experience and attracting the analogous method for the prediction of engineering geological processes influencing the stability of engineering construction.

References

Crouch Sl, Starfield A (1987) Boundary element method in solid mechanics. Mir, Moscow (in Russian)

Sherman SI et al (1983) The zones of faults' dynamic influence. Nauka, Novosibirsk (in Russian)

Starkov VI (1987) Tectonical deformations of the earth surface at the Rogun hydroelectric station site according to the results of instrumental meterages. Donisch, Dushanbe (in Russian)

Neotectonic and Mass Movements on the New Fez-Taza Highway (Northern Morocco)

17

Tabyaoui Hassan, Deffontaines Benoît, Chaouni Abdel-Ali, El Hammichi Fatima, Lahsaini Meriem, Mounadel Ahlam, Magalhaes Samuel, and Fortunato Gérardo

Abstract

The reflection initiated here, is in connection with an area poorly known from the geological point of view as it is composed of clayey, silty turbiditic sediments and structurally characterized by different geological domains (Saiss basin, Middle Atlas Mountain and Prerif ridges). Furthermore it is also characterized by the frequent mass movements of different nature and volume and a variety of erosion processes (such as rockfalls and rockslides, subsidence and collapse of roads, landslides, gully erosion and rain). We face here a unique chance to have a better geological and geomorphological control due to the construction of the new Fez-Taza highway in northern Morocco. Consequently, this study shows that mass movements are quite important especially in the Prerif ridge, where clayey rocks (clay-stones, marls, shales, muds, flyschs) newly outcrop on both side of this new highway. The factors that affect their occurrence are especially geology and soils, neotectonic movements and climate. The combination of satellite data interpretation, detailed mapping surveys, geophysical investigation, Digital Topographic Model, morphological and finally precise fieldwork campaigns, integrated within a Geographic Information System (GIS) helped to create geo-mapping documents for multi-source developer's in order them to pay attention to the potentially dangerous areas and to optimize safety works along such roads in some sections of the highway. At least such works lead us to differentiate Neotectonics and landslides (*s.l.*) and therefore simplify the passed and recent neotectonic history of the studied area.

Keywords

Mass movement • Neotectonics • Highway • Fez-Taza • Morocco

T. Hassan (✉) · C. Abdel-Ali · E.H. Fatima · L. Meriem · M. Ahlam
Faculté Polydisciplinaire de Taza, Université Sidi Mohamed Ben Abdellah, B.P. 1223, Taza-Gare, Taza 35000, Morocco
e-mail: hassan.tabyaoui@usmba.ac.ma

C. Abdel-Ali
e-mail: alichaouni@yahoo.fr

E.H. Fatima
e-mail: fatima.elhammichi@usmba.ac.ma

D. Benoît
Laboratoire de Géomatique Appliquée (ENSG-IGN), Université de Paris-Est Marne-La-Vallée, UPEM, Marne-La-Vallée Cedex 2 77450, France
e-mail: benoit.deffontaines@univ-mlv.fr

D. Benoît
Laboratoire International Associé ADEPT CNRS-NSC France-Taiwan, 5 Bd Descartes, Marne-La-Vallée Cedex 2 77450, France

M. Samuel · F. Gérardo
AlphaGéOmega, 62 rue du Cardinal Lemoine, 75005 Paris, France
e-mail: magalhaes_samuel@yahoo.fr

F. Gérardo
e-mail: gerardo.fortunato@hotmail.com

17.1 Introduction

Important part of the Moroccan highway system, the Fez-Oujda highway connection extends the highway Rabat—Fez to form eventually a large structuring West–East axis that integrates with the existing network and major roads projects as Taza—Al Hoceima connection and Oujda—Nador. It is also an important section of the Maghreb highway which originates in Nouakchott, capital of Mauritania, and serves the major cities of North Africa to reach Tobruk in Libya.

The drawing of the highway connects the two cities of Fez and Taza over a distance of 127 km through a difficult and mountainous area. The axis of the tracing was chosen to follow the crests of hills encountered, to avoid as much as possible crossing wadis and to ensure the presence of hydraulic works for the drainage of rainwater. However, even after as much as possible the natural terrain, this track involves great heights of cuts and fills exposing beautiful outcrops and exposures in this geologically poorly known area. The trace crosses many wadis, the main ones being: Oued Sebou, Hamri, Bou Zemlane, Matmata, Bou Hellou, Zireg, Inaouène (3 crossings).

The Fez-Taza highway is situated in a continental climate relatively temperate with an average annual rainfall varying from 390 mm to 740 mm. Rainfall events are generally brutal and the greatest rainfall are concentrated in few days of the wet season. The close succession of exceptional rainfall events, are sources of risks to such kind of environments.

The new Fez-Taza highway cross three geological domains: the Saïss basin, the Middle Atlas Mountain and the Prerif ridge. In the Prerif, landslides are mainly mudslides and flows landslides (flowslides) (Tribak et al. 2012). Their genesis is the combination of several factors: geometric (related to the inclination of the topographic surface), lithological (related to marly land beneath the superimposed hard rocks), and structural (complexity of the structure and their deformation). However, in the Middle Atlas, these movements are instead represented by rockfalls of Liassic sandstones, limestones and dolomites. In the Saïss basin, the extension of marl land is an obstacle to carriers and structure determination. The highway route offers a unique chance to have better geological and geomorphological control. The trenches show rich outcrops teaching since they permit to better constrain the differences between landslide and neotectonic.

The studies of mass movement and neotectonic in the Fez-Taza highway are considered through a successive process. Visual interpretation of satellite data (Landsat, as well as High resolution data 60 cm ground resolution such as GeoEye) allowed delineating the various types of mass movements (landslides, rock and debris falls, earth flows). The differentiation of rock and debris falls was done with the aid of ancillary data (geology and Digital Topographic Model). Draping all the images on the DTM surface has clarified zones of detachment from zones of accumulation. Also, by superimposing, in a Geographic Information System (GIS), the available geological maps (Taza, Tahala and Sefrou at 1/50.000 scale), on the processed images, discrimination between features and facies that undergo rock and debris falls was possible. Field truthing campaigns were raised to detect and delineate earth creep. The resulting study, comprising mass movements that were known before our study, plus those newly discovered, are presented herein. Their studies will be considered through the analysis of the processes and factors governing the instability.

In this paper, we present a general geology review and a detailed description of the different types of mass movement across the new Fez-Taza highway, and the factors that affect their occurrence.

17.2 Geological Framework

The Fez-Taza highway crosses from west to east, three important domains: (1) the Saïss basin, (2) the northern side of the Middle Atlas Mountain and (3) the Prerif ridges south of the Rif mountain front.

- The Saïss basin constitutes with the Gharb basins a large depression that extends eastward from the Atlantic to the Taza Strait. The Saïss basin opened in Late Miocene after the collapse of the northern edges of the Western Meseta and the Middle Atlas (Bargach et al. 2003). It behaved as a subsiding marine basin during the Late Miocene, and then was lacustrine during Late Pliocene and Quaternary (Taltasse 1953; Feinberg 1986). The Prerif thrust sheets are interfingered with Upper Miocene rocks in the eastern part of the basin, and with Pliocene rocks in its western part (Moratti and Chalouan 2006).

- In the northern extremity of the tabular Middle Atlas, sub-horizontal Lower Liassic dolostones and limestones are covered by unconformable Plio-Quaternary conglomerates and the lacustrine limestones formations of the Saiss Basin or by the Quaternary fluvial or travertine formations (Charrière 1990; Gourari 2001; Ahmamou 2002; Hinaje 2004).

- The Prerif thrust sheets form the frontal part of the Rif Cordillera. They are a tectonic-sedimentary complex, thrust over the South-Rif corridor (former foreland basin) or the Middle Atlas. The Prerif Ridges comprise elongated hills, formed mainly by Jurassic to Miocene rocks. The ridges are interpreted as part of the Meseta-Atlas cover of the foreland, involved in thrusts of the External Rif—due to the contraction of the African margin during compression from Late Miocene extending to Middle Pliocene (Faugères 1978; Zizi 1996; Moratti and Chalouan 2006).

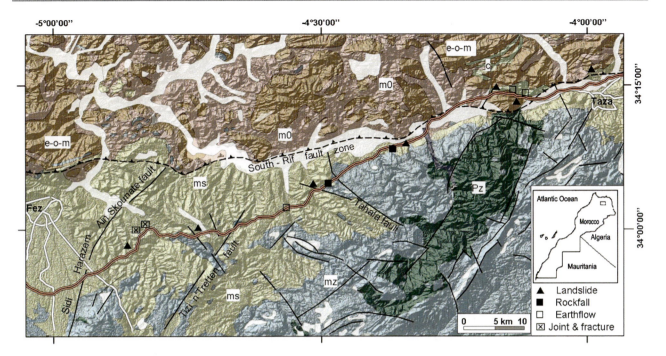

Fig. 17.1 Position of mass movements on the Fez-Taza highway on a geological and geomorphological background. *Pz* Palaeozoic, *mz* Mesozoic, *c* Cretaceous, *e-o-m* Eocene-Oligocene and Miocene, *m0* Middle Miocene, *ms* Upper Miocene

The structures recognized result of the effect of the overlapping tectonic layers recognized in the Rif area with folding-axis oriented WSW-ENE spilled southward and a brittle tectonic oriented NE-SW typical of Middle Atlas and associated with vertical synsedimentary movements, promoting subsidence and uplift of blocks (Sabaoui 1987). The major discontinuities correspond to (Fig. 17.1): (1) the south-Rif accident is an ENE-WSW-trending narrow elongated Prerif ridge, (2) the Tizi n'Tretten fault zone oriented NE-SW, (3) the Sidi Harazem—Ain Skounate fault zone is a set of parallel faults trending NE-SW (4) the Tahala fault oriented NW-SE confronts the liassic formation in the north to the upper Miocene marl cover in the south (Fassi Fihri and Feskaoui 1998).

17.3 Mass Movements Across the New Fez-Taza Highway

Landslides are complex phenomena that we attempt to explain by a number of relatively simple mechanisms resulting from topographic, geological, hydrological and climate condition. Some of these conditions may change over time. The analysis of these mechanisms in the Fez-Taza highway has led us to distinguish different types of slides. Figure 17.1 includes most mass-movement observed in the field. Some older, highlighted in the literature or by surveys conducted during construction are not included on this map.

In general, the Prerif nappes are the site of frequent mass movements. The factors that affect their occurrence are especially geology and soils, neotectonic movements and climate. Mass movements are quite important where clayey rocks (clay-stones, marls, shales, muds, flyschs) crops out. These rocks are potential shear planes as they retain water. The presence of these rocks types along the route of the Fez-Taza highway promotes various types of mass movements (Fig. 17.1).

17.3.1 Landslide

- The rotational slides (slumping) are very common in the Prerif and the Saiss sections of the highway (Fig. 17.1). The surface of these slides has a concavity upwardly facing with the possible presence of concentric cracks in plan and concave in the direction of sliding. This type of landslide concerns slopes with high steeply sloping (15–45°), consisting in particular of clay or marl materials. The instability is fostered by water infiltration in the most permeable layers or by the presence of swelling clays. This slide is quite common and attests to the renewal of marl and clay reliefs by artificial erosion (highway construction and canals).
- Translational slides are characterized by a shallow surface of rupture. This type of sliding interest slope formation in where predominant clay of Miocene or their

associated alteration coverage. Its mechanism is a translational movement of the field mass, more or less coherent, along a slightly inclined surface. The thicknesses of land set in motion are plurimetric and the debris spread to the base of the slope. These slides are very active and have been recognized in the Prerif at the western entrance of the city of Taza (Koudiate Zar Ramrama, 731 m). Their location is strongly influenced by water infiltration. An inventory of this type occurred during the rains period of February 2013 that showed that the distance could reach an horizontal length equal to two times the height of the slope where the break began.

- The clay flows dominate the reliefs of Koudiet Toumiyat located at the western entry of the city of Taza. These movements are the result of the evolution of deep rotational slides and originate in the downstream parts. This mass movement is recognized easily within the GeoEye images of Google Earth®. The clay flow is sometimes large and constitutes a risk due to the spread of debris on the northern side of the highway. The geological nature of the terrain represented by clays and gray marl late Miocene in age combined to steep slopes and to their water saturation, are the main factors triggering these movements and lead to a loss of structures slope stability on the side of the road. Field observations made on two field works (October 2012, February 2013), on the eastern part of Koudiet Toumiyat allowed us to see the growth and expansion of cracks running through the silty gravel formations based on the collapsed parts.

- The subsidence: corresponds to topographic depressions that are bowl-shaped due to the slow and gradual decline of soil without apparent breaks. They are manifested, by occasional vertical displacements of the ground surface in two different places in the Prerif and the Saiss basin. They are usually governed by the nature of clay and marl substrate (increase of the clays plasticity) during heavy rains.

17.3.2 Other Slope Movements

- Joints have been surveyed in the silt of Messinian in two sites in the South of Saiss basin (Region of Bled Cherada). The joints are predominantly oblique to the horizontal bedding and show preferred directions of N 130° E. They range in thickness from a few centimeters and in lengths up to 10 m. The average thickness of joints and the number per unit volume increases from west to east in the direction of digging of the Sebou River valley, suggesting that stress release of tension and alteration play an important role in fissure and joints genesis. However, the great majority of fissures and joints exhibit a matte surface texture without particles reorientation, with indications of brittle fractures rather than shearing.

- Gravitational displacement: the Jurassic formations of the Middle Atlas (limestones and dolomites rocks, and fractured joints), show traces of gravity moving rocks by translational motion without significant dissociation from the original material. These mountain slides are fossilized by post-Jurassic deposits.

17.4 Conclusions

The new Fez-Taza highway is characterized by frequent mass movements of different nature and volume. This section offers a unique chance to have better geological and geomorphological control. The trenches show rich outcrops teaching since they permit to better constrain the differences between landslide and neotectonic.

The combination of satellite data interpretation, detailed mapping surveys, geophysical investigation, DTM, morphological and finally precise fieldwork campaigns, integrated within a Geographic Information System (GIS) helped to create geo-mapping documents for multi-source developers. This instability corresponds in general to mass movement and dominates in the Prerif area. They are the result of geology and soils, recent and ongoing tectonic activity and climate.

Considering the risks of ground movement on some part of the Fez-Taza highway, the continuation and development of these researches is needed by actions such as the monitoring of movement, stability calculation, establishment of mechanical rupture models, modelling the motion path, research on protection techniques, development of files for each landslide. These actions will improve the understanding of these mass movements and attract the attention of developers on risk areas and potential or actual danger, and also the recommendation of stabilization works.

References

Ahmamou M (2002) Evolution et dynamique sédimentaires des carbonates fluvio-lacustres Plio-quaternaires dans le Saïss de Fès (Maroc). Thèse de Doctorat d'Etat, Université Mohamed V, Rabat, 230 p

Bargach K, Chalouan A, Galindo-Zaldivar J, Ruano P, Ahmamou M, Jabaloy A, Akil M, Sanz De Galdeano C, Chabli A, Benmakhlouf M (2003) Détermination de paléocontraintes à partir des galets striés des formations conglomératiques plio-quaternaires au front de la chaîne du Rif (Maroc) : la Ride de Trhat. Notes et Mémoires du Service Géologique, Maroc, 452, 99–108

Charrière A (1990) Héritage hercynien et évolution géodynamique alpine d'une chaîne intracontinentale : Le Moyen Atlas au sud-est de Fès (Maroc). Thèse d'Etat, Université Pau Sabatier, Toulouse III, France, 589 p

Fassi Fihri O, Feskaoui M (1998) Etude hydrogéologique de l'aquifère liasique du couloir Fès-Taza (Maroc). Karst Hydrology, IAHS Publ. no. 247, pp. 99–116

Faugères JC (1978) Les Rides sud-rifaines. Evolution sédimentaire et structurale d'un bassin atlantico-mésogéen de la marge africaine. Thèse Doct Etat. Univ. Bordeaux 1, n° 290, vol., 510 p., 11 tab., 119 fig., 42 pl

Feinberg H (1986) Les séries tertiaires des Zones Externes du Rif (Maroc). Biostratigraphie, paléoécologies et aperçu tectonique, Notes et Mém. Serv. Géol. Maroc 315, pp 192

Gourari L (2001) Etude hydrochimique, morphologique, lithostratigraphique, sédimentologique et pétrographique des dépôts travertino-détritiques actuels et plio-quaternaire du bassin karstique de l'oued Aggaï (Causse de Sefrou Moyen Atlas, Maroc). Thèse d'Etat, Univ. Sidi Mohammed Ben Abdallah, Fès, 476 p

Hinaje S (2004) Tectonique cassante et paléochamps de contraintes dans le Moyen et le haut Atlas (Midelt-Errachidia) depuis le Trias à l'Actuel. Thèse d'Etat Es-Sciences, Rabat 425 p

Moratti G, Chalouan A (2006) Tectonics of the Western Mediterranean and North Africa. J Afr Earth Sci (and the Middle East) 48(1):1

Sabaoui A (1987) Structure et évolution alpine de Moyen Atlas septentrional sur la transversale Tleta du Zerarda-Merhraoua. Thèse de 3ième cycle

Taltasse P (1953) Recherche géologiques et hydrogéologiques dans le bassin lacustre de Fès-Meknès. Notes et Mémoire, n° 115, 300 p. Service de géologie. Rabat, Maroc

Tribak A, El Garouani A, Abahrour M (2012) L'érosion hydrique dans les séries marneuses tertiaires du prérif oriental: agents, processus et évaluation quantitative. Rev Mar Sci Agron Vét (2012) 1:47–52

Zizi M (1996) Triasic-jurassic extentional systems and their Neogene reactivation in northern Morocco (the rides prérifaines and basin Guercif). Ph. D. thesis, Rice University, Houston

Importance of Geological Map Updates in Engineering Geology, Application to the Rif-Chain and Its Foreland (Northern Morocco)

18

Deffontaines Benoît, Tabyaoui Hassan, El Hammichi Fatima, Chaouni Abdel-Ali, Mounadel Ahlam, Lahsaini Meriam, Magalhaes Samuel, and Fortunato Gérardo

Abstract

This study focus on the importance of geological map updates in engineering geology. Presently we face such a huge quantity of data (1) acquisition of the earth surface (*s.l.*) such as multi-date remote sensing images, aerial photographs and Digital Terrain Model at various ground resolutions, High Resolution Data (HRD), and (2) lots of thematic data such as geological and geomorphological detailed works, thesis, report, mapping, etc… that it is needed and of major importance to gather and synthesize all these data in an up-dated numerical geological mapping. Therefore, we conducted a new multidisciplinary methodological approach; focus on Earth Sciences especially on the study of lithology, and structure. We use the contribution of all existing and available data such as digital terrain models (DTM), integrated of remote sensing space and airborne images. Recognition detailed geological materials and structures as well as the analysis and modeling of the physical processes controlling the deformation help also for the main focus of this research. The research is oriented in a first step towards the development of a Digital Terrain Model (DTM); the use of satellite images with multi-source, multi-resolution and multi-date (Radar-ERS, Landsat ETM 30 m, Aster 15 m and Quickbird 2.4 m up to Geo-Eye 60 cm) and aerial photographs; and Geological and geomechanical analysis of land facing the numerical simulation of tectonic deformation process, building on GIS software in order to update the geological map.

Keywords

Geological mapping update • Engineering geology • DTM • GIS • Rifian orogeny • Northern Morocco

D. Benoît (✉)
Laboratoire de Géomatique Appliquée (ENSG-IGN), Université de Paris-Est Marne-la-Vallée (UPEM), 5 Bd Descartes, 77450, Marne-la-Vallée Cedex 2, France
e-mail: benoit.deffontaines@univ-mlv.fr

D. Benoît
Laboratoire International Associé ADEPT 539 CNRS-NSC France-Taiwan, 5 Bd Descartes, 77450, Marne-la-Vallée Cedex 2, France

T. Hassan · E.H. Fatima · C. Abdel-Ali · M. Ahlam · L. Meriam
Polydisciplinary Faculty of Taza, Sidi Mohamed Ben Abdellah University, B.P. 1223Taza-Gare, Taza 35000, Morocco
e-mail: hassan.tabyaoui@usmba.ac.ma

M. Ahlam
e-mail: ahlammounadelfst@gmail.com

M. Samuel · F. Gérardo
AlphaGéOmega, 62 Rue Du Cardinal Lemoine, 75005, Paris, France

G. Lollino et al. (eds.), *Engineering Geology for Society and Territory – Volume 6*,
DOI: 10.1007/978-3-319-09060-3_18, © Springer International Publishing Switzerland 2015

18.1   Introduction

A geologic map displays the placement, distribution, characteristics, and age relationships of rock units and formation, along with structural features, on a two-dimensional base map. Everyone benefits directly or indirectly from ongoing geologic mapping such as local governments, industry, educators, and public depend on the information provided by geological maps to carry out their missions.

In morocco, Geological Survey of Morocco publishes the first geological map in 1927. Since then, the publications have continued, and to this day, it is not less than 1,600 articles (270 geosciences maps), are available to the public. Geological maps were made at scales from 1/500.000, 1/200.000, 1/100.000 and 1/50.000. Geological maps on paper format constitute three-quarters of the production of the Geological Survey, covering primarily the Northern Provinces.

Update of old geological maps is a good project involved in Morocco. Today, with the fast development of sensor techniques and computer methods, many kind of raster and vector based models for describing, modelling, and visualizing 3D spatial data are available. This development associated to the growth of engineering projects (construction of highways, urban development projects, etc.), the update of old geological maps in 2D and 3D become more and more important.

In this study, a multidisciplinary methodological approach is developed, in which the contribution of Earth Sciences is emphasized, as well as the contribution of digital elevation models (DEM) and integrated of remote sensing space and airborne images. Many works illustrate the importance of DEM and remote sensing in recognition of geological materials and structures as well as the analysis and modelling of the physical processes controlling the deformation (Douglas 1986; Klingebiel et al. 1988; Pryet et al. 2011). DEM are one of the most suitable tools for such kind of analysis. Although DEMs are currently being used for describing geological features related to geomorphology, hydrology and tectonics (Deffontaines 1990; 2000; Onorati et al. 1992; Seber et al. 1996; Spark and Williams 1996), they still have not become a common tool in geological mapping projects.

Modelling terrain relief via DTM is a powerful tool in GIS (Geographic Information System) analysis and visualization. In a GIS environment, three-dimensional images can be easily performed. The relief can be exaggerated and any type of 2D maps (Geological, Geophysical, etc.) can be draped over. With that, we can visualize the draped map of an area from any vantage-point.

The aim of this contribution is to illustrate the importance of the re-interpretation in the light of the new geological concepts and knowledge the DEM and remote sensing data combined with previous bibliographic works in providing valuable geological information that can be used as a guide in updating old geological map of Northern Morocco as a first step of a more regional project.

18.2   Methodology

Updating the geological map is the general purpose of this research. Several steps were undertaken.

1. First, available literature review was done for extracting information concerning the general geology and geomorphology of the study area. As a first application, the Prerif ridges, in northern Morocco have been chosen to illustrate the result of geological update. We look also forward to any drills, wells, seismic lines that may help to understand the 3D geometry.

2. In the second step, geological maps was scanned at 300 dpi using an A0 scanner, georeferenced in WGS84 projection in ERDAS Imagine 9.2, using a combination of latitude/longitude information and others references points on the map. Georeferencing was performed using a first order transformation with RMS-errors less than 0.3 m. The georeferenced map was then stored in Geotiff format at a scale of 1:50.000. The structural objects such as recognized fold axes, faults dip… are collected.

3. ASTER (Advanced Spaceborne Thermal Emission and Reflection Radiometer; METI/NASA 2009) Global Digital Elevation Model (GDEM) elevation data sets been used to perform the topographic analysis. The DEM data from this space mission cover most of the regions of the world and are publicly available (at no cost) at spatial resolutions of 1 arc second. The ASTER GDEM was resampled from 30 m (1 arc second) horizontal resolution to 15 m to minimize the terrain effect. The required study areas were, then extracted from DEM using there base map shape file in a WGS84 coordinate system. For this process, we used the software GLOBAL MAPPER 10v to crop the DEM with only required region's terrain data. Then, we used ArcGIS 10.2 software package to perform the spectral analysis over the DEM.

4. Correlation between DEM derived surfaces and geological maps: DEMs look at the surface from a strictly topographic point of view. Only the geological features reflected in the topography can be visible in DEM images. This fact allows DEMs to recognize geological structures, rock softness, and boundaries and drainage patterns (Ben Hassen et al. 2012; Duperet et al. 2003). The shaded topographic relief or hill-shadow of the DEM depicts relief by simulating the effect of the sun's illumination on the terrain. This technique was used for controlling and interpreting structural boundaries (Deffontaines 1990). These images are useful to display

Fig. 18.1 Geological setting and location of the studied area in northern Morocco. *1* Neogene and quaternary rocks, *2* Prerif ridges, *3* External prerif, *4* Internal prerif, *5* Mesorif, *6* Intrarif, *7* thrust nappes of intrarif origin, *8* Internal zones, *9* Flysch, *10* Middle atlas, *11* Meseta, *12* Location of the studied area

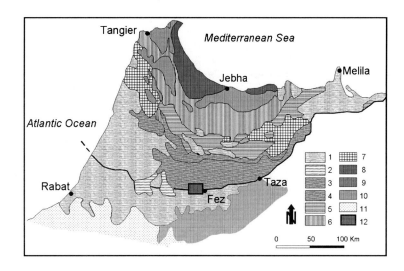

geological data related to landforms in terrains that show a close correlation between geology and topography. This initial exploration permits to further enhance lithological boundaries and to delineate areas where additional exploration is needed (satellites image analysis, field survey, etc.).

5. Data preparation and pre-processing of optical and microwave remotely sensed images, such as Landsat ETM + , ASTER 15/30 m and Quickbird 2.4 m were conducted using ERDAS software. Various image enhancement techniques were used to clearly visualize the images. Radiometric and geometric corrections were applied in order to remove the influence of the atmosphere and to get surface reflectance. Main values digital numbers (DN) of lithological units in individual Landsat and Aster bands were calculated and the values were plotted to see which band best discriminate the different lithologies. Based on the selected individual bands, image classifications were carried out using maximum likelihood classification. All the resulted images were subset into one projection system.

6. The geological map had then been redrawn and complete taking into account the previous items, our field experience, our field of competence (sedimentology as well as structural geology and engineering geology… and so on). For instance the legend and the existing lithologies had been revisited by field works and re-interpretation of the data.

18.3   Study Area

The Geological map of Fès-Ouest at 1:50.000 (Bruderer et al. 1950), communicates vast amounts of geological information. This map is edited by the Cherifienne Company of Oil for the Geological Survey of Morocco on 1950. It

shows the central part of the Prerif ridges. The Prerif thrust sheets form the frontal part of the Rif Cordillera. It corresponds to a tectonic-sedimentary complex, thrust over the South-Rif Corridor (Fig. 18.1).

To the south of this sheet, the Jbel Trhat area is chosen due to its complexity of folding and tectonic structure and history. This Prerifain ridge is a mountain area and corresponds to a fold with an E-W oriented axis, fewly deformed by faults. The estimate terrain elevation level is about 837 metres.

In this region, geological structures and rock unit boundaries show a strong correlation with relief. DEM and satellites images analysis show that geological boundaries may be mapped with detailed topographic analysis. Literature reviews (Bargach et al. 2004; Cherai et al. 2008; Lakhdar et al. 2004; Wernli 1978) and field works permitted to better understand the rock types, stratigraphy, structures and deformation history and then to update the geological map of this area Fig. 18.2.

This analysis show that Jbel Trhat is a complex anticlinal, with a kilometric length and periclinal ends, affected by thrusts of the Prerif nappe sheets. The fold is tighter toward the west, where layers of the southern side become overturned and a periclinal end can be observed. The fold asymmetry clearly indicates southward tectonic transport. This fold is probably active as it develops after Pliocene because conglomerates of this age are also overturned (Fig. 18.3).

In this portion of the map, the Prerif ridge shows a sedimentary sequence starting with a thick Jurassic sequence (J), with cancellophycus limestones of Aalenian and thick gray marls of Bajocian. The Neogene discordant on the Dogger starts with a calcareous and sandstone formation (60 m) of Lower Miocene (Burdigalian) (m1), surmounted by white marl (200–300 m) of Middle Miocene (m2). Unlike the geological map, our study confirms that the Middle

Fig. 18.2 Old and updated map of Jbel Trhat at 1/50,000 scale (work in progress). *J* Jurassic, *m1* Lower Miocene, *m2* Middle Miocene, *m3-1* Lower Tortonian, *m3a* Middle Tortonian, *m3 blue* and *gray* marls of Upper Tortonian—Messinian, *3-2* sandstone of Upper Messinian, *cl1* Pliocene, (cl3, qe & q1a) Quaternary

Fig. 18.3 Old and updated map of Jbel Trhat at 1/50,000 scale (work in progress). *J* Jurassic, *m1* Lower Miocene, *m2* Middle Miocene, *m3-1* Lower Tortonian, *m3a* Middle Tortonian, *m3 blue* and *gray* marls of Upper Tortonian—Messinian, (*3-2*) sandstone of Upper Messinian, (*cl1*) Pliocene, cl3, qe & q1a Quaternary

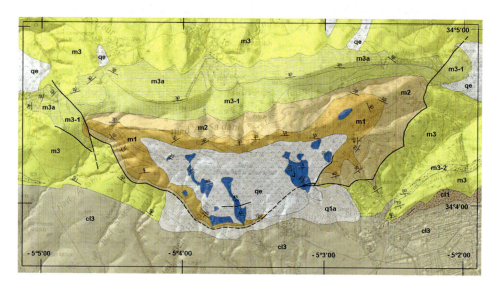

Miocene outcrops south of the Jbel Trhat. The Tortonian starts with bleu-gray marls (m3-1) and a cavernous yellow limestone (m3a). The Upper Tortonian—Messinian (m3) are represented by monotonous blue and gray marls series (800–900 m thick), intercalated by three sandy ferruginous beds, in turbidite character named "Grès I" and "Grès II" (Wernli 1978) (m3-2). The Pliocene correspond to polygenic conglomerates that crop out very locally (cl3). The quaternary is represented by variegated clays and silts (cl3). Alluvial fans and talus (qe) and mass rocks (q1a) are particularly abundant. They correspond to the accumulation of large blocks of calcareous or loam deposits supplied by clay.

The uses of satellites images and DEM permit to understand some typical scenarios of triggering or reactivation of mass movements and landslides (qe). The updated map shows some of theses phenomena (rockslide—landslide) that are the subject of detailed field investigations and geomechanical analyzes.

18.4 Results and Conclusion

DEM images have the ability to visualize structures and lithological boundaries reflected in the topography, especially those related to recent geological evolution. With DEM and satellites images, up-date of geological map become a dynamic process of gathering, evaluating, revising and up-dating geologic data and their implications and applications, both to the structure and to further exploration. Trend-topographic analysis in the Prerif ridge provides useful data about the role of doming processes and folding and faulting structures in the construction of the relief.

The results of this study support and come at a reference for any engineering further studies in this area. We are developing different types of maps of different scales (a) regional to local geological maps at a scale of 1:100.000 to 1/10.000 and (b) Engineering geological plan at even larger scale such as 1:5.000 scale. These documents are of great use to Individuals, consulting and civil engineering offices, and of course to Government administrations.

References

Bargach K, Ruano P, Chabli A, Galindo-Zaldi´Var J, Chalouan A, Jabaloy A, Akil M, Ahmamou M Sanz De Galdeano C, Benmakhlouf M (2004) Recent tectonic deformations and stresses in the frontal part of the Rif Cordillera and the Saïs Basin (Fes and Rabat regions, Morocco): Pure appl Geophys 161 (2004) 521–540

Ben Hassen M, Deffontaines B, Rebai N, Turki MM (2012) Contribution of drainage network and LANDSAT ETM + photo-interpretation for homogenization of the geological mapping coverage: Application to some examples in Southern Tunisia. Photo-Interpretation Euro J Appl Remote Sens 2012(3):31p

Bruderer W, Gouskov N, Gubler J, Jacquemont P, Levy R, Tilloy R (1950) Carte géologique régulière du Maroc au 1:100.000 : Fès Ouest. Soc. Cherif. Petroles, Notes et Mémoires Services Géologique Maroc, N°109

Cherai B, Charroud M, Lahrach A, Babault J (2008) Influence de la tectonique compressive et des mouvements verticaux d'origine mantellique sur l'évolution quaternaire du Saïs entre le Rif et le Moyen Atlas (Maroc). Actes RQM4. Oujda 2008:171–181

Deffontaines B (1990) Développement d'une méthodologie morphonéotectonique et morphostructurale; analyse des surfaces enveloppes, du réseau hydrographique et des modèles numériques de terrains; Application au Nord- Est de la France. Thèse, Univ. Paris VI, France, 230 p

Deffontaines B (2000) Formes et déformations de la surface terrestre: approche morphométrique et application. Thèse d'habilitation à diriger la recherche, Université Pierre et Marie Curie, Paris, 65 p

Douglas DH (1986) Experiments to locate ridges and channels to create a new type of digital elevation model. Cartographica 23(4):29–61

Duperet A, Deffontaines B, Passalacqua O (2003) Critères géomorphométriques issus des modèles numériques de terrain au service des applications hydrologiques. Bulletin Société française de photogrammétrie et de télédétection. ISSN-0244-6014. 172, pp 107–121

Klingebiel AA, Horvath EH, Reybold WU, Moore DG, Fosnight EA, Loveland TR (1988) A guide for the use of digital elevation model data for making soil surveys: U.S. geological survey open-file report 88–102, 18 p

Lakhdar A, Ntarmouchant A, Ben Abbou M, Ribeiro M L, El Hamzaoui 0, El Ouadeihe K, Hessane MA (2004) Minéralisations sulfurées et activité hydrothermale post miocène: impact sur la signature géochimique des eaux thermales de Moulay Yacoub (Bordure septentrionale du Sillon Sud Rifain Occidental, Maroc Septentrional). 4eme Coll 3MA, p 7 Agadir (Maroc)

METI/NASA (2009) ASTER global digital elevation model by Ministry of Economy, trade and industry of Japan (METI) and the National Aeronautics and Space Administration (NASA). Available at: http://asterweb.jpl.nasa.gov/gdem.asp and http://www.gdem.aster.ersdac.or.jp

Onorati O, Poscolieri M, Ventura R, Chiarini V, Crucillà U (1992) The digital elevation model of Italy for geomorphology and structural geology. Catena 19:147–178

Pryet A, Ramm J, Chilès JP, Auken E, Deffontaines B, Violette S (2011) 3D resistivity gridding of large AEM datasets: a step toward enhanced geological interpretation. J Appl Geophys 75(2011):277–283

Seber D, Vallvé M, Sandvol E, Steel D, Barazangi M (1996) Middle east tectonics: applications of geographic information systems. CGS Today, 7.2.1–6

Spark RN, Williams PF (1996) Digital terrain models and the visualization of structural geology. In: De Paor DG (ed) Structural geology and personal computers. Computer methods in the geosciences, v 15. Elsevier Science Ltd, pp 421–446

Wernli R (1978) Micropaléontologie du Néogène post-nappes du Maroc septentrional et description systématique des Foraminifères planctoniques. Notes et Mémoires du Service Géologique N° 331 207p

Disaster Awareness Education for Children in Schools Around Geological Hazard Prone Areas in Indonesia

19

Muslim Dicky, Evi Haerani, Motohiko Shibayama, Masaaki Ueshima, Naoko Kagawa, and Febri Hirnawan

Abstract

Geological disaster awareness has been increasing recently in Indonesia, especially since the great Aceh's Tsunami & Earthquake. It is important that all stakeholders have reasonable understanding on disaster response. School communities generally have limited disaster education opportunities and knowledge. This could imply a low level of disaster awareness. This paper aimed to examine the knowledge on disaster response and to highlight the needs to introduce earth science for school communities by exploring three areas: (i) disaster education, (ii) respond to disaster event, (iii) knowledge of earth science. The study had been started since 2006, targeted on schools located on the areas that experienced and/or potential to earthquake and tsunami in the future. Several schools around the coastline of Indian Ocean of Sumatera, Java, Bali & Lombok islands had been visited. In each school, we examined the curriculum, preparedness to face disaster, and activity of mitigation. Our posters and pamphlets were also distributed. Presentation and short drama were performed in the classroom to measure understanding of the contents. Result showed that disaster awareness were generally out of curriculum, due to limited knowledge of the curriculum development. For the respond to disaster event, most of participants are unaware what to do when disaster happens. Our visit had increased the curiosity of school communities to learn more about these disasters. These results suggest that dissemination of entry level of earth science is deeply needed, since there is no such subject (especially geology) in primary to secondary level schools in Indonesia.

Keywords

Disaster awareness • Earthquake • Tsunami • School children • Indonesia

19.1 Introduction

Since the aftermath of great Aceh's Tsunami & Earthquake in December 2004, geological disaster awareness has been increasing recently in Indonesia. It is important that all stakeholders including school communities have reasonable understanding on disaster response. School communities generally have limited disaster education opportunities and knowledge. This could imply a low level of disaster awareness.

M. Dicky (✉) · F. Hirnawan
University of Padjadjaran, Bandung 40115, Indonesia
e-mail: dicky.muslim@unpad.ac.id

E. Haerani
Faculty of Geology, University of Padjadjaran,
Jatinangor 45363, Indonesia

M. Shibayama · M. Ueshima · N. Kagawa
Natural Environmental Research Institute, Osaka 543-0045, Japan
e-mail: shibayama@themis.ocn.ne.jp

G. Lollino et al. (eds.), *Engineering Geology for Society and Territory – Volume 6*,
DOI: 10.1007/978-3-319-09060-3_19, © Springer International Publishing Switzerland 2015

This paper aimed to examine the knowledge on disaster response and to highlight the needs to introduce earth science for school communities by exploring three areas: (i) disaster education, (ii) respond to disaster event, (iii) knowledge of earth science.

19.2   Methodology

The activities had been started since 2006 and had targeted on elementary to senior high schools in several places, which located as the earthquake hazard and tsunami prone areas that experienced as well as potential areas of geological disaster in the future. Several schools around the coastline of Indian Ocean of Sumatera, Java, Bali & Lombok islands had been visited. In total, within 6 years there were more than 50.000 pamphlets and posters of earthquake and tsunami (Fig. 19.1) had been distributed in several schools and districts so far.

Participants in this study were stakeholders of education sectors in a remote area. For the purpose of this paper, the students & teachers were the focus of the study. The survey was piloted on a group of students & teachers from highest grade of each school.

Students & teachers involved in this research occupied public schools in the study area. The highest grade of each school was chosen as they have ability to read and write their opinion though in simple form. Teacher participants were chosen as they were the guardian of each classroom. We involved also the participation of the school principals as well as local government officers.

Pamphlets and posters for the tsunami and earthquake disaster prevention education are distributed directly as teaching materials. Pamphlets were prepared and printed in Osaka-Japan by several group of volunteers (Fig. 19.2), which then brought to Indonesia for this study. The posters are then posted in announcement board of each school, and the pamphlets are distributed to students and teachers in their classroom. Discussion session and short drama were performed in the classroom to measure understanding of the contents. The aim of discussion was to explore students & teachers' perceptions and knowledge of disaster education & response through a series of questions and answers (Shibayama et al. 2006).

A mixed method of descriptive and exploratory research design underpins this research. Descriptive & exploratory research designs are appropriate when little is known about the topic being investigated. Integration of quantitative and qualitative data which was generated from the survey, lends itself to the mixed method approach. According to Polit and Beck (2008) the greatest advantage of survey research for disaster issue is its flexibility and broadness of scope. Due to the limited amount of knowledge in the study area, the authors thought that a survey research would be more appropriate as it would generate a basic understanding of the phenomenon as well as reach a larger proportion of the population.

19.3   Result and Discussions

The word 'disaster' encompasses a myriad of occurrences and the meaning of the word is relative to each and every person experiencing the disaster. Generally speaking, a disaster or a major incident will overwhelm existing resources. However, in the context of this study it may be suggestive of a limited understanding as to what constitutes

Fig. 19.1 Pamphlets of earthquake (*left*) and tsunami (*right*) education for children in Bahasa Indonesia

Fig. 19.2 Activities to prepare the pamphlets of tsunami and earthquake by volunteers in Osaka, Japan

a disaster or major incident especially related with geological event or hazard such as earthquake & tsunami as they are major disaster in Indonesia (Anonymous 2006).

19.3.1 Disaster Education

When we came to each school, first we introduced ourselves and discussed with the school principal and teachers about disaster education in their school, especially related with disaster curriculum development and teaching materials for earth science. After introduction session in front of the classroom, participants were asked to identify their perceptions of their own level of disaster knowledge and experience of earthquake or tsunami in simple form. Although most of all participants stated they know about earthquake or tsunami events but many stated they have no ideas about kinds of natural disaster in their area. It is possible that the majority of participants answered since they have read newspaper or watch TV about the recent disaster events in Indonesia and elsewhere but unfortunately they don't have idea for their own area.

For the purpose of this study the term 'education' refers to any didactic formal education included in curriculum, where 'training' refers to practical hands on approach to disaster knowledge. Both of terms constitute activities such as lectures, desk top exercises, real-time exercises, etc. Many of participants had never attended specific disaster training or education outside their schools. Only a small amount of participants stated they had attended minor disaster specific courses in an extra-curricular activity such as "boy scout".

For the purpose of this study 'disaster specific courses' were considered as those that have been created specifically with the purpose of training common people in any aspects of disaster preparedness and response.

It was interesting to note that many of participants reported that they have experienced a disaster or major incident in their life. Some examples provided by the participants included examples such as house fire, flood, landslide, etc. The terms 'disaster' and 'major incident' were not qualified in the survey, however participant responses are suggestive of limited understanding of what constitutes a disaster or a major incident. Questions regarding the disaster event did not differentiate between -to some extent- predictable (i.e. house fire, flood, drought, etc.) and unpredictable (i.e. earthquake, tsunami, etc.).

Coastal area of Indian Ocean had experienced natural disasters with mass casualties due to earthquake & tsunami events. But it seemed from our study that there were no particular developments on the school's curriculum to include disaster education so far in the study area.

19.3.2 Respond to Disaster Event

Using our teaching materials in the classroom, we discussed with participants to rate their knowledge in their own level about disaster preparedness, especially about respond when an earthquake or tsunami happen. Majority of participants stated they do not know about what to do when a disaster occurs in their area. When participants were asked to rate their level of knowledge about how preparedness had been

Fig. 19.3 Classroom situation during discussion and short drama in Indonesia

constructed in their school and surrounding area, many of them do not know about school preparation to prepare for a disaster event.

Majority of the participants did not know about simple form of disaster preparedness for earthquake and tsunami events such as evacuation route, safety area, survival kit, communication tools, etc. This result is parallel to the above data regarding students & teachers' own level of education for disaster. In both instances more than half the sample had less than optimal confidence in their own disaster awareness. This suggests a feeling among students & teachers of limited preparedness to respond to a disaster event.

After a series of discussion, we then performed drama and/or story telling with our teaching materials to emphasize the need for understanding the subject of tsunami and earthquake hazard to all participants in the classroom (Fig. 19.3). Especially to stress out the need to be calm in panic situation to escape from disastrous event in time of earthquake or tsunami occur. It is worthy to note that pamphlets adopted from Japanese comic created and produced in Japan along with the appearance of foreign researcher in the classroom seemed had increased psychologically the enthusiastic attention from students and teachers in each school.

19.3.3 Knowledge of Earth Science

Based on the discussions with school principal, teachers & students as well as local education section office, it is interesting to note that based on national curriculum, earth science is included in the subject of Geography instead of Geology course in all level of elementary to high school. Even in a region where previous geological disaster had occurred, local content of curriculum for disaster is not developed yet so far. This might be due to limited knowledge and understanding of teachers in each school we visited.

An overwhelming of participants in the classroom stated that earth science education for students & teachers is very important. The form of earth science education and/or disaster training that most respondents believed would be beneficial for students & teachers were real-time exercises. Lectures provided by other competent institution (i.e. university, company) is the most stated by participants.

Education and training opportunities for participants in this study appear to be difficult to access due to their location and availability. While literature highlights the importance of disaster education and training for students & teachers but little appears to be understood about what type of education and training would be the most appropriate for a particular attendees (Duong 2009).

19.4 Conclusion

Previous disaster response experience and appropriate disaster or earth science education appear to be essential ingredients in providing a prepared and safe school. In a community where previous disaster response experience is limited, appropriate disaster education and training for students & teachers may increase the level of disaster awareness and help to make school community feel less vulnerable when having to face the unexpected (Aguilar and Retamal 2009).

Standardizing disaster and earth science education and making it more available may create a more cohesive and self-assured workforce. Further research needs to be conducted in order to close the gap in knowledge that exists in this area and to determine appropriate strategies for increasing disaster awareness among stakeholders.

References

Aguilar P, Retamal G (2009) Protective environments and quality education in humanitarian contexts. Int J Educ Dev 29:3–16

Anonymous (2006) National Action Plan of Disaster Risk Reduction 2006–2009, Indonesian Gov. Printing Office (in Bahasa Indonesia), p 196

Duong K (2009) Disaster education and training of emergency nurses in South Australia. Australas Emerg Nurs J 12:86–92

Polit DF, Beck CT (2008) Nursing research: generating and assessing evidence for nursing practice, 8th edn. Lippincott, Williams & Williams, Philadelphia

Shibayama M, Muslim D, Kagawa N, Shibakawa A, Hiraoka Y, Ueshima M, Kawamura D, Ota K (2006) Making of tsunami pamphlet for school children in Indonesia and Japan, and disaster prevention education. IGEO Poster Session Abstracts, GeoSciEd V (Bayreuth, Germany), p 8

Analysis of Recent Deformation in the Southern Atlas of Tunisia Using Geomorphometry

20

Mehdi Ben Hassen, Benoît Deffontaines, and Mohamed Moncef Turki

Abstract

In this work, we propose to locate, characterize and quantify some topographic deformations linked to the seismotectonic context, and anthropogenic actions in the southern Atlas of Tunisia. The analysis of morphometric parameters (Residual Topography, Hypsometric Integral, Drainage anomalies, Maximum curvatures and Roughness) has revealed that three structures in the study area, J. Ben Younes, J. Bou Ramli and J. El Abiod, are distinguished by a specific morphometric and anomalous response which may reflect an important morpho-dynamic activity caused principally by the numerous reactivation of the Gafsa fault.

Keywords

Geomorphometry • DEM • Structural landscape • Southern Atlas of Tunisia

20.1 Introduction

Tunisia is submitted to a long time convergence of both African and Eurasian crustal plates. The latter collision creates the northern Atlas orogene and the N–S present stress field creates many small topographic displacements on the major

M. Ben Hassen (✉) · B. Deffontaines
Bât IFG, Laboratoire de Géomatique Appliquée (ENSG, IGN), Télédétection et Modélisation des Connaissances, Université Paris-Est Marne-la-Vallée, 5 Bd Descartes, Marne-la-Vallée, Cedex 2, 77454, France
e-mail: mahdigeo2002@yahoo.fr

B. Deffontaines
e-mail: benoit.deffontaines@univ-mlv.fr

M. Ben Hassen · M.M. Turki
Faculté des Sciences de Tunis, Département de Géologie, Université de Tunis El Manar, Campus Universitaire 2092, El Manar Tunis, Tunisie
e-mail: mohamedmoncef.turki@fst.rnu.tn

B. Deffontaines
Laboratoire de Géomatique Appliquée, Ecole Nationale des Sciences Géographiques, Institut Géographique National, Et Laboratoire International Associé ADEPT France-Taiwan CNRS-NSC, 6-8 avenue Blaise Pascal, Marne-la-Vallée, Cedex 2, 77455, France

faulted structure. The seismotectonic database that we settle for the Tunisia has shown that the southern Atlas of Tunisia is strongly influenced by this geodynamic context and manifested by a relatively high seismic potential (Fig. 20.1).

Several methods may be used to describe active tectonic zones, among which "geomorphometry" is useful. For instance, this method may help describing, analysing and measuring the morphology of land surface (Deffontaines 1990; Pike and Dikau 1995; Pike 2002; Dehn et al. 2001; Bolongaro-Crevenna et al. 2005).

20.2 Identification of Recent Deformations

In this section, we will try to establish the relationship between morphometric parameters and the structural context of the study area.

Five major morphometric parameters were analyzed: the residual topography, hypsometric analysis, analysis of drainage anomaly, maximum vertical curvature and roughness of terrain.

The interpretation of specific signatures of morphometric indices allowed us to distinguish three structures of Gafsa chain: J. Ben Younes and J. Bou Ramli (showed in Fig. 20.2) and J El Abiod (showed in Fig. 20.3).

Fig. 20.1 Location, morphological and seismotectonic contexts of the studied area (superimposed on altitudinal hill shaded SRTM data : Azimuth = N315°E, Elevation = 45°): fault (modified from Hfaiedh et al. 1991); instrumental seismological data (for the 1975–2013 period) were provided by the National Meteorological Institute (Tunis)

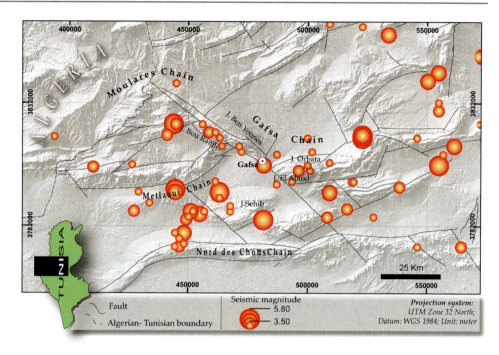

20.2.1 J. Ben Younes and J. Bou Ramli

Indeed, morphometric study shows that these reliefs are distinguished by many footprints of recent tectonic activity (Fig. 20.2):

- According to the catalog of the I.N.M., seven seismic events are occurred in this sector between 1977 and 2005 with a medium magnitude equal to 4. This reflects a relatively high seismic activity. The epicenters of these earthquakes are aligned along a NW-SE direction (Fig. 2.1).
- A high residual topography that can reach 470 m on the southern flank of J. Bou Ramli (Fig. 2.2).
- The Hypsometric Analysis in the Southern Atlas of Tunisia shows that the watersheds, located on the southwest flank of Ben Younes and Bou Ramli, have a higher hypsometric integral value (HI> 0.6) and a convex shape of the hypsometric curves. Their addition allows to delimit an area with immature landforms. The correlation with tectonic coverage of southern Atlas of Tunisia can be deduced that these high values of HI are located on the west of the Gafsa fault. Thus, the recent activity of this fault may explain the "youthfulness" of the landform in this area and the weakness of the climatic erosion compared to the active tectonic deformation (Fig. 2.3).
- The analysis of the hydrographic network of this sector shows a high frequency of drainage anomalies that may arise after the reactivation of one or more branch(s) of the Gafsa fault, inducing the deformation of the relief and setting up of these anomalies (Fig. 2.4).

- The ridges of these two structures show a very high maximum vertical curvature, while the NE flank varies from very low to medium (Fig. 2.5).
- The south-western flanks of these two structures have a significant surface rugosity expressed by a high roughness index. This roughness may reflect a strong present erosive potential (Fig. 2.6).

20.2.2 J. El Abiod

The area, located in the center of J. El Abiod (Gafsa chain) and whose height can reach 1,150 m, differs from other compartments of the landscape and it has special footprints of several morphometric parameters (Fig. 20.3):

- A relatively high concentration of earthquake epicenters reflecting a relatively high seismic activity (Fig. 3.1);
- A high residual topography (Fig. 3.2);
- South of J. Abiod, we clearly observed a "folding" of geological layers with a high roughness index, indicating a phase elastic compressive deformation that led to this structural configuration (Fig. 3.3);
- Disorder (multi-direction) of maximum curvature, thus defining a "disturbed area" in the center of the anticline (Fig. 3.4);
- A relatively high density of drainage (Fig. 3.5);
- A remarkable concentration of the drainage anomalies (Fig. 3.6).

On the other hand, the convex shape of the hypsometric curves and hypsometric integral value, relatively high

Fig. 20.2 Morphometric and seismic particularities of J. Ben Younes and J. Bou Ramli:
a Epicenters of earthquakes,
b Residual Topography,
c Hypsometric Integral,
d Drainage anomalies (*black*),
e Maximum curvatures,
f Roughness

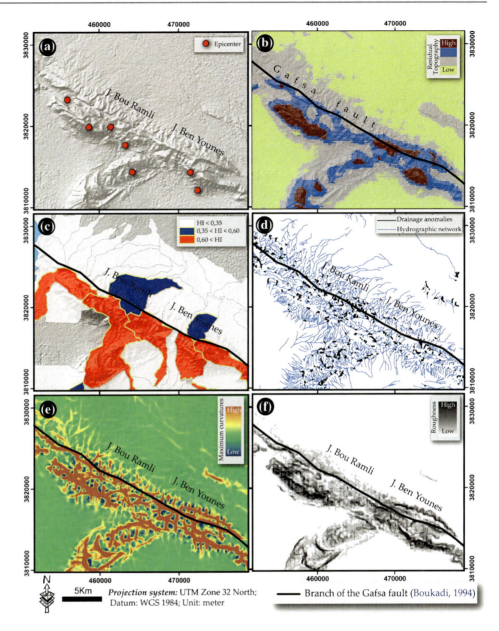

(IH> 0.5), watersheds located in this area reflect the immaturity terrain where erosion is still intense, reflecting the recent tectonic activity and/or low climate erosion.

20.3 Interpretation of Morphometric Indices

In this section, we will try to establish the relationship between morphometric parameters and the structural context of the study area.

The correlation of these indicators with the tectonic cover of the southern Atlas of Tunisia, can confirm the neotectonic reactivation of the Gafsa Fault as a major transpressive dextral strike-slip zone, under the influence of the N–S present stress fields. It should be noted that the Gafsa fault is a deep accident probably dating from late Mesozoic that has been reactivated, several times, during the Cenozoic (Zouari et al. 1990) creating the Gafsa elongated chain above the fault zone. Thus, during the Jurassic and lower Cretaceous, the Gafsa fault has acted as a normal fault with a strong normal component. Also, this earlier distensive tectonics caused the movement of salt material along this accident (Bedir et al. 1992; Boukadi 1994; Zouari 1995). At the end of the Cretaceous, the tectonic inversion in the tethyan region took place. The Gafsa fault appears as a N120°E tranpressive dextral strike-slip fault under the influence of the displacement to the north of the African continent due to the opening of the Atlantic Ocean and the Tethyan sea closure.

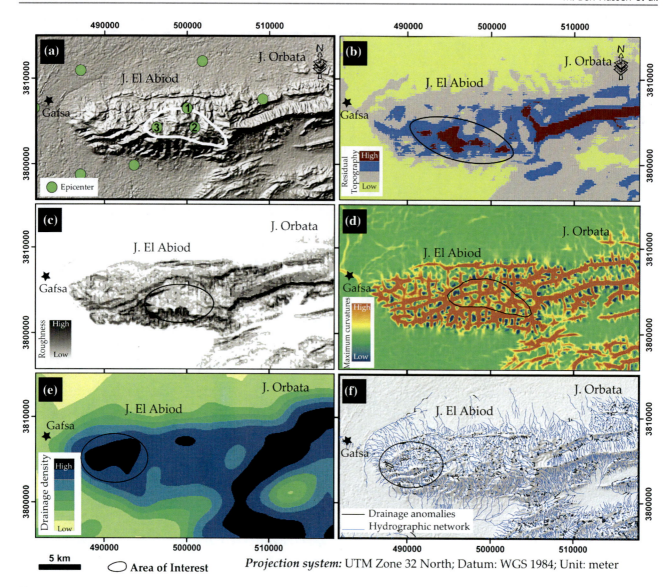

Fig. 20.3 Morphometric and seismic particularities of J. El Abiod: **a** Epicenters of earthquakes, **b** Residual Topography, **c** Roughness, **d** Maximum curvatures, **e** Drainage density, **e** Drainage anomalies (*black*)

But, it is difficult to distinguish between the tectonic and the halokinetic activities with only the morphometric methods. Thus, the tectonic activity of this fault can also be associated with the migration of the Triassic evaporitic material.

Indeed, the geological maps at 1/100 000 scale (produce by the ONM) and other studies, such as those of Zouari (1995), have indicated many local outcrops of Triassic materials in this region. Furthermore, the interpretations of seismic profiles have enabled Bedir (1995) and Hlaiem (1999) to show the presence of many Triassic salt bodies elongated several faults.

Thus, these two mechanisms (halokinesis and tectonics) have probable responsibility in the relief activity in this area and may explain their specific signature of the morphometric indices.

20.4 Conclusion

The identification of recent activity of faults (neotectonic) was performed in the Gafsa region (southern Atlas of Tunisia) based on the analysis of five geomorphometric parameters: the residual topography, hypsometric analysis, drainage anomaly, maximum vertical curvature and roughness of terrain.

The results of the morphometric analysis allowed us to confirm a probable recent activity of the Gafsa fault. Indeed, the kinematics of the Gafsa fault is clearly highlighted through the morphometric parameters by leaving its footprints on three highly immature landforms (J. Ben Younes, J. Bou Ramli and J. El Abiod) contrasting with the rest of the study area, located near the structural of Gafsa fault. On the contrary, it is difficult to distinguish and differentiate between the tectonic and the halokinetic activities.

Indeed, we particularly insist on the relationship between the movement of salt and changes in morphometric parameters, confirming the strong relationship that exists between the morphological structure of the surface and deep structures.

The correlation between morphometric indices and geological coverage has shown that some morphometric anomalies are related to the lithological variations in the study area. In the plains, the absence of apparent neotectonic activity is related to the soft and detritic nature of the rocks that can quickly fill the synclinal depressions and hide the traces of faults. However, these traces are easier to identify in the higher areas.

Thus, the morphometric analysis of the DEM and the drainage network has to provide a more accurate insight into the morphodynamics of the study area. This is explained by their sensitivity towards the perturbations caused by recent tectonic activity or the effect of the lithological cover of the terrain.

Acknowledgments We thank Pr. Fredj Chaabani (Faculty of Sciences of Tunis) and Dr. Ines Tagoug (University of Calgary) for their contribution in this study. We also thank the inhabitants of the Moulares city for their kind co-operation.

References

Bedir M, Ben Youssef M, Boukadi N, Slimane F, Zargouni F (1992) Les événements séquentiels méso-cénozoïques associés au système de coulissements de l'Atlas méridional de Gafsa. 3ème Journée de l'Exploration Pétrolière (ETAP). Tunis

Bedir M (1995) Mécanisme des bassins associés aux couloirs de coulissements de la marge atlasique de la Tunisie : simo-stratigraphie, seismo-téctonique et implications pétrolières. Thèse de doctorat en Sciences géologiques. Thèse de doctorat Doctorat es-Sciences, Université Tunis II, p 412

Bolongaro-Crevenna A, Torres-Rodriguez V, Sorani V, Frame D, Ortiz AM (2005) Geomorphometric analysis for characterizing landforms in Morelos State. Mexico, Geomorphol 67:407–422

Boukadi N (1994) Structuration de l'atlas de Tunisie : signification géométrique et cinématique des nœuds et des zones d'interférences structurales au contact de grands couloirs tectoniques. Thèse de doctorat en Sciences Géologiques, Univ. Tunis II, Faculté des Sciences de Tunis, 249p

Deffontaines B (1990) Développement d'une méthodologie morphonéotectonique et morphostructurale; analyse des surfaces enveloppes, du réseau hydrographique et des modèles numériques de terrains; Application au Nord-Est de la France. Thèse, Univ. Paris VI, France, 230 p

Dehn M, Gärtner H, Dikau R (2001) Principles of semantic modeling of landform structures. Comput Geosci 27(8):1005–1010

Hfaiedh M, Attafi K, Arsovski M, Jancevski J, Domurdzanov N, Turki MM (1991) Carte sismotectonique de la Tunisie. Editée par l'Institut National de la Météorologie

Hlaiem A (1999) Halokinesis and structural evolution of the major features in eastern and southern Tunisian Atlas. Tectonophysics 306 (1):79–95

Pike RJ, Dikau R (1995) Advances in geomorphometry. Z Geomorph N.F. Suppl Bd 101, 238 p

Pike RJ (2002) A bibliography of terrain modeling (Geomorphometry), the Quantitative Representation of Topography-Supplement 4.0. Open-File Report 02-465, U.S. Geological Survey

Zouari H (1995) Evolution géodynamique de l'Atlas centro-méridional de la Tunisie : stratigraphie, analyse géométrique, cinématique et tectono-sédimentaire. Thèse de doctorat d'état en sciences géologiques, 277p

Zouari H, Turki MM, Delteil J (1990) Nouvelles données sur l'évolution tectonique de la chaîne de Gafsa Bull Soc Géol Fr 8 VI, 621–629

Spatial Analysis of Remote Sensing Data in Early Stage of a Seismo-tectonic Research

Novakova Lucie

Abstract

In 1901, M4.7 earthquake hit the area of NE Bohemia, Czech Republic. The Hronov-Poříčí Fault (HPF) was found responsible for the event. Ongoing seismic monitoring proves the Hronov-Poříčí Fault Zone (HPFZ) is, in fact, the second most active area in the Bohemian Massif. Despite importance of the area, the HPFZ has not been described reasonably. Up to the moment, neither length of the HPFZ nor exact locations of its south branch are clear. Vagueness in length of the HPFZ causes large uncertainty in seismic risk assessment of the area. Integrated approach based on geographic information systems and remote sensing was employed to delineate lineaments in the wider HPFZ area. NASA provided Advanced Spaceborne Thermal Emission and Reflection Radiometer (ASTER) data was evaluated. ASTER digital elevation model provided basic topographic characteristics (surface curvature, slope and drainage systems). Edge detecting process was employed to define lineaments. Recent seismic activity and GPS monitored movements in the area were also assessed. The digital elevation model, extracted lineament, recent seismic activity, recent movements pointed out by GPS monitoring, were integrated and analysed in a geographic information system. Fault pattern suggested previously by various authors were compared with the GIS layers, and the extracted lineaments especially. Cross examination showed there are at least three possible variants of the south termination of the HPFZ. The spatial analysis also pointed out field tectonic mapping is necessary to describe the fault in detail and where to focus the survey.

Keywords

Active tectonics • Spatial analysis • Lineaments • Czech Republic

21.1 Introduction and Motivation

The knowledge of the range and continuation of a fault zone is very important for understanding the tectonic development and paleostress conditions of the study area, or for the estimation of earthquake hazard and risk. Previous works pointed out the magnitude of an earthquake is closely related to prolongation of the responsible fault (Wells and Coppersmith 1994).

Most studies focussed the Hronov–Poříčí Fault Zone (HPFZ) between the towns of Hronov and Trutnov. Despite the seismological importance of the HPFZ only the northern termination of the zone was described consistently so far. Assuming the southern termination of HPFZ is located in Poland or in Orlické hory Mts., the precise structural geological mapping of the brittle tectonic features has never been done within the whole range of the HPFZ area. Moreover, the tectonic conditions, its development and stress field remains poorly understand. The strongest earthquake was documented in 1901 with magnitude M4.7. Nowadays, it is supposed the HPFZ is about 30 km long.

N. Lucie (✉)
Department of Seismotectonics, Institute of Rock Structure and Mechanics AS CR, v.v.i, V Holesovickach 41, 18209 Prague, Czech Republic
e-mail: lucie.novakova@irsm.cas.cz

The maximum moment magnitude for such long fault might be M6.8. However, supposing the length three times higher, the maximum possible moment magnitude might be M7.4 (Wells and Coppersmith 1994). For this reason, it is highly important to specify length and range of the HPFZ precisely. The true length of the HPFZ should be considered when planning power plants, highways, tunnels and other engineering constructions in the broader area.

21.2 Geological and Tectonic Settings

The Hronov–Poříčí Fault Zone (HPFZ) is located in the easternmost part of the Trutnov–Náchod Depression. It is approximately 30–40 km long and up to 500 m wide system of parallel fractures, dividing two important structural units —the Intra Sudetic Basin and the Krkonoše Piedmont Basin. The NW-SE striking structure was formed due to the post-Cretaceous flexural folding and is filled with the Upper Cretaceous sediments and the Permo-Carboniferous volcano-sedimentary complex. The HPFZ had a complicated tectonic evolution, started in the late Paleozoic. Since then, several tectonic phases have taken place (Nováková 2014). The fault zone has been successively developed from an asymmetric anticline, whose steeply inclined SW part was axially disrupted due to the regional compression by a reverse fault. The main reverse fault (thrust) is accompanied by parallel or oblique normal or reverse faults. The reactivation of the HPFZ is recorded after the Upper Cretaceous sedimentation during the Late Saxonian tecto-genesis.

21.3 DEM Analysis

21.3.1 Method

Digital elevation model (DEM) was employed to assess major lineaments in the area in addition to thorough tectonics-focused review of previous mapping works. The Aster GDEM (LPDAAC 2013) of the HPFZ area and its surroundings was represented graphically as a shaded relief (Fig. 21.1a). Two orthogonal sunshine directions were utilized to avoid bias during later image processing. The Canny algorithm (Canny 1986) and Hough transformation (Duda and Hart 1972) were adopted. Image processing was carried out using the open-source image processor ImageJ (Schindelin 2012) implementing the Canny Edge Detector (Gibara 2011, Fig. 21.1b) and Hough Linear Transformation (Burger and Burge 2013) plugins respectively. We applied the Hough linear transformation to both whole (Fig. 21.1c) and partitioned images (Fig. 21.1d). The shaded relief images were divided into 20 (4 × 5) uniform square areas each. Directions of the linear structures identified in individual

partitions were displayed into rose diagrams. Finally, we confronted the automatic lineament identifications outputs to the tectonic mapping review.

21.3.2 Results and Discussion

Figure 21.1 shows ongoing results of the automatic lineament identification process. In the HPFZ area two major directions were found—NW–SE and WNW-ESE. These fault orientations are typical in the Bohemian Massif. The NW–SE direction is dominant (Fig. 21.2, left). In general, we found the lineament distribution provided by automatic processing similar to tectonic lines mapped during previous geological surveys (Fig. 21.2, right). Automatic lineament identification applied in individual partitions added two lineament orientations—W–E and WSW–ENE. Apparently, these two orientations are of local importance. It is obvious not all lineaments represent actual faults. Nevertheless the agreement between direction of the lineament and previously mapped fault direction (Nováková 2014) points out there is a link between terrain morphology and faults in the HPZF area. Moreover some of the lineaments identified in the square areas in actuality correspond to major faults including the HPFZ.

Four different parts can be distinguished due to spatial distribution of the lineament orientation (Fig. 21.1d and Fig. 21.3). These are stripes of the NW–SE orientation. Flat land on NE of the area represents the Sudetic Foreland. Lineaments here are ESE–WNW mainly. Morphologically significant stripe in the centre of the studied area is a Variscan relict mountain range. The NW–SE orientation dominates in this stripe. Strait line dividing these two parts is a clear demonstration of tectonic border of the Bohemian Massif in the area—250 km long the Sudetic Marginal Fault. SW end of the stripe corresponds to the HPFZ occurrence. W–E lineaments prevail in the SW part of the studied area. This terrain represents Permian-Carboniferous and Cretaceous basins. Fourth part could be identified in the SW corner where NW–SE orientation dominates again.

21.4 Lessons Learned

Spatial analysis of DEM data pointed out terrain lineaments in the HPFZ area and its vicinity. The identified lineament orientations correspond to geologically mapped tectonics. Moreover division due to local terrain lineament orientation matches main geological units in the area. Clearly, tectonics is responsible for many terrain lineaments in the HPFZ area. In addition, we showed open-source image processor ImageJ can be successfully utilized in spatial analysis of remote sensing data.

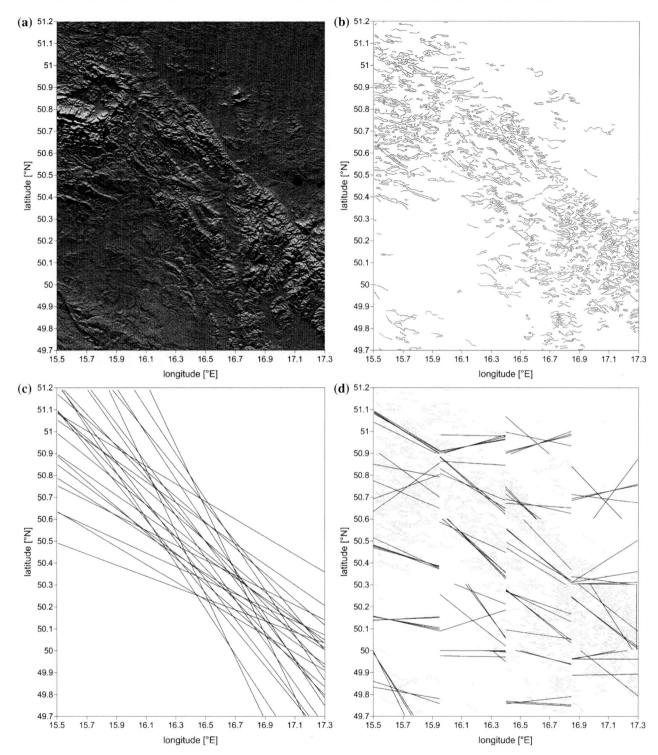

Fig. 21.1 Image processing. **a** Shaded relief of the HPFZ area. **b** Edge enhancement using the Canny algorithm. **c** Major lines identified using Hough algorithm in the whole area. **d** Major lines identified in partitions using Hough algorithm

Fig. 21.2 Dominant lineament orientations according to the automatic image processing (*left*) and fault orientations provided by geological mapping (*right*, Nováková 2014)

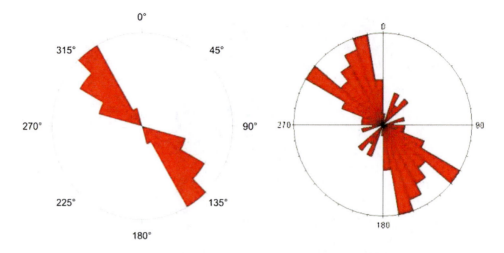

Fig. 21.3 Spatial distribution of the major lineament orientations in the HPFZ area

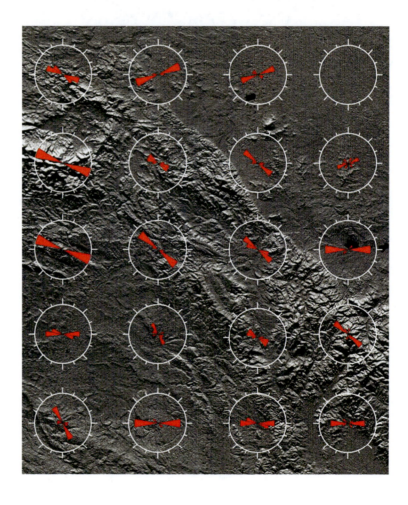

Acknowledgments The research was funded by GA CR (GA205/09/1244) and Institute's Research Plan No. A VOZ30460519. I am grateful to P. Novak for help with image processing. ASTER GDEM is a product of METI and NASA.

References

Burger W, Burge MJ (2013) Hough linear transformation plugin for ImageJ image processor, http://imagingbook.com/source

Canny JF (1986) A computational approach to edge detection. IEEE Trans Pattern Anal Mach Intell 8:679–714

Duda RO, Hart PE (1972) Use of the hough transformation to detect lines and curves in pictures. Commun ACM 15(1):11–15

Gibara T (2011) Canny Edge Detector plugin for ImageJ image processor. http://rsbweb.nih.gov/ij/plugins/canny/index.html

LPDAAC (2013) Global Data Explorer on-line application, http://gdex.cr.usgs.gov/gdex

Nováková L (2014) Evolution of paleostress fields and brittle deformation in Hronov-Porici Fault Zone, Bohemian Massif. Stud Geophys Geod 58:269–288. doi: 10.1007/sl1200-013-1167-1

Schindelin J, Arganda-Carreras I, Frise E, Kaynig V, Longair M, Pietzsch T, Preibisch S, Rueden C, Saalfeld S, Schmid B, Tinevez JY, White DJ, Hartenstein V, Eliceiri K, Tomancak P, Cardona A (2012) Fiji: an open-source platform for biological-image analysis. Nat Methods 9(7):676–682

Wells DL, Coppersmith KJ (1994) New empirical relationships among magnitude, rupture length, rupture width, rupture area, and surface displacement. B Seismol Soc Am 84(4):974–1002

Geomorphic Evidence of Active Tectonics: The Case of Djemila Fault (Eastern Algeria)

22

Youcef Bouhadad

Abstract

The Djemila active fault is a NE–SW trending and NW-dipping reverse fault. Historical seismicity around this fault is weak to moderate. However, field observations allowed us to observe differential uplifts of quaternary alluvial terraces on the hanging and foot walls. We interpret this as a strong evidence of recent tectonic activity. The uplifted terraces suggests an uplift rate of 0.14 mm/years. Therefore, the seismic potential of this fault may be greater than may be suggested by the historical seismicity of the region.

Keywords

Active reverse fault • Alluvial terraces • Uplift • Algeria

22.1 Introduction

At the outcrop scale field observations may provide useful information on the evidence of fault activity. Present-days landscapes result from the combined action of two factors the climate and tectonics (Molnar and England 1990; Burbank and Pinter 1999). Impact of tectonic in relief building has been suggested early and morphotectonics is now a widely recognized discipline for the study of morphological features related to tectonics (Scheidegger 2004). Active tectonics zones, mainly compressive context, exhibit several geomorphic features which may be used to understand the active faults history. Indeed, growth of faulted folds generates a response of the drainage system represented mainly by uplift/incision of alluvial fans and shift/diversion of drainage pattern (Mayer 1986; Shumn 1986; Merritts and Bulls 1989; Jackson et al. 1996; Pavlis et al. 1997; Bouhadad 2001; Lavé and Avouac 2001). Furthermore, trace of paleoearthquakes may be found in geomorphic records and then quantitative geomorphology provides rates of coseismic vertical and/or horizontal displacements that can be used as valuable tool for seismic hazard assessment (Merrits and Bull 1989; Keller et al. 1999; Martinez-Diaz et al. 2003). In this work we aim to discuss the geomorphic features related the Djemila active fault in northern Algeria (Harbi et al. 1999) (Fig. 22.1). In terms of geology, the studied area belongs to the Babor thrust sheet area formed by Jurassic-Cretaceous calcareous and marl. The known active faults of the region are: (i) the Kherrata fault which produced the February, 17, 1949 (Io = VII, MSK scale) earthquake and the youcef fault to the south and the Djemila fault studied herein.

22.2 Seismotectonics Setting

Northern Algeria belongs to the Africa-Eurasia plate boundary where tectonic plates are converging in the NW–SE direction (Anderson and Jackson 1988). The amount of shortening is about 4–6 mm/yr (De Mets et al. 1990). Consequently, the seismicity is relatively high and several majors earthquakes have been occurred in the past, particularly the Orleanville, September 9, 1954 (Ms = 6.5), the El-Asnam October 10, 1980 (Ms = 7.3) and the Zemmouri may 21st, 2003 (Mw = 6.9) (CRAAG 1994). Northern Algeria is NE–SW trending hills and mountains which forms the Tellean chain.

Y. Bouhadad (✉)
National Earthquake Engineering Center (CGS), B.P.252, Algiers, Algeria
e-mail: bouhadad_y@yahoo.com

G. Lollino et al. (eds.), *Engineering Geology for Society and Territory – Volume 6*,
DOI: 10.1007/978-3-319-09060-3_22, © Springer International Publishing Switzerland 2015

Fig. 22.1 Seismicity and active faults of the Setif (eastern Algeria) region. (*1* (6.0 ≥ Ms ≥ 5.0), *2* (5.0 ≥ Ms ≥ 4.0), *3* (4.0 ≥ Ms ≥ 3.0), *4* (Ms < 3.0)

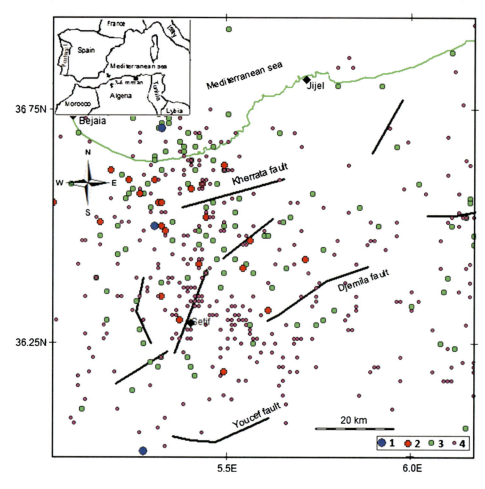

22.3 The Djemila Fault

The Djemila active fault shows a reverse mechanism and is trending NE-SW. It dips 50° toward the northwest. It constitutes a major geological structure of about 49 km of length (Vila 1980). The historical seismicity around this fault is weak to moderate. Nevertheless, field observations allowed us to identify a differential uplifts of quaternary alluvial terraces on the hanging and the foot walls. We interpret this as a strong evidence of recent tectonic activity. The uplifted terraces suggests an uplift rate of 0.14 mm/years for the hanging wall. Also, if we consider the length and the rate of terraces uplift, the seismic potential of the fault may be greater than may be suggested by the historical seismicity which extends back about two hundred years.

We interpret this as a geomorphic marker of the activity of the fault. The uplifted terraces suggests an uplift rate of 0.14 mm/years. Therefore, the seismic potential of this fault may be greater than may be suggested by the historical seismicity of the region. Hence, it is necessary to look in details to the seismic hazard implications of this fault.

22.4 Conclusion

Geomorphic features may help to recognize evidence of active tectonics at an outcrop scale. Field work undertaken in the Djemila active fault allowed us to observe differential uplift of alluvial terraces on the hanging and the foot walls.

References

Anderson H, Jackson J (1988) Active tectonics of the Adriatic region. Geophys JR Astr Soc 91:937–983

Burbank DW, Pinter N (1999) Landscape evolution: the interaction of tectonics and surface processes. Basin res 11:1–6

Bouhadad Y (2001) The Murdjadjo, western Algeria, fault-related fold: implications for seismic hazard. J Seismolog 4(5):541–558

CRAAG (1994) Les séismes en Algérie de 1365 à 1992. Publication du Centre de Recherche en Astronomie, Astrophysique et Géophysique, Département: Etudes et Surveillance Sismique, ESS, C.R.A. A.G, Alger-Bouzaréah

De Mets C, Gordon RC, Argus DF, Stein S (1990) Current plate motion. Geophys J Int 101:425–478

Harbi A, Maouche S, Ayadi A (1999) Neotectonics and associated seismicity in Eastern Tellian Atlas of Algeria. J Seismolog 3:95–104

Jackson J, Norris R, Yougson J (1996) The structural evolution of active fault and fold systems in central Otago, New Zealand: evidence revealed by drainage patterns. J Struct Geol 18:217–234

Keller EA, Gurroba L, Tierney TE (1999) Geomorphic criteria to determine direction of lateral propagation of reverse faulting and folding. Geology 27:515–518

Lavé J, Avouac JP (2001) Fluvial incision and tectonic uplift across the Himalayas of central Nepal. J Geophys Res 106 (11):26561–26591

Martinez-Diaz JJ, Masana E, Hernandez-Enrile JL, Santanach P (2003) Effects of repeated paleoearthquakes on the Alhama de Murcia Fault (Betic Cordillera, Spain) on the quaternary evolution of an alluvial fan system. Ann Geopysics 46 (5):775–791

Mayer I (1986) Tectonic geomorphology of escarpments and mountains fronts, In: Active tectonics, National academy press, Washington, DC, pp 125–135

Meritts D, Bull WB (1989) Interpreting quaternary uplift rates at the Mendocino triple junction Northern California, from uplifted Marine. Terraces—Geol 17:1020–1024

Molnar P, England P (1990) Late cenozoic uplift of mountain ranges and global mountain changes: chicken or egg? Nature 346:29–34

Pavlis TL, Hamburger MW, Pavlis GL (1997) Erosional processes as a control on the structural evolution of an actively deforming fold and thrust belt: an example from the Pamir-Tien Shan region, central Asia. Tectonics 16:810–822

Scheidegger AE (2004) Morphotectonics. 197 p, Edit. Springer, Berlin

Schumm S (1986) Alluvial river response to active tectonics: in Active tectonics. National academy press, Washington, pp 80–94

Vila J.M., 1980. La chaîne alpine d'Algérie orientale et des confins algéro-tunisiens. Thèse Sci. Univ. Paris, 665 p

Seismic Cycle of the Southern Apennine Deformation Front: The Taranto Gulf Marine Terraces Inputs and Implications

23

Benoît Deffontaines, Gérardo Fortunato, and Samuel Magalhaes

Abstract

Detailed tectonic analyses and geological mapping in muddy fold-belt front is a hard target. Using both fieldwork and GIS software associated to new soil datations (Sauer et al. in Soil development on marine terraces near Metaponto. Gulf of Taranto, Southern Italy, pp. 1–16, 2009) of the different marine terraces of the Taranto Gulf (Southern Italy), we were able to precise the location of the Metaponto-Pisticci staircase which appear to be situated above the first thrusting unit of the southern Apennine orogen corresponding to the deformation front. Consequently we can locate, characterize and quantify from an active tectonic point of view the deformation front of southern Apenines. By combining our data with (1) the AGIP seismic profile offshore the Taranto gulf that gives the precise geometry of the overthrusting sheet, (2) the known eustatic curve of the mediterranean sea along the southern Italian shore, and (3) the soil datations of the different terracic levels, we are now able to differentiate the signal of both active tectonic and eustatic processes and their related geomorphic features on the Taranto Gulf marine terraces. Therefore the observed seismic cycle of the Southern Apennine deformation front is revealed and characterized by a return period of about 261 ka, a vertical uplift of about 113 m and a shortening rate of 2.1 km by return period. Its deformation is not uniform and appear to be coherent with both a regular interseismic linear creep period (time = 251 ka/uplift = 71 m) and a rapid inferred cosismic uplift (time: lOKa/uplift = 42 m) that we interpret as a sismogenic period with probably numerous major earthquakes. Finally thanks to the marine terraces of the Metaponto-Pisticci staircase that lead us to better understand the deformation front of Southern Apenine and its associated landscape and to separate erosion and tectonic processes.

Keywords

Marine terraces • Soil datations • Seismic cycle • Deformation front • Taranto gulf (Southern Apennine, Italy)

23.1 Introduction and Aim of This Study

How to locate, characterize and quantify the deformation front of an orogen especially in muddy and silty areas? Classical structural geological and geomorphological methods are then at there boundaries as the absence of hard geological layers increase the difficulty to get dips and therefore to reveal faulted and folded structures. As a case example, we focus herein on the NNW–SSE trending southern Apennine deformation front in Basilicate (Southern

B. Deffontaines (✉)
Laboratoire International Associé N°536 ADEPT France-Taiwan CNRS-NSC, and Laboratoire de Géomatique Appliquée ENSG/ IGN and UPEM, Paris, France
e-mail: benoit.deffontaines@univ-mlv.fr

G. Fortunato · S. Magalhaes
« AlphaGeOmega », 62, rue du Cardinal Lemoine, 75005, Paris, France

Italy) where it is noteworthy the approximative inferred location of the orogenic deformation front (Ferranti et al. 2009), and we use herein the well preserved marine terraces NE–SW trending of the Taranto Gulf in order to precise both the geometry and its recent structural history.

The recent development of geomatic tools (e.g. Geographic Information System—GIS associated with photo-interpretation) validated with new numerous available data, as offshore AGIP seismic profiles, combined with terrace marine soil datations (Sauer et al. 2009) and the abundance of scientific studies of the place (Gignout 1913; Gigout 1960; Hearty and DaiPra 1992; Caputo 2007; Westaway and Bridgland 2007, 2009, among others), lead us to (1) re-interpret the structural geometry of this deformation front, and therefore to (2) precise both (1) the seismic cycle and the uplift and shortening rates of this southern Apennine deformation front. At least, we are now able to propose new assumptions on the origin of the formation of the marine terracic levels and on the passed geological and geomorphological history of the place.

23.2   Geographic, Geologic and Geomorphologic Presentation

The Pisticci-Metaponto pleistocene, holocene staircase area is situated in the central part of the Taranto gulf (southern Italy) and consist of well preserved and clearly expressed marine terracic levels that runs from Taranto (Apulia) to the NE up to Rocca Imperiale (Calabria) to the SW on more than 65 km long and a width of about 25 km (Fig. 23.l).

From the geological point of view the marrine terraces lithology above the erosional surface is filled up with gravel deltaic deposits (bottomsets, and foresets) and above windy loemy and loessy deposits silts, and sands.

From the structural point of view, the geometry of the Pisticci-Metaponto area is difficult to know as the soft muddy area is lacking bedding dips and due to the impossibility to preserve through time any fault plane. From the bibliography and the existing geological maps (scale 1/100,000), the southern Apennine deformation front is inferred west of the Pisticci-Metaponto staircase (20 km) which is not affected by cartographic faults. Nonetheless Bentivenga et al. (2004) describe the existence of one NE–SW normal fault seen and situated north of the Pisticci-Metaponto staircase. From this observation, Bentivenga et al. (2004) propose that the Pisticci-Metaponto staircase is made of a only one unique marine terrace offsetted by several normal faults, parallel to the shoreline, that explains the observed scarps. Contrasting to that assumption, (Caputo 2007; Westaway and Bridgland 2007) and others, demonstrate the exact prolongation of the buried sediments on both sides of the marine terrasses scarps, proving that these scarps

are not fault scarps. Unfortunately we were not able to find the Bentivenga stuctural observation during our field work. The tectonic regime of the studied area is controversy Westaway and Bridgland (2007, 2009) described an isostatic normal global extension, contrasting with Caputo and Bianca (2009) that describe a more classical compressional one.

From the geodynamic point of view, the roll-back of the Calabria off Sardinia (10–12 Ma) and the resulting opening of the Tyrrhenian sea, the northern part of the arc progressively collided with Adria to create the Apennine (while in the same time the southern part collided to Africa that forms the Maghrebides). The Lucanian Apennine of southern Italy is a fold-and-thrust belt that developed following the closure of the Mesozoic Tethys Ocean by the subduction of the Adriatic microplate (continental lithosphere) beneath the Southern Apennine. The belt is composed of various tectonic units derived from alternating basins and platforms that were located onto the western edge of the Apulian passive continental margin (D'Argenio and Alvarez 1980). Since Oligocene time these paleogeographic domains, namely the Apenninic Platform, the Lagonegro Basin and the Apulian Platform (Scandone 1972), experienced orogenic contraction and were stacked north-eastwards along low-angle thrust faults to produce a complex accretionary wedge (Roure et al. 1991; Monaco et al. 1998). The outer, eastern parts of the wedge correspond to a thin-skinned fold-and-thrust belt; its allochtonous units mainly consist of a detached Miocene flysch and a tectonic "melange", the so-called Varicoloured Clays (Casero et al. 1988), whose paleogeographic provenance has been debated in the literature (Lentini 1979; Pescatore et al. 1988; Monaco et al. 1998). The internal deformation of this melange has been referred to a generic post-Early Miocene time.

From the bibliography it appears that the outer province of the southern Apennine belt was structured in Quaternary time, i.e. up to Middle Pleistocene times (Pieri et al. 1996; Patacca and Scandone 2001), and there is no direct evidence for ongoing contractional deformation at the thrust front. It has been suggested that at around 650–700 ka active convergence turned into a regime of tectonic quiescence and generalised post-orogenic uplift (Ambrosetti et al. 1982; Westaway 1993; Patacca and Scandone 2001). Neogene contractional deformation was accompanied by deposition of syn-orogenic sediments, that occurred extensively within satellite basins located on top of the evolving fold-and-thrust belt, and within foredeep basins located ahead of the advancing thrust fronts (Hippolyte et al. 1994). It seems that the present thrust front is now buried below the foredeep deposits of the Bradanic Trough, that separate the outermost province of the belt from the relatively undeformed Apulian foreland, and has been inferred from subsurface data (Sella et al. 1988). The thrust front crosses the present coastline

Fig. 23.1 Location of the Taranto Gulf southeastern Italy, known geological features, quadrangle the studied area in Italy. One may note the obliquity of the NE–SW trending marine terraces toward the inferred NW–SE trending deformation front (*dashed black line*)

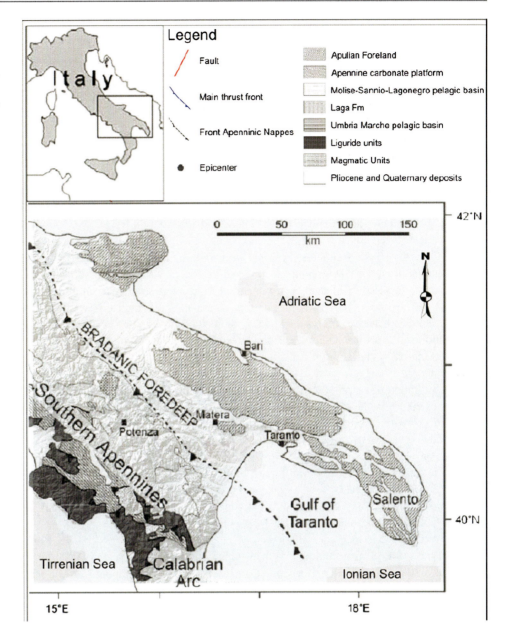

within the Taranto Gulf. In this area the bottom of the Ionian Sea is relatively steep and reaches a depth of more than 2,000 m below the sea level at around 150 km from coast line.

From the geomorphological point of view, many authors had been working on this area differentiating from 1 marine terracic level (Bentivenga et al. 2004), 6 (e.g. Mostardini et al. 1966; Vezzani 1967), 7 (Cotecchia and Magri 1967), to 11 (Bruckner 1980 PhD thesis and 1982). So great scientific divergence in the numbers despite the general agreement that the marine terraces are the results of both interaction of quaternary tectonics and glacio-eustatic sea level variations.

We distinguish by comparing the precise Digital Terrain Model (a mixed of SRTM, GRDEM-ASTER, and a 8 m ground resolution DTM) that we settled on the geological

maps (Geological Map of Italy 1:100,000—extract sheets 201-202-212) of the studied area on the Pisticci-Metaponto staircase from the Ionian Sea shoreline (east to west) 11 different terrassic levels separated by scarps or bluffs following the assumption of Bruckner (1980, 1982); Sauer et al. (2009).

In this paper, we will first have a closer look to the marine terraces sedimentology and their datations by a discussion about their validity, this will give us a curve: Altitude toward Time. Then in a second stage we will update the structural geology of the place by revealing the exact place of the Apennine deformation front situated 40 km east of what was inferred validated by numerous AGIP seismic lines and we will then deduce by comparison the seismic cycle differentiating the discontinuous uplift and shortening rates. At least, finally we will discuss some of those implications on the geomorphology and the recent history of these Pisticci-Metaponto marine terraces.

23.3 Datations of the Metaponto/Pisticci Marine Terraces

The sedimentology of the Pisticci/Metaponto marine terraces, Bruckner (1980), Sauer et al. (2009), present a characteristic vertical sediment sequence and may be resume as follow and described below from bottom to top: (1) marine sediments such as clays and silts deposited in deeper shelf; (2) intermittent layers of loam, gravels and sands that compose the terrace base sediments deposited in a middle shelf, sublittoral or lagoonal environment; (3) A main gravel sediment body deposited in a beach environment; (4) a fine continental textured cover sediments that is from lagoonal, alluvial, colluvial or aerian origins (e.g.: loess or loehm). The evolution of the sedimentation correspond generally to the uplift of the place starting from a marine to a continental environment.

To get a datation of a marine terracic level is a hard task (Sauer et al. 2009) on the technical and methodological as well as the geomorphic point of views. From the technical and methodological point of view, numerous methods and techniques are used each one with its own error bars well explained in Sauer et al. (2009). From the geomorphological–geological point of view, it is needed to know what geomorphic or sedimentological feature to date due to the great diachronism of the different deposits that compose a marine terraces. For instance tides and glacio-eustatic fluctuations of the sea level create and modify the geometry and datations of the marine terraces that lead to several different ages depending on the depth of the dated material.

First at depth it is possible to date (1) the erosionnal base surface that correspond to the angular unconformity in between the bedrock and the surficial geology. That corresponds to the first transgression or Rising Sea Level (RSL). It is also possible to date (2) the surficial geology corresponding to the marine deposits linked to the transgression highstand, maximum stillstand and beginning of the regression (that correspond to the age of the marine gravel body). Then third possibility (3) is to date the lagoonal environment and the continental deposits such as the fluvial, alluvial, swampy colluvional deposits which act as agradation processe or the glacio-eustatic regression. Then (4) it is also possible to date during the lowstand sea level, the associated loamy or loessic aerian deposits, and finally to date (5) any reactivation of the glacio-eustatic sea level fluctuations that changes this theoretical general terrace

marine sedimentological succession. The Marine Terrace 1 (Petrulla site) of the Metaponto-Pisticci staircase (Table 23.1) illustrate such discrepancies.

Due to the diachronism of the different sediments that compose the marine terrace, the question is then what geological/geomorphic object do we need to date for a marine terrace in order to get the neotectonic informations ? For our point of view, taking into account the existing bibliography (Sauer et al. 2009), it is needed to date the last maximum sea level. Terrassic bluffs appear then as the only geomorphic isochrone feature on the Pisticci-Metaponto staircase as it is the result of the erosion of a unique paleoshoreline. That is why it is needed to carefully date the terrace immediately up and down of each scarps/bluffs. One may note that even for the soils datation it is possible to get different ages due to glacio-eustatic fluctuations as it is possible to stack and superimposed different paleosoils. To date loess will give you indications of the more recent loess/loamy sediments deposits that settled during each maximum glaciar periods (or minimum sea level).

Table 23.2 resume and gather all available location (cf. black dots of Fig. 23.2a), altitude and ages of the selected terracic levels as well as the facies and the associated soils. It is then possible to draw the curve age versus altitude (Fig. 23.3) which is not a continuous linear feature. The resulting curve evidence colinear and equivalent time long oblique flats and steep ramps that is highly comparable to the general geometry of the creep/asismic corresponding to the red curve/oblique flat and earthquake return period corresponding to the orange curve (steep ramps) of the repeated sismic cycle...

Table 23.3 above resume the main characteristics of the sismic cycle revealed through the soil datations of the Pisticci-Metaponto marine terraces. It distinguish both component of the uplift rate and the shortening one. In the following paragraph we will focus on which cycle this could be.

23.4 Location of the Apennine Deformation Front in Basilicate and Taranto Gulf (SE Italy)

Field geologists know that where the muddy rather flat bedding dip had been preserved that usually correspond to the not deformed zone contrasting with those without bedding that appear to be highly deformed but without any trace of fault planes. That is why it is uneasy to get the structure in the fields of such areas without (or so few) bedding dips and fault plane due to the muddy lithology. Fortunately it exists indirect methods of structural geomorphology in order to decipher the structures based for instance in the drainage

Table. 23.1 Example of the diachronism of datations of the T1 Marine Terraces (Petrulla area), depending on the *surficial* geologic and geomorphic objects and the used datation technique (modify from Sauer et al. 2009; all *OSL ages from* Zander et al. 2003, 2006 in Sauer et al. 2009; 230Thl234U ages from Bruckner 1980)

Marine Terrace 1 (Petrulla site—modified from Sauer et al. 2009)

Technique	Depth (cm)	Age
OSL of loess	60 cm	16 ka BP
OSL of loess	90 cm	20.3 ka BP
OSL of loess	170 cm	24.9 ka BP
OSL of upper main gravel Body	nc	55.4 ka BP
OSL of terrace base	30 cm below surface	50.7 ka BP
U/Th of molluscs in upper mai gravel body	?	63 ± 3 ka BP
OSL of terrace base	90 cm below surface	73.8 ka BP
U/Th of molluscs in terrasse base	?	75 ± 7 ka BP
U/Th of molluscs in the lowest layer the main gravel body	?	110 ± 10 ka BP

Table. 23.2 Synthetic soil datations of the marine terraces of the staircase Pisticci-Metaponto. Til later than Brunhes/Matuyama boundary (780 ka) according to Bruckner (1980) (modified from Sauer et al. 2009)

Terrassic level Pisticci-Metaponto staircase	Coordinates (lat./long.) WGS84 Lat/lon	Altitude (m)	Age (ka)	MIS (mar. and fluvial sed.)	Sediments
TO (Lido di Metaponto)	40°21.05' N–16° 49,91' E	2	0,19	1	Aeolxan sand Beach sand
TO (Basentol/Metaponto)	40°21.80' N–16° 47,56' E	5	7	1	Alluvial sed.
T1 (Petrulla)	40°21.58' N–16° 46,40' E	22	80	5.1	Loess Marine gravel
T2 (San Teodoro I)	40°21.32' N–16° 44,68' E	43	100	5.3	Young Alluv. Fine sed. Older All. Fine sed Fluvial Grav Sed Marine Gravel body
T3 (San Teodoro II)	40°22.11' N–16° 44,75' E	61	120	5.5	Alluvial Fine Sed
T4 (Marconia)	40°21.02/20,99' N–16°41,09/10' E	96	195	7	Alluvial fine sed. Fluvial Grav. Sed. Marine Gravel Body
T5 (SE)	40°21.37' N–16° 40,56' E	120	310	9	Alluvial fine sed. Marine Gravel Body
T5 (NW)	40°21.73' N–16° 37,94' E	160	330	9	Alluvial fine sed. Fluvial Grav. Sed.
T6 (Tinchi I)	40°21.68' N–16° 37,11' E	196	405	11	Alluvial fine sed. Marine Gravel Body Marine Sand
T7 (Tinchi II)	40°21.87' N–16° 36,25' E	224	500	13	Alluvial fine sed. Marine gravel Marine sand
T8 (Pisticci)	40°22.14' N–16° 35,25' E	245	575	15	Colluvium Fluvial sediment + Tephra Layer
T9 (Rinne)	40°26.72' N–16° 36,38' E	320	670	17	Fluvial Sediment
T10 (Porcellini Brückner 1980)	40°26,844' N ?–16° 35,962' E ?	333 ?	730	19	?
T11 (?, Brückner 1980)	40°28'388' N ?–16° 33,891' E ?	427?	>780?	?	?

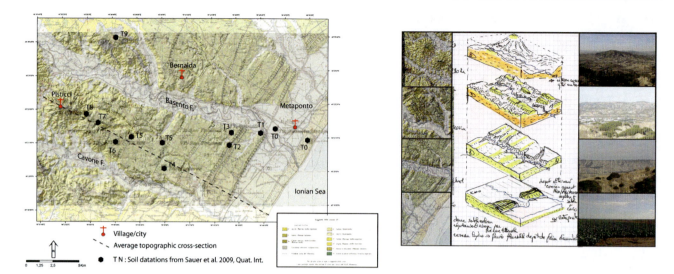

Fig. 23.2 Geomorphology of the Pisticci-Metaponto marine terraces. **a** Geological map superimposed above the hillshade DTM. *Black* large dots correspond to the location of the dated soils on the Metaponto-Pisticci marine terraces (modified from Sauer et al. 2009). Bluffs or scarps from Bruckner (1980) MNT SRTM, GRDEM-ASTER, MNT 208 m +intp of DTM (2,018 m). **b** Key examples of photographs, associated 3D sketch map and extract of geological map showing the increasing erosion of the older marine terracic levels (*bottom* to and *top* T9–11 respectively)

Fig. 23.3 Diagramm Age/Altitude of the Pisticci-Metaponto marine terraces (*green cross*); one may note the discontinuous uplift where we optimize in *red* (oblique flats) and in *orange* (steep ramps). The eustatic curve of the sea level during the last 450 ky (Holocene and Pleistocene time) is highlighted in *blue* below (Waelbroeck et al. 2002)

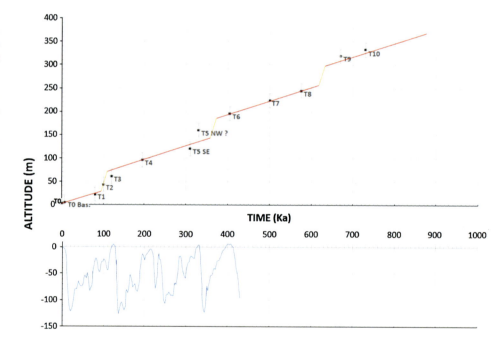

network geometry and its anomalies (Howard 1967; Deffontaines and Chorowicz 1991; Deffontaines et al. 1994, 2000). The circular and annular drainage NE of the Pisticci-Metaponto area contrast with the sub parallel one further north and lead us to trace the precise geometry of the deformation front which is situated close to the Matera city.

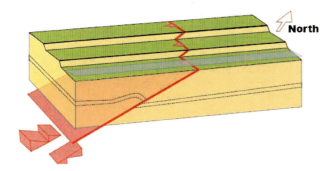

Fig. 23.4 3D view of the deformation of the Tarento gulf marine terraces (*Red plane*: Deformation front and thrusting Fault; in *green*: marine terraces and *brown* at the surface and terrassic scarps)

Numerous public parallel to the shoreline AGIP/ENI reflection seismic lines offshore Taranto Gulf are now public and available (http://unmig.sviluppoeconomico.gov.it/videpi/geografica.htm). The AGIP reflection seimic lines D484, F75_091, F75_089, F75_087 provide for instance the offshore Metaponto geometry of the rather flat ramp thrusting (6°TWT so real 3° "angle verging NNE dipping SSW) over the Adriatic continental lithosphere and validate the onshore geological interpretation above.

On the available AGIP offshore sismic reflection profile, one may note the presence of flat parallel reflection sismic layers of the bradanic trough ending to the south-west against a vertical misty zone which to our point of view characterize transpressive thick-skin strike slip faults. No thrust component is visible on the reflection profile and the vertical geometry is a clear tectonic feature. Therefore, we believe the present geometry of the deformation front is composed of both compression and transpressive faults revealing in the southern Apennine typical partitioning (external thrusting and internal strike-slip and normal faulting for the reajustement). Lentini et al. (2002) for instance highlighted numerous major NW–SE transpressive left lateral strike slip faults in the Lucanian area. Therefore contrasting the pure thin-skin model rootless nappes of the Apulian platform detached from basement (Mazotti et al. 2000), and contrasting also to the pure thick-skin model with basement involved thrusting (Menardi Noguera and Rea 2000), we believed that the southern Apennine face a mixed

in between both: thick (basement transpression) versus thin skin (decollement of the first thrusting units) tectonics which may better fit with the Steckler et al. (2008) teleseismic receiver data. The prevailing extension (Maggi et al. 2009) correspond to the western tectonic deformation and readjustment on the back of the thrusting and strike slip units.

The combination of both the seismic cycle with the offshore structural geometry lead us to propose the uplift and shortening rates situated in Table 23.3.

23.5 Discussion and Conclusions

Detailed soil datations of the different Pisticci-Metaponto marine terraces (Basilicata, Southern Italy) and the new geometry of the deformation front lead us to determine both its discontinuous uplift behaviour and therefore quantify the seismic cycle of the first unit situated above the deformation front of the southern Apennine orogenic belt following the tectonic model of Fig. 23.4. The sismic cycle correspond then to an alternance of asismic creep (246 ka, 71 m uplift, 5.5 mm/y shortening) and earthquake (15 ka, 42 m uplift and 53.5 mm/y shortening). They have strong implications in terms of oil exploration, maturation of the organic matter and on the tectonic history of the place (see for instance Jasvre et Pedley 2012).

Comparing the glacio-eustatic curve and the seismic cycle deduce above to the geological-geomorphological map, we may then be able to differentiate the glacio-eustatic or earthquake period origins of the scarps/bluffs situated above the Pisticci-Metaponto staircase. This give us the possibility to much better understand the passed and present evolution of Basilicate and related area.

If our reasoning is confirmed by further works and complementary datation (work in progress with Sauer et al.), one should assume that next rapid uplift very probably associated with major compressive and/or transtensive earthquakes will take place in 140 ka in Basilicata.

To our knowledge, it is the first time that scientists are able to determine interrelation in between both seismic and eustatic cycles and their related geomorphic features from a mapping point of view with this amount of confidence in

Table. 23.3 Characteristics of the seismic cycle: Marine terrace dotation versus altitude and seismic cycle, b, c, d = coordinate at the origin on the Y axis for the different straight lines with b = N.39; c = N.(−239); d = 0, with N = number of seismic cycle number. The shortening is deduced from the geometry of the thin skin decollement thrust (from the available offshore AGIP seismic section—see below)

Seismic cycle	Ages = X (ky)	Altitudes = Y (m)	Shortening (m)	Uplift rate (mm/y)	Shortening rate (mm/y)	Formulae
Oblique ramp = interseismic or creep	246	71	1354,76	0.28	5,507	Y = 0.2889X + b
Steep ramp = Cosismic Earthquakes return period	15	42	801,408	2.8	53,427	Y = 2.8193X + c
Average Seismic cycle	261	113	2156,168	0.43	8,261	Y = 0.4127X + d

such an active tectonic area. We hope this study will participate to better know the behaviour of the Apennine orogenic belt in Italy that will help to decrease the aggressivity against the civil protection commission after the recent Aquila earthquake episode.

Acknowledgments Authors express their warmest thanks to the Deffontaines's family who financed the whole study of the group since 2008, including common field works, as well as the two anonymous reviewers who improve this manuscript.

References

Ambrosetti P, Carraro F, Deiana G, Dramis F (1982) Il sollevamento dell'Italia centrale tra il Pleistocene inferiore ed il Pleistocene medio. Contributi conclusivi alia realizzazione della Carta Neotettonica d'Italia, Parte II, Pubbl. P.F. Geodinamica, vol 513. CNR, Rome, pp 219–223

Bentivenga M, Coltorti M, Prosser G, Tavarnelli E (2004) A new interpretation of terraces in the Taranto Gulf: the role of extensionnal faulting. Geomorphology 60:383–402

Bruckner H (1980) Marine Terrassen in Süditalien. Eine quartärmorphologische studie über das kiistentifland von Metapont. Düsseldorfer Geographische Schriften, 14, 235p

Bruckner H (1982) NF Suppl-Bd 43. Ausmab von erosion und akkumulation im verlauf des quartars in der Basilicata (Süditalien). Zeistchrift für Geomorphologie. Berlin, Stuttgart, pp 121–137

Caputo R (2007) Sea-level curves: perplexities of an end-user in morphotectonic applications. Glob Planet Change 57:417–423

Caputo R, Bianca M (2009) Comment on «late cenozoic uplift of southern Italy deduced from fluvial and marine sediments: coupling between surface processes and lower-crustal flow » by Westaway R. and Bridgland D. (Quaternary international 175:86–124), 204:98–102

Casero P, Roure F, Edignoux L, Moretti I, Muller C, Sage L, Vially R (1988) Neogene geodynamic evolution of the Southern Apennines. Boll Della Soc Geologia Ital 41:109–120

Cotecchia V, Magri G (1967) Gli spostamenti delle linee di costa quaternarie del mare ionico fra Capo Spulico e Taranto. Geologia Applicata e Idrogeologia 2:3–28

D'Argenio B, Alvarez W (1980) Stratigraphic evidence for crustal thickness changes on the southern Tethyan margin during the Alpine cycle. Geol Soc Am Bull 91:681–689

Deffontaines B, Chorowicz J (1991) Principle of drainage basin analysis from multisource data application to the structural analysis of the Zaire Basin. Tectonophysics 194:237–263

Deffontaines B, Lee JC, Angelier J, Carvalho J, Rudant JP (1994) New geomorphic data on Taiwan active orogen: a multisource approach, J Geophys Res, 99, B8: 20,243–20,266

Deffontaines B (2000) Formes et déformations de la surface terrestre: approches morphométriques et applications, Habilitation à Diriger des Recherches, Université Pierre et Marie Curie, P6,60p. + Annexes

Ferranti L, Santoro E, Mazzella ME, Monaco C, Morelli D (2009) Active transpression in the northern Calabria Apennines Southern Italy. Tectonophysics 476:226–251

Gignout M (1913) Les formations Marines pliocènes et Quaternaires de l'Italie du sud et de la Sicile, Annales de l'université de Lyon, NSI, 36, 693p. (in French)

Gigout M (1960) Sur le quaternaire marin de Tarente (Italie), Comptes rendus de l'Académie des Sciences, 250, 1094–1096, Lyon, (in French)

Hearty PJ, Dai Pra G (1992) The age and stratigraphy of Middle Pleistocene and younger deposits along the Gulf of Taranto (Southeast Italy). J Coast Res 8(4):882–905

Hippolyte JC, Angelier J, Roure F, Casero P (1994) Piggyback basin development and thrust belt evolution: structural and paleostress analyses of Plio-Quaternary basin in the southern Apennines. J Struct Geol 12:159–173

Howard AD (1967) Drainage analysis in geologic interpretation: a summation. Bull Am Ass Petr Geol Tulsa 51(11):2246–3428

Jarsvre EM, Pedley A (2012) Petroleum potential in the Gulf of Tarento, AAPG Annual Convention and Exhibition, Long Beach, California, 22–25 Apr 2012

Kaveh HF, Deffontaines B (2008) La limite slikke/schorre dans la baie du Mont Saint Michel (Golfe Normand Breton NO France), Photo-Interpretation, 3–4, pp 28–44 and 81–86

Lentini F, Carbone S, Di Stefano A, Guarnieri P (2002) Stratigraphical and structural constraints in the Lucanian Apennines (southern Italy): tools for reconstructing the geological evolution. J Geodyn, pp 141–158

Monaco C, Tortorici L, Paltrinieri W (1998) Structural evolution of the Lucanian Apennines. J Struct Geol 20:617–638

Mostardini F, Pieri M, Pirini C (1966) Stratigrafia del Foglio 212, Montalbano Ionico. Bolletino del Serv Geol d'Italia 87:57–143

Patacca E, Scandone P (2001) Late thrust propagation and sedimentary response in the thrust-belt—foredeep system of the southern Apennines (Pliocene-Pleistocene). In: Vai GB, Martini IP (eds) Anatomy of an Annali di Geofisica, Istituto Nazionale di Geofisica, Balogna, Orogen: The Apennines and Adjacent Mediterranean Basins. Annali di Geofisica, Istituto Nazionale di Geofisica, Bologna, pp 401–440

Pescatore, T, Renda P, Tramutoli M (1988) Rapporti tra le unita` lagonegresi e le unita` sicilidi nella media Valle del Basento, Lucania (Appennino Meridionale). Memorie Delia Soc Geologica Ital 41:353–361

Pieri P, Sabato L, Tropeano M (1996) Significato geodinamico dei caratteri deposizionali e strutturali della Fossa bradanica del Pleistocene. Memorie Della Soc Geologica Ital 51:501–515

Roure F, Casero P, Vially R (1991) Growth processes and melange formation in the southern Apennines accretionary wedge. Earth Planetary Sci Lett 102:395–412

Sauer D, Wagner S, Bruckner H, Scarciglia F, Mastronuzzi G, Stahr K (2009) Soil development on marine terraces near Metaponto (Gulf of Taranto, Southern Italy). Quaternary International, Gulf of Taranto, pp 1–16

Scandone P (1972) Studi di geologia lucana: carta dei terreni della serie calcareo-silico-marnosa e note illustrative. Bollettino della Societe´ dei Naturalisti in Napoli 81:225–300

Sella M, Turci C, Riva A (1988) Sintesi geopetrolifera della Fossa Bradanica (avanfossa della catena appenninica meridionale). Mem della Soc Geologica Ital 41:87–107

Vezzani L (1967) Depositi Plio-Pleistocenici del litorale Ionico della Lucania. Atti della Accad. Gioenia di Sci Naturale in Catania 6 (18):159–180

Waelbroeck C, Labeyrie L, Michel E, Duplessy JC, McManus JF, Lambeck K, Balbon E, Labracherie M (2002) Sea level and deep water temperature changes derived from benthic foraminifera isotopic records. Quat Sci Rev 21:295–305

Westaway R (1993) Quaternary uplift of southern Italy. J Geophys Res 98:21741–21772

Westaway R, Bridgland David (2007) improved uplift modeling of the gulf of Taranto marine terraces. Quat Int 210:102–109

Westaway R, Bridgland D (2009) Reply to comment by Caputo R, Bianca M., on «Late cenozoic uplift of sothern Italy deduced from fluvial and marine sediments: coupling between surface processes and lower-crustal flow» by Rob Westaway and David Bridgland

2007 improved uplift modeling of the gulf of Taranto marine terraces. Quaternary international 210: 102–109

Zander AM, Fulling A, Bruckner H, Mastronuzzi G (2003) Luminesznezdatierungen an litoralen sedimenten der terrassentreppe von Metapont, Süditalien. Essener Geogr Arbeiten 35:77–94 (in german)

Zander AM, Fulling A, Brückner H, Mastronuzzi G (2006) OSL dating of upper Pleistocene littoral sediments: a contribution to the chronostratigraphy of raised marine terraces bordering the gulf of Tarento South Italy. Geographfia Fisica e Dinamica Quaternaria 29:33–50

New Structural and Geodynamic Coastal Jeffara Model (Southern Tunisia) and Engineering Implications

Rim Ghedhoui, Benoît Deffontaines, and Mohamed Chedly Rabia

Abstract

Thanks to its geographical position in the western Mediterranean domain, Tunisia faces, since mid-Cretaceous (Aptian/Albian time period), to the inversion of the Tethys due to the northward African plate motion toward Eurasia. The coastal Jeffara is a part of the southern zone of deformation witness of the eastward migration of Tunisia to the Mediterranean Sea. We focus herein following Perthuisot (Cartes géologiques au 1/50.000 et notices explicatives des feuilles de Houmet Essouk, Midoun, Jorf, Sidi Chamakh, 1985) and others on the neotectonic of the coastal Jeffara (southern Tunisia) and its engineering implications. Based on the results of previous studies and new evidences developed herein, we propose a new structural and geodynamic coastal Jeffara model, influenced by the continuous post lower cretaceous northward migration of northern African toward the Eurasian plates. We herein study the Digital Elevation Model (issued from SRTM), which was checked with field surveys and 2D numerous seismic profiles at depth both onshore and offshore. All data were, then, integrated within a GIS Geodatabase, which showed the coastal Jeffara, as a part of a simple N–S pull-apart model within a NW-SE right lateral transtensive major fault zone (Medenine Fault zone). Our structural, geological and geomorphological analyses prove the presence of NNW-SSE right lateral en-echelon tension gashes, off shore NW-SE aligned salt diapirs, numerous folds offsets, en-echelon folds, and so-on… that are associated with this major right lateral NW-SE transtensive major coastal Jeffara fault zone that affect the Holocene and the Villafrachian deposits. These evidences confirm the fact that the active NW-SE Jeffara faults correspond to the tectonic accident, located in the south of the Tunisian extrusion, which is active, since mid-cretaceous, as the southern branch of the eastward Sahel block Tunisian extrusion toward the free Mediterranean sea boundary. Therefore this geodynamical

R. Ghedhoui (✉) · B. Deffontaines (✉)
Laboratoire de Géomatique Appliquée, Université Paris-Est
Marne-La-Vallée (ENSG-IGN and UPEM), 5 Bd Descartes,
77450, Marne-la-Vallée Cedex 2, France
e-mail: rimaigs@yahoo.fr

B. Deffontaines
e-mail: benoit.deffontaines@univ-mlv.fr

R. Ghedhoui · B. Deffontaines
Laboratoire International Associé ADEPT 536 Taiwan–France,
CNRS-NSC, Université Paris-Est Marne-La-Vallée, 5 Bd
Descartes, 77450, Marne-la-Vallée Cedex 2, France

R. Ghedhoui · M.C. Rabia
U.R. "Géomatique et Géosystèmes" (02/UR/10-01), Université de
la Mannouba, BP 95, 2010, Mannouba, Tunisia
e-mail: rabiamch@gmail.com

movement explains the presence, offshore, of small elongated NW-SE, N-S and NE-SW transtensive basins and grabens, which are interesting for petroleum exploration.

Keywords

Right lateral transtensive fault • Synthetic model • Seismic profiles • Geographic information system (GIS) • Coastal jeffara (S. Tunisia)

24.1 Introduction

Despite the presence of various studies (Perthuisot 1977, 1985; Bouaziz 1986, 1995; Bouaziz et al. 2002; Rabia 1998; Touati and Rodgers 1998; Jedoui 2000; Deffontaines et al. 2008; Gabtni et al. 2009, 2011, 2012; Bodin et al. 2010), the structure of the Jeffara is still subject of conflict. Besides, the authors do not all agree about the history of the deformation of the region, as the neotectonic context seems complex and multi-phase since the Lower Paleozoic (Burrolet and Desforges 1982; Ben Ayed 1986; Bouaziz 1995; Bouaziz et al. 1999, 2002; Gabtni et al. 2012).

The aim of this study is, therefore, to better constrain the structural geometry and the neotectonic history of the coastal Jeffara using geomatics, remote sensing data, structural geology, detailed field investigation and numerous transverse onshore and off shore petroleum seismic profiles in order to propose a new structural model confirming the existence of a NW-SE major dextral fault zone allowing the eastern extrusion of the Sahel block, due to the migration of the African plate towards Eurasia (Deffontaines et al. 2008).

24.1.1 Geographical Location

The Tunisian Jeffara basin has a sub-tabular, flat and low topography (Ben Ayed and Kessibi 1981), bounded to the north and to the east by the Gabes Gulf and the Mediterranean sea; to the west by the morphological boundary of the Dahar domain, the layers of which disappear under the Oriental Grand Erg (Ben Ayed and Kessibi 1981; Bodin et al. 2010), and in the south east by the "administrative Tunisian–Lybian border". The Jeffara is subdivided into two distinct and structural units: (1) to the SW, the continental Jeffara, which constitutes the transition zone of Mesozoic and Cenozoic outcrops of Dahar, and (2) the coastal Jeffara that parallel the shoreline consisting of the island of Jerba and the two peninsulas of Jorf and Zarzis (Perthuisot 1977) (Fig. 24.1a, b).

24.1.2 Structural Overview

From the bibliography, it seems that the deformation of the coastal Jeffara is complex. Perthuisot (1977, 1985), for example, showed the presence of two quaternary, tectonic phases which are responsible for the present morphology of Jerba island and the Jorf and Zarzis peninsulas: (1) an extensional Eo-thyrrhenian phase (slightly post-villafranchian) and (2) a compressive Tyrrhenian phase. The latter (1) reactivated the faults N045°E and N165°E in sliding movements at the level of the base and (2) led to the formation, on the cover, of layers of a large radius of curvature. Both tectonic phases affect the Villafranchian "crust" deposits which lead him to date these two tectonic episodes.

In contrast, Bouaziz (1995), suggests in his excellent "These d'état" a structure organized in "half-graben", with the presence of faults in a scale of a principal direction of N120°E to N160°E. At least, we should take also, into careful consideration the work of Touati and Rodgers (1998) who advocates the presence of: (1) a Mio-Pliocene extension of NE-SW direction, having reactivated the faults that are inherited from the E–W dextral sliding of Santonian age, of the NE-SW extension of the upper Jurassic—lower Cretaceous (syn-rift phase), and (2) a post-Villafranchian compression directed NNW-SSE responsible for the shaping of the present topography.

Therefore previous authors proposed contrasting structural history for the Jeffara deformations, which seems complex and poly-phase, since the cretaceous (Burrolet and Desforges 1982; Ben Ayed 1986; Bouaziz 1995; Bouaziz et al. (1999, 2002); Rabia 1998; Touati and Rodgers 1998; Gabtni et al. 2009, 2011, 2012).

24.2 SRTM Data Analysis

From the Digital Terrain Model SRTM (Shuttle Radar Topographic Mission 2000) altimeter data, characterized by a matrix spacing of 3″ (0.0008333°), hill-shading maps of azimuths N000°E, N045°E, N090°E, and N135°E, having fixed elevation of the light source (30°) and the vertical exaggeration (20x) were calculated (Fig. 24.2b, c).

The structural field analyses and the Digital Terrain Model and aerial photographs photo-interpretation, done using the methodology developed in Deffontaines (1990, 2000), Deffontaines and Chorowicz (1991), of the different hill-shading maps and of their associated roses directions confirm the presence, in the study area, of alignments of

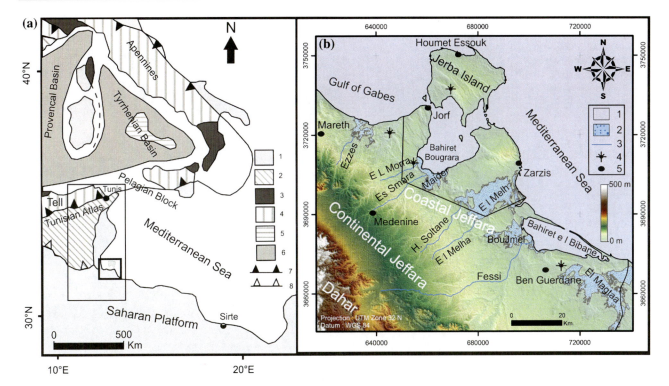

Fig. 24.1 **a** Geodynamic context of the western mediterranean domain (modified from Doglioni et al. 1999), Heavy black quadrangle in (**a**) correspond to the area (**b**) *1* continental platforms, *2* intracontinental fold belts, *3* crystalline massifs, *4* alpine chains, *5* neogene oceanic crust, *6* neogene thinned continental crust, *7* thrust fronts, *8* South Atlassic front; (**b**) geography of the studied area (*1* studied area, *2* sebkhas, *3* Oued/Talweg, *4* well, *5* town)

major faults NW-SE and WNW-ESE trending (Fig. 24.2a, d), as well asymmetric folds at the level of Jorf, Zarzis, and Jerba confirmed by the detail geological maps e.g. Perthuisot (1985) (see below).

24.3   Reinterpretation of the 1/50,000 Geological Maps and Contributions of Detailed Structural Field Work

The re-interpretation of the 1/50,000 geological maps in scale realized by Perthuisot (1985): Houmet Essouk (n°148), Midoun (n°149), Jorf (n°159), Sidi Chamakh (n°160) and Zarzis (n°160), associated with the hill shaded topography and fields observations show the presence of several examples of pedagogic structural cases which advocate the presence of an active NW-SE transtensive dextral strike-slip fault.

Though diachronic, the carbonated "Villafranchian crust" is an excellent indicator and marker of the current deformation highlighted by its vertical movements. Thus, the

Villafranchian and Mio-Pliocene deposits are here re-interpreted in tectonic structures with the help of the detailed Perthuisot (1985) geological mapping: the depressions within the Villafranchian crust are interpreted as tension joints and normal faults associated with a major dextral movement (Fig. 24.3). The first example shows that the Villafranchian deposits appear on both sides of the Mio-Pliocene clays and sands depression, with a conservation of the altitudes of the Villafranchian on both sides. The absence of the vertical component and the rectilinearity of the small valley suggest that is marked by a structural tension gashes directed N130°E (NNW–SSE), perpendicular to the axis of the principal minimum constraint σ3 (NNE–SSW) (Fig. 24.3d). The second example concerns the cartographic asymmetry of the Villafranchian crust. Field observations and the study of the DEM-SRTM, confirm the presence of a variation in altitude on both sides of the elongated narrow valley and therefore, of a vertical component of the deformation, characterized by a higher altitude in its western side. The cartographic asymmetry and the difference of altitude are the proof of a subsidence of the NE compartment, which

Fig. 24.2 (**a**) Coastal Jeffara structural scheme compiled from different D.E.M. hill-shading maps (*1* city, *2* fault, *3* fold, *4* Azimuth N315°E); (**b**) SRTM DEM hill-shading of the studied area map (*1* Azimuth N225°E, Elevation: 30°, Vertical exaggeration: 20x); (**c**) rose diagram of the major photo-interpreted alignments

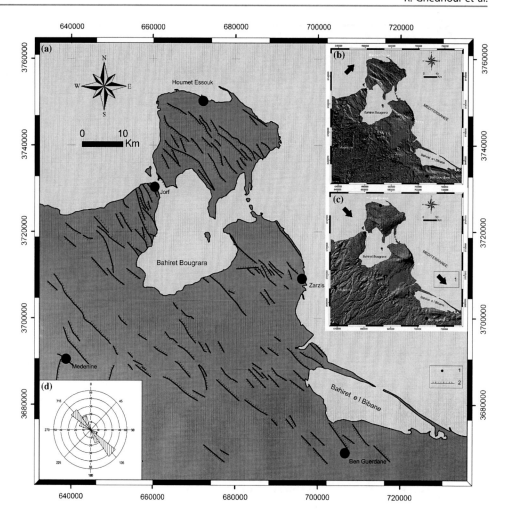

reveal therefore the presence of a normal fault (Fig. 24.3b). The sigmoid tectonic structures (in S), reveal a dextral strike slip going N120°E and N130°E, and are interpreted as "S" tension faults, associated with a dextral strike slip movement (Fig. 24.3b). Different cases correspond to structures in horst and graben, found between two normal faults with an opposite asymmetry on both sides (Fig. 24.3c).

continuity of seismic reflectors dipping towards the NE. It also shows the presence of a transparent and consequently a highly fractured zone in depth, characterized by NW-SE vertical faults. This geometry allows us to interpret them as strike-slip faults, supposed dextral, with a normal component toward the NE (Gabes gulf) and that are concentrated into several fault splays or slip corridors (Fig. 24.4).

24.4   2D Seismic Reflection Profiles

On shore 2D seismic reflection line (acquired by ETAP), NE-SW trending and perpendicular to the structures of the Jeffara is re-interpreted below: the profile P40 (NE on the right) is showing high fractured zones (reflector's dipping to the NE). It reveals an upper zone characterized by a big

24.5   Discussions and Conclusions

The analysis of the Digital Terrain Model (SRTM data) confirms the presence of: (1) alignments of major direction NW–SE, often mentioned in the bibliography (Perthuisot 1985; Ben Fergeni et al. 1990; Bouaziz 1995; Rabia 1998; Touati and Rodgers 1998; Gabtni et al. 2012), (2) directions

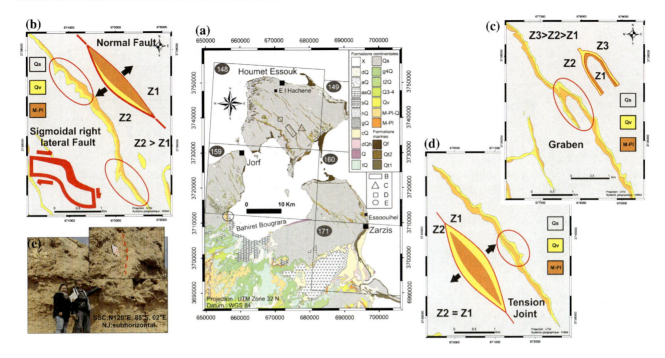

Fig. 24.3 (**a**) Mosaic of the detailed geological maps at 1/50,000 scale of Jerba island, Jorf and Zarzis (Southern Tunisia) (*148* Houmet Essouk, *149* Midoun, 159, Jorf: 160, Sidi Chamakh et 160: Zarzis); (**b**) Normal fault: (1) cartographical and lithological asymmetry, (2) difference in altitude; (**c**) graben structure bounded by two normal faults; (**d**) Tension joint: (1) cartographical and lithological symmetry, (2) altitude conservation on both sides of the tension joint; (**e**) Field photo showing a dextral strike slip fault on Pleistocene deposits (Boughara port, Bougrara, March 2102) affecting the Villafranchian

carbonated crust; Lithology: continental formations (*dQ* actual dunes, *aQ* actual alluviums, *sQ* filling of continental sebkhas, *tQ* historic terrace, *Qs* red silts, *Qv* villafranchian: sauman coloured crust, *M-Pl* Mio-pliocene: claystones and sandstones with conglomeratic thin layers); marine and costal formations (*Qf* Flandrian beach sandstones, *Qt1* Neo-tyrrhénian sand dune or marine deposits, *Qt2* Eu-tyrrhenian sand dune or marine deposits) (modified from the geological maps at 1/50,000 scale)

NNW-SSE to WNW-ESE not as clear than the previous one, as well as folds in dextral offset on Zarzis Peninsula. The reinterpretation of geological maps at 1/50,000 reveals that the tectonic structures present in the island of Jerba and the Jorf and Zarzis peninsulas, may be interpreted in terms of (1) normal faults directed N120°E; (2) an-echelon dextral tension gashes, aligned along the N120°E azimuth, (3) tension gashes N130°E trending, (4) "Sigmoidal" tension gashes and (5) horsts and grabens located along large faulted corridor, surrounded by major, N120°E to N130°E transtensive dextral strike-slip faults (Fig. 24.5b).

The interpretation of numerous parallel 2D seismic lines reveals the presence of a dense number of vertical strike-slip faults with normal component, interpreted as "strike-slip corridors". Jerba, Jorf and Zarzis are, thus, located above a

major dextral transtensive strike slip accident which is oriented N120°E to N130°E and which appears totally affected by highly frequent, vertical accidents. Those completely cut the costal Jeffara by corridors of many vertical faults oriented N120°E–N130°E with a small normal component.

In addition to the major NW-SE Gafsa-Tozeur fault, which stretches to the Jeffara and the NE-SW faults (Rabiaa 1998), evidence of dextral strike-slip movement was highlighted. These prove that the Jeffara fault corresponds to the major accident, bounding to the south, the Tunisian extrusion (Fig. 24.5a; see Deffontaines et al. (2008) this issue), helping the migration of the Sahel block towards the vacant border of the Mediterranean Sea. This also explains the presence, offshore, of NW-SE and NE-SW small oil basins, lying according to the opening of such fractures.

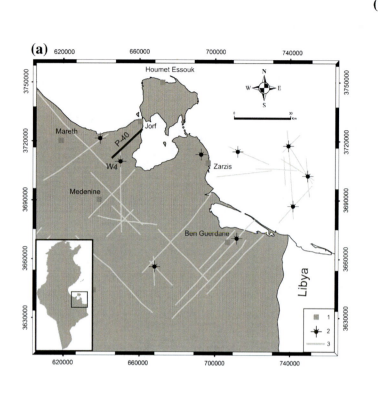

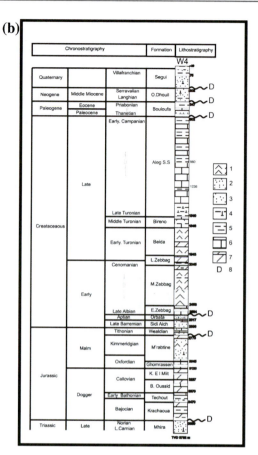

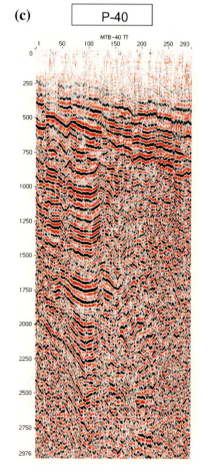

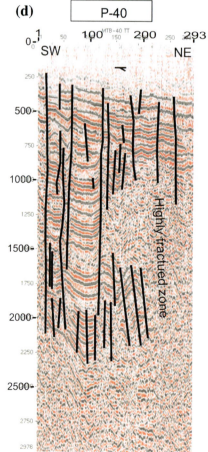

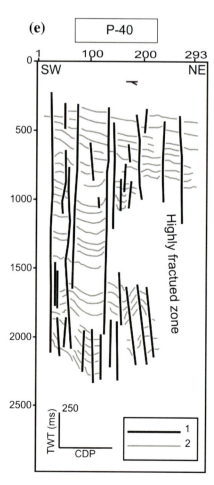

◄ **Fig. 24.4** (**a**) Block selected seismic lines (*1* town, *2* well, *3* seismic line; (**b**) lithostratigraphical units within W4 well (W4 geological report (ETAP), *1* gypsum and/or anhydrite, *2* sands, *3* sandstone, *4* marls, *5* claystone, *6* limestone, *7* dolomite, *D* unconformity); (**c**) seismic reflection profile (P40, ETAP); (**d**) transparency faires; (**e**) Interpretation of the line drawing (*1* fault, *2* reflector)

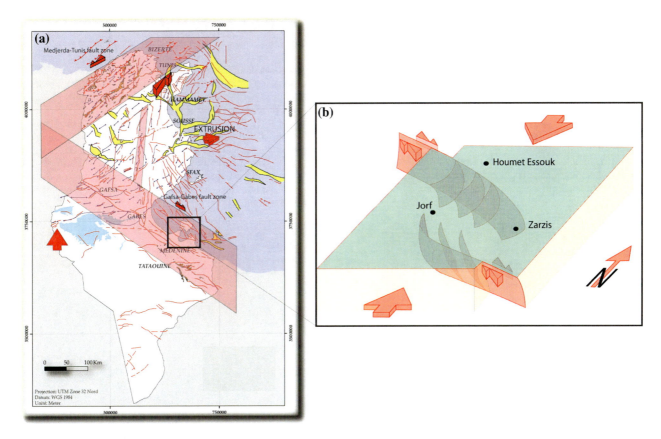

Fig. 24.5 (**a**) Geodynamic extrusion of Tunisia (Deffontaines et al. 2008, Potsdam); (**b**) 3 D synthetic block showing the geodynamic context of coastal Jeffara

Acknowledgments Authors would like to thank Ghedhoui family and Zetrini family, for their hospitality.

References

Ben Ayed N (1986) Evolution tectonique de l'avant pays de la chaine alpine de Tunisie du début du Mésozoique à l'actuel. Thèse Doctorat d'Etat, Université Paris-Sud, Centre d'Orsay, p 328

Ben Ayed N, Kessibi M (1981) Mise en evidence de deux couloirs de décrochement E-W dextre en bordure de la plateforme saharienne (Tunisie méridionale), Actes du 1er Congrés National des Sciences de la Terre, Tunis, Sept 1981, pp 291–301

Ben Ferjani A, Burollet PF, Mejri F (1990) Petroleum geology of Tunisia, Mem ETAP, vol 1. Tunisia, Tunis, p 194

Bodin S, Petitpierre L, Wood J, Elkanouni I, Redfern J (2010) Timing of early to mid-cretaceous tectonic phases along North Africa: new insights from the Jeffara escarpment (Libya—Tunisia). J Afr Earth Sci 489–506

Bouaziz S (1986) La déformation dans la plateforme du Sud tunisien (Dahar et Jeffara): Approche multiscalaire et pluridisciplinaire. Thèse 3ième cycle, Université de Tunis, 180 p

Bouaziz S (1995) Etude de la tectonique cassante dans la plateforme et l'Atlas sahariens (Tunisie méridionale): Evolution des paléochamps de contraintes et implications géodynamiques. Thèse de Doctorat, Université de Tunis II, p 486

Bouaziz S, Barrier E, Turki MM, Tricart P (1999) La tectonique permo-mésozoique (anté-vraconienne) dans la marge sud téthysienne en Tunisie méridionale. Bull Soc Geol France 45–56

Bouaziz S, Barrier E, Soussi M, Turki MM, Zouari H (2002) Tectonic evolution of the northern african margin in Tunisia from paleostress data and sedimentary record. Tectonophysics 227–253

Burrolet PF, Desforges G (1982) Dynamique des bassins néocrétacés en Tunisie. Mémoire Géologique, vol 7. Université de Dijon, pp 381–389

Deffontaines B (1990) Digital Terrain Model and Morpho-neotectonics, application in the Strasbourg area, Rhinegraben, France. Bull INQUA NC 13:58–59

Deffontaines B (2000) Formes et déformations de la surface terrestre: Approches morphométriques et applications, Habilitation à Diriger des Recherches (HDR). Université Pierre et Marie Curie, P6, p 60 +Annexes

Deffontaines B, Chorowicz J (1991) Principle of drainage basin analysis from multisource data, application to the structural analysis of the Zaire Basin. Tectonophysics 237–263

Deffontaines B, Slama T, Rebai N, Turki MM (2008) Tunisia structural extrusion revealed by numerical geomorphometry. Geol Soc Am spec workshop, Potsdam

Doglioni C, Fernandez M, Gueguen E, Sabat F (1999) On the interference between the early Apennines-Maghrebides back arc extension and the Alps-betics orogen in the Neogene geodynamics of the Western Mediterranean. Bol Soc Geol Ital 118:75–89

Gabtni H, Jallouli C, Mickus KL, Zouari H, Turki MM (2009) Deep structure and crustal configuration of the Jeffara basin (Southern Tunisia) based on regional gravity, seismic reflection and borehole data: how to explain a gravity maximum within a large sedimentary basin ? J Geodyn 142–152

Gabtni H, Jallouli C, Mickus KL, Turki MM (2011) Geodynamics of the Southern Tethysian Margin in Tunisia and Maghrebian domain: new constraints from integrated geophysical study. Arab J Geosci

Gabtni H, Alyahyaoui S, Jallouli C, Hasni W, Mickus KL (2012) Gravity and seismic reflection imaging of a deep aquifer in an arid region : case history from the Jeffara basin, southeastern Tunisia. J Afr Earth Sci 85–97

Jedoui Y (2000) Sédimentologie et géochronologie des dépots littoraux quaternaires: reconstitution des variations des paléoclimats et du niveau marin dans le Sud-Est tunisien. Thèse de Doctorat, Université Tunis II, p 338

Perthuisot JP (1977) Le Lambeau de Tlet et la structure néotectonique de l'ile de Jerba (Tunisie). Comptes Rendus des Seances de l'Academie des Sciences, Paris

Perthuisot JP (1985) Cartes géologiques au 1/50,000 et notices explicatives des feuilles de Houmet Essouk, Midoun, Jorf, Sidi Chamakh. Publications, Serv Geol Nat, Tunisie

Rabia MC (1998) Système d'Informations Géo-scientifique et Télédétection multi-capteurs: application à une étude mult-thèmes de la Jeffara orientale. Thèse de Docotrat, Université Tunis II, p 320

Touati MA, Rodgers MR (1998) Tectono-stratigraphic history of the Southern Gulf of Gabes and the hydrocarbon habitats. In: Proceedings of the 6th Tunisian petroleum exploration production conference, Tunis, Mai 1998, pp 343–370

Benoît Deffontaines, Mehdi Ben Hassen, and Rim Ghedhoui

Abstract

Neotectonics is revealed by numerous approaches such as a new re-interpretation and homogeneisation of the geological data set, earthquakes, field work studies and detailed numerical geomorphic analyses of the topography. Tunisia appear to be a case example for this kind of studies as it is affected by both active tectonics highlighted by few minor earthquakes and present numerous associated faults and folds. We develop first herein basic key geomorphic processing deduced from the bibliography and new indicators developed throughout this study, processed numerically through home made geodatabase, that lead us to propose a new structural scheme of extrusion revisiting the Tunisian neotectonic setting (Deffontaines et al. 2008, Postdam GFZ). We then update the Tunisian geological mapping by improving it with the high resolution existing Digital Terrain Model and remote sensing images. Fieldwork studies lead us to better solve mapping anomalies. At least we used numerous published existing seismic profiles that validate our model. We then propose herein an eastern extrusion model of Central Tunisia due to the northward migration of African plate toward Eurasia since mid cretaceous due to the Tethys inversion.

Keywords

Extrusion • Strike-slip fault • Transpression • Transtension • Tunisia

25.1 Introduction: State of the Art

Lateral extrusion along a collision zone has been proposed first in Tibet and Himalaya (Molnar and Tapponnier 1978; Tapponnier et al. 1983) and later extended to other mountain belts such as the Alps (Ratschbacher and Merle 1991) or in Taiwan (Angelier et al. 2009). Tectonic extrusion describes the lateral motion of structural units that move toward a weaker domain with respect to the mountain belt, in response to collision-induced shortening. The weak domain is often represented by a «free boundary» such as in analogue or numerical models. Such a mechanical exaggeration aims at producing clearer results. In the fields, the velocity and amplitude of extrusion are functions of (1) the shortening across the mountain belt, (2) the contrast in mechanical strength between the structural units within the belt and the adjacent weaker domains and (3) the existence of a

B. Deffontaines (✉) · M.B. Hassen · R. Ghedhoui
Laboratoire de Géomatique Appliquée, Université Paris-Est Marne-La-Vallée (UPEM), 5, Boulevard Descartes, Marne-La-Vallee, 77454, France
e-mail: benoit.deffontaines@univ-mlv.fr

M.B. Hassen
e-mail: mbenhass@ucalgary.ca

R. Ghedhoui
e-mail: rimaigs@yahoo.fr

B. Deffontaines · M.B. Hassen · R. Ghedhoui
Laboratoire International Associé ADEPT 536 CNRS-NSC, Aix-En-Provence, France

B. Deffontaines · M.B. Hassen · R. Ghedhoui
Laboratoire International Associé ADEPT 536 CNRS-NSC, Taipei, Taiwan

detachment level at depth that lead to the sliding of the overlying formations.

We aim at showing that despite the small size of the Tunisian Atlas mountain belt compared to Himalaya or the Alps, significant extrusion occurs at the eastern tips of the Tunisian collision zone since mid-cretaceous.

We focus herein on Tunisia as few recent structural studies prevail Fig. 25.1a, b) despite its tectonic activity (Boukadi 1994; Bouaziz et al. 2002; Hfaiedh et al. 1991). Furthermore Tunisia present a high quality of outcrops, as well as the huge quantity of remote sensing data and digital topographic data quite interesting to have a precise photo-interpretation (Deffontaines 1990, 2000 Deffontaines and Chorowicz 1991). Finally a quite good geological mapping (Perthuisot 1977, 1985 and so on.), microtectonic analzes (Bouaziz 1986, 1995, Bouaziz et al. 1999) as well as numerous available sismic reflection profiles and wells (e.g., and so on) that may be re-interprete and homogenize from North to South (Deffontaines et al. 2008; Ben Hassen et al. 2012).

We will first have a closer look to the extrusion southern branch so called Gafsa-Medenine (which are much more described in Ben Hassen et al. (2012 and this issue) of the transpressive context close to the Gafsa city and by Ghedhoui et al. (this issue) for the transtensive context. The Gafsa-Medenine fault zone is then inferred as a major right lateral strike-slip fault. At least we will propose following the work of Deffontaines et al. (2008) a global geodynamic model for the Tunisian extrusion that govern the present deformation since mid cretaceous time.

25.2 The Gafsa: Medenine Fault Zone—A NW–SE Transpression to Transtension Right Lateral Strike-Slip Major Fault Zone

25.2.1 The Gafsa-Medenine NW–SE Right Lateral Strike-Slip Fault Zone

Numerous studies had been done on that specific area of the Tunisian Atlas (Ben Ayed 1986; Zouari et al. 1990; Bedir et al. 1992; Bedir 1995; Zouari 1995; Touati et Rodgers 1998; Ghabtni et al. 2009, 2011; Bodin et al. 2010). A detail Tunisian geomorphometric approach detailed in Ben Hassen et al. (2012 and this issue) show (1) the different topography

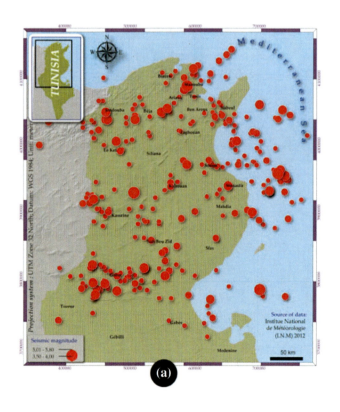

Fig. 25.1 **a** Earthquakes in Tunisia (M > 3.5) since 1970 (From INM data, in M.Ben Hassen, these UPEM, 2012); **b** Sismo-tectonic map of Tunisia (Sismic Magnitude and focal mechanisms parameters from the «Centre Seismologique International»; Topographic data from SRTM; bathymetric data from GTOPO30; faults from the sismo-tectonic map of Tunisia published by INM in 1991) from M. Ben Hassen these UPEM, 2012

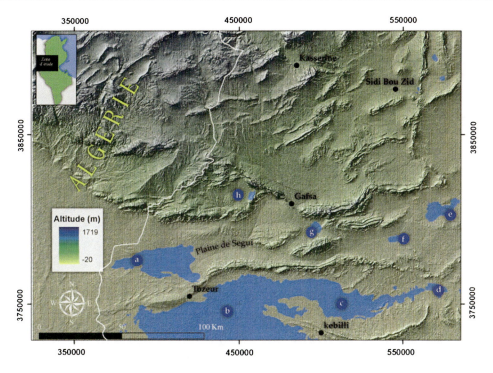

Fig. 25.2 Plains and depressions of the Southern Tunisian Atlas Tunisie: **a** Chott El Gharsa; **b** Chott El Jerid; **c** Chott El Fejaj; **d** Sebkhet El Hamma; **e** Sebkhet En Nouer; **f** Sebkhet Sidi Mansourr; **g** Sebkhet Guettar; **h** Garaat Ed-Douza

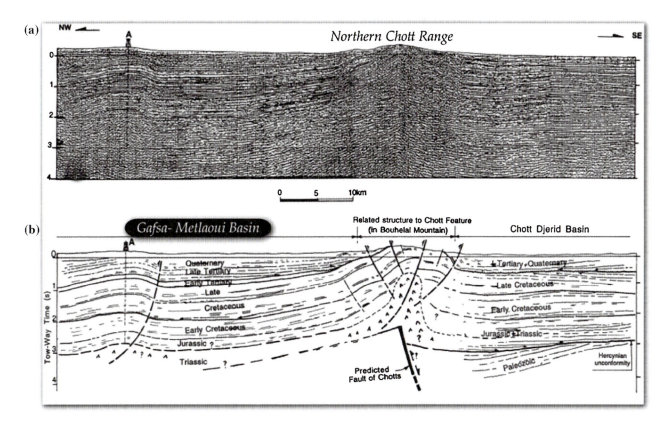

Fig. 25.3 Sismic reflexion profile highlighting the diapiric salty Jurassic to triassic formations (Mestaoua?) situated above a paleonormal Tethyan inverted normal fault on the Gafsa-Metlaoui basin (from Hlaiem 1999 in M. Ben Hassen PhD thesis UPEM, 2012)

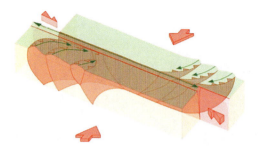

Fig. 25.4 4A: 3D structural model of the Gafsa Medenine fault zone: a transpressive NW-SE right lateral strike-slip fault zone close to Gafsa (West of the Gabes gulf—on the *left*) contrasting with the transtensive

component close to Medenine (South of the Gabes gulf—on the *right*) see M.Ben Hassen et al. and R.Ghedhoui et al. (this issue)

(Fig. 25.2), the structures and their right lateral offsets, a quite simple N–S published seismic profile Fig. 25.3 (from Hlaiem 1999), and the interpretation of the area in terms of re-activation of a major NW–SE right lateral strike-slip transpressive fault zone (see the strike-slip Gafsa fault model on Fig. 25.3a). Whereas further SE on the coastal Jeffara area, south of the Gabes gulf, and close to the Medenine city, this Gafsa-Medenine fault zone taking into account the bibliography (Bouaziz 1986; Rabia 1998) is newly described by Rim Ghedhoui et al. this issue. The Medenine fault zone act as a transtensive right lateral strike-slip fault where pull apart prevail that create the Boughrara depression as illustrated by the Fig. 25.4b. The Boughrara may be filled locally by salt diapir(s) (see description of the origin of such below and in Ghedhoui et al. this issue).

25.2.2 The N–S Axis

The well known N–S axis acts therefore as a graben situated in between a transpressive and a transtensive tectonic domains. Therefore the N–S axis appear to be a N–S neutral elongated area» in between two distinct tectonic and topographic domains. If the western one corresponds to a transpressive high plateau (see Fig. 25.5) that contrasts to the transtensive eastern part of the N–S axis that correspond to the Sahel block or plain. The very thick buried Mestaoua formation Jurassic in age (upper Lias to Aalenian) composed of salt and evaporites, then fills in the faults zone and the gaps and creates different kind of diapirs.

- (1) some diapirs west of the N–S axis are situated in a N–S compressive environment above the former E–W Tethysian normal fault that are then inversed due to the African/Europe convergence;
- in other place the diapirs act as (2) transcurrent along the major fault zones that correspond to the left lateral Diapir fault zone zone NE-SW trending (Tunis Medjerda) and the NW-SE trending Gafsa-Medenine right lateral fault zone;

- and at least (3) diapirs may fill in N–S and oblique grabens at sea/offshore in a normal transtensive environment. These therefore have great oil implications especially on the reactivation of the diapirs and on the very particularly geometry of the different offshore oil traps that are situated close to the diapir itself.

We then may summarize the deformation in Tunisia as follow.

25.3 Discussion and Conclusions

On this Tunisian extrusion model deduced from structural drainage analyses and anomalies, drainage network classifications, specific and optimized analyses of the Digital Terrain Model (DTM-SRTM), summit level surface analyses. integrated within a GIS (Deffontaines 1990; Deffontaines et al. 1994; Deffontaines 2000; Ben Hassen et al. 2012, Ghedhoui et al. this issue.), and that take into consideration previous works (Bouaziz et al. 2002), we propose that the well known diapir fault zone (Medjerda-Tunis fault zone) correspond to a major left lateral transpressive to transtensive (from west to east respectively) trending northeastward acting as the northern major boundary of the Central Tunisia extrusion. It appears in the fields as a compressive structure linked to the behaviour of the continuous uplifting of the elongated NE-SW salt diapirs that parallel this major transcurrent extrusion tectonic zone. It is associated with the NW-SE trending well known Gafsa-Gabes fault zone (Ben Hassen 2012; Ben Hassen et al. 2012 and this issue, and Ghedhoui et al. (2013) and this issue) which is characterized by numerous en-echelon folds acting as a major right lateral fault zone that bounds the southern part of the central Tunisia extrusion (Deffontaines et al. 2008). Within the extruded central Tunisia, the North–South axis also known as Al Abiod N–S fault zone appear to be a reactivated graben closely associated with the eastern extrusion of central Tunisia and differentiating high Atlasic and low eastern tunisian Domain (Bouaziz et al. 2002). The N–S axis acts as

Fig. 25.5 Geodynamic extrusion of Tunisia (modified from Deffontaines and et al. Potsdam, 2008): The Tunis-Medjerda fault zone act as a transpressive to transtensive *left* lateral fault zone wheras the Gafsa-Medenine fautl act as a Transpressive to transtensive fault zone. The neutral area is highlighted by the N–S axis which behaves as a major graben infilled by the elongated diapir made of the *thick* salty Mestaoua serie (400–800 m thick)

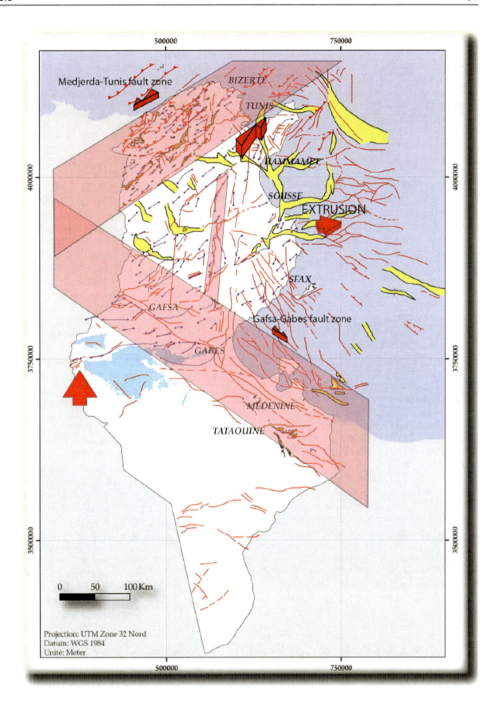

the neutral zone in bounding the transpressive domain on the West and the transtensive domain on the East. Therefore the N–S axis act as a normal fault/graben fills in by halocinetic processes which highlight outcropping compressive structures. We interpret this Tunisian extrusion active as a mature feature as it has been developing all along the Tethysian closure since mid-cretaceous (Bouaziz 1986). Our study therfore highlights that the different tectonic regimes that prevail since the inversion of the Tethys (mid-Creataceous Aptian-Albian?) is a mixed of both compression and strike slip tectonic regime affected by halocinesis local phenomenons. We therefore may reinterprete completely differently the passed history of Tunisia and this gives several major indications for engineering geology such as oil indications, sismic hazards and so on.

Acknowledgments Authors would like to thank Ghedhoui's family, Ben Hassen's family, and Deffontaines's family who financed the major part of this research.

References

Angelier J, Chang TY, Hu JC, Chang CP, Siame L, Lee JC, Deffontaines B, Chu HT, Lu CY (2009) Does extrusion occur at both tips of the Taiwan collision belt? Insights from active deformation studies in the Ilan Plain and Pingtung Plain regions. Tectonophysics 466:356–376

Bedir M (1995) Mecanisme des bassins associes aux couloirs de coulissements de la marge atlasique de la Tunisie: simo-stratigraphie, seismo-tectonique et implications petrolieres. These de doctorat en Sciences geologiques. These de doctorat Doctorat es-Sciences, Universite Tunis II, 412 p

Bedir M, Ben Youssef M, Boukadi N, Slimane F, Zargouni F (1992) Les evenements sequentiels meso-cenozoiques associes au systeme de coulissements de l'Atlas meridional de Gafsa. 3eme Journee de l'Exploration Petroliere (ETAP). Tunis

Ben Ayed N (1986) Evolution tectonique de l'avant pays de la chaine alpine de Tunisie du debut du Mesozoique a l'actuel. These Doctorat d'Etat, Universite Paris-Sud, Centre d'Orsay, p 328

Ben Hassen M (2012) Analyse de la deformation recente dans l'Atlas meridional de la Tunisie par geomorphometrie et Interferometrie Radar (DInSAR). These de doctorat en Sciences de l'Information Geographique, Universite: Paris-Est Marne-la-Vallee/ France; et en geologie, Universite: Tunis El Manar/Tunisie. 341p

Ben Hassen M, Deffontaines B, Turki MM (2012) Contribution de la Geomatique (SIG et teledetection) pour la mise a jour de la couverture geologique a l'echelle du 1/ 100 000eme du sud de la Tunisie, Photo-Interpretation European Journal of Applied Remote Sensing, N° 2012/3, 31 p

Bodin S, Petitpierre L, Wood J, Elkanouni I, Redfern J (2010) Timing of early to mid-cretaceous tectonic phases along North Africa: New insights from the Jeffara escarpment (Libya—Tunisia). J Afr Earth Sci 58:489–506

Bouaziz S (1986) La deformation dans la plateforme du Sud tunisien (Dahar et Jeffara): Approche multiscalaire et pluridisciplinaire. These 3^{ieme} cycle, Universite de Tunis, 180 p

Bouaziz S (1995) Etude de la tectonique cassante dans la plateforme et l'Atlas sahariens (Tunisie meridionale): Evolution des paleochamps de contraintes et implications geodynamiques. These de Doctorat, Universite de Tunis II, p 486

Bouaziz S, Barrier E, Turki MM, Tricart P (1999) La tectonique permo-mesozoique (ante-vraconienne) dans la marge sud tethysienne en Tunisie meridionale. Bull Soc Geol France 45–56

Bouaziz S, Barrier E, Soussi M, Turki MM, Zouari H (2002) Tectonic evolution of the northern african margin in Tunisia from paleostress data and sedimentary record. Tectonophysics 357:227–253

Boukadi N (1994) Structuration de l'atlas de Tunisie: signification geometrique et cinematique des nreuds et des zones d'interferences structurales au contact de grands couloirs tectoniques. These de doctorat en Sciences Geologiques, Univ. Tunis II, Faculte des Sciences de Tunis, 249 p

Deffontaines B (1990) Developpement d'une methodologie morphoneotectonique et morphostructurale; analyse des surfaces enveloppes, du reseau hydrographique et des modeles numeriques de terrains; Application au Nord-Est de la France. These, Univ. Paris VI, France, 230 p

Deffontaines B (2000) Formes et deformations de la surface terrestre: Approches morphometriques et applications, Habilitation a Diriger des Recherches (HDR), Universite Pierre et Marie Curie, P6, p. 60 + Annexes

Deffontaines B, Chorowicz J (1991) Principle of drainage basin analysis from multisource data, Application to the structural analysis of the Zaire Basin. Tectonophysics 194:237–263

Deffontaines B, Slama T, Ben Hassen M, Rebai N, Turki MM (2008) Tunisia structural extrusion revealed by numerical geomorphometry. Geol Soc Am spec workshop, Potsdam, June 2008

Gabtni H, Jallouli C, Mickus KL, Zouari H, Turki MM (2009) Deep structure and crustal configuration of the Jeffara basin (Southern Tunisia) based on regional gravity, seismic reflection and borehole data: How to explain a gravity maximum within a large sedimentary basin? J Geodyn 47:142–152

Gabtni H, Jallouli C, Mickus KL, Turki MM (2011) Geodynamics of the Southern Tethysian Margin in Tunisia and Maghrebian domain: new constraints from integrated geophysical study. Arab J Geosci 6:271–286

Hfaiedh M, Attafi K, Arsovski M, Jancevski J, Domurdzanov N, Turki MM (1991) Carte sismotectonique de la Tunisie. Editee par l' Institut National de la Meteorologie

Hlaiem A (1999) Halokinesis and structural evolution of the major features in eastern and southern Tunisian Atlas. Tectonophysics 306 (1):79–95

Molnar P, Tapponnier P (1978) Active Tectonics of Tibet. J Geophys Res 83:5361–5375

Perthuisot JP (1977) Le Lambeau de Tlet et la structure neotectonique de l'ile de Jerba (Tunisie). C. R. Acad Sci, Paris

Perthuisot JP (1985) Cartes geologiques au 1/50.000 et notices explicatives des feuilles de Houmet Essouk, Midoun, Jorf, Sidi Chamakh, Publications, Serv. Geol. Nat. Tunisie

Rabia MC (1998) Systeme d'Informations Geo-scientifique et Teledetection multi-capteurs: Application a une etude mult-themes de la Jeffara orientale. These de Docotrat, Universite Tunis II, p. 320

Ratschbacher L, Merle O (1991) Lateral extrusion in the Eastern Alps, part 2: structural analysis. Tectonics 2:245–256

Tapponnier P, Peltzer A, Le Dain Y, Armijo R, Cobbold P (1983) Propagation extrusion tectonics in Asia: new insights from simple exper-iments with plasticine. Geology 10:611–616

Touati MA, Rodgers MR (1998) Tectono-stratigraphic history of the Southern Gulf of Gabes and the hydrocarbon habitats. Proceedings of the 6th Tunisian petroleum exploration production conference, Tunis, Mai, pp 343–370

Zouari H (1995) Evolution geodynamique de l'Atlas centro-meridional de la Tunisie: stratigraphie, analyse geometrique, cinematique et tectono-sedimentaire. These de doctorat d'etat en sciences geologiques, 277 p

Zouari H, Turki MM, Delteil J (1990) Nouvelles donnees sur revolution tectonique de la chaine de Gafsa. Bull Soc Geol Fr 8(4):621–629

The Extrusion of South-West Taiwan: An Offshore-Onshore Synthesis

26

Benoît Deffontaines, Liu Char-Shine, and Chen Rou-Fei

Abstract

Neotectonic Interpretation of the different marine surveys (ACT, …) swath bathymetry and different onshore-offshore sismic profiles combined to classical structural fieldwork, geodetic, seismological and interferometric studies lead us to propose a global structural scheme and confirm the regional escape tectonics affecting both onshore-offshore of SW Taiwan. First, it is highlighted here the difficulty to only interprete the swath bathymetry even in the northern tip of the Manila accretionnary prism which is a rather simple geological context but affected by both (1) a strong amount of sedimentation due to the Taiwan mountain belt erosion, and (2) to the submarine erosion of the giant Penghu canyon. Second point, is the importance of the seismic interpretation in order to get the offshore bedding and structural data combined with the swath bathymetry and to the photointerpretation of the digital Terrain Model combined to the accurate geological maps to precisely delineate the blocks that is inferred to be submitted to a classical escape tectonic. Third, the precise study of the two new major structural boundaries Fangliao and Young-An structures which guides the SW Taiwan extrusion. Combined with onshore studies (e.g. interferometry (DINSAR), geodetic, seismology and field work) which gives (1) locations, characterization and quantification on the interseismic displacements and (2) lead us to modify our view of the global tectonic structures of the SW Taiwan. To conclude, it is highly recommended to combine both approach on and offshore geology in order to better understand geology and active structures in that part of the world.

Keywords

Extrusion • Neotectonics • Structural geology • Interferometry • Sismic profiles • SW Taiwan

B. Deffontaines (✉)
Laboratoire de Geomatique Appliquee (LGA),
Université Paris-Est Marne-La-Vallee (UPEM), 5 Bd Descartes,
77454, Marne-La-Vallee Cedex 2, France
e-mail: benoit.deffontaines@univ-mlv.fr

B. Deffontaines
Ecole Nationale des Sciences Geographiques (ENSG),
6, et 8 Avenue Blaise Pascal Cité Descartes—Champs-sur-Marne,
77455, Marne la Vallée Cedex 2, France

L. Char-Shine
National Taiwan University, Taipei, Taiwan ROC

C. Rou-Fei
Academia Sinica, Taipei, Taiwan ROC

26.1 Introduction

South of Taiwan, the Philippine plate, bounded by the Luzon volcanic arc, overrides towards the West the South China Sea oceanic crust into the Manila trench. This subduction is transformed northward and onshore into a continent subduction (Lallemand and Tsien 1997) or what is commonly called collision between the Philippine and Eurasian plates. In other words, Taiwan appears to be a large onshore accretionary prism. The convergence between the two plates provides a unique example in the world of the subduction of a continent under an oceanic crust. Thus, it is interesting to analyse in detail the ongoing processes related to plate convergence in Taiwan, as a key example for geological studies of mountain building in the world (e.g., Suppe 1984; Suppe et al. 1985). Within this context, the transition in space and time between subduction and collision deserves particular attention. For this reason, the aim of this paper is to describe the structural and displacement pattern in SW Taiwan, where the transition from the offshore northernmost segment of the Manila Trench to the southernmost segment of the fold-and-thrust Taiwan belt occurs. Such a study implies the combined use of offshore and onshore sources of information.

The obliquity between the Chinese continental margin (Eurasian Plate) and the convergence of the Luzon volcanic arc (Philippine Sea Plate) involves progressive migration towards the South of the active collision (e.g. Biq 1972; Bowin et al. 1978; Wu 1978; Suppe 1984; Suppe et al. 1985; Ho 1986). In the Taiwan area, a present 82 mm/yr towards azimuth N310°E according to GPS measurements (Yu et al. 1995, 1997) reveal the high tectonic activity and the natural hazard importance of this study.

As indicated by Mouthereau et al. (1996) and confirmed by Yu et al. (1997) based on geodetic (GPS) studies, SW Taiwan is the region where maximum deformation occurs. However, few geologists have studied the detailed geometry and mechanisms of structures in SW Taiwan, mainly because of the scarcity of outcrops of suitable rock nature as mudstones prevail, abundance of flat lowlands and absence of differentiate reliefs, as well as the luxuriant vegetation. Sun (1964) published an interesting detailed photogeologic interpretation of the whole area using Howard's (1967) photointerpretation methodology, which takes into careful account the vertical offsets suggested by the topography and the detailed drainage network patterns (see also Deffontaines and Chorowicz 1991). In the coastal plain area of SW Taiwan, Sun (1964) therefore recognised numerous large, flat domes, which he interpreted as horsts and grabens. Later, Deffontaines et al. (1993) proposed the presence of oblique vertical transfer zones that affect the Foothills of Taiwan, probably inherited from the Eurasian passive margin and reactivated by the Plio-Quaternary Penglai orogeny. Deffontaines et al. (1997, 2000) moreover revealed the location of the deformation front of the orogen in the SW Taiwan, as well as the geometry of the pop-up structure for the Tainan anticline. Mouthereau et al. (2002), based on balanced geological cross-sections, confirmed the tectonic styles all along the western foothills of the orogenic belt of Taiwan, from North to South Taiwan confirming both the location of the deformation front in the SW Taiwan and oblique structures to the belt.

In this paper, we compile the structural information provided by recent offshore and onshore surveys to provide a spatially continuous reconstruction of the deformation and structure in SW Taiwan, from sea to land. We then propose a structural scheme of the onshore-offshore SW Taiwan. And using both GPS and DinSAR we propose to locate characterize and quantify the active structures. We use therefore a variety of tools, among which the major ones are the seismic reflection profiling at sea, the field work and mapping on land, and the morphological analysis both on land and at sea, as well as interferometric results of ALOS (JAXA) (DINSAR).

26.2 Tectonic Escape

If the theory of tectonic escape or extrusion has been documented close to collision zone since the 80s (Molnar and Tapponnier 1978; Tapponnier et al. 1983, see also Deffontaines et al. this issue), its application to SW Taiwan is uneasy mainly due to the small extension of the area, the lack of fields outcrop and the few geophysic evidences (no relief, few outcrops, lack of seismicity). Nevertheless Lu (1994, 1996) suggest to apply the phenomena to both side of the Taiwan orogen. More recently (Angelier et al. 1999; Lacombe et al. 2001) suggest a rigid blocks model where four blocks escape to the southwest along major structural discontinuities.

26.3 The Extrusion of SW Taiwan

Onshore seismic profiles have been shot in onshore SW Taiwan in the 50s and 60s. Few remain, apparently because of the difficult paper storage conditions under humid tropical climate. More recently, Suppe et al. (1985) and Yang et al. (2003) interpreted the Nanliao and Kuantzulin structures as underlain by stacking antiform and duplexes. Huang et al. (2005) also proposed a new structural interpretation of SW Taiwan, proposing a "triangle or thrusting wedge model" for the geometry of this complex area. Comparing two NW-SE offshore seismic sections and a single onshore seismic section and the corresponding line drawing interpretations, they suggested the presence of a "propagating intercutaneous nappe".

Despite the interest of such 2D structural approaches, due to the obliquity of the belt, major lateral component of the displacement exist so 3D representation was missing. This component is however clearly evidenced by both the GPS measurements (Yu et al. 1997), and the DinSAR results (Deffontaines 2000, 2004, 2005; Fruneau et al. 2001; Pathier et al. 2003; Mouthereau et al. 2003) and confirmed by "microtectonic" analyses in the field (Angelier et al. 2009). The work by Huang et al. (2005) thus provides a good example of the intrinsic limitations of the 2-D seismic interpretations when the deformation is 3-D in nature. Is worth noting that in a context where strike-slip and oblique components of deformation play an important role geometrical pitfalls affect, and biased structural constraints may result from, balanced geological cross-sections. To summarise, except where structures and motion vectors have a common vertical plane of symmetry, structural reconstruction is a typical 3-D task and can hardly be resolved using 2-D means. This is the case in SW Taiwan. The aim of this contribution is therefore to use marine data with a variety of seismic profiles trends, onshore field data and full consideration of the morphological features of structural significance on both sides of the shoreline.

From the geographic point of view, the onshore SW costal plain of Taiwan appears to be an alternance of NNE-SSW trending elongated topographic ridges and trough (clearly seen on the hypsometric image of the 40 m ground resolution DTEM (Deffontaines et al. 1997), interpreted as the erosional remnance of both harder and softer rocks toward the strong erosion that affect this muddy area. Therefore it is difficult from classical photo-interpretation to decipher the exact places of anticlines and synclines axes even looking at drainage anomalies following the Deffontaines and Chorowicz (1991) methodology. Studies in the fields were confirmed by existing accurate geological mapping done by CPC and the Geological survey of Taiwan (MOEA) which were of great help.

From the lithological point of view, the foreland sequence of SW Taiwan is composed of a more than 5 km very thick, massive and monotonous undercompacted mudstone (e.g. Chou 1991; Hsieh 1972; Lacombe et al. 2001; Brusset et al. 2003; Huang et al. 2005). etc. As shown for instance by the TN-1 wells, drilled in the northern part of the Tainan structure in 1968 and described by Hsieh (1972). Above the thick mudstones was recognized the "Tainan formation" composed of yellowish gray, very thin fine grained sands containing few fossils. Its lower part contains light gray fine grained sands with gray clay rich in mollusks. Then A formation which is composed of bluish gray mudstone intercalated with light gray, very fine grained calcareous siltstone which may be related to Erchunghi formation of northern Taiwan. Then develops the *Upper Gutingkeng* (gray sandy mudstone or sandy siltstone intercalated with gray muddy sandstone 700 m thick is equivalent of the upper part of the Cholan formation) and the *Lower Gutingkeng* formations (bluish gray massive mudstone, more than 3,800 m thick and corresponding to the Cholan formation in northern Taiwan), and ends the wells due to technical problems by the Niaotsui formation which is more than 300 m thick and consist of dark gray massive mudstone.

From the structural point of view, several major domains onshore and offshore (see Liu et al. 2005) were distinguished and highlighted by faults, whatever there activity, which are coarsely described below: The Tainan Basin (TB) is situated on the southern edge of the Eurasian continental plateau showing extensional features far from SW Taiwan and reactivated in a compressionnal environment close to the deformation front (Deffontaines et al. 1997; Lacombe et al. 1999, 2001; Mouthereau et al. 1996, 1999, 2001). His formation is due to the N–S stretching of the Eurasian margin in Middle Eocene to Oligocene time and to the NW-SE stretching due to the opening of the South China Sea. (2) The Pingtung Plain (PP) is situated in between the eastern low metamorphic Central Range bounded by the Chaochou fault (CCF) and the SW Foothills where the Kaoping river flows guided by a N–S elongated diapiric ridge (Huang et al. 2005), and at least the accretionnary prism Lundberg et al. (1991), and Reed et al. (1991) first described from seriated seismic profile (R/V Mona-Wave) the structure of the wide Manila accretionary prism close to Taiwan. *The deformation front (DF)* onshore Deffontaines et al. (1997) and offshore (Liu et al. 1997) is situated west of the Tainan ridge.

So after a global presentation of the analysed and interpreted data used first offshore, where we will focus on the two major structures that bounds the potential escape zone in order to retrace the arguments second we will look for onshore arguments using fieldwork, geodetic measurements and simologic approach, before the discussion, conclusion and perspectives (Figs. 26.1 and 26.2).

Difficulties occurred to interpret the geology at some places where submarine large erosion occur (the Penghu Canyon) and where steep slopes make structural interpretation difficult. However, these studies revealeds two major discontinuities: the Fangliao Fault Zone and the Young-An Fault Zone, shortly described below.

1. **The Fangliao fault zone and associated submarine landslides** Numerous marine surveys had been done close to the Hengchun peninsula and the topographic and geomorphic description of have already been done (You et al.). We herein interpret these data in terms of structural geology, proposing that a major left lateral strike-slip fault characterises the Fangliao canyon, as shown by the structural interpretation of the parallel seismic profiles illustrated below.

The detailed analysis of the bathymetry (data from Liu Char-Shine ad co-workers) provides evidence of two major landslides on the western flank of the Hengchun peninsula. These sliding masses fill in the Fangliao canyon, inducing a typical "horse foot" trace in the bathymetry. They correspond to the sliding of the missing western flank of the anticline located west of Hengchun (see Fig. 26.3).

2. **The Yung-An Fault Zone and the Penghu canyon** Numerous seismic sections acquired during the Taiwan-France ACT survey crosscut the complex Penghu canyon. The seismic, 3,5 Khertz interpretation of this domain was difficult because of the presence of deep slopes and intense erosion of the complex Penghu

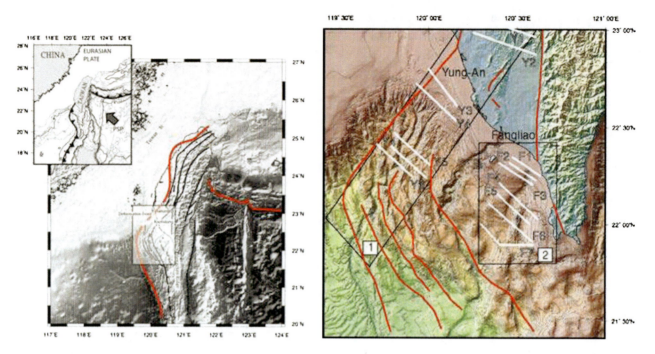

Fig. 26.1 Geodynamics of Taiwan and quadrangle location of studied area. **a** *Left* one may note the onshore-offshore situation of the studied area situated in the incipient collision zone (Lallemand et al. 1995) in between Manila subduction zone to the south and the collision zone onshore Taiwan. **b** *Right* location of the offshore seismic (sismic reflection profiles Fangliao fault zone (F1 to F7) from north to south respectively, Young-An fault zone: Y1 to Y5 from North to South also

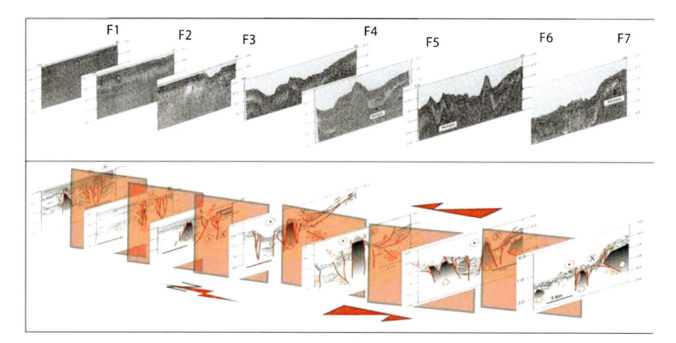

Fig. 26.2 The Fangliao left lateral fault zone SE boundary of the SW Taiwan extrusion. One may note the left lateral transtensive strike-slip fault zone which is injected by an elongated mud diaper south of the neutral zone (North to the *left*, *red lines* correspond to faults)

canyon system. Whereas a classical hill shading (see method described in Deffontaines et al. 1993) highlights NE-SW morphological discontinuities within the Manila accretionnary prism, the demonstration of the existence of a major structural discontinuity is given by the detailed seismic profile analysis and interpretation (especially location of faults, folds and dip of layers). Therefore, the Young-An fault zone appears to be a positive flower

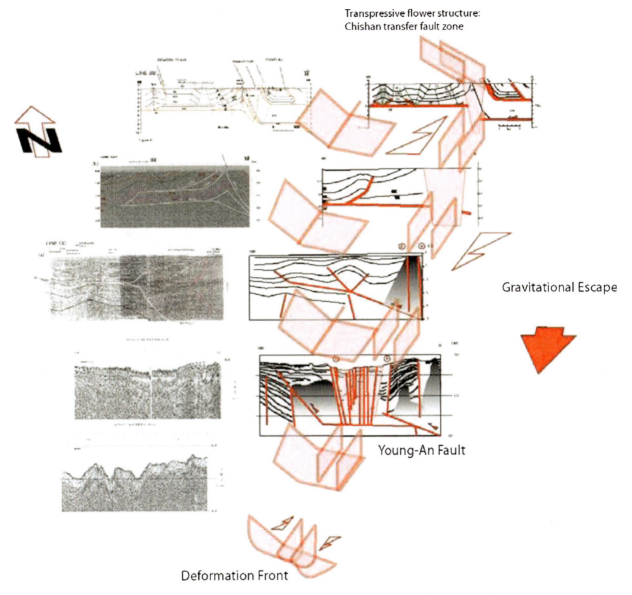

Transpressive flower structure:
Chishan transfer fault zone

Gravitational Escape

Young-An Fault

Deformation Front

Fig. 26.3 The western boundary of the SW Taiwan extrusion: the Yung-An alignment (Transfer, vs. strike-slip, and role of the erosion) that correspond to tectonic partitioning

structure marked locally by an elongated pressure ridge above the fracture zone. This major fault zone is right-lateral, as shown by the numerous offset of folds on both sides of it. Also note that its displacement at depth, close to the deformation front, is less important than the displacement close to the shoreline, because of reactivation during the geological evolution (Figs. 26.4 and 26.5).

From the GPS point of view, In the first approximation, rather than involving heterogeneous displacement, the interseismic deformation of the Pintung basin mainly consists of a relatively continuous deformation, as shown by the surprisingly regular pattern of displacement data recorded at numerous GPS stations. The offshore boundaries of the basin however corresponds to two major structural discontinuities, the Fangliao canyon that we interpret as a major left-lateral fault zone (associated with submarine landslides affecting the western flank of Hengchun anticline, west of the peninsula) and the Young-An fault zone that we interpret as the major right-lateral fault zone facilitating the offshore extrusion of the Pintung basin. We suspect the existence of another right-lateral fault zone northwest of the YAFZ, which would affect the Penghu canyon as marked by submarine drainage anomalies. These major offshore fault zones are not well-preserved onshore, which is not surprising in light of the active erosion of SW Taiwan and of the soft lithology in the studied area. Numerous hazards are related

Fig. 26.4 Thrusting component
in the concrete dyke close to the
Tsenwenghsi (river), the
shortening is clearly seen by the
pipe completely squeezed within
the thrust fault plane (B.
Deffontaines slide)

Fig. 26.5 General deformation
model of the SW Taiwan

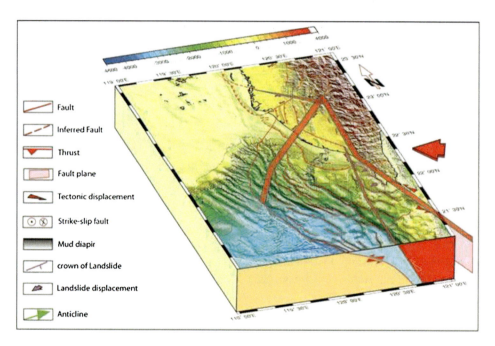

to these major structural fault zones, including possible
earthquakes (despite the seismic gap or lack of present
earthquake activity) and tsunamis that could be triggeres by

submarine landslides such as those evidenced close to
the Hengchun peninsula or above the deformation front
(Deffontaines et al. 1999).

26.4 Discussion and Conclusion : The Model of Extrusion of SW Taiwan

Several structural key points arises resulting of our study in the SW Taiwan: The backstop (eastern boundary correspond to the Chaochou fault the indentor is the Peikang high and the western boundary that correspond to the continental slope of the eurasian plate and the Tainan basin. Highlighted by the shelf break before the erosion by the Penghu canyon. The relative sismic gap is linked to the soft marly gutingken formation that gives a rather continuous deformation at the surface revealed by both the interferometric wok as well as the GPS data.

The two lateral boundaries of this SW Taiwan extrusion are to the west the Young-An fault zone ending in the city of Chishang onshore which act as a triple junction. The Fangliao/chaochou onshore split offshore into two distinct major fracture zones the fangliao canyon and the Hengchun fault.

The mud diapirs offshore (Liuchyu island act as grabens that parallel to the extrusion displacement which highlight the transtensive component of the deformation. The neutral zone correspond to the Fengshan transfer fault zone already described by Deffontaines et al. 1993 and 1999. This fault correspond to a graben like structure which differenciate two distinct tectonic domain a transpressive one on the NE and a transtensive one in the SW. Many perspectives arise in terms of gas hydrates as soon as the structural geometry is much better know in that place.

Acknowledgments The authors would like to thank deeply the LIA ADEPT N°536 France-Taiwan and the Sino-French collaboration and more especially Jacques Angelier, Hu Jyr-Ching, Erwan Pathier, Benedicte Fruneau, Johann Champenois, Lu Chia-Yu, Lee C.-T., Brusset S., Chu Hao-Tsu, Lee T. Q, Lee Jian-Cheng, Jo Deramond, Rau Ruei-Juin, Li Fung Chun, Hsu Shu-Kun, Hsu Hsi-Hao, Chang Chung Pai, and Chang Tsui-Yu for so many fruitful discussions and friendship for more than two decades.

References

Angelier J, Barrier E, Chu HT (1986) Plate collision and paleostress trajectories in a fold-thrust belt: the Foothills of Taiwan. Tectonophysics 125:161–178

Angelier J, Chu HT, Lee JC (1997) Shear concentration in a collision zone: kinematics of the Chihshang Fault as revealed by outcrop-scale quantification of active faulting, Longitudinal Valley, eastern Taiwan. Tectonophysics 274:117–143

Angelier J, Yu SB, Lee JC, Hu JC, Chu HT (1999) Active deformation of Taiwan collision zone: discontinuities in GPS displacement field. In: Proceedings of the international symposium on Subduction et collision active dans le SE asiatique, Fourth colloquium on sino-French cooperation program in Earth sciences, Montpellier, France

Biq C (1976) Western Taiwan earthquake-faults as miniature transform faults. Bull Geol Surv Taiwan 25:1–8

Biq C (1989) The Yushan-Hsuehshan megashear zone in Taiwan. Proc Geol Soc China 32:17–20

Bonilla MG (1975) A review of recently active faults in Taiwan. USGS Open-report, pp 1–55 (75-41)

Bonilla MG (1977) Summary of quaternary faulting and elevation changes in Taiwan. Mem Geol Soc China 2:43–56

Byrne T, Crespi J, Pulver M, Spiker E (1999) Deformation of the Taiwan hinterland: strike-slip faulting, extrusion or both? In: Proceedings of the international symposium on subduction et collision active dans le SE asiatique, Fourth colloquium on sino-French cooperation program in Earth sciences, Montpellier, 9–12 May 1999

Chen H (1984) Crustal uplift and subsidence in Taiwan: an account based upon retriangulation results. Spec Publ Central Geol Surv 3:127–140 (in chinese)

Davis D, Suppe J, Dahlen FA (1983) Mechanisms of fold-and-thrust belts and accretionary wedges. J Geophys Res 88(B2):1153–1172

Deffontaines B, Lee JC, Angelier J, Carvalho J, Rudant JP (1994) New geomorphic data on Taiwan active orogen: a multisource approach. J Geophys Res 99(B8):20243–20266

Deffontaines B, Lacombe O, Angelier J, Chu H-T, Mouthereau F, Lee C-T, Deramond J, Lee J-F, Yu M-S, Liew P-M (1997) Quaternary transfer faulting in Taiwan Foothills: evidence from a multisource approach. Tectonophysics 274:61–82

Deffontaines B, Liu C-S, Angelier J, Mouthereau F, Lacombe O (1999a) Quaternary tectonic evolution of offshore SW Taiwan. In: Proceedings of the international symposium on Subduction et collision active dans le SE asiatique Fourth colloquium on sino-French cooperation program in Earth sciences, Montpellier, 9 ± 12 May 1999

Deffontaines B, Chu H-T, Angelier J, Lee C-T, Mouthereau F, Li F-C, Brusset S, Chang C-P, Sibuet J-C, Chang T-Y, Lacombe O, Liu C-S, Deramond J, Lee J-C, Lu C-Y, Hu J-C, Yu M-S, Lee J-F, Bureau D (1999b) Onshore ± offshore Taiwan earth science data base within a geographical information system. In: Proceedings of the international symposium on Subduction et collision active dans le SE asiatique, Fourth colloquium on sino-french cooperation program in Earth sciences, Montpellier, 9–12 May 1999

Fruneau B, Pathier E, Raymond D, Deffontaines B, Lee CT, Wang HT, Angelier J, Rudant JP, Chang CP (2001) Uplift of Tainan foreland (SW Taiwan) revealed by SAR interferometry. Geophys Res Lett 28:3071–3074

Fuh S-C, Liu C-S, Wu M-S (1997a) Migration of the Canyon systems from Pliocene to Pleistocene in area between Hsyning structure and Kaoping slope and its application for hydrocarbon exploration. Pet Geol Taiwan 31:43–60

Fuh S-C, Liu C-S, Lundberg N, Reed D (1997b) Strike-slip faults offshore southern Taiwan: implications for the oblique arc-continent collision processes. Tectonophysics 274:25–39

Ho CS (1986) A synthesis of the geologic evolution of Taiwan. Tectonophysics 125:1–16

Hu JC, Angelier J, Yu SB (1997) An interpretation of the active deformation of southern Taiwan based on numerical simulation and GPS studies. Tectonophysics 274(1–3):145–169

Huang C-Y, Wu W-Y, Chang C-P, Tsao S, Yuan P-B, Lin C-W, Kuan-Yuan X (1997) Tectonic evolution of accretionary prism in the arc-continent collision terrane of Taiwan. Tectonophysics 281:31–51

Hung J-H, Wiltschko DV, Lin H-C, Hickman J-B, Fang P, Bock Y (1999) Structure and motion of the southwestern Taiwan fold-and-thrust belt. TAO 10(3):543–568

Hsu T-L, Chang HC (1979) Quaternary faulting in Taiwan. Mem Geol Soc China 3:155–165

Lacombe O, Mouthereau F, Deffontaines B, Angelier J, Chu H-T, Lee CT (1999) Geometry and quaternary kinematics of fold-and-thrust units of SW Taiwan. Tectonics 18(6):1198–1223

Lacombe O, Mouthereau F, Angelier J, Deffontaines B (2001a) Structural, geodetic and seismological evidence for tectonic escape in SW Taiwan. Tectonophysics 333:323–345

Lallemand SE, Tsien H-H (1997) An introduction to active collision in Taiwan. Tectonophysics 274:1–4

Lee TQ, Angelier J (1995) Analysis of magnetic susceptibility anisotropy of the sedimentary sequences in southwestern foothills of Taiwan and its tectonic implications. In: Proceedings of the International conference and Third sino-French symposium on Active Collision in Taiwan, Abstracts volume, 22–23 March 1995, Taipei, Taiwan, pp 213–218

Liu CS, Huang IL, Teng LS (1997) Structural features off southwestern Taiwan. Mar Geol 305–319

Lu CY (1994) Neotectonics in the foreland thrust belt of Taiwan. Petrol Geol Taiwan 29:15–35

Lu CY, Malavieille J (1994) Oblique convergence, indentation and rotation tectonics in the Taiwan Mountain belt: insights from experimental modelling. Earth Plan Sci Lett 121:477–494

Lu CY, Angelier J, Chu H-T, Lee J-C (1995) Contractional, transcurrent, rotational and extensional tectonics: examples from northern Taiwan. Tectonophysics 246:129–146

Lacombe O et al (2001b) Tectonophysics 333:323–345

Lu CY, Jeng FS, Chang KJ, Jian WT (1998) Impact of basement high on the structure and kinematics of the western Taiwan thrust wedge: insights from sandbox models. TAO 9(3):533–550

Lundberg N, Reed D, Liu C-S, Lieske J (1997) Forearc-basin closure and arc accretion in the submarine suture zone south of Taiwan. Tectonophysics 274:5–23

Mouthereau F, Angelier J, Deffontaines B, Lacombe O, Chu HT, Colletta B, Deramond J, Yu MS, Lee JF (1996) Cinematique actuelle et recente du front de chame de Taiwan. CR Acad Sci, t. 323, II, Paris, pp 713–719

Mouthereau F, Angelier J, Deffontaines B, Brusset S, Lacombe O, Chu H-T, Deramond J (1998) Folds and fault kinematics and tectonic evolution of the southwestern thrust belt of onshore Taiwan. In: Annual meeting of geological society of China, Chungli, 20–21 March 1998, p 140

Mouthereau F, Lacombe O, Deffontaines B, Angelier J, Brusset S (2001) Deformation history of the southwestern Taiwan foreland thrust belt: insights from tectono-sedimentary analysis and balanced cross-sections. Tectonophysics 333:293–322

Ratschbacher L, Merle O (1991) Lateral extrusion in the Eastern Alps, part 2: structural analysis. Tectonics 2:245–256

Rau RJ, Wu FT (1998) Active tectonics of Taiwan orogeny from focal mechanisms of small-to-moderate earthquakes. TAO 9:755–778

Reed D, Lundberg N, Liu C-S, Kuo B-Y (1992) Structural relations along the margins of the offshore Taiwan accretionary wedge: implications for accretion and crustal kinematics. Acta Geol Taiwan 30:105–122 (Science reports of the National Taiwan University)

Rocher M, Lacombe O, Angelier J, Chen H-W (1996) Mechanical twin sets in calcite as markers of recent collisional events in a fold-and-thrust belt: evidence from the reefal lime-stones of southwestern Taiwan. Tectonics 15(5):984–996

Sengor AMC, Gorur N, Saroglu F (1985) Strike-slip faulting and related basin formation in zones of tectonic escape: Turkey as a case study, vol 37. SEPM Special Publications, pp 227–264

Seno T, Stein S, Gripp AE (1993) A model for the motion of Philippine Sea Plate consistent with NUVEL-1 and geological data. J Geophys Res 98:17941–17948

Sun SC, Liu CS (1993) Mud diapirs and submarine channel deposits in offshore Kaohsiung-Hengchun, southwest Taiwan. Petrol Geol Taiwan 28:1–14

Suppe J (1984) Kinematics of arc-continent collision, dipping of subduction, and back-arc spreading near Taiwan. Mem Geol Soc China 4:67–90

Suppe J, Hu CT, Chen Y-J (1985) Present-day stress directions in western Taiwan inferred from borehole elongation. Petrol Geol Taiwan 21:1–12

Tapponnier P, Peltzer A, Le Dain Y, Armijo R, Cobbold P (1983) Propagating extrusion tectonics in Asia: new insights from simple experiments with plasticine. Geology 10:611–616

Teng LS (1990) Geotectonic evolution of late Cenozoic arc-continent collision in Taiwan. Tectonophysics 183:57–76

Tsai Y-B (1986) Seismotectonics of Taiwan. Tectonophysics 125: 17–37

Wu F, Rau R-J, Salzberg D (1997) Taiwan orogeny: thin-skinned or lithospheric collision? Tectonophysics 274:191–200

Yeh YH, Barrier E, Angelier J (1991) Stress tensor analysis in the Taiwan area from focal mechanisms of earthquakes. Tectonophysics 200:267–280

Yu SB, Chen HY (1996) Spatial variation of crustal strain in the Taiwan area, Sixth Taiwanese geophysical meeting, Chiayi, November 1996, abstract volume, pp 659–668

Yu SB, Chen HY (1998) Strain accumulation is Southwestern Taiwan. TAO 9(1):31–50

Yu S-B, Chen H-Y, Kuo L-C (1997) Velocity field of GPS stations in the Taiwan area. Tectonophysics 274:41–59

Yu S-B, Kuo L-C, Punongbayan RS, Ramos EG (1999) GPS observation of crustal deformation in the Taiwan-Luzon region. Geophys Res Lett 26(7):923–926

Active Tectonic Risk Assessment—Problems with Soil and Soft Sediment Deformation Structures

27

Philip E.F. Collins

Abstract

Many sites feature soils and sediments that have undergone syn- or post-depositional deformation. Typical forms of deformation include load casts, wedges, involutions and diapirs. In studies on geologically-young (Quaternary) soils and sediments in much of northwest Europe and some other areas, such forms are frequently attributed to the former action of non-glacial freezing and thawing during past cold periods. Similar features are found in arctic and high altitude areas. They are used to help reconstruct past climate and associated ground thermal regimes, including the extent of permafrost. The identification of the features as being relict with a low likelihood of such intense freeze-thaw processes being replicated has implications for site risk assessment and, consequently, design. Features that reflect syn- or post-depositional deformation elsewhere in the world, or in pre-Quaternary rocks, are frequently attributed to strong ground motion, and are used to help reconstructed past earthquake histories, both for regions and for individual potentially-active tectonic structures. There are direct implications for risk assessment and design. In theory, since the processes related to ice growth/decay and cyclic ground motion might be expected to be different, it should be possible to differentiate between them using diagnostic criteria. Unfortunately, either because different researchers are unconsciously biased by their training (perhaps causing the difference in interpretation depending on geographical region and the age of the features), or because the criteria are inadequately defined and are equally applicable to different processes, problems remain for fully understanding site risk. This paper compares the current state of diagnostic features used for interpreting deformations and uses these to explore possible interpretations and implications of a range of features found in apparently 'cryogenic' and 'palaeoseismic' settings. It ends with a call for a renewed focus on developing a robust and testable diagnostic toolkit for site investigation of soil and soft sediment deformation structures so that the features can be understood in terms of either "fossil" ice processes or active tectonic risk.

Keywords

Quaternary • Deformation • Risk • Infrastructure

P.E.F. Collins (✉)
Brunel University, Kingston Lane, Uxbridge, London,
UB8 3PH, UK
e-mail: philip.collins@brunel.ac.uk

27.1 Introduction

Unexpected ground conditions are a major area of risk for construction and infrastructure management. Frequently, carefully executed site investigation can successfully identify difficult ground and provide a framework for mitigation. A limitation can, however occur, where subsurface features are unclear, ambiguous, or where the investigator has limited experience of differentiating apparently similar features. Soft sediment and soil deformation structures are a good example of this as they may be missed in typical borehole and trial pit sampling. The deformation mechanism can also be difficult to determine. Finally, the investigator may interpret deformation features based on prior experience, and education.

A better understanding of soft sediment deformation structures is needed as this will permit both a better understanding of long term risk (e.g. do the structures represent potentially active or relict soil processes), and current ground conditions (e.g. different processes may create different three dimensional structures that may affect macro-scale shear, consolidation and drainage).

27.2 Potential Causes of Soft Sediment and Soil Deformation Structures

A wide range of processes are known to create deformation structures. These include sub-marine slumps, glacial action, burrowing by animals, roots, freeze-thaw and earthquakes. The last two will be considered here as they can both potentially affect large areas and, as discussed below, can produce similar features. Interestingly, the they also link to two groups of investigators—those educated principally in the impact of climate change (particularly former periglacial conditions), and those educated in seismic impacts.

As soil temperatures pass below 0 °C, the soil water progressively freezes, resulting in compression and heave. Localised variability is caused by soil materials with different thermal properties, and varying amounts of water. Very low temperatures can cause a reversal of the ice expansion, inducing tension, and sometime cracking. As soil thaws, localized settling may occur, associated with plastic creep. High porewater pressure may be induced by the downwards migration of a freezing front, or by rapid thaw. This can induce liquefaction.

During a seismic event, horizontal and vertical pressure waves can result in oscillations between compression and tension. Dry soil may crack, buckle or disintegrate. Wet soil may plastically deform, shear or liquefy. This is particularly the case during larger magnitude events, and where a cohesive surface layer prevents the rapid vertical release of porewater.

27.3 Comparison of Selected Soft Sediment and Soil Deformation Features Attributed to Periglacial and Seismic Action

One of the most commonly-cited criteria to differentiate periglacial and seismic deformations is geographical extent. If cold climate conditions were responsible, then similar forms should, in principle, be found across a broad region. In reality, the formation of deformations by frost action is dependent upon the type of soil/sediment, water availability and local variations in geothermal flux. Theoretically, at least, deformations with a seismic origin should be focused around the responsible faults, with a change in deformation style away from the epicenter. Again, exactly what deformation occurs is determined by factors such as event magnitude/depth, plasticity, liquefaction potential and depth of burial. Ambreysis (19xx) found evidence for liquefaction many tens of kilometres from fault ruptures.

As a result, it is sensible to consider whether the detailed form of deformation structures can be used to differentiate causal mechanisms. In theory, different rates and directions of deformation should be associated with freeze-thaw and shaking. Unfortunately, a comparison of deformation features described in the scientific literature indicates a significant amount of overlap. Some examples of these are given in Table 27.1.

27.4 A Quaternary Case Study from South-Central England

Mid-Quaternary fluvial sands and gravels, unconformably resting on moderately permeable Eocene sands underlie an almost flat river terrace at Eversley Common, Hampshire, UK. The site occurs beyond the maximum glacial limits and will have experienced repeated cold-warm climate shifts

Table 27.1 Selected deformation structures attributed to periglacial and seismic processes

Periglacial		Seismic	
Form	Process and key features	Form	Process and key features
Vertically oriented clasts	Differential freeze-thaw heave between a clast and the soil/sediment matrix leading to vertical movement and alignment. May occur at a uniform depth across a unit/site	Aligned clasts	Clasts aligned to flow of liquefied material. Localised
Fissure fill/ Sediment wedge	Gravity driven (i.e. downwards) infilling of thermal contraction or mass movement-induced fissure. Typically wider at top. May show stratification	Dyke—Neptunian (formed under water)/Fissure fill (sub-aerial)	Gravity driven (i.e. downwards) infilling of seismically-induced fissure. Typically wider at top. May show stratification
		Dyke—injection	Pore fluid pressure driven infilling of fissure (principally upwards). Typically narrower at top. Particles may be graded (fining up)
		Sill—injection	Pore fluid pressure driven infilling of fissure (principally lateral)
Flame structures	Plastic upwards deformation under cyclic freeze-thaw stress	Diapiric structures	Plastic upward deformation under seismic load stress. Grades into injection dykes if soil/sediment becomes fully liquefied. Reflects deeper unit having a lower dynamic viscosity than overlying soil/sediment. Margins may feature micro-faults, with upwards displacement. Adjacent areas may show evidence of subsidence
Load cast, pillow structures	Isolated mass of sediment that has sunk into an underlying unit that has experienced freeze-thaw induced plastic deformation, liquefaction and/or localised consolidation producing density changes. Range of sizes (<cm to >m)	Load cast, pillow structures	Isolated mass of sediment that has sunk into an underlying unit that has experienced hydro-plastic deformation, liquefaction and/or localised consolidation due to cyclic shear density changes. Range of sizes (<cm to >m).
Involutions	Plastic deformations resulting from ice growth and decay. May also result from rapid thaw settlement and liquefaction	Sismoslump/ Involutions	Convoluted sedimentary structures reflecting in situ deformation due to cyclic lateral seismic loading. Under and overlying strata may be intact or show grading (increasing deformation towards surface). Structures may show no sign of compression due to subsequent burial

over ~0.5–0.6 Ma. It is also close to the Variscan Front, a major East-West fault zone that runs approximately parallel to the former ice maximum limit (Fig. 27.1). Only one nearby small historical earthquake is known (Musson 1994).

Soft-sediment deformation structures at the site, including load casts, flames/diapirs, possible dykes, homogenites, oriented clasts and 'kink' structures. A boudinage-like structure indicative of high pore-water pressure and lateral migration was observed in one exposure (Fig. 27.1). In some exposures, there is evidence for an initial significant deformation event or phase, with later modification by slowly acting processes, including limited vertical realignment of stones.

If periglacial action was responsible for the larger structures, their large size suggests a deep seasonal thaw layer. Regional permafrost has been hypothesized (Hutchinson and Thomas-Betts 1990), though there is no unambiguous local evidence for its presence. If it was present, this might explain features that would require a reduction in the permeability of the underlying Tertiary sands.

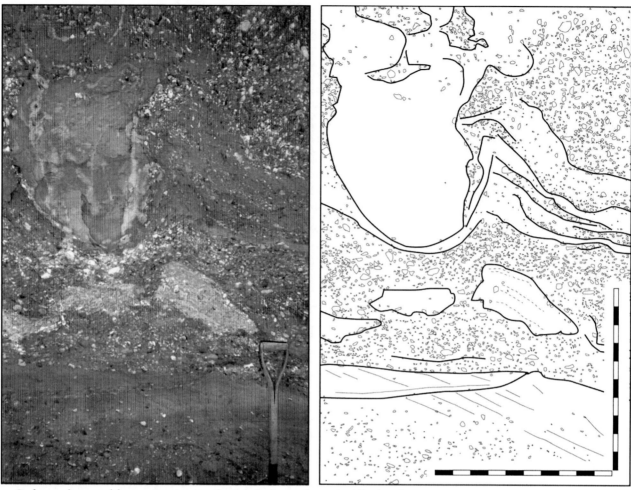

Deformed ground from a Quaternary fluvial site in south-central England

Fig. 27.1 Soft sediment deformation structures, Eversley Common, UK. Note load cast, 'boudinage-like' structure and angular 'kink'

The absence of significant historical seismicity might suggest an earthquake origin is unlikely. However, the deformations are large and, if they are seismically-formed, would represent a large and perhaps infrequent event. In addition, the 'kink' structure shown in Fig. 27.1 may indicate abrupt compression occurring above a partially liquefied layer. This raises the probability of a large seismic event affecting the site at some point in the past, though does not necessarily disprove a periglacial hypothesis.

27.5 Conclusions

The apparent similarity of soft sediment and soil deformation structures presents a significant challenge for site (and indeed regional) risk assessment. Careful examination may allow causes to be determined, but site investigators need to be aware of the bias introduced by their own training. Significant problems exist if the evidence is in the form of narrow borehole records and samples, especially if the samples are disturbed.

Further work is needed on deformation structures. In particular, robust criteria for differentiating rapid and slow deformation in different types of material need to be developed.

References

Hutchinson JN, Thomas-Betts A (1990) Extent of permafrost in southern Britain in relation to geothermal flux. Q J Eng GeolHydrogeol 23:387–390

Musson R (1994) A catalogue of British earthquakes. BGS technical report no. WL/94/04

Formation of Earthquake Faults by the Fukushima Hamadori Earthquake and an Estimation of Displacement Distribution Around the Faults Using Airborne LiDAR Data

28

Shunsuke Shinagawa, Shuji Anan, Yasuhito Sasaki, Sakae Mukoyama, Shin-ichi Homma, and Yoko Kobayashi

Abstract

A series of earthquake faults was produced by the 2011 Fukushima Hamadori Earthquake (Mw 7.1), which broke out one month after the Great East Japan Earthquake. Two lines of earthquake faults appeared extending 30 km in length. They were inferred to be located partly on the active faults and partly on its extended line. The purpose of this presentation is to analyze these faults. We made use of a new method of airborne LiDAR data in the investigation of the displacement distribution by the earthquake. The results obtained fit generally well with the distribution data of net fault displacement produced by field investigation. Furthermore, the results suggest that the earthquake caused block tilting and land subsidence. Our new method makes available information difficult to obtain by the conventional method of field research of displacement distribution and it is useful for investigation immediately after the earthquake.

Keywords

Earthquake fault • Fault displacement • Airborne lidar • Differential analysis

28.1 Introduction

The Fukushima Hamadori Earthquake occurred on April 11, 2013, one month after the Great East Japan Earthquake of March 11. It formed earthquake faults near the Itozawa Fault (14.7 km in length) and Yunodake Fault (15.8 km in length). At the Itozawa Earthquake Fault, which is situated near the hypocenter, the maximum vertical displacement is 220 cm, while at the Yunodake Earthquake Fault which is farther from the hypocenter, it is 80 cm (e.g. Toda and Tsutsumi 2013).

The earthquake faults actually appeared not where they were assumed to be by the past topographical interpretations. Thus improving the accuracy of active fault survey methods is now an imminent task for the applied geological research. It is necessary to confirm the location of emerged earthquake faults to compare with the results of various other survey methods.

In order to estimate the locations where earthquake faults would emerge, and the magnitude of displacement which they would cause, investigation is generally done in situ, but this method is extremely time-consuming and costly. In recent years, however, airborne LiDAR surveying has pervaded rapidly. It is now possible to make use of this technology for our purpose doing the measurements before and after an earthquake. In this research, first, DEM differential analysis was conducted making use of LiDAR surveys performed at two different times, before and after the earthquake, and the earthquake fault displacement distribution was assumed. Then the data were compared with the in situ survey results; the effectiveness of the new method was verified.

S. Shinagawa (✉) · S. Anan · Y. Sasaki
Public Works Research Institute, Tsukuba, Japan
e-mail: sinagawa@pwri.go.jp

S. Mukoyama · S.-i. Homma · Y. Kobayashi
Kokusai Kogyo Company Limited, Tokyo, Japan

28.2 Differential Analysis of Results of the Two Airborne LiDAR Surveys

28.2.1 Principle and Method of the Differential Analysis

The value of difference of elevation can be easily calculated from bi-temporal DEM pair. However, in order to get the accurate distance and direction of the ground movement, it is necessary to obtain the three-dimensional locus of the point.

In this study, we applied the technique of image matching analysis to the measurements of the ground displacement (Japan Patent No. 4545219). Since the visual image is two-dimensional planar surface, and the directly measured distance is a horizontal element, the elevation value corresponding to each planar coordinate point has to be calculated by the three dimensional vector. However, because the image made from DEM has the elevation value at each grid point, the vertical component is available by interpolation of the DEM elevation value around the endpoint of the calculated vector. A remarkable advantage of this technique is that it does not require neither the measurement for mapping nor selection of specific characteristic for tracking. Moreover, random point cloud data provide with areal and spatial quantitative movement.

High resolution DEM that can be made of the airborne LiDAR survey is useful as it enables us to produce measurable image from a terrain model. In case there is a wide area crustal movement caused by an earthquake etc., to get more accurate ground displacement rate, it is necessary to obtain measurement data for correction. For a good image matching analysis, it is preferable that the image should show geomorphic quantity without the azimuthal anisotropy. The PIV method was selected for the image matching analysis in this study. The movement was identified by the image correlation, and the movement vectors were calculated by the average of several movement parts in the search area.

28.2.2 Measurement Precision and Precautions

For the images used in this study, 1 pixel is the single grid size of 2 m × 2 m DEM. And the search window size is 64 pixels × 64 pixels (128 m × 128 m). In image matching, sub-pixel interpolation was employed, so it is normally possible to calculate displacement of about 1/10 pixel size. Since 2 m

grid of topographical data was set to be 1 pixel, it is assumed that the reliable displacement is ±20 cm.

28.2.3 DEM Used for the Analysis

The research site covers the area where earthquake faults were formed by the Fukushima Hamadori Earthquake. The DEM used was 2 m grid data based on the airborne LiDAR data of before and after the Fukushima Hamadori Earthquake (Table 28.1).

28.2.4 The Impact of Wide Area Crustal Movement on DEM and Its Cancellation

In the study area, crustal movement caused by the Great East Japan Earthquake and the Fukushima Hamadori Earthquake overlaps. Therefore it is necessary to cancel the displacement caused by other than our target earthquake. No displacement data are available of the Great East Japan Earthquake of March 11, nor of the after slip from March 11 to April 11 when Fukushima Hamadori Earthquake occurred. Therefore a planar distribution model of displacement based on a primary polynomial approximation was prepared making use of the daily values at April 9, produced by three GNSS-based Control Stations around the survey region. The values of each grid provided by the model were assumed to be as the sums of the displacement caused by the earthquake of March 11 and the after slip during the period from March 11 to April 11.

28.2.5 Analysis Procedure

(1) Displacement distribution before and after the earthquake

First, shaded slope maps before and after the earthquake were prepared making use of DEM. Then horizontal displacement was obtained for each 128 m^2 area by image matching analysis, and the difference of average elevations of matched area was taken as the vertical displacement. The centers of the search windows were shifted in the south-north and east-west direction by 64 m each to draw the vectors of displacement (Fig. 28.1).

(2) Displacement distribution along the earthquake faults

In order to clarify the details of the displacement distribution across the earthquake faults, the differences between

Table 28.1 DEM used for the analysis

	Date of surveying	Grid size (m)	Note
Before the earthquake	Sep., 2006 to Feb., 2007	2	Taken by Kokusai Kogyo Co. Ltd
After the earthquake	May 11 to June 10, 2011	2	Provided by geospatial information authority of Japan[a]

[a] High precision elevation data for restoration and disaster prevention measurement

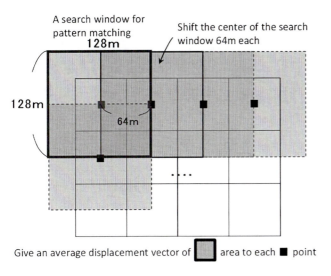

Fig. 28.1 The procedure to obtain the vectors of displacement (Shinagawa et al. 2013)

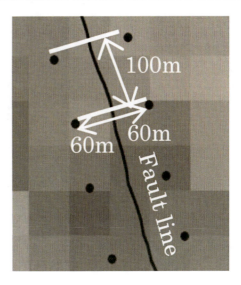

Fig. 28.2 The procedure to obtain the displacement values along the earthquake fault (modification of Shinagawa et al. 2013)

the two grids were sought, each covers the upheaved and subsided points with an earthquake fault in between. Given the displacement values of the measured point of 64 m mesh explained above as well as positional information of the measurement points, the displacement value of meshes including measuring points were given to each measurement point. The differential of the set of points 120 m apart straddling the fault line was presumed to be the earthquake displacement which was estimated by differential analysis (Fig. 28.2). Regarding the vertical displacements, the relative rise on the northeast side was given a positive value. As to the horizontal displacements at right angles with the strike of the faults (N55°W for the Yunodake Earthquake Fault and N20°W for the Itozawa Earthquake Fault), displacements were estimated by calculation. Positive values were given to the right lateral displacements and negative values to the left lateral displacements respectively.

28.3 Comparison of Results of Differential Analysis with Those of In Situ Survey

Figure 28.3 shows the vertical displacement distribution and horizontal displacement vector of the entire area covered by differential analysis.

28.3.1 Vertical Displacement

The comparison between the results of our study and those of in situ survey indicates that the area where vertical displacement was confirmed to have occurred conforms with the distribution of ground surface earthquake fault almost

exactly. By and large our estimates of vertical displacement give somewhat larger values than those of the in situ survey data.

Although our study site does not adequately encompass the earthquake affected area, our results suggest the land block enclosed by the Itozawa Earthquake Fault and the Yunodake Earthquake Fault has inclined to north-east. It also indicates that the land block has subsided remarkably on the west side of the Itozawa Earthquake. These observations are in harmony with those of the crustal movement study based on the interference SAR (Geospatial Information Authority of Japan 2011).

We see in Fig. 28.3 that subsidence has occurred in the distribution area of the valley bottom plain. Since alluvial deposits cover the valley bottom plains, it is possible that earthquake motion caused consolidation subsidence. The estimated subsidence varies from point to point as indicated in Fig. 28.3, but any subsidence greater than 20 cm was detected in this study.

28.3.2 Horizontal Displacement

In our study site, displacement from north-east to east-northeast is dominant. Estimated displacement near the electronic reference point conforms almost perfectly with the magnitude and direction of displacement that the Geospatial Information Authority of Japan measured immediately after the earthquake.

The horizontal displacement near Itozawa Earthquake Fault generally runs in the direction of east-northeast at right angle with the earthquake fault, but the direction and magnitude of land surface displacement change beyond the fault.

Fig. 28.3 Vertical and horizontal displacement distribution estimated by geomorphic image matching analysis using airborne LiDAR data before and after the 2011 Fukushima Hamadori Earthquake (modification of Shinagawa et al. 2013)

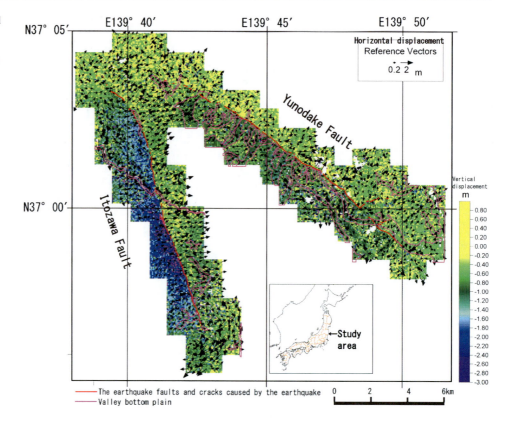

Near the south end of the Itozawa Earthquake Fault, there is a relatively large displacement band with a fixed width which runs in the direction of south–west. This band of displacement corresponds with the course of aircraft that performed the measurement before the earthquake. The fact that airborne LiDAR were performed at different times with this vicinity as the boundary of two surveys may suggest this unique trend was caused by an error in the process of amalgamating two different sets of data taken at different times.

28.3.3 Verification of the Results of Displacement Analysis of the Earthquake Faults

Figure 28.4 is an example (the vertical displacement of Idosawa Fault) of the results of differential analysis in comparison with those of in situ survey. The direction and magnitude of displacement in vertical as well as horizontal component vary greatly by the measured point. The moving average of the 1,000 m sections along the line of the fault

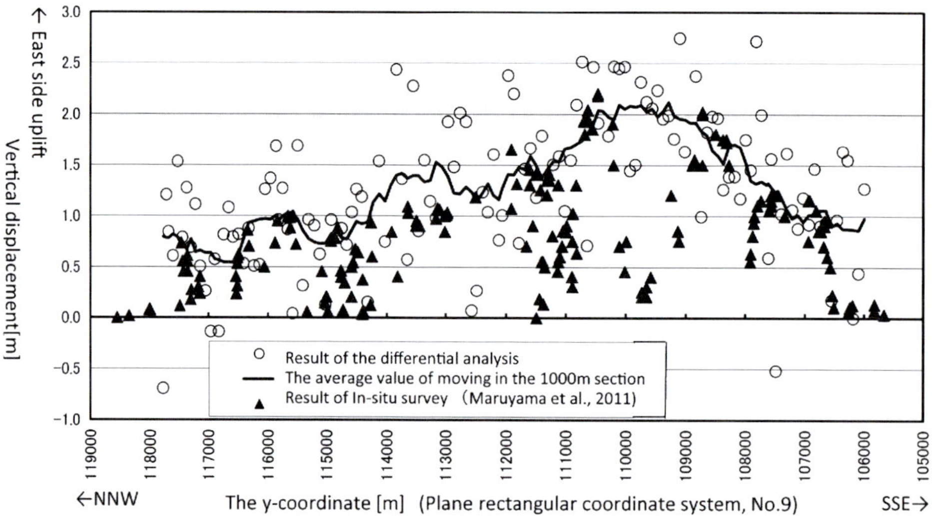

Fig. 28.4 The vertical displacement of the Itozawa Fault (Shinagawa et al. 2013)

generally encompasses the values of the larger side of in situ survey data. In in situ surveys the measured value of displacement at the fault is taken as "displacement". However, since displacement of land near the fault often accompanies flexural deformation of land, in situ data tend to be underestimated. Take this into account the moving average of the 1,000 m sections would produce correction effect. There are cases where the data of in situ and differential analysis differ greatly. There are three possible factors that cause such large differences; first, the deformation occurred which did not accompany a clear earthquake fault. Second, the earthquake motion has caused ground settlement hence subsidence around the fault. Third, a matching error may have been caused by an incorrect identification in similar topographical sites.

28.4 Conclusion

The method of DEM differential analysis employed to investigate the earthquake faults caused by the 2011 Fukushima Hamadori Earthquake clearly indicates ground surface displacement distribution around the earthquake faults. However we need to cautiously exclude extreme values in the data. The specific causes of such extreme values are unknown.

The ground surface displacement data produced by the DEM differential analysis overlaps with the upper side of in situ measurement data. One advantage of the differential analysis is that it produces information with respect to a ground settlement phenomenon which accompanies not only fault displacement, but tilt motion and earthquake motion over the area. This is not available through the conventional in situ survey.

References

Geospatial Information Authority of Japan (2011) http://www.gsi.go.jp/cais/topic110425-index-e.html

Maruyama T et al (2011) http://unit.aist.go.jp/actfault-eq/Tohoku/report/idosawa/idosawa.pdf

Shinagawa S et al (2013) J Jpn Soc Eng Geol 53:271–281

Toda S, Tsutsumi H (2013) Bull Seis Soc Am 103:1584–1602

Part IV

Applied Geology for Infrastructure Projects

Convener Eng Svetozar Milenkovic—*Co-convener* Zoran Berisavljevic

This session aims to enlighten and emphasize the importance of applied geology as the one of the key factors in rational designing and construction of large-scale infrastructure projects such as motorways and railways. Contributions should range over a wide spectrum of the applied geology involvement through development of site investigations and characterization, empirical and numerical modelling, presenting illustrative and innovative case studies of infrastructure projects, and to underline needs and benefits of cooperation between geologists and civil engineers, which seems to be the weakest link in the designing process. Whether it is needed to perform bearing capacity and settlement calculations, slope stability or some other analysis, particular emphasis is placed on determining appropriate soil and rock parameters as they are the basis of every geotechnical design. The experiences and lessons learnt from past projects are highly appreciated and warmly welcome in the session.

Field Monitoring of the Behavior of Pile-Net Composite Foundation in Oversize-Deep-Soft Soil

29

Yu-feng Wang, Qian-gong Cheng, and Jiu-jiang Wu

Abstract

For the observation of the behavior of the pile-net composite foundation in oversize-deep-soft soil, some cross sections in the Chaoshan railway station were selected for field monitoring. According to the recorded data, an analysis on the behavior of this structure was conducted with the following conclusions reached: (1) At the beginning of filling, the values of the earth pressure on pile tops and in the soil between piles both increased abruptly. And the growth rate of the earth pressure on pile tops is higher than that of soil between piles. When the depth of fill hits a certain value, the pile-soil stress ratio will reach its maximum value with an arching effect appeared. (2) As the increase of filling depth, pore water pressure increased with excess pore water pressure generated. After the filling, the distribution of excess pore water pressure disappeared gradually. (3) The axial force and the skin resistance of tube pile present a close relationship with the properties of each stratum. (4) The settlement value of each layer is also relevant to its depth and properties with its settlement ratio directly proportional to the increase of filling depth. (5) As the increase of filling depth, the tensile displacement of geogrid presents an increasing trend. The increase of the tensile displacement of geogrid in soil between piles presents a lag effect with its lag time about 50 days. (6) The generation of lateral displacement is accompanied with the filling of embankment. Nearer the surface, the value of lateral displacement was higher. And as the increase of depth, it decreased gradually. For different layers, the value of lateral displacement also varies.

Keywords

Oversize-deep-soft soil • Pile-net composite foundation • Field monitoring • Bearing capacity • Settlement characteristics

29.1 Introduction

Recent years, pile-net composite structure, as a new treatment technology, has been widely used in construction for its advantages. This technique emerged in the world since 1970s. However, the first theoretical study was just appeared in 1990,

which is the present of the empirical formula of pile-soil stress ratio proposed by Jones et al. After then, Low et al. studied the arching effect of soil between piles and the effect of geogrid in theory and experiment (Low et al. 1994). The real sign for the worldwide usage of this structure should be the appearance of British Standard 8006 presented by the U.K. in 1995, which fills up the blank in the design of pile-net composite foundation. So far, the research of pile-net composite structure has achieved fruitful results (Helwany et al. 2003; Graeme et al. 2005; Huang et al. 2005; Abusharar et al. 2009). But, there is still no uniform cognition for the complex of this structure, especially in the usage of soft soil subgrade (Bergado et al. 2000; Han and Gabr 2002; Chen et al. 2008; Kousik 2010; Borges and Oliveira 2011).

Y. Wang (✉) · Q. Cheng (✉) · J. Wu
Department of Geological Engineering, Southwest Jiaotong University, Chengdu 610031, Sichuan, China
e-mail: wangyufeng1987118@126.com

Q. Cheng
e-mail: chengqiangong@home.swjtu.edu.cn

Here, in order to learn the behavior of pile-net composite foundation in soft soil further, authors take the oversize-deep-soft soil (its area reaches $2.5 \times 10^5 \, \mathrm{m}^2$) in the Chaoshan railway station as a case study to observe the deformation of foundation strengthened with pile-net composite structure. During the research, the field monitoring of the surface settlement, lateral displacement, pile-soil stress ratio, pore water pressure, axial load and skin resistance of tube pile, and stretching of the geogrid was conducted. And then, based on the analysis of these data, the bearing capacity and settlement mechanism of the pile-net composite foundation in oversize-deep-soft soil is discussed to provide insight into fundamental understanding of such cases.

29.2 Experiment Setup

29.2.1 Geological Background

The studied area locates at the Chaoshan railway station in Chaoan County, Guangdong Province. In this area, the strata are mainly composed of muck, muddy silty clay, and silty sand deposited in the alluvial-lacustrine (Q_4^{al}) and marine-continental (Q_4^{mc}) environment of the Quaternary, with high water content, high liquid limit, high compressibility, high porosity, underconsolidated, low strength, etc. So, the bearing capacity of this area is very poor. The detailed geological profile of the testing section and the CPT data are illustrated in Fig. 29.1.

29.2.2 Instrument Setup

According to the field condition, the section located at the mileage of DK207+373 was chosen for field monitoring. In this section, the soft soil foundation was strengthened with

tube piles (type: PHC 500 A 100-12; diameter: 0.50 m; thickness: 0.1 m; concrete grade: C60), pile caps, ground beams (adopted C35 concrete and were cast-in-place), cushions, and geogrids, as shown in Fig. 29.2. After the construction process of tube piles and ground beams, a ballast mattress layer with both geogrid layers interbedded was paved on the top. And the total thickness of this layer is 0.6 m. The relevant design value of bidirectional tensile strength of geogrides is 80 kN/m.

The filling process of this section was conducted from March 29 to April 27, 2010 with its final depth of fill reached 2.2 m. And the installation of the monitoring instruments, including earth pressure cells, wire stress gauges, displacement transducers, settlement gauges, inclinometers, and pore pressure gauges, is finished during the filling process. Figures 29.3 and 29.4 show the layout of each instrument. The information of the instruments is listed in Table 29.1.

29.3 Results and Discussion

29.3.1 Pile-Soil Stress Ratio

According to the monitoring data, the variations of the average earth pressures on the pile top and in the soil between piles were obtained (Fig. 29.5) with their ratio shown in Fig. 29.6. The monitoring began on March 15, 2010, and ended on October 9, 2010, lasting more than 6 months. From Figs. 29.5 and 29.6, the following trends were observed:

1. The distributions of earth pressure and pile-soil stress ratio can be divided into three stages as the variation of filling height.

The first stage is the construction period from March 15, 2010 to April 26, 2010. In this process, the earth pressure on pile top presents a gently linear increasing trend with its

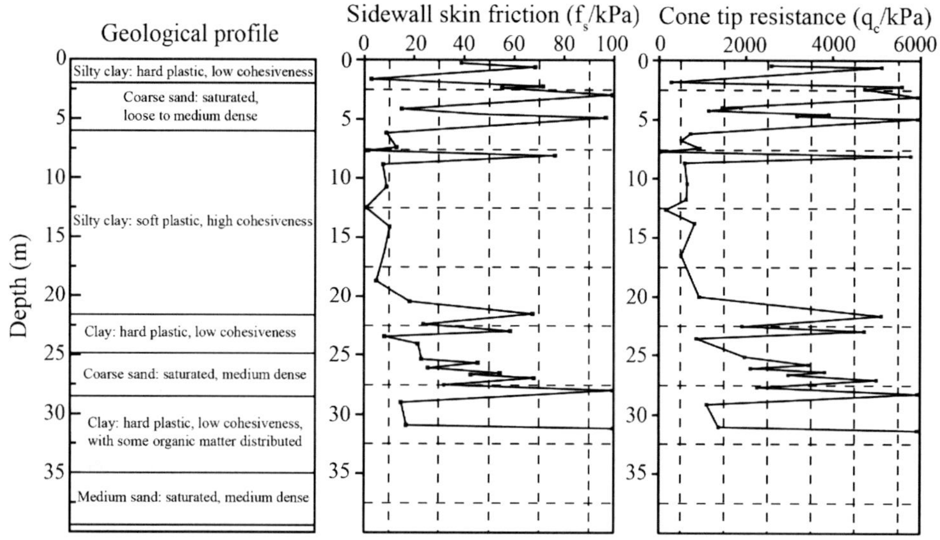

Fig. 29.1 Geological profile of the testing section and the CPT data

Fig. 29.2 Cross-sectional view of the strengthening scheme (unit: m)

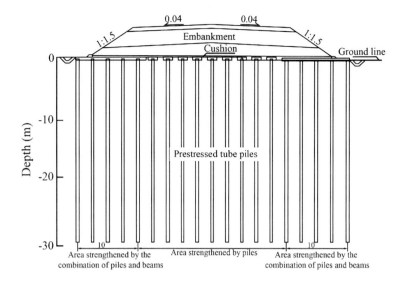

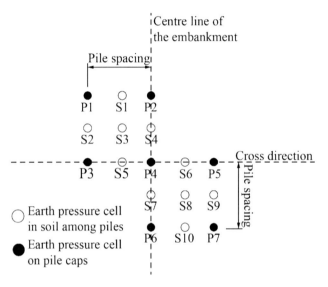

Fig. 29.3 Layout of the earth pressure cells

growth rate at 0.97 kPa/d during the first 15 days. From the 16th to the 42nd day, its growth rate increased to 1.75 kPa/d. For the earth pressure in soil, it also presents an increasing trend from the 1st to 37th days with its average growth rate at 0.45 kPa/d. However, from the 38th to the 42nd days, the earth pressure in soil presents a decreasing trend as the increase of filling height, which may be induced by the generation of soil arching effect. The pile-soil stress ratio displays an increasing trend with its changing rate reached 0.09/d.

The second stage is the stabilization phase in the post construction period, which lasts for 136 days. During this stage, the earth pressure on pile top continued to increase with its growth rate decreased to 0.197 kPa/d. The pressure acting on the soil between piles continued to decrease at a changing rate of 0.024 kPa/d. The pile-soil stress ratio still

shows an increasing trend with its increasing rate only being 0.017/d which is much smaller than that in the first stage.

The final stage lasted from July 19 to October 9, 2010, which is the unloading process of construction. As the excavation of the filled layer, the earth pressure and pile-soil stress ratio both show a downward trend. However, during the 194th to 203rd days, the pile-soil stress ratio increased slightly due to the continuous adjustment of the cushions on the load borne by piles and the soil between piles.

2. The earth pressure in soil between piles is much smaller than that on piles. After the filling, the pile-soil ratio was always larger than 1. At the end of the monitoring, it even reached 7.6, which indicates that soil between piles barely supports the load and the main load is borne by piles.

3. From Fig. 29.5, around on the 35th day of filling, the earth pressure in soil between piles reached its maximum value 0.1747 MPa, which indicates the generation of soil arching effect after this point. At the same time, the load borne by the soil between piles began to transfer to piles. The specific filling height is approximately 1.85 m, i.e., the height of the soil arch is 1.85 m, a little larger than the theoretical value.

29.3.2 Pore Water Pressure

The monitoring of pore water pressure is from April 10 to October 6, 2010. For comparison, data measured by PPG02–PPG05 were plotted in Fig. 29.7a, and that from PPG9 and PPG10 were plotted in Fig. 29.7b. Others were damaged during the construction without data obtained.

As shown in Fig. 29.7, the followings can be obtained:

1. During the first 9 days from March 29, 2010, there is no obvious increase for pore water pressure as the increase of filling height.

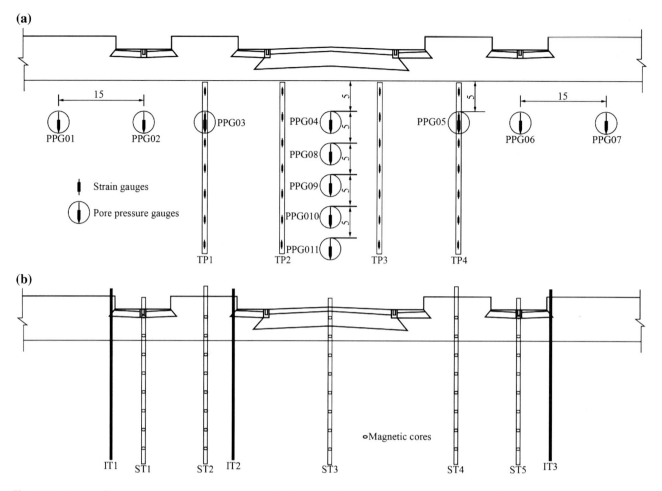

Fig. 29.4 Layout of the substructure instruments (unit: m). **a** Layout of the pore pressure gauges and the wire stress gauges, **b** Layout of the settlement tubes *ST* and the inclinometer tubes *IT*

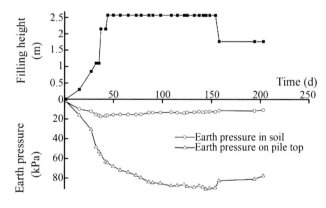

Fig. 29.5 Variations of the earth pressure

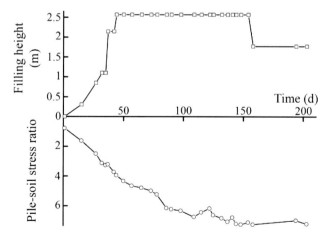

Fig. 29.6 Variations of the pile-soil stress ratio

2. The pore water pressure increases with excess pore water pressure generated during the filling stage and decreases with excess pore water pressure dissipated during the interruptions in construction. For example, during the period from the 5th to the 6th day, the pore water pressure measured by PPG04 increases from 0.0421 to

0.0502 MPa as the filling height increases from 0.85 to 1.1 m. However, the pore water pressure decreases from 0.0502 to 0.0463 MPa during the intermission from the 7th to the 9th day.

Fig. 29.7 Pore water pressure
with filling height and time

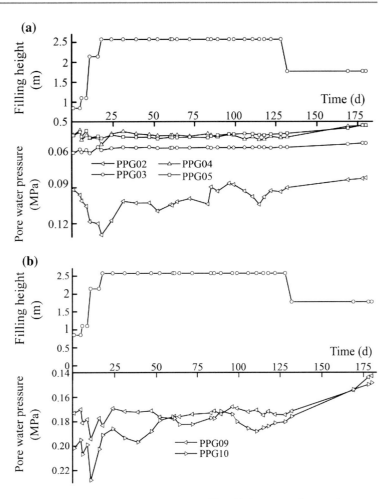

3. As the finish of filling, the value of pore water pressure goes into stable with slight fluctuations observed, such as, the value measured by PPG03 which fluctuated between 0.0563 and 0.0587 MPa. Besides the PPG02, the others also display similar trend. Different from others, the value measured by PPG02 presents a wide fluctuation due to the influence of construction machine with its value fluctuated from 0.09 to 0.13 MPa.

4. On the 132nd day, due to the need for construction, the filled layer was excavated 0.8 m depth. And during the excavation, the water pressure decreased correspondingly. For example, the value measured by PPG03 decreased from 0.0563 to 0.0531 MPa, and that of PPG04 reduced from 0.176 to 0.149 MPa. This indicates that the variation of pore water pressure is closely related to the overlying load.

29.3.3 Axial Force of Piles

During construction, some wire stress gauges installed on TP1 were destroyed. So, the data monitored by the rest on other piles (TP2, TP3, and TP4) were used for analysis as shown in Fig. 29.8. From Fig. 29.8, some observations are obtained as follows.

1. As the variation of depth, the axial force displays in nonlinear with its maximum value appeared at 21 m depth. That is to say, the axial force increases with depth at the range of 3–21 m, and decreases with depth at the range of 21–36 m.

2. The load transfer of the axial force is closely related to the properties of layers where the tube pile distributed. Take the TP4 as example, it can be observed that, at the depths of 3–6 and 21–36 m where the coarse sand, medium sand, and silt layers distributed, the axial force increases faster than that at the depth of 6–21 m where the clay and soft clay layers distributed.

3. With increasing time and overlying load, the axial force of piles all increased with different fluctuations. At the depths of 3–9 and 30–36 m, the growth rate of the axial force is slower than that at the middle part.

4. The distribution of the axial force on each pile reveals that the layers composed of clay and coarse sand shared most of the overlying load. And the axial force at the bottom of pile is small but not zero. All these indicate that the tube piles in this section appear to be an end-bearing friction pile.

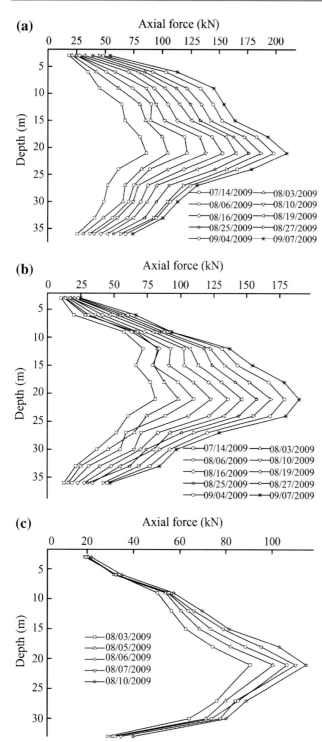

Fig. 29.8 Curves for the axial force as a function of depth. **a** Measured by TP2. **b** Measured by TP3. **c** Measured by TP4

29.3.4 Layered Settlement

The ST4 was destroyed during construction, so the data recorded by the magnetic cores installed on ST1, ST2, ST3, and ST5 were analyzed in this paper. The variations of

layered settlement with filling height and time are plotted in Fig. 29.9. From these figures, the following trends are obtained.

1. As the variations of time and filling height, the settlements at different depths all present obvious increases. And in the cross-sectional direction of the subgrade, the settlements also differ, i.e., the nearer the settlement tube away from the center line of the subgrade, the larger the settlement.

2. When the filling speed is relatively fast, the growth rates of the layered settlements at different depths also increase. Take the example of ST2, during the 77th–104th days, as the filling height of the monitoring section changed from 0.85 to 2.75 m rapid, the data measured by magnetic cores on ST2 present a relatively high rate of increase with a steep decrease in the settlement curves, as shown in Fig. 29.9b.

3. After the rapid increase of settlement in each layer due to the fast increase of filling height, the settlement then decreases at different rates during the following 10 days, which can observed in each subfigure. For example, on the 104th day of construction, the settlement measured by the magnetic core at the depth of 0.82 m on ST5 was 111 m, however, it fell to 101 mm on the 125th day with 10 mm decreased.

4. The settlements in the layers composed of soft soil are relatively larger than others. And after the filling, the layered settlements at different depths almost tend to steady with little fluctuations. During whole monitoring process, the growth rate of the settlement is almost directly proportional to the filling speed.

29.3.5 Behavior of Geogrid

During construction, a total of five displacement transducers were installed on the bottom of the first layer of geogrid in the testing section with their labels as FDT1, FDT2, FDT3, FDT4, and FDT5, respectively. Among them, FDT1, FDT3, and FDT5 were installed at the center of pile caps in plane, and the rest were at the center of the soil between piles. After installation, just FDT2, FDT3, and FDT5 worked normally with their data shown in Fig. 29.11. From Fig. 29.11, the followings can be observed (Fig. 29.10).

1. As the increase of filling height, the elongation of the geogrid also increased, especially for FDT2 located in the soil between piles. And the elongation of the geogrid overlays on the soil between piles is larger than that on the pile tops. When the monitoring is finished, the elongations measured by FDT3 and FDT5 on the pile tops are 0.36 and 0.53 mm, respectively; however, that measured by FDT2 overlays on the soil between piles reaches 1.17 mm, which is 2–3 times larger than the

Fig. 29.9 Settlement variations with filling height and time. **a** Measured by the magnetic cores on ST1. **b** Measured by the magnetic cores on ST2. **c** Measured by the magnetic cores on ST3. **d** Measured by the magnetic cores on ST5

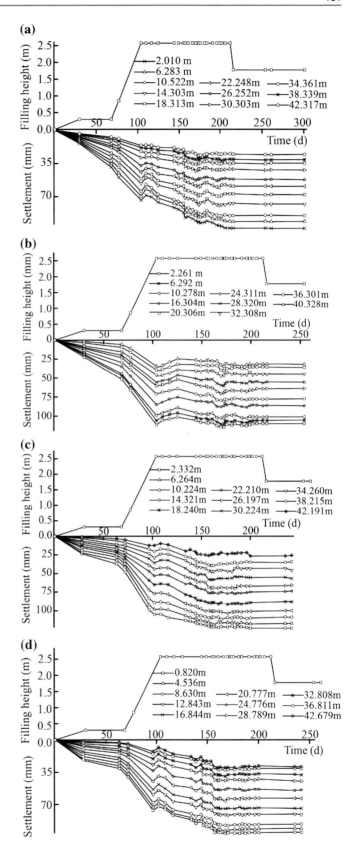

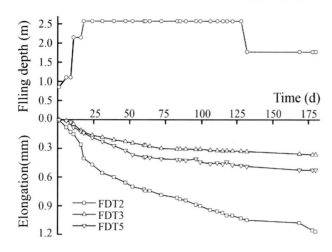

Fig. 29.10 Elongation variations with time and filling height

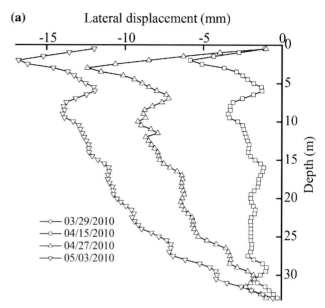

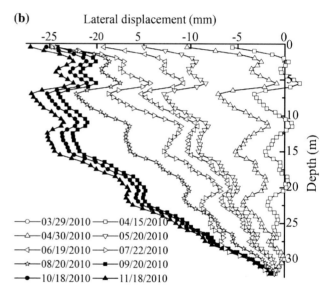

Fig. 29.11 Lateral displacement with time and depth. **a** Recorded by IT3. **b** Recorded by IT2

former. It can be inferred that the elongation rate and the tension of the geogrid overlaid on soil between piles are both larger than those on pile top, but the overall elongation and tension of the geogrid are not necessarily large.

2. The curve recorded by FDT2 reveals that the elongation variation of the geogrid overlaid on the soil between plies presents a lag effect as the increase of filling height. The delay time is about 50 days. In the first 11 days of filling, the average changing rate of elongation is 0.014 mm/d. However, in the following 8 days from the 16th to 24th of filling, the average changing rate of elongation increased to 0.026 mm/d. The reason for the display of this phenomenon is that, as the filling develops, the settlement difference between soil and pile increased gradually. So, the geogrid overlaid on soil between piles began to play its role with elongation increased.

29.3.6 Lateral Displacement

After the layout of inclinometer tubes, the IT3 was destroyed. And the IT1 was also abandoned some days later due to the need of construction. Hence, only part of the data recorded by IT1 and the whole recorded by IT2 were used for analysis with their variations shown in Figs. 29.11 and 29.12. From these figures, the following observations can be obtained.

1. The lateral displacement emerged as soon as the filling process began and shows a decreasing trend as the increase of depth. The lateral displacements at the bottom of both piles are both near zero and fluctuated slightly with time and over loading. The maximum values of both piles at different days all emerged at the depth of 0–10 m.

2. As the filling and the consolidation of soil proceeding, the lateral displacement of the embankment increases accordingly. In Fig. 29.12, it can be observed that, during

the filling, the growth rate of the lateral displacement is obviously higher than that during the post filling period. For example, during the fill period from March 29, to April 27, 2010, the maximum growth rate of the lateral displacement measured by IT2 reaches 0.247 mm/d. However, during the post filling period from April 27 to November 18, 2010, the maximum growth rate of the lateral displacement measured by IT2 is only 0.075 m.

3. The lateral displacement of the embankments under loading also varies in different soil layers. Generally, the lateral displacement of the embankment in soft soil layers, such as the silty clay layers or clay layers, is relatively larger than that in other layers. Furthermore, the

Fig. 29.12 Maximum lateral displacement with time and depth

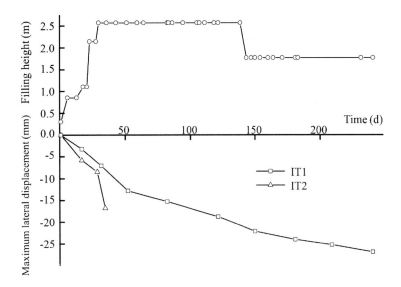

Table 29.1 Types and quantities of instruments in the testing section

Category	Earth pressure cells	Pore pressure gauges	Wire stress gauges	Magnetic cores	Displacement transducers	Inclinometer tubes
Amount	17	11	52	110	5	3
Embedded time	03/13/10	11/08/09–11/17/09	08/03/09	11/08/09–11/17/09	03/20/10	11/08/09–11/17/09
Measuring range	0.4 MPa	0.4 MPa	Tension range: 20 kN Pressure range: 10 kN	Accuracy: 1 mm Maximum: 50 m	30 mm	Accuracy: 0.1 mm Maximum: 40 m
Working principle	Vibrating string	Vibrating string	Vibrating string	Magnetic type	Vibrating string	Resistive
Locations	7 on pile caps, 10 in soil between piles	1 hole with 1 gauge	13 in each testing piles	22 in each settlement tube	2 on pile caps, 3 in soil between piles	3 in near the slope of the embankments

lateral displacement around the dividing line of both soil layers is also relatively larger.

4. The maximum values of the lateral displacements recorded by both inclinometer tubes are all smaller than 30 mm, which indicates that, the layout of the pile-net composite structure strengthens the stability of the subgrade effectively.

29.4   Conclusions

The objective of this paper is to gain insight into the behavior of pile-net composite foundation in oversize-deep-soft soil. From the study, the following conclusions can be drawn.

1. As the variation of filling height, the distributions of earth pressure and pile-soil stress ratio can be divided into

three stages. The earth pressure in soil between piles is much smaller than that on pile tops. After the filling, the pile-soil stress ratio is always larger than 1 with its maximum reached 7.6, which indicates that the tube piles play a good role with most of the load shared.

2. Due to most of the load is borne by tube piles, there is no obvious increase for pore water pressure as the increase of filling height. In the coarse sand layer and the medium sand layers, the release of the excess pore water pressure is faster than that in the silt layer and the silty soil layer.

3. As the variation of depth, the axial force displays in nonlinear with its maximum value appeared at 21 m depth. And the load transfer property of the axial force is closely related to the properties of layers where the tube pile distributed. At the depth where the coarse sand, medium sand, and silt layers distributed, the axial force

increases faster than that at the depth where the clay and soft clay layers distributed.

4. During whole monitoring process, the growth rate of the layered settlement is almost directly proportional to the filling speed. When the filling is rapid, the growth of settlement will increase fast with a decrease phase occurred after the fast filling. The variation of settlement is also related to depth and the properties of the soil layers. Besides, differential settlement also occurs at different locations along the cross-sectional direction of the subgrade.

5. The elongation of the geogrid increases with filling, and the elongation rate and the tension of the geogrid overlaid on the soil between piles are both larger than those on the pile tops. Besides, there is a lag effect on the elongation variation of the geogrid overlaid on the soil between plies as the increase of filling height. And the delay time is about 50 days.

6. The lateral displacement emerged as soon as the filling process began and shows a decreasing trend as the increase of depth. Similar to the display of axial force, the lateral displacement also varies in different soil layers.

Acknowledgements This research was supported by the National Natural Science Foundation of China (No. 41172260, 41372292, 51108393), the National Basic Research Program of China (973 Program) (No. 2008CB425801) and the Specialized Research Fund for the Doctoral Program of Higher Education (20110184110018).

References

Abusharar SW, Zheng J-J, Chen B-G et al (2009) A simplified method for analysis of a piled embankment reinforced with geosynthetics. Geotext Geomembr 27(1):39–52

Bergado DT, Teerawattanasuk C, Youwai S et al (2000) Finite element modeling of hexagonal wire reinforced embankment on soft clay. Can Geotech J 37(6):1209–1226

Borges JL, Oliveira MD (2011) Geosynthetic-reinforced and jet grout column-supported embankments on soft soils: numerical analysis and parametric study. Comput Geotech 38(7):883–896

British Standard 8006 (1995) Code of practice for strengthened/reinforced clays and other fills. British Standards Institute

Chen RP, Chen YM, Xu ZZ (2008) A theoretical solution for pile-supported embankments on soft soils under one-dimensional compression. Can Geotech J 45(5):611–623

Graeme D, Skinnera R, Kerry R (2005) Design and behaviour of a geosynthetic reinforced retaining wall and bridge abutment on a yielding foundation. Geotext Geomembr 23(3):234–260

Han J, Gabr MA (2002) Numerical analysis of geosynthetic-reinforced and pile-supported earth platforms over soft soil. J Geotech Geoenvironmental Eng 128(1):44–53

Helwany SMB, Wu JTH, Froessl B (2003) GRS bridge abutments—an effective means to alleviate bridge approach settlement. Geotext Geomembr 21(3):177–196

Huang J, Han J, Collin JG (2005) Geogrid-reinforced pile-supported railway embankments—A three-dimensional numerical analysis. Transp Res Rec 1936:211–229

Kousik D (2010) A mathematical model to study the soil arching effect in stone column-supported embankment resting on soft foundation soil. Appl Math Model 34(12):3871–3883

Low BK, Tang SK, Choa V (1994) Arching in piled embankment. J Geotech Eng 120(11):1917–1938

Deformation Behavior of Excavated High Loess Slope Reinforced with Soil Nails and Pre-reinforced-Stabilizing Piles

30

Qian-gong Cheng, Yu-feng Wang, and Jiu-jiang Wu

Abstract

For the research of the deformation behavior of excavated high loess slope reinforced with the combination system of soil nails and pre-reinforced-stabilizing piles. An open cut slope is chosen as an example for field monitoring with the following results obtained: (1) As the excavation depth increasing, the pre-reinforced-stabilizing piles all moved toward outside due to the existence of earth pressure behind piles. (2) As the increase of excavated depth, the value of earth pressure before piles presents a nonlinear variation with one maximum point appeared along depth. Differing from that before piles, the fluctuation of the earth pressure behind piles obviously increased with both maxima points distributed along depth. (3) Similar to the display of earth pressure, the distribution of the reinforcement stress along piles also fluctuated as the increase of excavation. (4) The axial nail load presents an obvious increase during excavation, which indicates that the layout of soil nails played an effective role for the stabilization of excavated slope. (5) Based on the distribution of axial nail load during excavation, it is inferred that the potential sliding surface of slope soil above piles displays in an arc-shape with its toe of slip surface located at the top of pile.

Keywords

High loess slope • Soil nails • Pre-reinforced-stabilizing piles • Field monitoring • Deformation

30.1 Introduction

The technique of soil nailing has been used worldwide in recent years for its simple usability and economic efficiency in stabilizing slope and supporting excavation (Schlosser 1991; Chen 2000; Li et al. 2008; Da Costa and Sagaseta 2010; Gong et al. 2011). Based on this technique, some other supporting measures, such as, composite soil-nailing walls, composite structure with piles and soil nails, composite structure with scattered row piles and soil nails, etc., have also been widely applied in construction (Yang et al. 2005; Liu et al. 2010; Zhang et al. 2011). However, the application of these composite structures in loess slope is still scare, especially in high loess slope (>30 m) (Hu et al. 2010).

Here, in order to learn the deformation behavior of excavated high loess slope reinforced with the combination system of soil nails and pre-reinforced-stabilizing piles further, an open cut slope, located at the entrance of Guanyintang tunnel in Shan County, is chosen as an example for field monitoring, which is meaningful in improving the design level of high loess slope along railway to guide practices in the future. During the research, the field monitoring of the lateral displacement, earth pressure, reinforcement stress, and axial nail load, etc., was conducted. And then, based on the analysis of these data, the supporting mechanism of the combination system of soil nails and pre-reinforced-stabilizing piles in high loess slope is discussed to provide insight into fundamental understanding of such cases.

Q.-g. Cheng (✉) · Y.-f. Wang · J.-j. Wu
Department of Geological Engineering, Southwest Jiaotong University, Chengdu, 610031 Sichuan, China
e-mail: chengqiangong@home.swjtu.edu.cn

30.2   Field Monitoring Setup

30.2.1   Excavation

The studied area locates at the entrance of Guan Yintang
Tunnel in Shan County, Henan Province, with its open cut
slope composed of loess. The excavation of the slope was
divided into seven stages. And the detailed information of
stages 1–6 is shown in Fig. 30.1. The last stage is the
refilling of soil after the finish of the open cut tunnel construction, which is not exhibited in this figure.

30.2.2   Instrument Setup

According to the field condition, the piles, numbered $8^{\#}$ and
$10^{\#}$, were chosen for monitoring. Along the vertical reinforcing steel bars distributed near the front and back sides of
both piles, four inclinometers were installed with their
numbers $8^{\#}A$, $8^{\#}B$, $10^{\#}A$, and $10^{\#}B$. In addition, 32 earth
pressure cells and wire stress gauges were also installed with
2 m spacing in vertical. Take the example of pile $8^{\#}$, the
detailed locations of these instruments are presented in
Fig. 30.2.

Fig. 30.1 Excavation profile

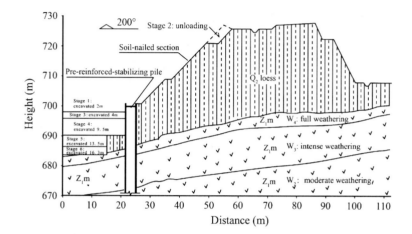

Fig. 30.2 Layout of monitoring
instruments along pile $8^{\#}$ (unit:
cm)

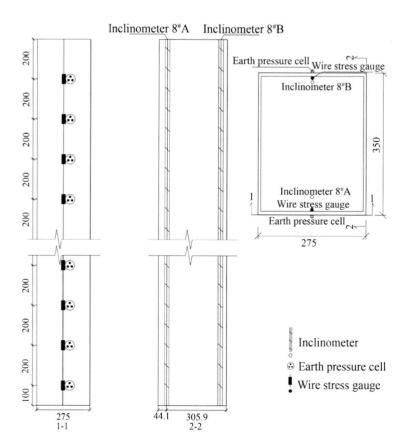

Fig. 30.3 Layout of soil nails used for monitoring

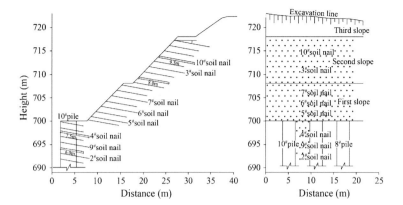

Besides the monitoring of the behavior of piles, some instruments were also installed along soil nails to record the interaction between soil and nails. Figure 30.3 shows the layout of the soil nails used for monitoring. All of the soil nails were installed at 10° from horizontal with different lengths. Along each of the nails, numbered $2^{\#}$, $3^{\#}$, $4^{\#}$, $5^{\#}$, $6^{\#}$, $7^{\#}$, $9^{\#}$, and $10^{\#}$, wire stress gauges were instrumented with 1.5 m spacing. The diameter of each nail is 25 mm. In Fig. 30.3, the $1^{\#}$ and $8^{\#}$ nails were not presented, which are 3 m long.

30.3   Results and Discussion

30.3.1   Deformation Behavior of Piles

30.3.1.1   Horizontal Displacements of Piles

The horizontal displacements of piles are recorded by inclinometers. Based on the readings, the deflection-depth curve of pile and the horizontal displacement-time curve of pile top were obtained, respectively. For the displays of these curves of piles $8^{\#}$ and $10^{\#}$ are similar, just the data of pile $8^{\#}$ are presented in this paper as shown in Figs. 30.4 and 30.5. Here, the negative value indicates the displacement is outward of the slope.

From Figs. 30.4 and 30.5, we may see the followings.

(1) During excavation, the pre-reinforced-stabilizing piles moved toward outside due to the existence of earth pressure behind piles with their maximum horizontal displacements appeared at the top surfaces. For the A-side of pile $8^{\#}$, the maximum horizontal displacement reaches 91.8 mm.

(2) Along piles, the shapes of the deflection-depth curve of the A-sides are parabolic type with regular variations.

(3) Before the excavation reaching 4 m, the variation of pile top horizontal displacement is great, which indicate that, at the beginning of excavation, the slope moved outward rapid. When the excavated depth reaches 9.5 m, the variation of pile top horizontal displacement becomes gently with the corresponding value of $8^{\#A}$ fluctuated between −86 and −92 mm. This indicates that, when the excavated depth reaches certain value, the earth pressure behind piles tends to steady with a gentle variation of pile top displacement occurred.

(4) During the unloading stage, the horizontal displacement of pile displays a decrease tendency, which exhibits that unloading can promote the stability of slope.

All these natures indicate that the horizontal displacement of pile can be divided into instantaneous displacement and secondary displacement. The former occurs in a short time after excavation or unloading. And then, the later occurs due to the variation of earth pressure induced by soil rheology.

30.3.1.2   Earth Pressure Distribution Along Piles

According to the data recorded by the earth pressure cells along pile $8^{\#}$, the variations of earth pressures with time and depth are showed in Figs. 30.6 and 30.7, from which we may see the followings.

(1) For the A-side of pile $8^{\#}$, the earth pressure increases with depth ranging from 0 to 16 m and turns into decrease when depth exceeds 16 m. That is to say, the maximum value of earth pressure distributed at the depth of 16 m with its value reached 388.27 kPa.

(2) The variation of earth pressure along the B-side of pile $8^{\#}$ also displays in nonlinear as the variation of depth with both maxima points distributed. The values at both maxima points are 235 and 334 kPa, respectively.

(3) As the variation of time and excavation depth, the earth pressure along the A-side of pile $8^{\#}$ decreases gradually. That is because, with excavation, the pile began to deform with the earth pressure gradually transformed into active soil pressure from earth pressure at rest.

(4) Along both sides of pile $8^{\#}$, the decrease of earth pressure during the stages 1 and 2 is obviously faster than that of other stages, which indicates that the earth pressure tends to stable with time.

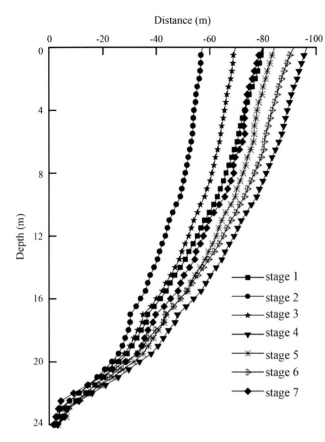

Fig. 30.4 Deflection of the pile

30.3.1.3 Reinforcement Stress Along Piles

For the monitoring of the internal stress of piles, some wire stress gauges were installed on the concrete reinforcing bars distributed at the both sides of piles 8[#] and 10[#]. Here, the data monitored by the wire stress gauges on pile 8[#] were used for analysis as shown in Figs. 30.8 and 30.9. From both figures, some observations are obtained as follows. Here, positive value indicates the tensile stress.

(1) In stages 1 and 2, the reinforcement stress along the A-side of pile 8[#] is negative indicating the existence of compressive stress. However, in the following stages, the value turns into positive gradually, which indicates the appearance of tensile stress. Differing from the A-side, the reinforcement stress along the B-side of pile 8[#] is always in negative.

(2) During the first stage, the variation of the reinforcement stresses along both sides of pile 8[#] is obviously greater than that in other stages, especially at the very beginning of stage 1. That is because, at the beginning of excavation, the concrete of piles was just poured several days with very low strength. The earth pressure loaded on the piles was mainly shared by the concrete reinforcing bars distributed in piles. So, the growth rate of reinforcement stress in the first stage is fast. In the following stages, the strength of the poured concrete of piles increased greatly with more earth pressure borne. Hence, the growth rate of reinforcement stress in the following stages reduced rapidly.

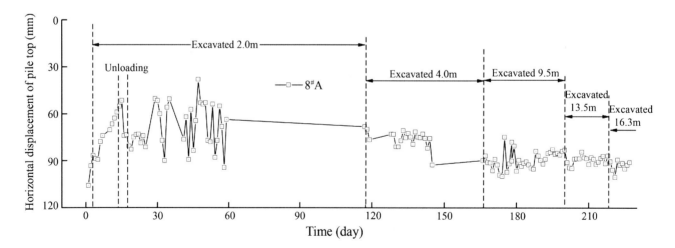

Fig. 30.5 Horizontal displacement-time curve of pile top with excavation and time

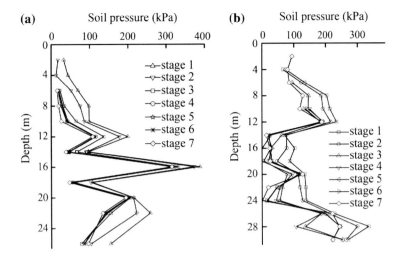

Fig. 30.6 Earth pressure variations with depth and excavation

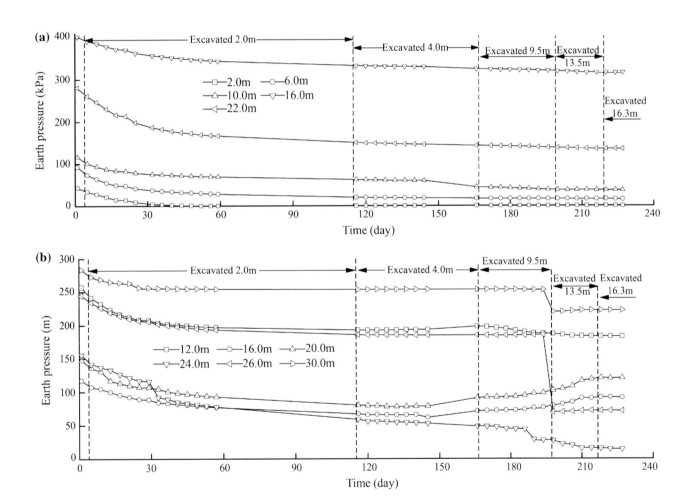

Fig. 30.7 Earth pressure variations with excavation and time

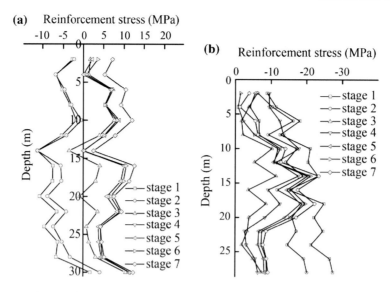

Fig. 30.8 Reinforcement stress variations with depth

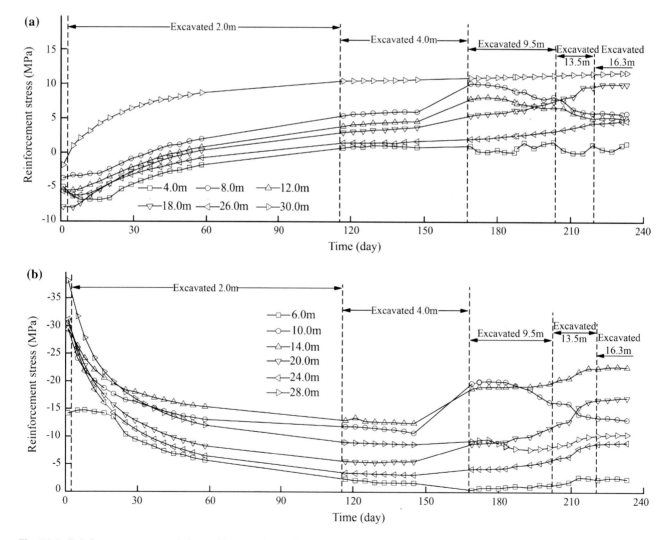

Fig. 30.9 Reinforcement stress variations with excavation and time

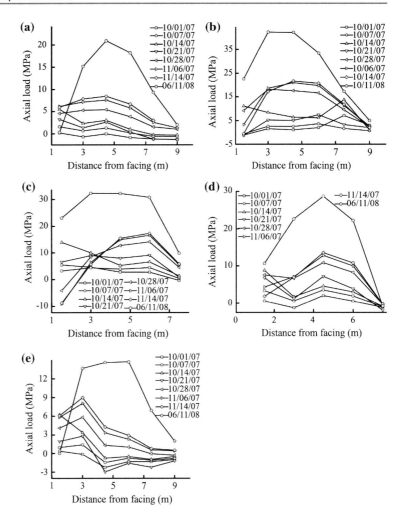

Fig. 30.10 Stress-time-nail length curves of soil nails installed in the soil above piles. **a** 3# soil nail. **b** 5# soil nail. **c** 6# soil nail. **d** 7# soil nail. **e** 10# soil nail

(3) For the variation of the reinforcement stress along the A-side of pile 8#, an obvious fluctuation can be observed during the stage 4 with following characteristics presented.

- At the depth of 0 to 4 m, the A-side of pile 8# is vacant, the value of reinforcement stress is nearly zero.

- At the depth of 6–12 m, the reinforcement stress firstly presents a rapid increase as excavation, and then decreases gradually as the adjustment of earth pressure loaded on piles.

- At the depth of 14–26 m, the reinforcement stress increases continuously as the increase of bending moment suffered by piles.

- At the depth of 28–31 m, the piles inserted into bedrock with minor displacement occurred. The reinforcement stress in this part tends to stable.

(4) For the variation of the reinforcement stress along the B-side of pile 8#, it also displays obvious fluctuation in the stage 4 with some characteristics similar to that of the A-side presented.

30.3.2 Behavior of Soil Nails

Besides the monitoring of pre-reinforced-stabilizing piles, the axial load in each instrumented soil nail was also obtained from wire stress gauge readings. Based on the monitored data, the behaviors of soil nails installed in different sections of the slope are obtained as shown in Figs. 30.10 and 30.11.

From these figures, it can be concluded as follows.

(1) For the nails installed in the soil above piles, the nearer the nails away from the toe of the slope, the larger the axial load. Hence, the maximum value among these soil nails appears on the 5# soil nail which reaches 42 MPa.

(2) At the very beginning of the installation of soil nails, the interaction between the cement paste around nails and soil is incomplete, so the axial loads along soil nails display irregular variations with their maximum distributed near the facing. As the improve of the interaction between nails and soil, the distribution of

Fig. 30.11 Stress-time-nail length curves of soil nails installed in the soil between piles. **a** 2$^{\#}$ soil nail. **b** 4$^{\#}$ soil nail. **c** 9$^{\#}$ soil nail

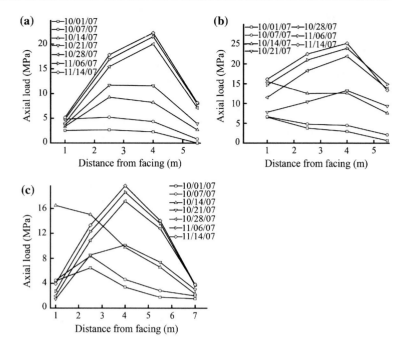

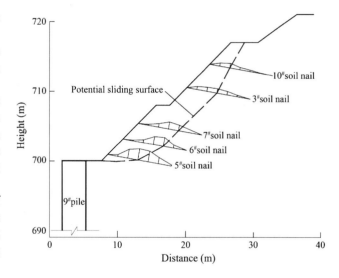

axial load along each nail shows in an parabola form with its maximum registered at the mid of nails.

(3) For the nails installed in the soil between piles, the positions of their maximum values are similar, which are distributed 4 m away from the facing.

(4) As the excavation work processed, the axial loads along nails increases continuously with its growth rate decreasing, which indicates that the soil nails play an effective role in the control of slope deformation. There is no obvious failure occurred in slope during excavation.

(5) Rainfall plays an important role in the deformation of slope, especially for the part near the facing. Due to the slope facing where the soil nails distributed was not reinforced with guniting, the axial loads near the facing increase obviously during rainfall.

Based on the display of the stress-time-nail length curves, the potential slip plane in slope above piles can be inferred as shown in Fig. 30.12.

Fig. 30.12 Location of the potential sliding surface in the slope above piles

30.4 Conclusions

The objective of this paper is to gain insight into the deformation behavior of excavated high loess slope reinforced with the combination system of soil nails and pre-reinforced-stabilizing piles. From the study, the following conclusions can be drawn.

(1) During excavation, the pre-reinforced-stabilizing piles all moved toward outside due to the existence of earth

pressure behind piles. The maximum horizontal displacement of the A-side of the pile 8$^{\#}$ reaches 91.8 mm.

(2) As the increase of excavated depth, the value of earth pressure before piles presents a nonlinear variation with its maximum appeared at the depth of 16 m. Differing from that before piles, the fluctuation of the earth pressure behind piles obviously increased with both maxima points distributed at the depths of 12 and 28 m, respectively.

(3) Similar to the display of earth pressure, the distribution of the reinforcement stress along piles also fluctuated as

the increase of excavation. During the first stage, the variation of the reinforcement stresses along both sides of pile $8^{\#}$ is obviously greater than that in other stages, especially at the very beginning of stage 1.

(4) According to the recorded data of axial nail load, it is revealed that the axial nail load presented an obvious increase during excavation, which indicates that the layout of soil nails played an effective role for the stabilization of excavated slope.

(5) Based on the distribution of axial nail load during excavation, it is inferred that the potential sliding surface of slope soil above pre-reinforced-stabilizing piles displays in an arc-shape with its toe of slip surface located at the top of pile. Instead, the potential sliding surface of soil between piles is present in an polyline type.

Acknowledgments This research was supported by the National Natural Science Foundation of China (No. 41172260, 41372292, 51108393), the National Basic Research Program of China (973 Program) (No. 2008CB425801) and the Specialized Research Fund for the Doctoral Program of Higher Education (20110184110018).

References

Bin L, Min Y, Zhiyin Y et al (2010) Design and practice of composite structure with scattered row piles and soil nailing for pit-protection. China Civil Eng J 106–114

Chen Z (2000) Soil nailing application in foundation engineering. China Building Industry Press, Beijing (in Chinese)

Da Costa A, Sagaseta C (2010) Analysis of shallow instabilities in soil slopes reinforced with nailed steel wire meshes. Eng Geol 113:53–61

Gong C, Cheng Q, Yang L, Yuan GB, Wang YF (2011) Case study of deformation behaviour of high loess slope in excavation. J China Railway Soc 32:120–124

Hu J, Xie Y, Wang W (2010) Study of stability characteristics of multi-stair high cut slope in loess highway. Chin J Rock Mech Eng 3093–3100

Li J, Tham LG, Junaideen SM et al (20080 Loose fill slope stabilization with soil nails: full-scale test. J Geotech Geoenvironmental Eng 134:277–288

Schlosser F (1991) Soil nailing recommendations. Presses de l´ENPC, French

Yang ZY, Zhang J,Wang K (2005) Development of composite soil-nailing walls. Chin J Geotech Eng 27:153–156

Zhang LL, Zhang QX, Ma QX (2011) Soil nail's axis force of pile-anchor composite soil-nailed wall. J Beijing Univ Technol 1338–1342

Effects of Alkali Silica/Aggregate Reaction on Concrete Structures in Bundelkhand Region, Central India

Suresh Chandra Bhatt and Bhuwan Chandra Joshi

Abstract

The Bundelkhand massif covering 26,000 km^2 area forms the northern segment of Indian shield. It mainly consists of Bundelkhand gneissic complex (3.1 Ga) and granitic complex (2.5 Ga). These gneisses and granitoids are mainly traversed by E–W and NE–SW trending shear zones. Betwa a tributary of Yamuna is marked by number of concrete dams which are major resources of water in this region. The rock fragments of granites, gneisses, migmatites, basic, ultrabasics and metasedimentaries have been used as coarse aggregate for construction of dams. The main constituents of these rocks are strained quartz, altered feldspar and elastic (mica) minerals. The progressive development of reaction rims of alkali-aggregate and silica reactions were observed in three concrete dams; Sukuwa Dukuwa, Kamala Sagar and Saprar dams. The petrological studies of coarse aggregates (deformed gneisses, schist and granites) reveal that the higher percentage of strained quartz (50–60 %), altered feldspar (25–35 %) and elastic (micaceous) minerals with clayey matrix were found most deleterious reactive agents to cause alkali aggregate/silica reactions. It also indicates that the low-alkali cement and supplementary cementing materials have not been used with reactive aggregates to prevent AAR and ASR reactions.

Keywords

AAR/ASR • Dams • Bundelkhand • India

31.1 Introduction

Numerous concrete structures such as dams and bridges have been suffering from alkali silica (ASR) and alkali aggregate reactions (AAR) since long time. The ASR and AAR has been considered deleterious reactions between alkalis in cement and certain variety of silica and altered feldspar present in aggregate. The ASR and AAR reactions commonly produce expansion by exerting pressure on surrounding matrix and cause extensive distress and cracking in concrete structures (Pan et al. 2012). Normally the higher the alkali content of the cement and the higher the cement content of the concrete, the greater the rate of expansion and cracking. The cracking of the concrete creates channels for water movement in the concrete, which may lead to an increase in saturation. It is universally accepted that the strength of concrete will directly depend on strength and engineering properties of aggregates (Popovics 1979). The low degree of weathering and good strength would be strong quality of an aggregate to resist abrasion and degrading in mixture of Portland cement, which effectively withstand the forces applied by traffic, reservoir and running water during service life of pavements, dams, bridges and other concrete structures (Balbaki et al. 1992; PCCM Manual 1989). Since the identification of AAR (1940), the petrological and mineralogical studies of AAR susceptible aggregates along with their chemistry and laboratory tests to predict swelling and cracking in concrete, have been considered important tools to cope up with this problem.

S.C. Bhatt (✉) · B.C. Joshi
Department of Geology, Institute of Earth Sciences, Bundelkhand University, Jhansi, India
e-mail: geoscb@yahoo.com

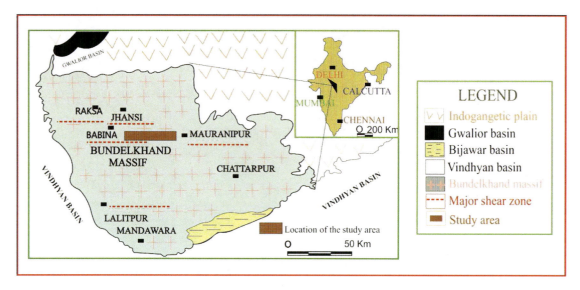

Fig. 31.1 Geological map of Bundelkhand region showing location of the study area

Betwa a major tributary of Yamuna is marked by number of concrete dams in Bundelkhand region of Central India (Figs. 31.1 and 31.2). In this paper the observations on petrographic examinations of aggregate and progressive development of reaction rims have been observed in Sukuwa Dukuwa, Kamala Sagar and Saprar dams (Fig. 31.3). The Saprar, Jamini and Dhasan are major tributaries of Betwa, across which these dams have been built up (Fig. 31.2). Rajghat, Matatila, Sukuwa-Dukuwa, Paricha, Kamla sagar etc. are other major dams in this terrain. Most of them are constructed on the river Betwa. Present paper mainly focuses on the identification of optical and mineralogical characters of reactive strained quartz, altered feldspar and other reactive micaceous minerals present in most of coarse aggregates used in these concrete structures.

31.2 Geomorphological and Geological Set Up

The study area occupying the northern territory of Bundelkhand massif is represented by rugged and undulating topography (Fig. 31.2). The low to moderate hills ranging 200–300 m height mainly constitute different types of granites and linear quartz reefs. The rivers of the area characterized by meandering (curved) and straight courses are assumed to be flowing through fractures/faults (Fig. 31.2). These rivers originating from high ranges of Vindhyans, are flowing SW–NE in this granitic country.

Geologically three major groups of rocks are recognised in the study area (Fig. 31.1). The TTG gneisses (3.3 Ga), gneisses (3.2 Ga) with inclusions of amphibolites and migmatites and mylonitised gneisses constitute gneissic

complex. The metasedimentary and metavolcanics (2.56 Ga) forms the metasedimentary complex while the porphyritic hornblende and grey granite (2.5 Ga), pink granite, mylonitised granite, quartz reefs and mafic dykes belong to the granitic complex (Basu 1986, 2007; Roday et al. 1995; Bhatt and Hussain 2008; Bhatt and Gupta 2009; Bhatt and Mahmood 2012). The E–W to ENE–WSW and NE–SW trending major brittle ductile crustal shear zones are also traced in the investigated area (Fig. 31.1).

The banded light to dark grey and fine to medium grained multiply folded (F_1–F_3) mafic gneisses are exposed in and surroundings of Ghisauli, Badera, Jaunpur and Chaurara villages located in the south of Babina town (Fig. 31.1). Thes rocks (3297+83 Ma; Mondal et al. 2002) associated with mica schists, migmatites, quartzites, amphibolites, ultramafics and mafics are also exposed near Kuraicha and Baragaon to the south of Mauranipur (Fig. 31.1). They also show sheared contact with pink granites in southeast of Kuraicha and south of Chituad in Mauranipur sector. The northern margins of the massif, compact granitoids are demarcated by the pink granitic gneisses. The fine grained, greyish to dark green mafic and ultramafic lensoidal bodies associated to gneissic complex are also traced (3249+ Ma). The fine grained and sheared fuchsite quartzite occurring as isolated patches within the older mafic gneissic terrain, are noticed near Sukwan Dukwan reservoir in the south of Babina (Fig. 31.1).

Based on field setting and petrogical characters five types of granites are reported in Bundelkhand region. (1) The greyish green hornblende granite and (2) light to dark grey coarse grained leucogranite constituting large porphyries of quartz and feldspar (mainly orthoclase) are occurring in the form of small hillocks and in lensoidal bodies of this massif

Fig. 31.2 Physical relief map and location of dams

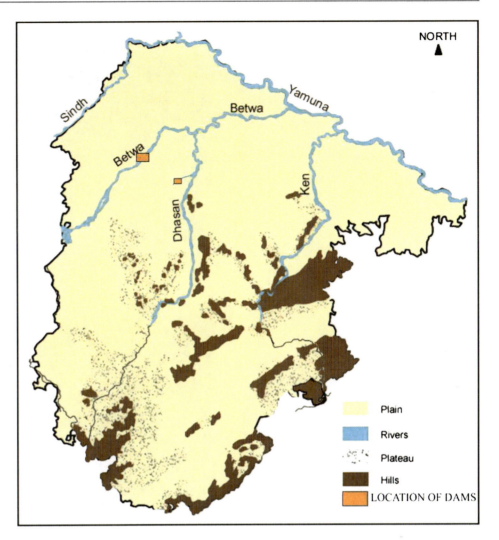

(Fig. 31.1). (3) The large tracts of massive, compact, pinkish to dark red and coarse grained pink granites are widely distributed in the northern and southern parts of the massif (Fig. 31.1). (4)The huge outcrops of fine to medium grained, light to dark pink granitoids and (5) fine to medium grained grey granites are appeared as dome shaped hillocks in this massif (Fig. 31.1). The sheared contact between gneisses and granites is also marked in the south of Mauranipur (Fig. 31.1).

The major quartz reefs exhumed parallel to NE–SW trending shear zones are recorded as spectacular tectonic features. These reefs are offsetting older gneissic complex, iron formation and granitites at several places. The white to milky white recrystallised quartz are found as main constituents of these reefs. The dark grey, fine to medium grained and NW–SE trending dolerite dykes (Rao 2000–2150 Ma) are considered last magmatic phase in this massif.

Petrographic and Microstructural study of coarse Aggregates: The petrographic analysis rocks used as coarse aggregates in concrete were carried out under polarising microscope and discussed below.

Gneisses: The fine to medium grained mafic gneisses constitutes subhedral to anhedral phenocrysts of quartz, feldspar (mainly orthoclase and plagioclase) and muscovite with accessories of biotite and chlorite (Fig. 31.4b). The earlier formed host quartz grains are tabular to elongate and about 50–60 % are strained. The foliation (gneissosity) is defined by preferred orientation of elongated and ribbon quartz and feldspar grains along with flakes of muscovite and biotite. The quartz grains exhibit strong undulose extinction, deformation lamellae and fractures (Fig. 31.4b, d, e). The eqigranular recrystallised quartz grains formed due to granular effects are found on the margins of host grains. The medium grained and cleaved orthoclase are sometimes seems to altered microcline. The plagioclases are twinned and exhibit kinking effects. At places brown flakes of biotite appeared as altered green chlorite. Magnetite occurs as inclusions.

The streaky and granite gneisses are sheared and predominantly constitute rotated phenocrysts of quartz and k-feldspar (Fig. 31.4d). The asymmetrical and rotated porphyroclasts of quartz and feldspar are dominantly displaying sinistral sense of shear movement.

Fig. 31.3 a Kamla Sagar Dam
showing effects of alkali silicate
and alkali aggregate reactions.
b Saprar Dam showing signatures
of alkali silicate and alkali
aggregate reactions. **c** Sukuwa-
Dukuwa Dam showing signatures
of alkali silicate and alkali
aggregate reactions

Mylonitised Gneisses: Three types of mylonite (proto-mylonite, mylonite and ultramylonite) zones have been distinctly identified. The protomylonite mainly constituting large phenocrysts (0.2–04 mm) of quartz and K-feldspar (50–90 %), are characterized by strong undulose extinction and deformational lamellae (Fig. 31.4d). The recrystallised quartz grains are widely occurring in margins of host quartz. The progressive growth of mylonitic foliation is represented by preferrely oriented fabrics of elongated and ribbon of quartz (strained) and feldspar grains (Fig. 31.4e). Due to excessive reduction in grain size and dynamic recrystallisation, the proto, blasto and S–C mylonites were progressively

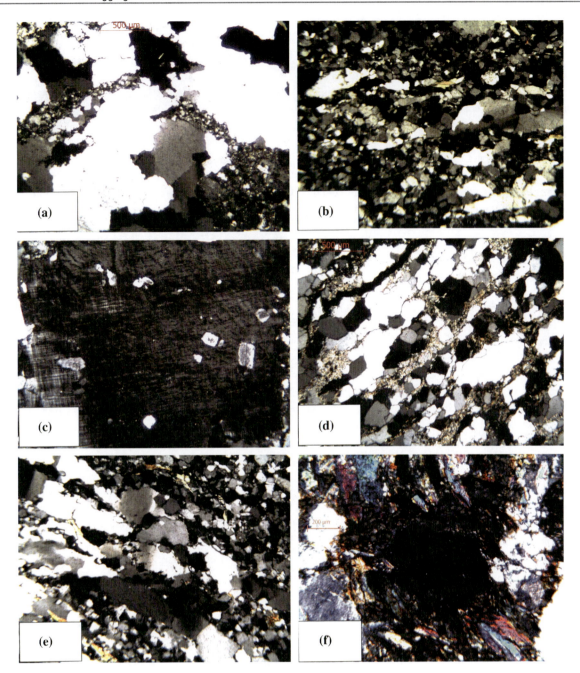

Fig. 31.4 **a** Photomicograph showing polygonal host quartz grains with rare recrystallised (strained) quartz grains in grey granite. **b** Photomicrograph showing elongated strained quartz grains displaying unulose extinction and recrystallisation in gneisses. **c** Orthoclase showing alteration effects with inclusion of recrystallised quartz in granite gneiss. **d** Mylonitic foliation represented by preferred orientation of elongated quartz and feldspar grains in mylonitised gneisses. **e** Photomicrograph showing hornblende, pyroxenes and flakes of flaky minerals. **f** Photomicrograph showing S–C planes in sheared gneisses.

evolved (Fig. 31.4d, e). The alteration effects (microclinisation) and kink bands are also seen in few phenocrysts of orthoclase (Fig. 31.4c) and plagioclase respectively.

Schist: The interclations of mica-schist are also noticed within these mafic gneisses near Kamla Sagar dam. The schistosity is represented by alternating quartz and micaceous bands. Quartz, feldspar, muscovite found as important minerals. Few shreds of mica, zircon, biotite and magnetite are also reported as accessories.

Mafic/Ultramafics: The peridotite and pyroxenite mainly constitute olivine, plagioclase, pyroxenes, amphiboles (mainly hornblende), talc and chlorite as main mineral constituents. The colourless faint green and anhedral grains of olivine contain inclusion of iron. Orthopyroxenes dominantly consisting of enstatite, occasionally enclose olivine to form poikilitic texture. Among amphiboles green to greenish brown hornblende are dominantly found whereas tremolite and chromite are rarely occurred.

Migmatites: The grey to dark green well folded lensoidal bodies of migmatites are found in the downstream of Kamla Sagar dam. The leucocratic (quartz and feldspar) and dark bands (mafic minerals) are characteristic features of these rocks. Quartz, feldspar (orthoclase, microcline and plagioclase) biotite, muscovite, chlorite are observed as chief mineral constituents.

Quartzite: Few outcrops of quartzites are occurred as small lenses within the sheared gneissic complex to the north of Mauranipur. Quartz, sericite, chlorite, rutile and tourmaline were examined as main minerals constituents. The medium to coarse grained quartz grains (angular to subhedral) appeared as lenses. The small streaks of biotite with rare plagioclase and magnetite occur in quartz laminae. The fine grained fuchsite quartzite mainly constitutes alternating bands of quartz (0.5–1 cm) and fuchsite (1.5–3 cm). The fuchsite mainly appears as dark green shreds with bands of quartz.

Grey Granite: Quartz, K-feldspar, plagioclase and mica are found as main mineral constituents. Plagioclase appeared with altered core, is followed by outward and a relatively free zone. The tabular to polygonal quartz grains (subhedral to euhedral) constitute 20–30 % of total composition of rock. The secondary recrystallised quartz grains found in the margins of host grains. Anhedral to euhedral phenocrysts of orthoclase constitute 40–50 % of total composition. Microcline showing perthetic growth and perfect cleavage is also observed in few thin sections. Magnetite and recrystallised quartz grains are seen as inclusions. Brown to greyish brown laths of biotite are also seen.

Pink Granite: Pink granite displaying porphyritic texture contains large phenocrysts of quartz, orthoclase, microcline and plagioclase. Magnetite, zircon, chlorite and rutile occurred as accessories. Subhedral to euhedral quartz grains embedded within recrystallised matrix and plagioclase showing twining and perfect cleavage are also observed. The medium to coarse grained and well cleaved orthoclase forms 70–75 % of total rock composition. The perthetic growth in microcline with inclusion of quartz is also noticed. Magnetite and recrystallised quartz occur as inclusions. Biotite is found in small shreds.

31.3 Discussion

The petrographic observations done on coarse aggregates reveal that the mylonitised (sheared) gneisses, granites, quartzite and schist containing higher percentage of strain

quartz (60–80 %), altered feldspar (30–35 %) and micaceous minerals may be considered more susceptible to alkali silicate and alkali aggregate reactions. It was also observed that the size of quartz may have adverse influence on deterioration of concrete. However, the average grain size do not show any discernible effect on expansion of concrete. It is also inferred that the microstructural features like cleavage, fractures, kinks, deformation lamellae, undulose extinction, dynamic recrystallisastion and twining examined in quartz, feldspar and micaceous minerals may have serious effects on strength of rock aggregate and corresponding concrete. Apart from this, the crystalline forms of silica (chert, chalcedony and cryptocrystalline quartz) found in various rock aggregates may also be responsible to cause alkali silica or alkali silicate reactions. The feldspar, mica and clayey minerals may cause kaolisation, oxidation and other carbonic reactions and eventually accelerate the alkali silicate reaction. Therefore, the shape, size and microstructural characters of rock aggregate along with the percentage alkali present in cement paste may be considered important parameters for determination of strength of concrete.

The most of the dams in the study area particularly Sukuwa Dukuwa and Saprar dams are more than 50 years old. The reaction rims and minor cracks observed in these concrete dams indicate that the ASR and AAR reactions have become more progressive since last few years and may further cause extensive cracks and great damage to these structures. Since, there are no proper preventive measures to stop deleterious ASR and AAR reactions in these dams. Therefore, it is suggested that such dams may be abandoned. It is also advisable that the low alkali content cement should be used and potentially reactive aggregates have to be identified before construction.

References

Balbaki W, Balbaki M, Benmokrane B, Aitcin PC (1992) Influence of specimen size on compressive strength and elastic of high performance concrete. American Society of Testing Materials, p 113–118

Basu AK (1986) Geology of parts of the Bundelkhand granite massif. Central India: Rec Geol Survey India 17(2):61–124

Basu AK (2007) Role of the Bundelkhand Granite Massif and the Son Narmada mega fault in Precambrian crustal evolution and tectonism in Central and Western India. J Geol Soc India 70:745–770

Bhatt SC, Hussain A (2008) Structural history and fold analysis of Basement Rocks around Kuraicha and adjoining area Bundelkhand massif, Central India. Geol Soc India 72:331–347

Bhatt SC, Mahmood K (2012) Deformation pattern and microstructural analysis of sheared gneissic complex and mylonitic metavolcanics of Babina-Prithipur Sector, Bundelkhand Massif, Central India. Indian J Geosci 66(1):79–90

Bhatt SC, Gupta MK (2009) Tectonic significance of shear indicators in the evolution of Dinara Garhmau shear zone, Bundelkhand Massif, Central India. In: Kumar S (ed) Magmatism tectonism and mineralization. Macmillan Publishers India Ltd., New Delhi, pp 122–132

Mondal MEA, Goswami JN, Deomurari MP, Sharma KK (2002) Ion microprobe Pb^{207}/Pb^{206} age of Zircon from the Bundelkhand massif, northern India for crustal evolution of Bundelkhand Aravalli protocontinent. Precambr Res 117:85–100

Pan JW, Feng YT, Wang JT, Sun QC, Zang CH, Owen DRJ (2012) Modelling of alkali-silica reaction in concrete: a review. Front Struct Civ Eng 6(1):1–18

Popovics S (1979) Concrete-making materials. McGraw Hill Book Company, New York, p 370

Rao JM, Rao GVSP, Widdowson M, Kelly SP (2005) Evolution of proterozoic mafic dyke swarms of the Bundelkhand Granite Massif, Central India. Curr Sci 88(3)

Roday PP, Diwan P, Singh S (1995) A kinematic model of emplacement of quartz reef and Indian subsequent deformation patterns in central Bundelkhand batholith. Proc Ind Acad Sci (Earth-Planet Sci) 104(3):465–488

Applied Engineering Geology Methods for Exemplar Infrastructure Projects in Malopolskie and Podkarpackie Provinces

Zbigniew Bednarczyk and Adam Szynkiewicz

Abstract

More than 90 % of landslides in Poland in number of 35,000 is located in the Malopolskie and Podkarpackie provinces. They are causing serious economic loses every year. In 2010, these reached 2.9 bln EUR during the flood in southern Poland. The research was connected with EU financed projects for road reconstructions. The mass movements size of 0.4–2.2 mln m^3 had low rates of displacement from few mm to over 5 cm a year. Colluviums built from saturated flysch soil–rock mixtures required usage of specific multidisciplinary engineering geology methods. Ground conditions were difficult for in situ and laboratory test. The investigation methods included high quality core sampling, GPR scanning, laboratory index, IL oedometer, triaxial and direct shear tests. Mapping by GPS-RTK method was employed for actualization of landslide morphology. Inclinometer and piezometer monitoring measurements were performed by the period of over 6 years. The new real-time early warning system the first of its kind in Poland was installed in 2010 in Beskid Niski Mountains. Obtained data provided parameters for LEM and FEM slope stability analysis. It allowed control of landslide behavior before, during and after stabilization works. The research proved that chosen investigation methods helped in remediation works. The results of the study reveal that large flysch landslides were difficult for counteraction and remediation methods should be considered very carefully. Comprehensive monitoring and modelling before the counteraction stage could lead to a better recognition of landslide remediation possibilities and early warning.

Keywords

Engineering geology site investigations • Landslide monitoring

32.1 Localization and Landslide Geology

The research was performed on the area of southern Poland, where damages to infrastructure, transportation networks or private properties caused by the landslides are the most common in the country. Investigations were conducted on 23 landslides with serious threats to the public roads or other

Z. Bednarczyk (✉)
Poltegor-Institute, Parkowa 25, 51-616, Wrocław, Poland
e-mail: zbigniew.bednarczyk@igo.wroc.pl

A. Szynkiewicz
Kart-Geo, Bacciarellego 39/1, 51-649, Wrocław, Poland
e-mail: adam.szynkiewicz@gmail.com

important infrastructure. Mass movements localization and instrumentation in Beskid Niski is presented on Fig. 32.1. Landslides were formed in marine flysch deposits folded during Alpine Orogenesis. Intensive erosion in river valleys and high groundwater level, during the Holocene era, characterized by thick weathering zones activated huge numbers of landslides (Gil et al. 1974; Raczkowski and Mrozek 2002). The main reason for landslides occurrence were high slope inclination combined with flysch type geology. Heterogeneous mixture of soils and rocks were involved in creep processes. Mass movements were reactivated in wet periods many times. Saturated claystones had mechanical parameters as weak cohesive soils. Sandstones usually occurred as thin layers with different degree of digenesis. It allowed water infiltration together with seepage due to many crack and joints. Movements were usually localized on slopes of inclination varied from 15° to 35° with shallow groundwater levels from 0.5 to 2.0 m. Colluviums depths were changing from few to dozens of meters. Intensive rainfalls together with floods, erosion in river valleys, snow melting and variations of the pore pressure inside soil layers were enhancing the sliding activity (Bednarczyk 2008, 2010, 2012). The problem has intensified in May/June 2010, after intensive rainfalls and flood in southern Poland (Fig. 32.2).

32.2 Field Investigations

Landslides built from rock-soil flysch type deposits required site-specific investigations. These needs site inspection, engineering geology investigations, actual water balance as well as measurements of various specific trigger parameters. Presented investigations included over 600 m of core drillings diameter of 132 mm, depths of 9–30 m together with NNS sampling. Ground Penetration Radar scanning length of 10 km were carefully scaled by boreholes. GPS-RTK method supplemented by conventional geodesy methods was found an effective way of landslide mapping. Over 500 m of inclinometer casings and piezometers were installed for slopes instrumentation. Landslides sizes, depths, failure mechanisms, lithology and engineering geology parameters and groundwater conditions were recognized this way. Implementation of different types of ground movement, groundwater level and pore pressure monitoring allowed

identification of landslide triggers. A new on-line instrumentation was built at Szymbark landslides. Obtained data was applied for modeling of slope stability and design of landslide stabilization works (Fig. 32.3).

32.3 Laboratory Tests

Laboratory tests included index tests (grain size, moisture content, Attenberg limits, bulk density, density of soil particles, content of organic/bitum. material), direct shear, oedometer and triaxial CID tests. Soils represented silty loams, silty clays to claystones, had high moisture content of 18–37 %, liquidity index of 0.1–0.8, eff. cohesion of 6.5–10 kPa, eff. angle of shearing resistance of 11–15° and a very high compressibility. The highest moisture content of ≈30 % and plasticity index of 30–40 % were observed in samples taken near the sliding surface (Fig. 32.4). The slip surface depths of 2–16 m were in good agreement with index test results.

32.4 Landslide Monitoring

Identification of landslide triggers were performed using standard and real-time monitoring methods. Validation of the monitoring results for 23 sites in the network was possible due to continuous site inspection program. Groundwater table depths and pore pressures were measured in

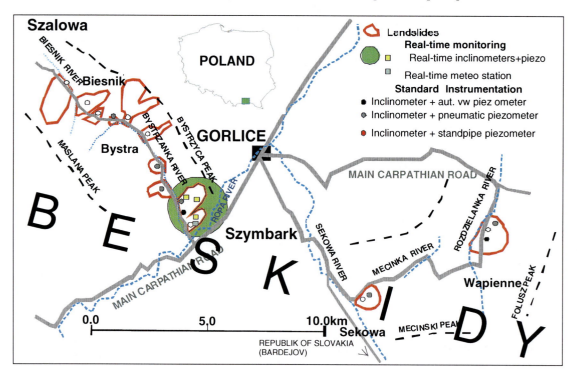

Fig. 32.1 Landslide localization

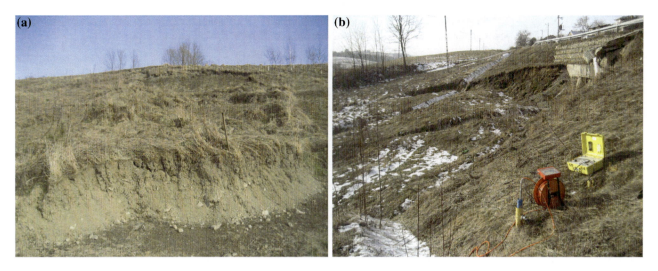

Fig. 32.2 Landslide activation. **a** Szymbark. **b** Strzeszyn—monitoring measurements

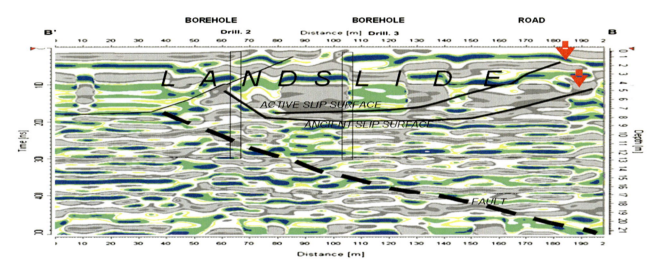

Fig. 32.3 GPR profile Sekowa landslide

Fig. 32.4 Index laboratory tests, Szymbark landslide

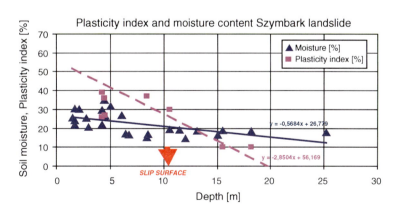

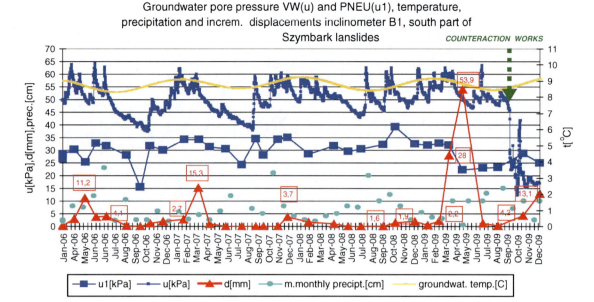

Fig. 32.5 Monitoring measurements Szymbark landslide

order to estimate effective stresses using standpipe, pneumatic and automatic vibrating wire piezometers. Subsequent monitoring surveys performed by the period of over 6 years, every 30–45 days determined the magnitudes, depths, directions and rates of displacements. The displacements varied from several up to 400 mm (after damage of inclinom. casing).

The movements usually occurred after the pore pressure reached 50–100 kPa on the slip surface. Comparison of displacements, pore pressure, groundwater level depth, temperature and precipitation data shows, that the largest displacements at Szymbark landslides occurred when the pore pressure decreased after high value periods in May–June 2006 (9.6–11 mm), during stabilization works in Dec. 2006 (49 mm) and May–June 2010 (Fig. 32.5). The first in Poland, real-time monitoring system was installed in May 2010 in Szymbark. Installation was realized just before record high precipitations of 100 mm/m^2 in 3 h time. Four on-line stations installed over the public road included two 3D inclinometers (12 and 16 m, totally 56 tilt sensors), one in-place inclinometer (14 m, 3 IP uniaxial sensors), automatic pore pressure transducers and automatic weather station. Total deformation sqrt $(x^2 + y^2)$ reached 34–50 mm during 34 months time (Fig. 32.6).

32.5 Numerical Modeling

The slope stability analyses based on assumption of rotational failure were realized using LEM Janbu, Bishop methods and FEM methods (Figs. 32.7, 32.8). The applied

approach allowed analyzing influence of high pore water pressure on the shear strength reduction. Proposed counteraction methods were tested using classical LEM methods. At Sekowa landslide, values of relative factor of safety Fs, calculated by Bishop method, were slightly above Fs = 1.13 before stabilization and Fs = 1.58 after it. Expected displacements were calculated by FEM methods, SoilVision codes and linear elastic models. Monitoring data were included in boundary conditions definitions. The expected displacements on Sekowa landslide are presented on Fig. 32.8. The final FEM mesh indicated that landslide was active and dangerous for the public road.

32.6 Design and Control of Stabilization Works

The results of exemplar counteraction works together with chosen investigation, monitoring and stabilization methods are presented in are presented in Table 32.1. The basic landslide remediation questions were addressed to the effective mitigations needed to maintain stability. At Sekowa remediation included gabion wall built on micropiles foundation along the river (Fig. 32.9). Second retaining wall on micropiles foundation was installed above the road. Drainage system lead down groundwater to the river and to a new culvert under the road. At Szymbark landslide partial stabilization at the front landslide part was decided. It included gabion walls along the river, surface and internal drainage system, two new culverts under the road, horizontal drainage boreholes with filters, anchors 20 m long to the bedrock, and

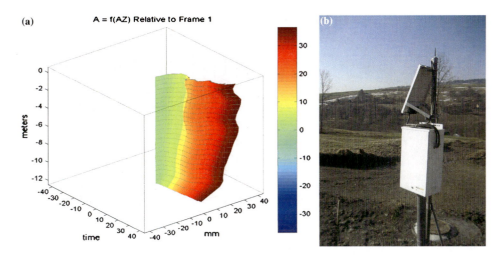

Fig. 32.6 On-line monitoring. **a** 3D displacements plot. **b** Field Station

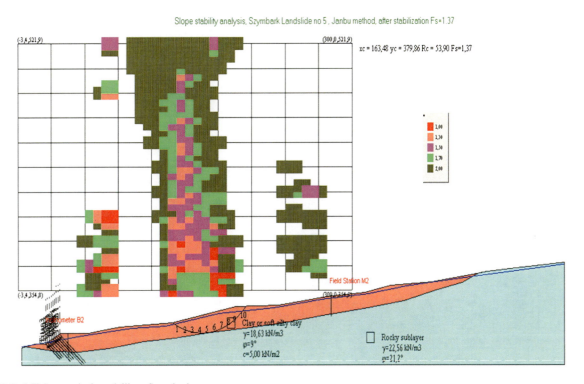

Fig. 32.7 LEM numerical modelling, Szymbark

Geobrugg high tensile wire mesh. The pore pressure values after counteraction dropped in both areas (from 45 to 30 kPa at Sekowa and from 70 to 15–50 kPa at Szymbark). Groundwater level depths were lowered from 0.5–1.3 to 2.2–3.0 m. The movements before counteraction works varied from 26 to 138 mm. In some cases they increased to over 60 mm during remediation, but after usually were lowered to ± few mm (Fig. 32.10). However, on Szymbark landslide, displacements up to 10 mm were recorded after remediation works. In May-June 2010 they increased after intensive rainfalls up to 18 mm, what was detected by early warning system.

Fig. 32.8 FEM numerical
modelling, Sekowa

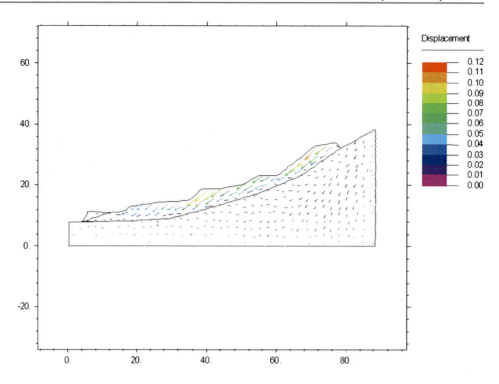

Table 32.1 Control of mitigation measures

Landslide	Vol. (m³)	Depth (m)	Displacements, (mm) before during after stabilization			Investigation methods	Counteraction methods
Sekowa	0.4	2.7–5.1	26	61	4–5	Boreholes, lab. tests, GPR, GPS-RTK, inclinometers, pore pressure transducers piezometers	Retaining gabion wall on pile foundation, micropiles, surface drainage system
Szymbark	2.2	1.3–5.0	138	19	8–13	Boreholes, lab. tests, GPR, GPS-RTK, inclinometers, pore press. stand + on-line	Gabion walls, anchors, high tensile wire mesh, internal and surface drainage

Fig. 32.9 Stabilization of Sekowa landslide

Fig. 32.10 Control of stabilization works, Sekowa

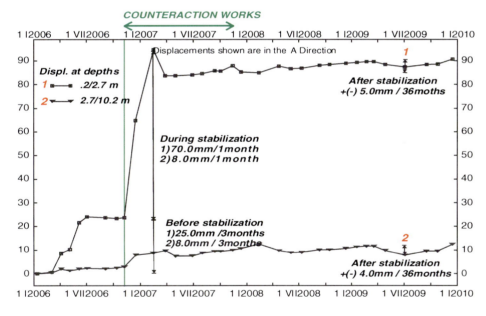

32.7 Conclusions

The research proved that chosen investigation, monitoring and modelling methods were useful for infrastructure projects. The continuous site inspection programme detected displacements from few mm to 40 cm to the depths of 1.3–18.0 m. The main triggering factors were connected with precipitations and changes of pore pressure. Proposed stabilization methods were fully effective at Sekowa landslide

and four other landslides. In Szymbark the remediation limited displacement ranges. The results of the study reveal that large flysch landslides were difficult for counteraction and remediation should be considered very carefully. Comprehensive monitoring and modelling methods before the counteraction stage could lead to a better recognition of remediation possibilities and early warning. Implementation of a new real-time monitoring techniques enabled continuous observation of landslide behavior to apply appropriate management to the sites of highest risk to the road network.

References

Bednarczyk Z (2012) Geotechnical modelling and monitoring as a basis for stabilization works at two landslide areas in Polish Carpathians. In: Proceedings of 11th International & 2nd North American Symposium on Landslides. Canadian Geotechnical Society, Banff, Canada, Balkema, Taylor and Francis, pp 1419–1425

Bednarczyk Z (2010) Soil-structure interaction on three stabilized flysch landslides in polish carpathians. In: *The Proceedings of the First International Conference on Advances in Multiscale Mechanics AIMM'10*. KAIST, South Korea, Techno-Press, pp 1389–1397

Bednarczyk Z (2008) Landslide geotechnical monitoring network for mitigation measures in chosen locations inside the SOPO Landslide Counteraction Framework Project Carpathian Mountains, Poland The First World Landsl. Forum, Tokyo, ICL, UN, pp 71–75

Gil E, Gilot E, Kotarba A, Starkel L (1974) An early Holocene landslide in the Niski Beskid and its significance for paleogeographical reconstructions. St Geom Carpat-Balc 8:69–83

Raczkowski W, Mrozek T (2002) Activating of landsliding in the Polish Flysch Carpathians by the end of 20th century. Studia Geom Carpatho Balcanica 36:91–111

A Brief Overview of the Typical Engineering Characteristics of Tropical Red Soils

33

George Brink

Abstract

Tropical soils in many ways have unique characteristics that can mainly be ascribed to the compositions and micro-structures of a material developed under hot, wet soil-forming conditions. For the purpose of this article, the term "tropical soil" will loosely refer to soils formed under such conditions, predominantly through chemical weathering processes. Most of these soils contain abundant iron and aluminium oxides due to the rapid breakdown of feldspars and ferromagnesian minerals, the removal of silica and bases and the concentration of iron and aluminium oxides or sesquioxides (sesquioxide = three atoms of oxygen to two atoms of another element i.e. Al_2O_3 or Fe_2O_3). Due to the high iron content, these soils are more often than not red in colour. The unique processes and conditions of soil formation in a tropical environment directly result in material compositions and structures that influence the engineering properties of tropical soils and the determination of these properties. Tropical soils do not behave in a similar fashion to temperate zone soils and consequently do not behave as expected when using conventional laboratory testing methods. This includes the susceptibility of the material to physical and mechanical breakdown, as well as a change in material properties due to cementing of the available sesquioxides. It is the aim of this paper to summarise and discuss the established unique engineering properties associated with soils formed in a tropical environment as summarised in the existing literature.

33.1 Introduction

The title of this article implies that tropical soils are sufficiently distinctive to be evaluated separate from conventional soils from more temperate climatic regions. Tropical soils in many ways have unique characteristics that can mainly be ascribed to the compositions and micro-structures of a material developed under hot, wet soil-forming conditions. For the purpose of this article the term "tropical soil" will loosely refer to soils formed under such conditions, predominantly through chemical weathering processes and separate from those soils formed in areas subject to more temperate climates with distinct seasonal variations. Throughout the available literature, tropical soils can broadly be subdivided into (i) tropical red

soils and (ii) tropical black soils. Due to the significant differences in material properties and engineering behaviour of the two identified general tropical profiles, emphasis will predominantly be placed on the discussion and investigation of tropical red soils in this paper.

Tropical red soils are often erroneously referred to either as 'laterites', 'laterite soils', 'latosols' or 'lateritic soils'. The existing chemical, geological and geotechnical data concerning laterites confirm that a universal, standardised terminology used to describe these materials has not been established and numerous inconsistencies have developed in the identification, classification and nomenclature between tropical red soils and laterites. For the purpose of this article, the term "laterite" refers solely to those materials or soil horizons which either represent laterite formation in the form of hardened pedogenic horizons or contain distinct evidence of laterization in the form of nodules, concretions or distinct discolouration, and is excluded from the tropical red soils category discussed here.

G. Brink (✉)
Exxaro Resources, PO Box 9229, Pretoria 0001, South Africa
e-mail: george.brink@exxaro.com

G. Lollino et al. (eds.), *Engineering Geology for Society and Territory – Volume 6*,
DOI: 10.1007/978-3-319-09060-3_33, © Springer International Publishing Switzerland 2015

33.2 Unique Engineering Characteristics of Tropical Red Soils

The usefulness of conventional index and strength tests for the identification and engineering classification of tropical soils has been questioned in the past, primarily due to the dependence of such tests on the sample preparation process and the variation of the natural soil structure.

33.2.1 Index and Physical Properties

The well-known global classification systems, such as the Unified Soil Classification System (USCS), are all based on the grain size distribution and Atterberg Limits for soils from temperate climatic environments. The clay minerals in these soils are generally stable, allowing for the relatively accurate determination of the index properties and allowing the development of numerous relationships between index properties and engineering behaviour. Due to the instability of clay minerals in tropical soils and the susceptibility of changes in material properties on drying, the relationships between the results of classification tests and the engineering behaviour are not as easy to obtain for tropical red soils (Mitchell and Sitar 1982). Further, it should also be kept in mind that drying not only occurs during sampling and testing, but also depends on the local climatic conditions, drainage and profile position. This may result in significant heterogeneity in the established regional soil characteristics (Fig. 33.1).

Northmore et al. (1992) established that index property values of tropical red soils are extremely sensitive to pre-treatment of the samples prior to testing, particularly to pre-drying and, for plasticity and shrinkage tests, to the degree of manipulation or mixing of the test sample. Drying of tropical red soils typically result in irreversible changes in the physical properties of the material. This initiates aggregate

formation and cementation by sesquioxides and the irreversible loss of water from the structures of hydrated clay minerals (Table 33.1).

The unique features of weathering and pedogenesis in tropical environments result in compositions and structural development that directly influences the engineering properties of these soils and their determination. Cementation of particles by the sesquioxides, together with the hydrated states of some minerals, lead to high void ratios, low compressibility and potentially high permeabilities compared to the high plasticity values and clay contents.

It is generally assumed that the permeabilities of in-situ tropical red soils are high (up to 10^{-2} cm/s), but as was the case with characterizing the strength of these materials the available data on the permeabilities of undisturbed tropical soils are very limited. Northmore et al. (1992) recorded values of 1.1×10^{-3}–5.8×10^{-6} cm/s for the Kenya specimens and 5.6×10^{-4}–2.2×10^{-6} cm/s for the specimens from Indonesia. This roughly equals the established permeabilities of silty materials or very fine sands from more temperate environments. As the permeability of a material is highly dependent on its structure, there is a significant variation in the available permeability results of undisturbed specimens and those permeabilities determined on remoulded specimens. The permeability of remoulded and compacted tropical red soils was found to vary between 1×10^{-6} and 1×10^{-8} cm/s (Saunders and Fookes 1970), with Northmore et al. (1992) reporting a "hundred to three hundred fold" decrease in the permeability of remoulded specimens compared to the values established for undisturbed specimens.

The effect of sub-surface biotic activity on the overall permeability of tropical red soils should not be neglected. Blight (1991) has highlighted the influence of termite activity on the overall permeability of the soil, as well as that of open channels from plant roots and other organisms. The presence of these cavities may increase the overall permeability by up to several orders of magnitude.

33.2.2 Mechanical Properties

The available published data on the strength testing of undisturbed samples of tropical red soils is limited, with the majority of available test results reflecting the shear strength parameters obtained from remoulded specimens. During their study, Northmore et al. (1992) determined the shear strength characteristics for a number of undisturbed and remoulded samples of tropical red soil from Kenya and Indonesia. Values of effective cohesion (c′) were found to be highly variable, ranging from 0 to 97 kPa and values of effective angle of internal friction (ϕ′) in the range of 11° and 44° for those tests completed on undisturbed samples. These represent a big variation in values and are generally higher

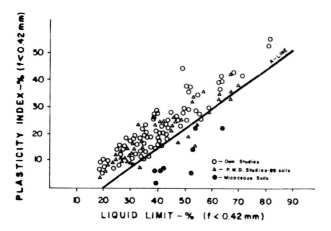

Fig. 33.1 Established variability in Atterberg limits tropical and lateritic soils from Ghana (Gidigasu 1974)

Table 33.1 Effect of drying on index properties of hydrated clay from Hawaiian Islands (Gidigasu 1974)

Property	Natural moisture content	Partially air dried	Completely air dried
Sand (%)	30	42	86
Silt (%)	34	17	11
Clay (%)	36	41	3
Liquid limit (%)	245	217	NP
Plastic limit (%)	135	146	NP
Plasticity index (%)	110	71	NP

NP = Non plastic

Table 33.2 Summary of available shear strength characteristics of undisturbed samples of tropical red soils

Reference	Average dry unit weight (kg/m^3)	Friction angle (°)		Cohesion (kPa)	
		Range	Average	Range	Average
Vargas (1974)	–	23–33	28	0–59	24
Tuncer and Lohnes (1977)	1,300	27–57	42	48–345	163
Fos (1973)	1,150	36–38	37	22–28	25
Northmore et al. (1992) (Indonesia)	836	11–36	22	2–97	32
Northmore et al. (1992) (Kenya)	1,112	12–44	25	0–55	24

than those soils similar plasticity or clay fraction but formed in temperate environments. Wesley (1988) comments that test results of allophane-rich soils generally reflect soils of higher shear strength values than those of halloysite-rich soils (Table 33.2).

The residual friction angles of remoulded or destructured materials were found to vary between 18° and 41° (mostly 28°–38°), with effective cohesion values ranging between 0 and >50 kPa (De Graft-Johnson et al. 1969; Gidigasu 1974; Saunders and Fookes 1970). In an effort to establish the effect of remoulding on the engineering behaviour of the soil, Northmore et al. (1992) also determined the shear strength characteristics of two sets of destructured and compacted samples to compare with the results obtained from the testing of the undisturbed samples. The test results revealed largely similar stress-path and strength behaviour despite the large structural differences of disturbed and destructured materials. In both cases the compacted specimen revealed higher effective friction angles than the undisturbed samples.

Northmore et al. (1992) concluded from their testing of tropical red soils from Indonesia and Kenya that there appears to be a positive relationship between the effective shear strength of tropical soils and its plasticity index and moisture content. This is the opposite of the established relationship in clayey soils from temperate climatic regions. A general positive relationship further exists between the increases in the effective strength of the material with an increase in depth below natural ground level, most likely due to either the development of material structure.

Consolidation test data from the literature indicates highly variable initial void ratios, typically varying between 0.92 up to 5.36. It has generally been assumed that the tropical red soils are fairly incompressible compared to clays from temperate climates, but the available consolidation test results on undisturbed samples reflect very high rates of primary consolidation (indicating high rates of short term settlement), making the calculation of the consolidation coefficient (C_v) very difficult. When evaluating the rate of consolidation according to the percentage of consolidation in the first minute of testing, it is evident that allophone-rich materials generally experience a reduction in the consolidation rate with increasing stress, whereas halloysite-rich samples tend to maintain a high rate of consolidation throughout. C_v values typically range between 6.3×10^{-6} and 2.5×10^{-2} cm^2/s. The consolidation coefficients of compacted or remoulded tropical red soils were found to typically range between 1×10^{-1} and 1×10^{-3} cm^2/s (Saunders and Fookes 1970). The compressibility results (M_v) ranges from high to very low, with an overall decrease in compressibility with an increase in the applied stress. Secondary consolidation (C_a), used as an indication of the long term consolidation behaviour of the soil, was also found to generally increase with an overall increase in the applied stress (Northmore et al. 1992).

33.3 Conclusions

The unique processes and conditions of soil formation in a tropical environment directly result in material compositions and structures that influence the engineering properties of tropical soils and the determination of these properties. The available undisturbed and remoulded results discussed here reflect a significant variation in the values in the shear strength characteristics of tropical red soils and, as was found to be the case with the index properties, the heterogeneity of the material characteristics, degree of weathering and cementation are the most likely all reasons for the variations. It is therefore recommend that the data from index tests alone should not be used to distinguish between soils formed in different climatic regimes and from different parent rock types. However, it may be possible to make use of the classification tests, gradings and Atterberg Limits to establish a correlation with the expected engineering behaviour for specific application and based on local experience.

Test results of undisturbed and remoulded soil specimens confirm that tropical red soils are highly sensitive to structural breakdown and manipulation. Manipulation of the material using heavy equipment may result in marked changes in characteristics and the ease with which it is handled on site. Further, the sensitivity of tropical red soils to drying was found to result in changes (frequently irreversible) in the physical characteristics of the material, mainly due to aggregate formation and cementation by sesquioxides and the irreversible loss of water from the structures of hydrated clay minerals.

References

Blight GE (1991) Tropical processes causing rapid geological change. In: Forster A, Culshaw MG, Cripps JC, Little JA, Moon CF (eds) Quaternary engineering geology, Engineering geology special publication, vol 7. Geological Society, London, pp 485 490

Buchanan F (1807) A journey from Madras through the Countries of Mysore, Canara and Malabar, vol 2. East Indian Company, London, pp 436–460

Fermor LL (1911) What is laterite? Geol Mag 5(8):453–462

De Graft-Johnson JWS, Bhatia HS, Gidigasu MD (1969) The engineering characteristics of the lateritic gravels of Ghana. In: Proceedings of special session engineering properties of lateritic soils, 7th international conference soil mechanics foundation engineering, Mexico, vol 1, pp 117–128

Gidigasu MD (1974) Degree of weathering in the identification of laterite materials for engineering purposes—a review. Eng Geol 8:213–266

Joachin AWR, Kandiah S (1941) The composition of some local soil concretions and clays. Trop Agric 96:67–75

Lacroix A (1913) Laterites of Guinea and alteration products associated with them. Nouveau Arc Mus Hist Nat 5:255–356

Marques JJ, Schulze DG, Curi N, Mertzman SA (2004) Major element geochemistry and geomorphic relationships in Brazilian Cerrado soils. Geoderma 119:179–195

Martin FJ, Doyne HC (1930) Laterite and lateritic soils in Sierra Leone. J Agric Sci 20:135–143

Millard RS (1962) Road building in the tropics. J Appl Chem 12: 342–357

Mitchell JK, Sitar N (1982) Engineering properties of tropical residual soils. Engineering and construction in tropical and residual soils. In: Proceedings of the ASCE geotechnical engineering division speciality conference, Honolulu, Hawaii, pp 30–58

Northmore KJ, Culshaw MG, Hobbs PRN, Hallam JR, Entwisle DC (1992) Engineering geology of tropical red clay soils: summary findings and their application for engineering purposes. British geological survey technical report WN/93/15

Pendelton RL (1936) On the use of the term laterite. Am Soil Surv Bull 17:102–108

Saunders MK, Fookes PG (1970) A review of the relationship of rock weathering and climate and its significance to foundation engineering. Eng Geol 4:289–325

Uehar G (1982) Soil science for the tropics. Engineering and construction in tropical and residual soils. In: Proceedings of the ASCE geotechnical engineering division speciality conference, Honolulu, Hawaii, pp 13–27

Wesley LD (1988) Engineering classification of residual soils. In: Proceedings of the 2nd International conference on geomechanics in tropical soils, Singapore, vol 1, pp 77–84

Remote Analysis of Rock Slopes with Terrestrial Laser Scanning for Engineering Geological Tasks in Reservoir Planning

34

Hieu Trung Nguyen, Tomás M. Fernandez-Steeger, Hans-Joachim Köhler, and Rafig Azzam

Abstract

Building the lower basin of a pumped-storage hydropower plant in an active limestone quarry without any sealing is a challenging project. A proper site investigation is crucial to increase the overall efficiency of planning as well as realisation by anticipating potential problems and sticking points. The local geology comprises a complex system of reef structures and bedded sequences that is intersected by faults. Sets of faults can be identified on different scales and the fracturing patterns vary vertically and horizontally. Both the ongoing quarrying as well the sheer size and steepness of the outcrop prevent any extensive data acquisition using traditional methods. Therefore, terrestrial laser scanning (TLS) was used to map the quarry and to create a high resolution digital elevation model (HRDEM). The HRDEM allows to perform spatial analysis with respect to the distribution of geohydraulic and geotechnical properties of the rock mass in the quarry. The major advantages of this approach are the increased level of detail, a substantial improvement of documentation and synergetic effects that arise from the multiple different applications of scan data e.g. for analysis, interpretation, planning and solving geohydraulic and geotechnical issues. It has shown in practice that this multi-facetted usage of the collected data outweighs the initial efforts of data collection and processing by far.

Keywords

Pumped-storage hydropower • Rock mass characterization • Terrestrial laser scanning • Blautal

34.1 Introduction

For the planning approval of the pumped-storage hydro-power plant (PSH) Blautal, commissioned by the municipal utilities Ulm (SWU) and EDUARD MERKLE GMBH & CO. KG, extensive geohydraulic and geotechnical site investigations are mandatory. Located about 4 km of the spring Blautopf in the karst landscape of the Swabian Jura's southern edge in Southern Germany, the PSH shall have an installed capacity of 60 MW by a built-in reservoir of 1.1 million m^3 and a height difference of 170 m. Initial planning has stipulated an unsealed lower basin situated on a mountainside inside the quarry as the preferred option of construction. The absence of any sealing means that the basin will be incorporated within the natural groundwater. Due to the subsequent use, the mine plan provides that the limestone will be quarried to 10 m below the groundwater table, creating a steep, 95 m high rock slope next to the eastern bank of the basin. As there are environmentally sensitive biotopes nearby, the anticipated impact on natural groundwater conditions must be limited in space. Therefore, two major aims have been defined for the site-investigation: To specify the potential influence on the groundwater during excavation, PSH construction and operation time on the one hand. On the other hand first estimations of the expected dynamic

H.T. Nguyen (✉) · H.-J. Köhler
Dr. Köhler & Dr. Pommerening GmbH, Am Katzenbach 2, 31177 Harsum, Germany
e-mail: hieu.nguyen@koehler-pommerening.de

T.M. Fernandez-Steeger · R. Azzam
Department of Engineering Geology and Hydrogeology, RWTH Aachen University, Lochnerstraße 4-20, 52064 Aachen, Germany

Fig. 34.1 Tripod mounted
instrument, scanning a 65 m
height slope. *Light-colored lime
—and marlstone, complex
geological features and different
rock face conditions due to
ongoing quarrying complicate the
rock classification*

stress changes within the rock are to be given, since pore water pressure can significantly reduce slope stability (Köhler et al. 2013). To meet these goals, a detailed geohydraulic and geotechnical rock mass description is essential. The obtained information serves as input data for a detailed geohydraulic rock mass model with special emphasis on the location and spatial distribution of discontinuities as well as their degree of karstification. This will allow a hydraulic characterization of the rock mass and the identification of potential failure mechanisms due to increased buoyancy or sliding wedge formation. The local geology comprises a complex system of reef structures and bedded sequences that is intersected by faults. Sets of faults can be identified on different scales and the fracturing patterns vary vertically and horizontally. Both the ongoing quarrying as well the sheer size and steepness of the outcrop prevent any extensive data acquisition using traditional methods (Fig. 34.1). In the presented study area, traditional manual recording would be very laborious and time-consuming and therefore economically not feasible, as outlined above. Thus, remote sensing by terrestrial laser scanning (TLS) was used to map the quarry and subsequently used to create a high resolution digital elevation model (HRDEM).

34.2 Methodology

Terrestrial laser scanning utilizes reflected laser pulses emitted from a tripod-mounted scanning instrument (Fig. 34.1) to determine distances to targets of interest. The scanned surfaces are recorded as point data with accuracy ranging from 4 to 3 cm. Thus, TLS data provide the possibility to measure and visualize topographic relief down to centimeter resolution for close-up digital inspection.

Provided that the exact scanner position is known, each point can be georeferenced and used to compute a digital elevation model (DEM). The newly generated DEM in turn can be registered with color photographs and serve as the basis for engineering geological rock mapping. The entire quarry has been time-efficiently scanned from different positions by a terrestrial laser scanner with average point spacing from 10 to 15 cm, including additional scans of selected slopes with 2 cm point spacing. Each scanner position has been determined by the use of Real Time Kinematic (RTK) satellite navigation. In addition, a photographic mapping has been applied to assist data analysis and interpretation. Applying the data preparation and processing after (Nguyen et al. 2011), a DEM of the quarry in 1 m resolution and HRDEMs of different slope exposures in 2 cm are used to create up to date topographic maps, cross-sections and virtual outcrop models. Thus, different—otherwise laborious—tasks can be completed in a `virtual lab', even for inaccessible areas.

34.3 Applications: A Spatial Analysis Approach

The following descriptions focuses on how to transform structural information from a HRDEM of a jointed rock mass into an organized discontinuity-network. For this purpose, the characterization of a rock mass has been exemplarily executed on a virtual outcrop model of the steep eastern slope of the investigation site. Due to low coloring diversity and the complex geological structure with its multiple reef formations and interconnected bedding, an interactive data interpretation is required for the rock mass characterization. In the beginning, an orthogonal projection of the HRDEM allows to precisely visualize the morphology

Fig. 34.2 Manual rock map by using the HRDEM. It forms the basis to locate geological features and to determine their attitude

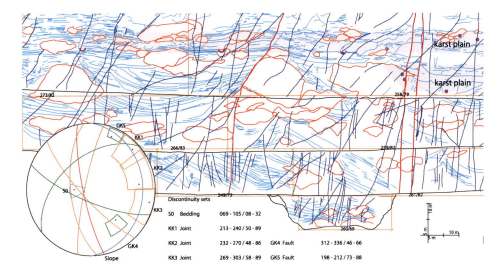

of vertical and overhanging rock faces with an adaptable map scale for an adequate display. This enables different applications, e.g. the traditional mapping by simply tracing geological structures, the measuring of geometric features and the estimated localization of karst plains and shafts (Fig. 34.2). Identified features are from top to bottom: thin calcareous marlstone layers with underlying massive and clustered biohermal limestone interlocked with reworked framework elements and stratified limestone in the intermediate basins. All these structures are bound together by secondary encrustation or cementation. Apart from the bedding-planes, the fracture network consists of two conjugate sets of steep closely spaced joints and two conjugate sets of widely spaced faults with normal down-dip displacement. At different levels of elevation, karst features are exposed. However, it was difficult to clearly separate different lithological units during manual pre-allocation: As the joint patterns of the examined outcrop are complex and intersecting, tracing certain discontinuities often became unclear and ambiguous. Thus, analyzing the HRDEM in a virtual lab has to be repeated under different display conditions (angle of view, angle of light, color shade, photographic overlays). In a later stage, an adjustment by secondary data, e.g. calculated or estimated discontinuity frequency, has proved to be very helpful. To this extent, the framework of a discontinuity-network has been established by a heuristic description, encompassing the discontinuities in a rock mass. For the next step, using the orientation of each computed vertex or face of the HRDEM, a TLS data set can be automatically compiled into local dip and strike direction. This enables the researcher to map the populations of structural slope facets, to carry out a stereographic analysis and then ultimately identify discontinuity sets. For a time-efficient approach, the HRDEM has been divided into several partitions prior to structural analysis in order to decrease the noise generated from the spatial variability of

identical fracture patterns over the entire outcrop. In order to identify individual sets, overpopulated steep west inclined facets resulting from manmade slope expositions have to be removed before transforming the orientation data into density patterns. By using the frequency distribution of spatial orientations, several present discontinuities can be characterized from the density pattern by successively decimating dominant sets to emphasize subdominant orientations (Köhler 2013). With the possibility to color and visualize identified sets within the HRDEM, it shows that the analyzed orientation data can be related to flat-angled east dipping bedding surfaces, to steep NNW-SSE and WSW-ENE striking conjugate joints and to subdominant steep WNW-ESE striking faults and a few NE-SW striking faults (Fig. 34.2).

Application continued with extracting traces of geological features by using an algorithm which sets closely spaced scanlines onto the partitions of the HRDEM and utilizes a moving search window for corner detection. By computing the intersection points between the scanlines and traces, resulting discontinuity frequency and their spatial pattern can be visualized into a grid, suitable for designating homogeneous areas (Fig. 34.3a). Here, discontinuity spacing showed distinguishable areas of closely to widely spaced joints. Again, a cross validation with manual pre-allocation is necessary, as scan resolution and rock faces determine the quality of traceable features and spacing data (Fig. 34.3b). However, as the HRDEM depicts the in situ condition of the slope, blasting of ongoing mining operations significantly increases the joint density, creating a disordered network that strongly differs from the original, tectonically induced, system of fractures. These areas were treated separately for evaluating the discontinuity connectivity in the rock mass (Fig. 34.3b). By displaying those features with other surface characteristics in engineering geological plans and other visual methods of displaying field data like cross-sections or

Fig. 34.3 a Narrow spaced scanlines shows spatially different joint density, separating massive reef bodies and thinly bedded marlstone layers. At the time of the recording, different blasting techniques altered traceable joint densities. **b** Translated into density classes for a practical use. As low resolution areas, fresh blasting works and anthropogenic features influence the classification, the results has to be field-validated

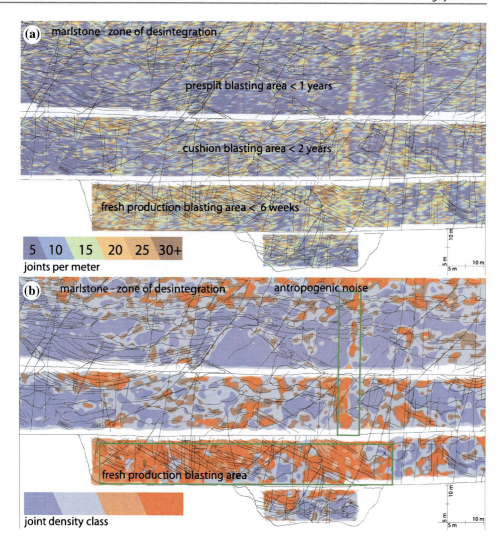

3D models, a high horizontal and vertical precision of the different information locations was secured. Based on this data and in combination with the results from prior applications, a discontinuity network can finally be organized as sets of discontinuities with identified spacing, persistence and attitudes. The major advantage of this procedure is the generation of detailed multiple-layered thematic 3D maps which provided a substantial improvement of data, knowledge representation and documentation, which in turn can be used as a template for ongoing applications as rock slope stability analysis.

34.4 Conclusion

The use of remote sensing with TLS shows that maps and cross sections generated from TLS data serve as a record of the spatial location of actual data and are suitable tools for characterizing discontinuities in a rock mass in detail over a large area. Operational experiences prove that primer higher efforts for data collection and processing will be compensated by far due to multi-use of data and flexibility in data supply, in terms of quantity and quality. The key advantage is that according to the investigation process and demand, data can be extracted and aggregated at varying levels of detail to respond quickly and economically to changing project requirements.

References

Köhler H-J, Hennings S, Nguyen HT, Fernández-Steeger TM (2013) Structural analysis for permeability and stability assessment by use of TLS for the pumped storage hydropower plant Blautal.—19. Tagung für Ingenieurgeologie, München, 13–15 März 2013, pp 395–400

Nguyen HT, Fernández-Steeger TM, Wiatr T, Rodrigues D, Azzam R (2011) Use of terrestrial laser scanning for engineering geological applications on volcanic rock slopes an example from Madeira Island (Portugal). Nat Hazards Earth Syst Sci 11:807–817

Dynamically Loaded Anchorages

Santoro Federica, Monia Calista, Antonio Pasculli and Nicola Sciarra

Abstract

The present work regards the study of soil-structure interaction, in particular during pull-out tests on anchorages. Due to the complexity of the system, numerical analyses carried out by a commercial code, based on the Finite Difference Method (FDM), were performed. In order to calibrate the overall selected approach, first of all, a simple two-dimensional model of the system was firstly study. The preliminary results and suggestions for incoming 3D modeling improvements are discussed. Furthermore attention has been focused on the nature of the impulsive impacting force due to a virtual debris flow or avalanches striking on the structure, including the anchorages, in order to correctly simulate the test of pull out. Accordingly, a simplified, preliminary model of the anchorages-net system, including the effect of the mass deposition, is proposed and discussed.

Keywords

Anchorages • Impulsive phenomena • Numerical modeling • Pull-out test • Soil-structure interaction

35.1 Introduction

Currently the anchoring techniques are undergoing major development; in many cases these may offer a solution to the problems of stability at different depths (Hobst and Zajic 1983; Mashimo and Kamata 2002). At the same time the use of numerical modeling codes, performed on well defined geological and geotechnical model, becomes important in the design phase and the prevention of geological risks, as for example phenomena of collapse. In the case of numerical modeling which provide for the study of the interaction between soil and structures, the problem becomes more complex. For this reason it is necessary to start with simplified models, in order to control and manage data better, and then proceed with gradually more complex models that can provide results as close as possible to reality. In this context also criteria, which are used for the evaluation of the forces transmitted by debris flow or avalanche impact against structures, become important. As regard, it must be said that in most cases these criteria neglect the actions that occur in the initial phase of the phenomenon; also the results of experiments reported in the literature do not provide useful information for the design of structures for protection, because measured pressures at the impact have larger dispersion. Typically to estimate the actions exerted by the fluid flows, a modified value of the hydrodynamic pressure exerted by a fluid in permanent motion is considered, or a multiple of the hydrostatic pressure. Actually, the pressures that arise at the impact can reach extremely high values, although the duration of the actions of maximum intensity is very short. So the impulsive nature of the phenomenon is evident (Federico and Amoruso 2005, 2008). Actually the anchorages object of this study are part of a barrier system in which they are connected through break devices to a frame and a net. Furthermore in this work we explore the feasibility to develop a semi-analytical model of the system based on the elasticity of the net. In addition, the elastic and dissipative behavior of bolts and wire-rope dissipation are also

S. Federica (✉) · M. Calista · A. Pasculli · N. Sciarra
INGEO Department, University of Chieti—Pescara, Via dei Vestini 31, 66100, Chieti, Italy
e-mail: f.santoro@unich.it

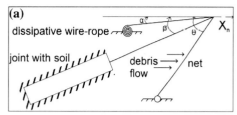
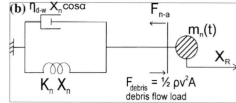

Fig. 35.1 **a** Simplified conceptual 2D system. **b** Scheme of the mechanical model

included. Moreover, as the consequence of the debris or avalanche interaction with the net is a mass deposition, during the time transitory, the mass of the system increases at a rate proportional to the debris flux and to the solid fraction. In order to study the overall response, as first step, a collection of point like (zero-dimensional), interacting elements, is assumed to be suitable in catching the mean dynamical features of the phenomena. Accordingly, the related mechanical model is based on concentrated masses, interacting to each other by elastic rings and dash-pots. The mass of the net is variable and increasing with time. The 'air bag' effect that smoothes the impulsive load due to debris impact may be observed. Actually the momentum transferred during the impact between the debris and the net frame system is delayed. As a consequence the intensity of the transmitted forces to the bolts is reduced. This semi-analytical model to calculate the impulsive force that is necessary to simulate pull out tests is currently under study by our research team. We are focusing our attention on the behavior of the anchorages and their interaction with surrounding soil. The achieved results will be discussed in a forthcoming, more detailed paper. As well as for other studies (Pasculli et al. 2006; Sciarra et al. 2011) conducted by our research group, also this study was performed by a commercial computer code, FLAC2D 6.0, based on FDM approach (Finite Differences Method). The selected code is commonly used to analyze static and dynamic problems, including dry and wet conditions. Moreover the simulations of structural elements, for example piles, anchorages, etc. are allowed as options (Itasca Consulting Group Inc 2008). Anchorage-net system: first modeling attempt.

In this paper only a preliminary description of the mathematical model of the anchorage-net system we are performing is given. Thus, to investigate how the anchorages work and to study the dynamical loads to which they are subject, we assumed that the simplified two-dimensional system, sketched in Fig. 35.1a, is a reasonable representation of the actual barrier system. As first approach, we considered the net as a point mass affected by elasticity (elastic spring) linked to the dissipative wire rope (dash pot) by a 'parallel' connection (same displacements), as reported in Fig. 35.1b. The mass of the net is assumed to increase with time because of the deposition of the solid fraction of the debris. The tensile force due to the anchorage and the load of the debris

flow are assumed to be the external forces acting on the net-system. In order to study the dynamics of the system, as first step, we applied the balance of the forces only along the direction parallel to the slope topography, where the selected anchorage is located. Balance of force momentums were not considered, as the net is assumed to be only hinged. Thus through a simple inspection of the 'conceptual' model of the sub-system shown in Figs. 35.1 and 35.2, the following scalar equation follows:

$$\frac{d}{dt}\left[m_n(t)\frac{dx_n}{dt}\right] = -F_{n-a}\cos\beta - \eta_{d-w}\frac{dx_n}{dt}\cos\alpha - k_n x_n + F_{debris}.$$

This can be arranged in the following expression:

$$m_n(t)\frac{d^2x_n}{dt^2} = -F_{n-a}\cos\beta - \left[\eta_{d-w}\cos\alpha + \frac{dm_n(t)}{dt}\right]\frac{dx_n}{dt} - k_n x_n + F_{debris}$$

(35.1)

where 'n' stands for net; 'a' anchorage; 'd' dissipation; 'w' wire-rope; $m_n(t)$ is the increasing mass associated to the net, x_n the displacement coordinate of the net along the slope topography (see Fig. 35.1); $F_{n-a}\cos\beta$ and $\eta_{d-w}\frac{dx_n}{dt}\cos\alpha$ are respectively, the projection along the $\hat{x}_n$ direction (see also Fig. 35.1) of the tensile force exerted by the anchorage on the net and the dissipative force of the wire rope for which a linear law was assumed; η_{d-w} is the 'equivalent' dissipation coefficient; $k_n x_n$ includes the elasticity of the net and F_{debris} is the load exerted by the solid fraction of the debris. It is worth to note that the debris mass deposition on the net affects the dynamics of the system like a further viscous force.

As a simple assumption we set $\frac{d}{dt}m_n(t) = \rho \cdot v$ where ρ and v are, respectively, the density of the solid fraction and the velocity of the debris. Assuming, for the sake of simplicity, ρ and v constant in time, it follows: $m_n(t) = m_n + \rho v t$. The debris flow load is introduced considering the dynamical pressure exerted against the net, thus according to the kinetic term of the Bernoulli law, the following expression yields: $F_{debris} = 1/2(\rho v^2 \cdot A \cdot sen\theta)$ where $A \cdot sen\theta$ is the projection of the area A of the net on a plane perpendicular

Fig. 35.2 **a** Distribution of displacement vectors after the application of load and before failure; **b** geometrical model of interfaces; **c** shear stress increment after failure; **d** exaggerated grid after failure

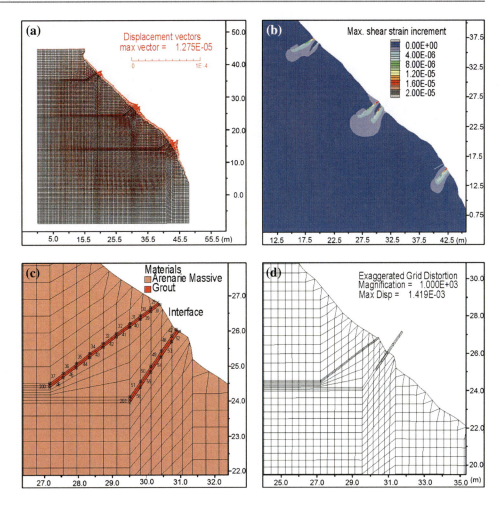

to the debris flow direction. More details about the model will be given in further works. In particular the term F_{n-a} will be computed including also the dynamical response of the soil-anchorage system.

35.2 Test Case

The study area is located in the central part of Italy, in the western area of the Abruzzo Region, along the A24 highway. The two sections of interest are located next to the left side of Varri valley (Rome direction) and to the right side of a near valley that belongs to a tributary of Turano river (L'Aquila direction); They are included between 800 and 900 m above sea level. The two artificial slopes, that have been modeled, are very steep, and at places vertical.

A detailed geological and geomechanical survey allowed to define the geological setting and the geotechnical features of the area, that is part of the margin of Monti Carseolani, belonging to the carbonatic Laziale-Abruzzese platform

(Parotto and Praturlon 1975; Mariotti 1992), which is connected to the fore deep of the Salto-Tagliacozzo basin.

In particular, in the two studied slopes we can see the following deposits:

- Massive Sandstones (Messinian): coarse particle size, interbedded with decimetric strata with fine particle size (thickness 30 m).
- Detritic-sandstone deposit: thickness 0.5–2 m (Slope-L'Aquila direction (highway progressive km 63 + 350– 65 + 600).
- Stratified sandstone (Messinian): thinly stratified, associated with sandstones with coarse particle-size in decimetric strata (thickness 30 m).

Physical and mechanical parameters that were used to develop geotechnical model employed in the subsequent steps of numerical modeling are shown in Table 35.1. The two sections were modeled using Mohr Coulomb constitutive model, in order to focus the attention on the yielding and the successive failure of soil surrounding the anchorages. We chose to simulate soil and structure with grid and interface elements that connect soil and anchorages.

Table 35.1 Geotechnical parameters of deposits outcropping in the two slopes and geotechincal parameters of interfaces and physical—mechanical parameters of anchorage

Geotechnical slopes features		Geotechnical features—interfaces	
Massive sandstones		Normal stiffness	1E7–1E6 kN/m^3
Density	22 kN/m^3	Shear Stiffness	1E7–1E6 kN/m^3
Friction angle	35–40°	Cohesion	50 kN/m^2
Effective Cohesion	150/200 kN/m^2	Physical-mechanical parameters—anchorage	
		Density	2.4 kN/m^3
Debris		Bulk modulus	6.833E7 kN/m^2
Density	20 kN/m^3	Shear modulus	6.4E6 kN/m^2
Friction angle	32°	Diameter	0.1 m
Stratified sandstones		Length	3–4 m
Density	20.8 kN/m^3		
Friction angle	30–35°		
Effective Cohesion	100–150 kN/m^2		

In the first phase it was necessary to work on a two-dimensional simplified model, in order to understand how to manage and control steps and results. So a square mesh was used, assuming a simple vertical structure (also simulated by square mesh) inserted near the symmetry axis of the soil grid. Between the soil grid and the structure some interface elements were inserted, depending on the mesh size, and they were monitored in terms of shear force and shear displacements. Starting from the simplified model we continued using a progressively more complex model, up to the sections described previously.

The two sections were represented by a grid with square mesh (about 0.35 m for side). In order to simulate pull out tests over the structure (also simulated with grid elements) variable loads, ranging from 10^3 N up to 10^5N were assumed. The instability of the system, by a numerical computational point of view, occurred when the load of 10^5N was applied. Both anchorages slipped and the max displacement vector measured was 2.07×10^{-4} m. Also it was possible to monitor and correlate shear stress and shear displacements of different point located on the interface (Fig. 35.2).

35.3   Conclusions

In this article the analysis set up and the initial phase of the work in which a static variable force is applied to a simple two-dimensional model in order to calibrate and verify first results were discussed. These first results are encouraged, accordingly a 3D analyses are under developing, taking into account not only the application of the static variable force, but also the dynamic time history load, derived from the considerations explained in this article about the impulsive

phenomena. The simple mathematical-mechanical modeling describing anchorage system, preliminarily proposed in this paper, will be furthermore developed.

References

Federico F, Amoruso A (2005) Numerical analysis of the dynamic impact of debris flows on structures. Proceedings of ISEC-0-3rd International structural engineering and construction conference, Shunan, Japan

Federico F, Amoruso A (2008) Simulations of mechanical effects due to the impact of fluid—like debris flows on structures. Italian J Eng Geol Environ 1:5–24

Hobst L, Zajic J (1983) Anchoring in rock and soil. Publisher Elsevier, Amsterdam, pp 311–320

Itasca Consulting Group Inc. (2008) Dynamic analysis, FISH in FLAC, getting started, structures, users guide, example application, verification problem—FLAC Version 6.0. Minneapolis, Minnesota USA

Mariotti G (1992) Note introduttive alla geologia dell'Appennino centrale. AA. VV. (Eds.) VSimposio Paleoecologia Delle Comunità Bentoniche - Guida all'escursione, pp 30–45

Mashimo H, Kamata H (2002) Experimental investigation the effect of rock bolts on tunnel stability in sandy ground. Physical modelling in geotechnics ICPMG '02, Rotterdam, Balkema, pp 797–801

Parotto M, Praturlon A (1975) Geological summary of the central appennins. Quad. de La Ricerca Scientifica, 90:257–311, CNR, Roma

Pasculli A, Calista M, Mangifesta M (2006) The effects of spatial variability of mechanical parameters on a 3D landslide study. Proceedings of the 4th international FLAC symposium on numerical modeling in geomechanics, Madrid, Spain, May 29–31 2006. Paper: 01–05 (ISBN 0-9767577-0-2)

Sciarra N, Calista M, Marchetti D, D'amato Avanzi G, Pochini A, Puccinelli A (2011) Geomechanical characterization and 3d numerical modeling of complex rock masses: a slope stability analysis in Italy. Proceedings of the 2nd International FLAC/DEM symposium, February 14–16 2011, Melbourne, Australia

LiDAR and Discrete Fracture Network Modeling for Rockslide Characterization and Analysis

Matthieu Sturzenegger, Tim Keegan, Ann Wen, David Willms, Doug Stead, and Tom Edwards

Abstract

On November 25, 2012, a rockslide occurred along the Canadian National Railway tracks, approximately 150 km northeast of Vancouver, Canada. The volume of the slide was approximately 53,000 m^3. It caused four days of service disruption and the collapse of a rock shed protecting the tracks. This paper studies the triggering factors and the failure mechanism based on a combination of airborne and terrestrial LiDAR data, site investigation and discrete fracture network (DFN) modeling. This work provided input parameters for subsequent run-out analysis and design of a railway protection structure (rock shed).

Keywords

Failure surface • LiDAR • Discrete fracture network • Finite element analysis

36.1 Introduction

On November 25, 2012, Canadian National Railway (CN) had a major rock and debris slide cover the track along the Fraser Canyon, approximately 25 km north of Boston Bar, British Columbia, Canada (Fig. 36.1). The slide was approximately 70 m wide, with 9 m of debris covering the track, resulting in a service disruption of more than four days. The rock slide resulted in the collapse of a rock shed, protecting the track from this historically unstable rock slope (Fig. 36.2).

Keegan (2007) shows that ground hazards are CN's third most costly type of railway hazard and that amongst ground hazards, "rock" landslides contribute to 4.3 % of the annual cost related to train accidents; rock landslide may also transform into "debris" landslides, which constitute 5.8 % of the annual accident cost.

This paper presents a rockslide investigation, which incorporates both remote sensing techniques and discrete fracture network (DFN) modeling in order to optimize the characterization and back-analysis of the rock instability. The paper describes the triggering mechanism, geometry, and failure mechanism of the rockslide. This study formed the first phase of a combined rockslide hazard assessment and protection structure design project.

M. Sturzenegger (✉) · A. Wen · D. Willms
Klohn Crippen Berger Ltd., Vancouver, BC, Canada
e-mail: mSturzenegger@klohn.com

A. Wen
e-mail: awen@klohn.com

D. Willms
e-mail: dwillms@klohn.com

T. Keegan
Klohn Crippen Berger Ltd., Edmonton, AB, Canada
e-mail: Tkeegan@klohn.com

D. Stead
Department of Earth Sciences, Simon Fraser University, Burnaby, BC, Canada
e-mail: dstead@sfu.ca

T. Edwards
Canada National Railway, Edmonton, AB, Canada
e-mail: tom.edwards@cn.ca

36.2 Site Conditions and Geological Setting

36.2.1 Climate and Seismicity

In the area, the daily average temperature ranges between −2 and 21 °C, in January and July, respectively. The annual

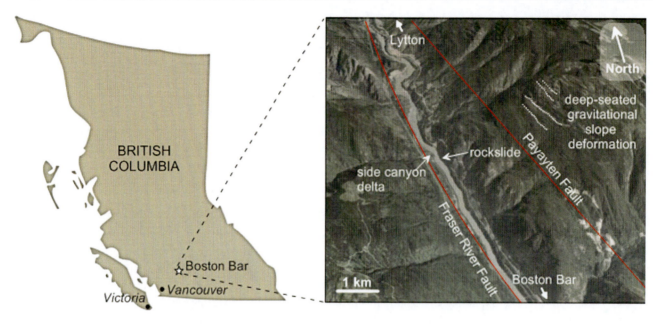

Fig. 36.1 Rockslide area location, geomorphology and regional faults

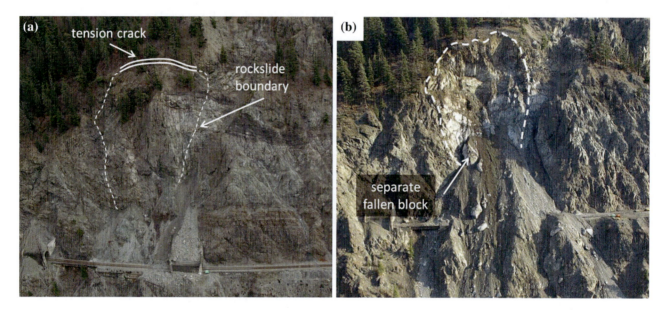

Fig. 36.2 Rockslide. Pre-slide (**a**), and post-slide (**b**), topography

precipitation is 432 mm. From November to March, half of precipitation is snowfall (Environment Canada 2013). The Jackass Mountain weather station located near the rockslide indicated a first series of freeze–thaw cycles preceded by three rainfall periods of 2 to 5 days at the end of October and beginning of November 2012. During mid-November, three days of high-intensity rainfall were recorded, with an average and maximum daily precipitation of 13.1 and 21.5 mm, respectively; these three days of precipitation represent more

than the 90th percentile of the long-term November record, and can be defined as an "extreme weather event" (IPCC 2007). The night of November 24 2012 marked the onset of a second series of freeze–thaw cycles.

A review of seismic data indicated that earthquakes are unlikely to have destabilized the rock slope, and consequently, it appears that the likely trigger factor was the succession of freeze–thaw cycles, accompanied by heavy precipitation.

36.2.2 Geological Setting

The rockslide area is located on the steep eastern bank of the Fraser River between Lytton and Boston Bar, approximately 300 m above the river level and 120 m above the railway tracks (Figs. 36.1 and 36.2). The river parallels the right lateral Fraser River Fault, which forms the boundary between the Coastal Belt and the Intermontane Belt of the Canadian Cordillera. The Pasayten Fault is located upslope, on the east side of the valley.

Post-glacial river erosion has resulted in the formation of terraces in many places along the river bank. However, terraces are absent near the rockslide, and a small side-canyon delta on the west side of the Fraser Canyon, upstream of the site, deflects the river flow against the east side of the canyon, accelerating erosion at the base of the slope (Piteau 1977). There is evidence for mountain-scale deep-seated gravitational slope deformation on the east flank of the Fraser River valley, including ponds, talus, and antislope scarps (Fig. 36.1b). The recent rockslide occurred within the boundary of a paleo-landslide (Fig. 36.3a).

The bedrock consists of the Jackass Mountain Group, a thick succession of Cretaceous shallow-water deltaic sedimentary rocks (MacLaurin et al. 2011). Rocks exposed along the track comprise greyish brown and highly fractured argillite, with locally well developed, steeply dipping cleavage. The rock overlying the argillite and exposed mid-slope above the track is a sandstone or sandy siltstone. At the top of the slope, a massive, thick-bedded pebble conglomerate forms a line of prominent cliffs (Fig. 36.3a). The bedding dips moderately into the slope.

Three seismic refraction lines done on the slope directly above the rockslide back scarp indicated that competent bedrock lies 20–22 m below the surface, progressively decreasing to 12 m upslope. Competent bedrock is overlain by a layer of fractured and weathered bedrock; this layer is approximately 15 m thick directly above the failure surface, progressively decreasing to 7 m upslope. Approximately 5 m of overburden overlies the fractured bedrock.

Two local faults were observed at the rockslide location (Fig. 36.3). Fault 1 (F1) is sub-vertical and strikes NW–SE, slightly oblique to the rock slope. It appears below the tunnel at Mile 109.4 and its trace extends along the base of the rockslide through a gully, where it can be recognized as a surface with calcite coating. A black shale layer shows a 20 m offset along the gully suggesting an apparent normal displacement (Fig. 36.3). Fault 2 (F2), which is characterized by very smooth surfaces and calcite coating, can be observed along the north side of the failure surface where the black shale layer is folded. It is an apparent reverse fault with an apparent offset of a few meters.

36.3 Rockslide Characterization and Analysis

The rockslide occurred on November 25, 2012. Prior to the event, tension cracks (Fig. 36.2a) were observed, which would eventually form the back scarp of the failure surface. A rockslide volume of 53,000 m^3 was estimated based on the comparison of pre- and post-slide digital elevation models (DEMs).

Rock slope characterization (Table 36.1) was undertaken using a combination of aerial imagery, terrestrial and airborne LiDAR, field mapping and seismic refraction. Terrestrial and airborne LiDAR data proved to complement each other to capture both the steep rock faces and moderately dipping talus zone of the terrain.

It was observed that the failure surface (Domain 2 in Fig. 36.3a) consists of three sectors. In Sector A (to the north), the failure surface mostly follows the fault F2. In

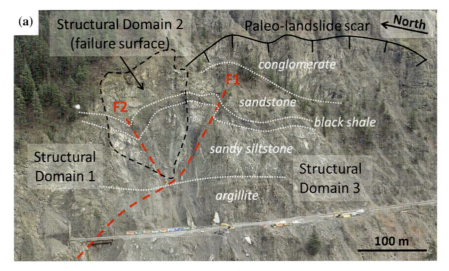
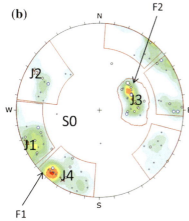

Fig. 36.3 Rockslide area. **a** Local geology and structure. **b** Stereonet (lower hemisphere) for Structural Domain 2 (S0 represents the bedding)

Table 36.1 Domain 2 (failure surface) discontinuity parameters

Set	No. of joints	Dip (°)	Dip direction (°)	Joint length (m)			Joint spacing (m)		
				Min	Mean	Max	Min	Mean	Max
J1	22	88	69	3.0	12.1	22.0	0.3	2.5	6.6
J2	17	85	120	3.0	9.9	28.0	2.4	7.2	12.1
J3	20	44	252	5.0	11.5	25.0	1.4	5.6	9.0
J4	19	84	31	3.0	8.0	26.0	0.9	3.3	9.0

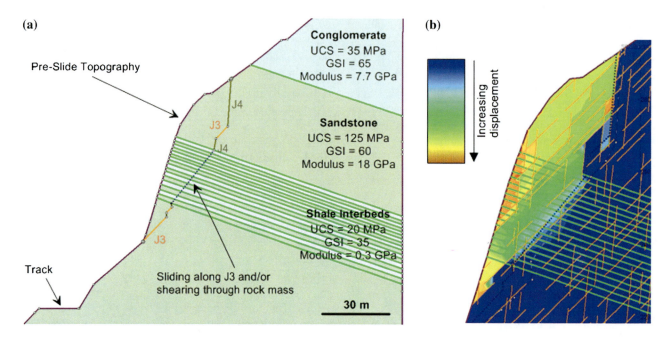

Fig. 36.4 Rockslide back-analysis. **a** Cross section. **b** Relative displacement obtained with *Phase2* with DFN model

Sector B (along the center), the failure surface follows a combination of joint sets J3, J4, and likely some component of intact rock bridges. In Sector C (to the south), the failure surface appears to be quite irregular and composed of a combination of joint sets J2, J3, J4 and intact rock bridges. The failure mechanism is complex, with a major component of two-dimensional sliding in Sectors A and B, and lateral release in Sector C. At the top of the three sectors, a combination of J1, J2 and J4 provided upper release through open sub-vertical joints.

A separate block failed along the northern part of the main failure surface (Fig. 36.2b), probably by sliding along joint set J3 and/or fault F2. This generated a massive block of sandstone having a volume of approximately 400 m³.

Based on rock slope characterization, a cross section was selected for finite element (FE) modeling using the software *Phase2* (RocScience) (Fig. 36.4). The Hoek-Brown rock mass criterion and a Mohr-Coulomb constitutive model for the discontinuity shear strength were assumed. Several back-analysis scenarios, with varying persistence, friction angle, and cohesion along a stepped failure surface were modeled.

Only the most realistic case, which incorporates a DFN model to reproduce the failure surface, is illustrated in this paper.

DFN generation in *Phase2* requires input and calibration of joint persistence and spacing (joint sets J3 and J4, see Table 36.1) so that the DFN matches the post slide profile (Fig. 36.4). A back analysis showed that in order to achieve a critical strength reduction factor (SRF) of 1.0, a 44° friction angle was required.

36.4 Discussion and Conclusion

The rockslide location coincides with the presence of two local faults, small-scale folds of the sedimentary layers, and a paleo-landslide scar (Fig. 36.3a). A daylighting joint set also played a major role in destabilizing the slope (Fig. 36.3b). Other major rockslides have been reported to coincide with structural features (e.g., Brideau et al. 2009; Pedrazzini et al. 2011; Sartori et al. 2003) and/or the location of a paleo-landslide (e.g., Mathews and McTaggart 1978).

These structural and morphological characteristics should be used as criteria for the identification of future potential rockslides along infrastructures, such as railway tracks.

The results of the finite element analysis suggest that the incorporation of a DFN model can realistically reproduce the step-path geometry of the failure surface. Consequently, DFN models may become increasingly valuable in the modeling of similar cases (Gischig et al. 2011; Oppikofer et al. 2011; Sturzenegger and Stead 2012). However, it should be noted that finite elements modeling may not always be appropriate to simulate absolute displacement along a discrete surface. In this study, the reliability of the results was constrained using an initial limit equilibrium analysis, which is more commonly used in the industry.

References

Brideau M-A, Yan M, Stead D (2009) The role of tectonic damage and brittle rock fracture in the development of large rock slope failures. Geomorphology 103:30–49

Environment Canada (2013) Canadian climate normals 1971–2000 (Lytton). http://climate.weatheroffice.gc.ca/climate_normals/index_e.html

Gischig V, Amann F, Moore JR, Loew S, Eisenbeiss H, Stempfhuber W (2011) Composite rock slope kinematics at the current Randa instability, Switzerland, based on remote sensing and numerical modeling. Eng Geol 118:37–53

Intergovernmental Panel on Climate Change (IPCC) (2007) 4th report, climate change 2007

Keegan T (2007) Methodology for risk analysis of railway ground hazards. Ph.D. thesis, Department of Civil and Environmental Engineering, Edmonton, Alberta

MacLaurin C, Mahoney J, Haggart J, Goodin J, Mustard P (2011) The Jackass Mountain Group of south-central British Columbia: depositional setting and evolution of an early Cretaceous deltaic complex. Can J Earth Sci 48:930–951

Mathews W, McTaggart K (1978) Hope rockslides, British Columbia, Canada. In: Voight B (ed) Rockslides and avalanches. Elsevier, Amsterdam, pp 259–275

Oppikofer T, Jaboyedoff M, Pedrazzini A, Derron M-H, Blikra LS (2011) Detailed DEM analysis of a rockslide scar to improve the basal failure surface model of active rockslides. J Geophys Res 116 (F2):22 pp

Pedrazzini A, Jaboyedoff M, Froese C, Langenberg W, Moreno F (2011) Structural analysis of Turtle Mountain: origin and influence of fractures in the development of rock slope failures. In: Jaboyedoff M (ed) Slope tectonics. Geol Soc London, Spec Publ 351:163–183

Piteau D (1977) Regional slope-stability controls and engineering geology of the Fraser Canyon. Geol Soc Am Rev Eng Geol 3:85–111

Sartori M, Baillifard F, Jaboyedoff M, Rouiller JD (2003) Kinematics of the 1991 Randa rockslides (Valais, Switzerland). Nat Hazards Earth Syst Sci 3:423–433

Sturzenegger M, Stead D (2012) The Palliser rockslide, Canadian Rocky Mountains: characterization and modeling of a stepped failure surface

Experimental Study on Water Sensitivity of the Red Sand Foundation in Angola

Wei Zhang, Zhenghong Liu, Jianguo Zheng, Sumin Zhang, and Yongtang Yu

Abstract

In Angola's capital Luanda and its surrounding areas, there distributing abundant brownish red sand called Quelo, which possesses the characteristics of softening and collapsibility when immersed with water. So how to improve the red-sand foundation reasonably and economically is of vital importance and urgency for the infrastructure construction of the area. Based on certain housing project in a typical red sand site in Luanda, a large in-situ water immersion experiment on the site was performed, which simulate the foundation treatment of replacement cushion, foundation dimension and load value of actual buildings. This paper records this experimental study and aims to investigate into the effectiveness of foundation treatment method, this paper records and studies the water-immersion deformation of the foundation after treatment and the rules of water permeation. The results show that: (1) the special formation structure of the red sand site determined a wider impacted area after water immersion; (2) the mudstone distributed under the red sand layer may expand slightly after water immersion; (3) the indoor test tends to exaggerate the hazard of red-sand collapsibility; (4) the bearing capacity of red-sand after softening with water-immersion should be adopted in the foundation design of multi-storied residential buildings on the red-sand site, and (5) the foundation treatment of replacement cushion is effective and reasonable.

Keywords

Quelo sand • Collapsible soil • Softening with water immersion • Foundation treatment • Permeability

37.1 Introduction

In Luanda, the capital city of western African country Angola, and its surrounding areas, there is wide distribution of a brownish red silty sand soil called Quelo (or Muceque), which possesses collapsibility and deteriorated mechanical properties after immersed with water. This unique physical and mechanical feature of the soil often poses much jeopardy to local construction projects.

According to existing literature, Quelo sand is thought to be Pleistocene marine sediments which were initially composed of fine sand and medium sand particles, but later, were remodeled and laterized under the terrestrial environment, producing clay mineral compositions of kaolinite, illite and iron oxide (hematite and goethite). The iron oxide contained in the soil is the main reason of its being red in color.

The collapsibility of Quelo sand has been discovered and proved by Novais-Ferreira and Silva (1961) adopting the method of plate loading test (PLT), and by Silva (1970) using consolidation test method. The results of these tests show that with low moisture content, the subsoil was able to withstand high load (up to 800 kPa) with small settlement; but once saturated, the bearing capacity will decrease due to the bonding failure of particles.

W. Zhang · Z. Liu (✉) · J. Zheng · S. Zhang · Y. Yu
China JK Institute of Engineering Investigation and Design, Xi'an, China
e-mail: 22232508@qq.com

G. Lollino et al. (eds.), *Engineering Geology for Society and Territory – Volume 6*,
DOI: 10.1007/978-3-319-09060-3_37, © Springer International Publishing Switzerland 2015

Currently, as a number of infrastructure projects are under construction in the Quelo sand distributed areas in Luanda, how to design the foundation of the buildings has become focal attention. Usually the design follows two schemes: one is to follow the foundation design for collapsible soil, which requires either manual replacement of part or all of the collapse soil with non-collapsible soil by way of certain measures, or using piles that penetrate the whole collapsible soil layer; the other is to design according to the bearing capacity of the subsoil after its softening with water-immersion, which, in this case, can meet the bearing capacity with simply replacement of the compacted cushion. The former scheme can ensure the safety of the buildings but require relatively high cost, while the latter one, though with low cost, needs further research on its safety. Based on one of the house construction projects in Luanda, this study probes into the rationality of designing foundations on Quelo sand site where the soil possesses property of softening and collapsibility after water-immersion. For this purpose, water immersion experiment was carried out aiming to investigate the prototype foundation of five-storied buildings to obtain data revealing rules of water permeation in the foundation soil and the additional deformation of the foundation when soaked with water.

37.2   Engineering Geological Conditions of the Test Site

Located in southern part of Luanda city, about 14 km to the Atlantic on the west and about 15 km to the Kwanza River on the south, the testing site possesses the stratum structure shown in Table 37.1 and Fig. 37.4, which resembles the typical structure of Quelo sand distribution. The brownish red silty sand② layer is the typical Quelo sand layer and contains low moisture in tree-covered district. Both the in-situ testing index in the borehole (SPT and dynamic penetration test) and the collapsibility coefficient in the indoor test are impacted greatly by water content within the soil. The silty sand③$_2$ layer, which takes on mixed-color, is mainly made up with white, yellow and red soil blocks. The silty sand③$_1$ layer, serving as the transitional layer between the layers of silty sand② and silty sand ③$_2$. Both the sandstone④$_1$ layer and the mudstone④$_2$ layer are of weak diagenesis, and some of the mudstone samples have shown swelling during the indoor test.

Soil samples are taken from the borehole to conduct indoor test, and the physical and mechanical property indexes (average) of each layer of the foundation soil have been obtained and is shown in Table 37.1. The collapsibility coefficient is obtained from confined compression test, and

the test pressure, for the depth within 10 m underground the pressure being 200 kPa, and for the layer below 10 meters underground the pressure being the saturated deadweight of the overlying soil layer. As shown in Table 37.1, the silty sand layers of ②, ③$_1$ and ③$_2$ are found with high collapsibility coefficients according to the indoor test results, and the self-weight collapse settlement under the self-weight stress below the compacted cushion in Fig. 37.4 is calculated at 173 mm according to formula 1.

$$\Delta_{zs} = \sum_{i=1}^{n} \delta_{zsi} h_i \qquad (37.1)$$

Where δ_{zsi} means the self-weight collapsibility coefficient of layer i, while h_i refers to the thickness of the ith layer in mm.

And no ground water of any kind has been found within 30 m underground at the testing site.

37.3   Scheme of the In-Situ Test

37.3.1   Design of the Building Foundation

The construction project supporting the in-situ test includes 424 five-storied residential buildings with the height of 15 m. Main body of the buildings adopts brickwork structure, and the foundations are built in area where the Quelo sand is the main bearing stratum. According to the PLT, the characteristic value of bearing capacity of the Quelo sand is between 100 and170 kPa, averaging at 120 kPa when softened with water immersion. If the basic design of the foundation adopts this bearing capacity, the follow scheme can be generated: (1) adopting the reinforced concrete strip foundation base, with the bottom width of 1.0–1.6 m, and the buried depth being 1.0 meter; the average additional stress being 120 kPa; (2) replace the soil within 1.0 meter depth under the reinforced concrete base with refilling compacted cushion of brownish red Quelo sand material.

37.3.2 Arrangement of the Test Pit

The purpose of the full-scale foundation immersion test is to prove the validity of the above foundation design scheme in terms of safety, so the foundation treatment, the buried depth and the form of the foundation base, as well as the apron are all simulations of the real construction engineering arrangement. As shown in Figs. 37.1 and 37.4, the plane of the water immersion test pit is a rectangular, two foundation

Table 37.1 Physical and mechanical property index of the foundation soil

Layer no.	Soil type	t/m	w/ %	ρ_d / g·cm^{-3}	d_s	E_{s1-2} / MPa	δ_s	δ_{zs}	Soil color
②	Silty sand	7.3	5.9	1.66	2.67	13.4	0.020	0.014	Brownish red
③₁	Silty sand	2.3	6.1	1.72	2.68	16.6	0.021	0.021	Variegated, mainly red
③₂	Silty sand	2.9	8.6	1.80	2.68	17.2	0.014	0.013	Variegated (white, yellow and red)
④₁	Sandstone	3.0	10.2	1.70	2.68	16.1	0.009	0.009	Variegated, mainly off-white
④₂	Mudstone		15.2	1.71	2.73	18.9			Grayish green

t layer thickness, w water content, ρ_d dry density, d_s specific weight of soil particle, E_{s1-2} compression modulus under pressure of 100–200 kPa, δ_s coefficient of collapsibility, δ_{zs} self-weight coefficient of collapsibility
Soil classification is based on China's *Code for Investigation of Geotechnical Engineering* (GB 50021–2001, 2009 Edition)

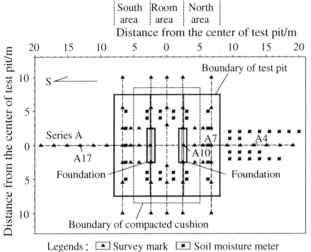

Fig. 37.1 Layout of the test pit

37.3.3 Testing Process

The test includes the following steps: (1) Apply load of 147 t (the first loading) on the foundation before the immersion test, with the corresponding additional contact stress of the concrete base being 123 kPa, close to 120 kPa, the average additional stress of the building foundation; (2) immerge exclusively the north area of the test pit with water, see Fig. 37.2a, to stimulate and test the immersion range and the deformation feature of the foundation when the outside of the building were fully immerged with water; (3) immerge the whole pit with water, see Fig. 37.2b; (4) increase the load to 191 t (the second loading), with the corresponding additional contact stress of the base increasing to 159 kPa; (5) Again immerge the whole pit with water.

37.4 Field Test Results

37.4.1 Volume of Immersion

The test injected a total of 5,530 m³ of water into the pit with the duration of 111 days, which was divided into the three periods of a, b, and c as shown in Fig. 37.3. The three periods include the period of immersing the northern area of the site exclusively, the period of full immersion of the construction site after the first load and the period of full immersion of the site after the second load, with immersion volumes of 948, 3,570 and 1,012 m³ respectively. Due to limitation of the outside conditions, water-break appeared in the process of immersion (see Fig. 37.3).

strips with 5 m long and 1.2 m wide are placed in the center of the test pit, and the basic axis distance is 4.8 m. Below the foundation is a whole sheet of compaction replacement cushion, 1.0 m thick. From north to south, the test pit can be divided into three areas: "the north area", "the room area" and "the south area".

To monitor the deformation of the foundation and subsoil in the process of testing, 58 deformation observation marks have been arranged in and around the test pit (Fig. 37.1). To monitor the rule of water permeation in the process of testing, 46 (Fig. 37.1) soil moisture meters have been buried in and outside of the testing pit with buried depth being 2–10 m below the bottom (lowest point) of the testing pit.

Fig. 37.2 Scene of immersion testing site. **a** Test pit with only the north area immersed. **b** Test pit with full water immersion

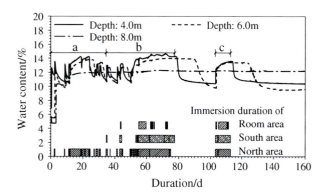

Fig. 37.3 Water content changing with time in north area

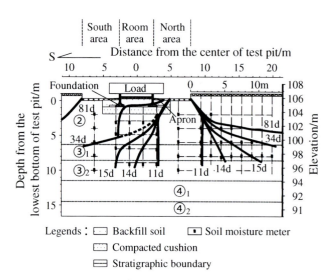

Fig. 37.4 Soaking range at different duration days

37.4.2 Rule of Water Permeation

By way of calibration and through the readings of the moisture meters buried in the subsoil, the real-time moisture content of the soil can be obtained. Figure 37.3 shows the moisture content (mass fraction) curve changing with time of the north area (outside the compaction cushion) monitored at the respective depth of 4.0, 6.0 and 8.0 m of the test pit. It can be seen from Fig. 37.3 that when water reached the place where the moisture meters were buried, water saturation of the soil increased obviously, and correspondingly, the readings of the meters also changed. By analyzing the changes of the readings of all the 46 moisture meters, the real-time

immersion range of water in the process of the test can be acquired. Figure 37.4 shows the immersion range curve at different times after the test beginning.

As can be seen from Fig. 37.4, the immersion test has made the all natural Quelo sand in the pit soaked by water (the 81st d curve of Fig. 37.4). Figures 37.3 and 37.4 also reveal the following phenomena:

(1) When soaked with water, moisture content of the silty sand② layer experienced a sharp increase at first, and then gradual increase; But when the water was stopped,

the moisture content experience of the layer was a sharp decrease first, which was then followed by gradual decrease (Fig. 37.3).

(2) Moisture content of the subsoil of the silty sand② layer is closely related with water supply. As shown in figure 37.3, water content at the depth of 4.0 and 6.0 m both experienced a process of first increasing and then decreasing each time when the subsoil was immerged in water and when the water was cut off.

(3) For the mixed-colored silty sand ③₂ layer and the subsoil of lower part of ③₁ layer, their moisture content remained constant for a long time once it was increased (Fig. 37.3).

(4) During the early stage of immersion, the direction of water penetration was mainly vertical, with the lateral infiltration distance less than 2 m; but as the soaking continues till the water infiltrating surface reached the silty sand③₂ layer, water invasion range gradually expanded to the lateral side; Approaching the ending period of the immersion, the minimum angle between the saturation line and the horizontal plane was about 8° (the 81st d curve of Fig. 37.4).

The above phenomena indicate that, the silty sand② layer possesses good water permeability (with larger permeability coefficient), and the silty sand③₂ layer and the underlying sandstone and mudstone possess poor permeability (with smaller permeability coefficient). So in the process of immersion, water can penetrate very far in the lateral direction. Therefore, when the thickness of the Quelo sand was poor, the ground water can easily increase the moisture content in the Quelo sand layer under the foundation, and form upper perched water that submerges the Quelo sand layer.

37.4.3 Deformation of the Foundation

During the process of the test, precision leveling instrument was used to monitor the pre-set deformation observation marks regularly every day to acquire the additional deformation data of the foundation after water immersion. The curves of the deformation changing with time are shown in Fig. 37.5, recorded by the most representative A4, A7, A10 and A17 observations marks (plane position shown in Fig. 37.1). Figure 37.6 shows the settlement section connected by the deformation marks of the above A series at different times (i.e. along the north and south axis of shown in Fig. 37.1) after the test started with first immersion.

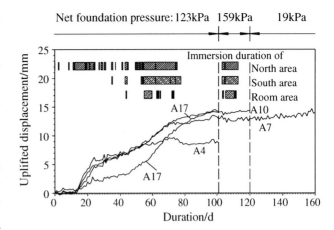

Fig. 37.5 Deformation curve of the most representation marks changing with time

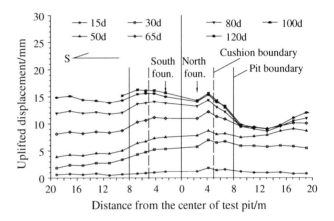

Fig. 37.6 Settlement section connected by A series of deformation marks at different times

Analyses on the monitoring results show that certain amount of uplift rather than obvious settlement occurred after water immersion on each of the deformation observation marks. Other marks (for example, A7, A10 and A17 in Fig. 37.5), except the marks set to the north of the test pit, have lifted continuously in the test. While the marks to the north of the test pit was also witnessed with continuous uplift at the starting period of the test, slight settlement appeared during the ending period of the first loaded water immersion (the 74th d after the test started). Comparing

Figs. 37.5 and 37.4, it is clear that within 11d after the test started, though layers of silty sand②, ③$_1$ and ③$_2$ were all wetted with water, no obvious deformation appeared on any of the observation marks; but with the on-going of the immersion, uplifting gradually appeared on all the marks. This phenomenon indicates that the deformation is mainly caused by the weak expansion of the mudstone④$_2$ layer after being wetted with water. During the late period of the first loading (100d after the test started), the uplift deformation readings of the marks were between 9.0 and 17.0 mm, with the average of 14.3 mm. Figure 37.6 reflects the deformation of the foundation from another perspective. As can be seen from Fig. 37.6, that (1) the deformation of the foundation has no obvious difference with its surrounding foundation soil after immersion though the base has been loaded with 123 and 159 kPa; (2) a "deformation groove" can be seen in the ground north of the pit, and the reason may be that it is located in area covered by trees, where the moisture content of Quelo sand is low and where the collapse intensity is bigger, so certain collapse settlement formed when the place was immersed with water. If the depth of the groove were regarded as the collapsible settlement of the Quelo sand after immersion, the collapsibility volume is relatively small (less than 6.3 mm).

Analyzed comprehensively, the overall deformation value after water-immersion is small, and no significant collapse settlement appeared in the Quelo sand. This means that water-immersion is not enough to cause inhomogeneous deformation that may affect the safety of the building. This also indicates, therefore, that for multi-storied civil buildings, it is feasible to design the foundation on the basis of the bearing capacity of the Quelo sand after being softened with water immersion, given its unique engineering properties.

37.4.4 Discussion

According to the indoor test results, the collapse settlement of the subsoil of the test site could be 173 mm under self-weight pressure alone after water immersion when calculated with formula 1. Its settlement value would be much larger if additional stress was taken into account. However, the actual collapse settlement measured in the in-situ test was very small. So this indicates that evaluation of the foundation soil engineering property generated from indoor tests exaggerates the risks of the foundation soil, and the designs based on data thus obtained tends to be too conservative.

Similar deviation in the volume of collapse settlement obtained through the in-door calculated value with formula 1 and through the in-situ measured value was also found existing in construction projects in China's loess distribution area, the typical collapse site in the world. It is believed that the contradiction is mainly caused by the differences of soil structures and stress conditions existing between the indoor test soil samples and the actual construction site. Therefore, the current *Code for building construction in collapse loess regions* practiced in China has stipulated that in evaluating the collapsibility, loess with collapse coefficient smaller than 0.015 is regarded as non-collapse loess, and shall not taken into account when calculating the collapse settlement. Besides, the calculation of self-weight collapse settlement value is also based on formula 1, but with a correction coefficient between 0.5 and 1.5 (the lowest one used for the southeast region; and the largest one for northwest region). For the Quelo sand, similar effect may also exists in the calculation of collapse settlement, where exaggerated settlement value may be resulted if calculated through indoor test by way of formula 1.

37.5 Conclusions

On the basis of an actual house construction engineering project and simulating the cushion replacement method, foundation dimensions and loading value adopted in actual project, this paper reported an in-situ water immersion test conducted on full-scale foundation in typical Quelo sand areas, and the comprehensive research on the foundation reaction after water immersion and on the rule of its water permeation. The following findings have been reached:

(1) Typical Quelo sand site consists of a top layer of Quelo sand of high permeability, and a lower layer of mixed-colored sand, sandstone and mudstone with low permeability. This unique stratum makes water permeate far laterally when the ground keeps being soaked in water, leading wider plane of immersion range.

(2) The mudstone under the test site expands slightly after immersion, which causes the upward deformation of the foundation but with limited value.

(3) Though indoor tests show that the Quelo sand is with collapsibility, the in-situ test results have shown collapse settlement value much smaller than that obtained through the indoor test. So the indoor test has exaggerated the risks of the Quelo sand. This phenomenon is in line what has been discovered in some of China's loess (typical collapsible soil) site.

(4) In the in-situ immersion test, there is no obvious difference in the deformation between the soaked foundation and those in other districts. Therefore, in designing the foundation of multi-storied house-buildings, the bearing capacity of the red sand can adopt that after the soil being soaked and softened, and it is feasible to improve the Quelo sand foundation soil with cushion replacement.

Reference

Novais-Ferreira H, e Silva CAF (1961) Soil Failure in the Luanda Region. Geotechnique Study of These Soils. Proc. 5th Int. Conf. Soil Mech. Found. Eng, Paris

Silva JAH (1970) Engineering geology and geotechnical behaviour of expansive and collapsing soils of Angola. Tese do Estágio Realizado no Imperial College of Science and Technology, Londres

Geological Characterization and Stability Conditions of the Motorway Tunnels of Arrangement Project of the NR43, Melbou (W. Béjaïa)

38

Nassim Hallal and Rachid Bougdal

Abstract

The oriental coastal region of Béjaïa city is situated in the mountainous chain of the Babors. The relief is steep as well as slopes which are very abrupt. In the North, the Babors coastal massif dominates the Mediterranean Sea forming the Cap of Aokas and Djebel Djamaa N' sia in the city of Melbou. The recent realized tunnels (T1, T2) cross the limestone massif of Djebel Djamaa N' sia dated of the Jurassic era, which shows an intensive deformation due to the fracturing as well as to the karstification. Several instabilities have been observed at the time of their realization. Both tunnels (T1, T2) are excavated by mining in two sections; Skullcap and stross. Basing on the geologic survey, the geotechnical interpretation and the geomechanical classification of Bieniawski are the basis for what the retaining type were stopped (Flies, anchoring bolts or the special processes for particular cases).

Keywords

Babors • Tunnels • Instability • Support systems • Geomechanic classification

38.1 Introduction

In the framework of the extension of NR43 (Béjaïa), two tunnels (T1, T2) were executed for the aim of crossing the limestone massif of Djebel Djemaa N' sia. The total length of the both tunnels is about 1,497 m (Fig. 38.1).

The thickness of the skullcap cover is approximately 170 m for T2 and 135 m for T1. These two tunnels were excavated by mining in two sections, skullcap and stross.

38.2 Presentation of the Work

The arrangement project of the NR43 consists in the realization of two tunnels, which are monotube with two ways, of a width of 11.10 m, to 7.23 m of height and a 2 × 4 m

N. Hallal (✉) · R. Bougdal
Scientific University of Algiers USTHB, Algiers, Algeria
e-mail: geoaokas@yahoo.fr

R. Bougdal
e-mail: bougdalr@yahoo.fr

road. A quoted pavement of 80 cm in Width on the both sides (sea and mountain) in every tunnel is also planned.

38.3 Geological Setting

The tunnels T1 and T2 on the NR43 which cross the massif of Djebel Djamaa N' sia are situated in the Melbou municipality. The geological formations observed in both tunnels have the same nature of the Jurassic limestone (Coutelle 1979).

38.4 Geology of the Studied Area

The massif of Djebel Djemaa N' sia is featured by hard limestone rocks (Figs. 38.5 and 38.6), grey color and brown patina (Leikine 1971). Most of the outcrops show an intense deformation due to the paroxysmal phase of the Miocene (kirèche 1993), that is expressed by fractures more or less closed papered with calcite and iron oxides deposits. The stratification plans are observed in several points on the extreme of the massif; they are characterized by directions

Fig. 38.1 Situation map of both tunnels (T1, T2)

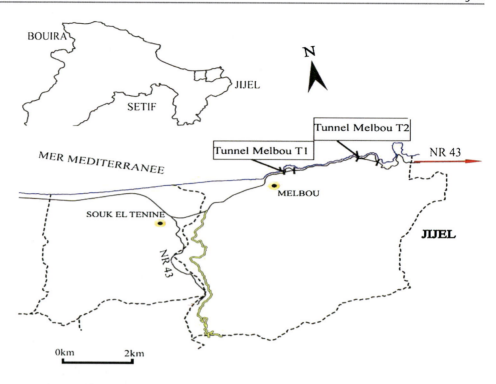

Fig. 38.2 Karstic cavity in the tunnel T1

Fig. 38.3 Karstic cavity in the tunnel T2

going from N040° to N070° and dips from 40° to 75° southward.

A big vertical fault (N070°) is observed on the Djebel Djemaa N'sia upstream side. Secondary faults, which break off in the contact of the major fault, are observed along the NR43.

Along the borders of this massif, limestone breaches few to well concreted and masses landslides with reddish clay matrix are observed, they mask the limestone feet of cliffs at around depressions from the East tunnel portal T02 (PK 0 +900) until the Melbou exit of (PK 1+700), where we find strengthened sand beds which are surmounted by a layer of reddish clays.

From the tunnel T02 (PK 0+900) to the slope of the ravine, schist formations (crossed by the engineering work) appear, the first one is of marly nature and yellowish to brunette color which is in unconformity on a gray to black clay formation, crossed by calcite veins of very variable thicknesses (Obert 1981).

Fig. 38.4 a Stereographic diagrams of the measures inside tunnel T1. Lower hemisphere. **b** Stereographic diagrams of the measures inside tunnel T2. Lower hemisphere

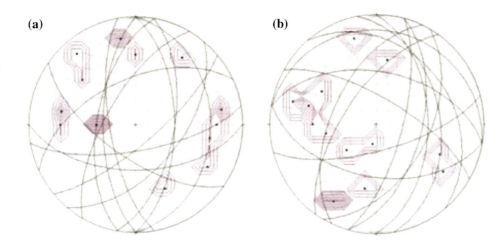

(a)   **(b)**

38.5 Main Causes of Instability

38.5.1 The Fracturing

The limestone massif is crossed by a big vertical fault (N070°), Observed on the upstream side of Djebel Djemaa N' sia. The continuation of this fault toward the southeast controls the geomorphology of limestone massif bounds. Secondary fault systems, which break off in the contact of the major fault, are observed along the NR43. Three sets of faults were registered along these tunnels during the digging works. The first one presents a direction of N025° to N040°, the second presents a direction of N050° to N075° and the third is characterized by a N090° direction.

A systematic measurement of the fracturing was effectuated during the progress in both tunnels (Fig. 38.4). As a consequence, data have been collected and statistically treated by a stereographic projection. These analyses confirm the existence of three fractures families.

Fractures observed at both tunnels are the result of the fracturing engendered by the tectonic movements of compression undergone by the rocks' formations during geologic history (Kirèche 1993). Three families of discontinuities are observed. They present, generally, the same orientation as the families observed in the Tunnel of Cap Aokas (Hallal and Bensafia 2011; Bougdal 2009). The spacing and the thickness of the discontinuities are very variable, they contain mostly brunette to yellowish clays, with sometimes alternations of thin layers blackish. These zones are mostly the seat of detachment of dihedral due to the bad holding of the rock.

38.6 Karstification

The limestone massifs of the Babors Lias are known by the existence of numerous dissolution and subterranean canal cavities which can develop on several kilometers.

In the calcareous massif of Djebel Djamaa N' sia, the karstic cavities were met throughout the drilling of both tunnels T1 and T2. They are characterized of a simple clayey filling of in the typical open karstic canal, passing by a karstic levels or sealed cavities. The majority of these phenomena develop on fracturing plans.

For the tunnel T1, several karstic spaces and caves were met, The most important were met in PK 0+121 (Fig. 38.2) and in PK 0+223 (Fig. 38.3), where the first one requires 49 m^3 of the concrete for its filling and the second space presents a total volume of 2,199.92 m^3 situated on highly-rated upstream of the tunnel axis. Besides, several karstic space were also been met, they are characterized by clayey fillings and intense fracturing associated with breaches. Between the Pk 0+760.90 and 0+780.80 a karstic system developed. Indeed, this part of the tunnel is characterized by fillings furniture established by pebbles and blocks of limestone wrapped in a brown and yellowish clayey matrix. This training presents a very precarious stability. None of the recut karstic space is really active from the point of view of in-rushes of water. For the tunnel T2, a karstic zone was met during the digging between Pk 0+169 and Pk 0+114. It is established by blocks centimeter in metrics wrapped in a reddish clayey matrix. The most important were met at the level of Pk 0+100, Pk 0+157 and in Pk 0+177 underlined by the presence of stalactites and stalagmites. The cave met at the level of Pk 0+177 presents a stream of low debit and an alluvial deposits testifying of a former strong flow.

38.7 Hydrogeology

38.7.1 General Framework

The fracturing and the karstification confers to the calcareous massif of Djebel Djemaa N' sia (massif of Babors) a sufficient permeability for the infiltration and the accumulation of subterranean waters (Obert 1981). The hydrogeologic situation of both tunnels is identical; it is represented in both cases by the Jurassic permeable limestone formations. The flows of groundwater met are those of karstic type through a discontinuity network and karstic channels.

The tunnel T2 is different from the tunnel T1 in this case by the presence of a small ravine of an important debit during the rainy periods, So it emerges from this hydrogeologic situation that more or less important water in-rushes could arise during the drilling of the tunnel and infiltrations of water can be expected in the tunnel T2 during the drilling.

38.7.2 In-Rushes of Water in Tunnels

38.7.2. Tunnel T1
- No coming water occurred during the works of excavation;
- A water seepage resulting from fractures in direct contact with karstic space;
- A water seepage is especially observed in skullcap, they appeared near the karstic space and to the contact of the faults;
- Numerous seepages are observed in tectonic zones, with very low debit.

38.7.2. Tunnel T2
- The temporary nature of the comings waters shows that it is mainly percolations waters till the surface of the rock massif in rainy period;
- Seepage and coming of water were more frequent near the karstic cavity, they appeared skullcap to the right of fractures. In the karstic space, met at the level of Pk 0 +177, a stream of weak debit was met and sediment trainings witness of a high debit of water.

In both tunnels, during heavy rain, the coming of water were systematically listed and gauged when their debit allowed it.

38.8 Behavior of the Rock Massif in the Progress

During the digging works, the observations made during the progress to estimate the holding of the rock after mining and to recover extra profiles in the unstable zones are proved true difficult because of the works of marinating and bleeding, Which followed directly mine blasting, and fast pose of a protection coat of concrete thrown in skullcap and partially in facings.

38.8.1 Skullcap

The unsticking of the dihedrals is inexistent in skullcap during the progress of both tunnels, a little lesser in tunnel T2, situated in the faults zones.

The holding of the rock in skullcap has been proved globally good to very good, rarely bad or very bad in intensely broken or karstified zones.

38.8.2 Facings

For both tunnels, the holding of the rock in facing has been proved globally good to very good, with some noticed instabilities. A bad to very bad holding in intensely broken zones, in the zones of weaknesses as well as in the karstic spaces.

38.8.3 Attack Front

Faces presented a good behavior, with some limited cases of bad holding due to the fracturing and to the karstification of the rock and to the nature of filling.

38.9 Geotechnics

38.9.1 Geomechanical Characteristics of Rocks

- According to the standard IN ISO 14689-1, the calcareous massif of DJ Djemaa N' sia presents a resistance to compression with a rocky matrix estimated as high to very high.
- The results of mechanical tests made on the calcareous massif within the framework of the study of each constraint met in the progress (fractured and karstic zones) revealed low resistance in the compression with a rocky matrix globally less resistant.

38.10 Application of the Method of BIENIAWSKI in the Crossed Massif

The method applied for the classification of the rock massif of DJ Djemaa N' sia is the method of BIENIAWSKI (Bieniawski 1968; Barton et al. 1974), in order to arrive at a global evaluation of its quality.

The Figs. 38.5 and 38.6 summarize the evolution of the variation of the RMR and the retaining structure set up according to the digging in both tunnels.

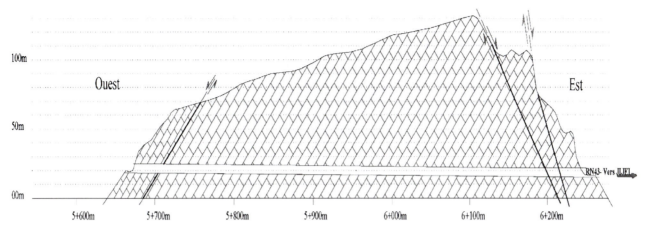

Fig. 38.5 Geological and geotechnical cup of the tunnel T2

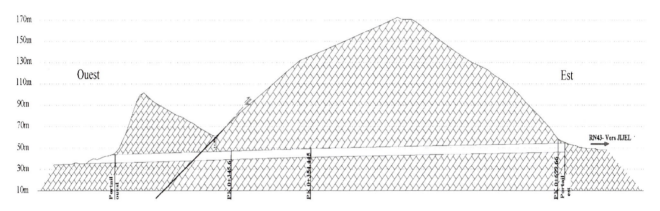

Fig. 38.6 Geotechnical and geological cup of the tunnel T1

RMR									
MIN	28	41	44	33 33	25	37	49		45
MAX	39	49	50	43 50	37	45	52		56
Soutènnement réalisé		C		A	C	C	C		C

0+123.75 0+175.00 0+230.00 0+665.00 0+699.00

Fig. 38.7 Variation of RMR and type of retaining structure realized along T2

RMR																						
MAX	50	53	42	52	36	51	40	47	42	53	53	50	40	48	42	40	50	45	52	45	50	45
MIN	37	40	27	42	28	40	30	40	29	42	42	42	34	42	37	42	40	34	40	35	42	37
Soutènnement réalisé	C	C	B10	C	C	B10		B10	B10	C	C	C	C	C	C	C C	C			C	C	C

0+080.20 0+142.60 0+223.60 0+279.10 0+301.60 0+395.10 4+450.40 0+542.90 0+612.40 0+649.90 0+685.90 0+760.90 0+803.00 0+884.60 0+928.40

Fig. 38.8 Variation of RMR and type of retaining structure realized along T1

These figures show that the geotechnical characteristics of the met rock formations are variable. The bad characteristics were met in the zones of strong fracturing and the karstic zones which coincide with the zones of faults. The variation of the RMR along tunnels T1 and T2 is given in the following Figs. 38.7 and 38.8.

38.11 Conclusion

The study and the follow-up geologic works made along both tunnels (T1, T2), allowed us to collect a significant number of data which found the following utilities:

From a geologic point of view, the massif crossed by tunnels is essentially represented by limestones dolomitic massifs of the lower Jurassic.

From a hydrogeologic point of view, no ground-water was revealed in the massif and no coming of water was observed in the tunnel T1, except for the inside of the tunnel T2 in Pk 0+177, a stream of weak debit was met.

The analysis of the in situ tests results and those in laboratory, realized within the framework of the study of each met constraint and according to the standard IN ISO 14689-1 used for the classification of calcareous massif of DJ. Djemaa N' sia, the geomechanical characteristics of the crossed massif are summarized as follows:

- A compression resistance with a rocky matrix estimated as high to very high for the calcareous massif of Djebel Djemaa N' sia.
- The geomechanical application of the classification of Bieniawski shows that the geological and geotechnical characteristics of the crossed massif vary along traces, It shows a succession of zones classified in the categories II (good), III (average), IV (bad) and V (very bad).

By basing itself on the Bieniawski classification, the retaining structures adopted for every classification in every tunnel are the following ones:

In the category II (C)
- A coat of concrete thrown by thickness 50 mm;
- Bolts of anchorings of 4 m length with a stitch of 1.5 m × 1.5 m.

In the category III (B10)
- A coat of welded fatigues dress;
- A coat of concrete thrown by thickness 10 cm;

- Bolts of anchorings of 4 m length with a stitch of 1 m × 1.5 m.

In the category IV (A)
- A coat of concrete thrown by thickness 150 mm;
- A coat of welded fatigues dress;
- Arch reticule (H = 94 mm), spaced of 1 m;
- A coat of welded fatigues dress;
- Bolts of anchorings of 4 m length with a stitch of 1 m × 1 m.

In the category V
- Bend umbrella;
- A coat of welded fatigues dress;
- Arch reticule (H = 162 mm), spaced of 1 m;
- A coat of welded fatigues dress;
- A coat of concrete thrown by thickness 200 mm;
- Bolts of anchorings of 4 m length with a stitch of 1 m × 1 m.

References

Barton N, Lien R, Lunde J (1974) Engineering classification of rock masses for the design of tunnel support. Rock Mech 189–239

Bougdal R (2009) Doublement du tunnel de Cap Aokas, synthèse des données géologiques et géotechniques. Rapport géologique (Document interne)

Coutelle A (1979) Géologie du Sud-Est de la grande Kabylie et des Babors d'Akbou. Thèse doctorat sciences, Paris

Hallal N, Bensafia W (2011) Dédoublement du tunnel d'Aokas: Géologie et caractérisation géotechnique du massif traversé. Stabilité de l'ouvrage. Mémoire MASTER. USTHB. Alger

Kirèche O (1993) Evolution géodynamique de la marge tellienne des Maghrébides d'après l'étude du domaine parautochtone schistosé. Thèse doctorat sciences, Université d'Alger (USTHB)

Leikine M (1971) Etude géologique des Babors occidentaux (Algérie). Thèse de doctorat, Université de Paris

Obert D (1981) Etude géologique des Babors orientaux (Domaine tellien, Algérie). Thèse de Doctorat, Université Pierre et Marie Curie, France

A Modified Freeze-Thaw Laboratory Test for Pavement Sub Soils Affected by De-icing Chemicals

Assel Sarsembayeva and Philip Collins

Abstract

De-icing chemicals are the most effective and cheap method to prevent winter slipperiness on pavement surfaces in urban settings. Analysis of existing experimental studies and theoretical methods shows that solutions of de-icing chemicals run off surfaces and are deposited in adjacent soils. However, there is a lack of knowledge about the effect of de-icing chemicals on the engineering properties of pavement sub soils, where the de-icing agents may penetrate beneath the pavement surface. This is a particular issue as pavement surfaces typically cool and warm at a faster rate than surrounding areas. In particular, during seasonal freeze-thaw cycles, the lower freezing point of the chemical solution may induce moisture migration toward a freezing front within the sub soil, leading to increased heave potential, and subsequent thaw collapse. The paper describes an experimental study that simulates the movement of water and de-icing solution from the deposition area upward to the highway's sub base. The chemical effect of de-icing solutions on the strength characteristic, water and chemical content of each 10 cm layer of tested soil column will be assessed by in situ measurement and post-experiment analysis. In the long run the impact of the de-icing chemicals precipitation on the bearing capacity and deformation of sub base soils of roads will be evaluated.

Keywords

Roadside soil • De-icing chemicals • Engineering properties • Moisture migration

39.1 Introduction

Since the 1950s de-icing chemicals have been the most popular and effective method to keep highways surfaces clean and safe from winter slipperiness as the freezing temperatures of de-icing chemicals are much lower than pure water (Bing and Ma 2011). This freezing point can be determined from the eutectic diagram of each type of de-icing chemical.

De-icing chemicals are the cheapest way to prevent ice formation and usually guarantee its quick removal from the pavement surfaces. There are various types of inorganic de-icing agents, the freezing temperature of which can be determined from the eutectic diagram (Dreving 1954; West 1982). Widely used agents include chlorides, which are cheap but corrosive, and organic compounds like acetates, formats and urea, which are not so aggressive to crops and metals, but more expensive and may case an unpleasant smell of vinegar during use.

De-icing salts can be delivered to soil near roads by runoff and spray and cause the secondary salinization of the roadside area. There are numerous reports on simulation works and field observations of the deposition and redistribution of de-icing chemicals in the roadside environment (e.g. Norrström and Jacks 1998; Blomqvist and Johansson 1999; Lundmark and Olofsson 2007; Lundmark and Jansson 2008).

A. Sarsembayeva (✉) · P. Collins
Brunel University, Kingston Lane, Uxbridge, Middlesex, UB8 3PH, London, UK
e-mail: assel.sarsembayeva@brunel.ac.uk

G. Lollino et al. (eds.), *Engineering Geology for Society and Territory – Volume 6*,
DOI: 10.1007/978-3-319-09060-3_39, © Springer International Publishing Switzerland 2015

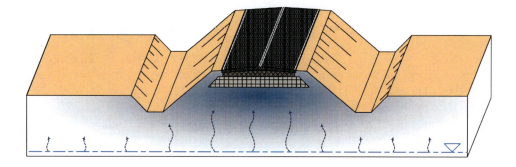

Fig. 39.1 The moisture migration in the highway basement reviewed without the usage of de-icing chemicals

Deposition of de-icing salts in the roadside inevitably changes the chemical content and moisture regime of the subsoil. This may have a great effect on its bearing capacity, especially during seasonal freeze-thaw cycles. Furthermore, it keeps the moisture of the nearby soil in a liquid state (Kane et al. 2001).

As the thermal conductivity of pavement materials is far and away higher than the adjacent soils the thermal gradient causes significant migration of moisture upward the freezing front that is where pavement layers (O'Neill 1983; Othman and Benson 1993; Giakoumakis 1994; Kane et al. 2001; Vidyapin and Cheverev 2008). Figures 39.1 and 39.2 describe the moisture migration toward the freezing front in winter period.

In the spring, when the swelled mass of sub soils starts to thaw and drain, a loss of bearing capacity occurs, as shown in Fig. 39.3 (Othman and Benson 1993; Simonsen and Isacsson 1999; Qi et al. 2006; Vidyapin and Cheverev 2008).

We hypothesis that the ability of de-icing chemicals to keep moisture in a liquid state at low temperature may enhance moisture migration toward the highway subsoil in

cold periods. Progressive, but slow freezing to below the eutectic point, such as occurs in areas like Central Asia and Siberia, is likely to induce cryosuction of the unfrozen soil moisture, inducing frost heave (Hoekstra 1966; O'Neill 1983; Othman and Benson 1993; Qi et al. 2006; Bronfenbrener and Bronfenbrener 2010a, b; Matsumura and Yamazaki 2012). This will enhance thaw-related loss of bearing capacity of highways in the spring.

In situ measurement of the migration, and cyclic freezing and thawing of de-icing agents is problematic. Instead, a modified frost heave test methodology has been developed.

39.2 Development of a Laboratory Model

The laboratory method for frost heave and thaw weakening susceptibility observations of soils is based on ASTM D5918-06 Standard, (2007). The installation scheme of this test simulates the movement of water and de-icing solution from the deposition area upward to the highway's sub base. A series of identical insulated soil columns within a

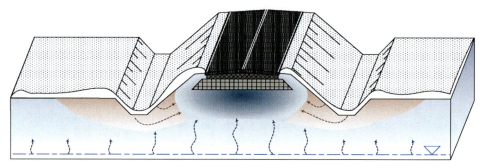

Fig. 39.2 Migration of salt-water solution in the highway basement during the freezing

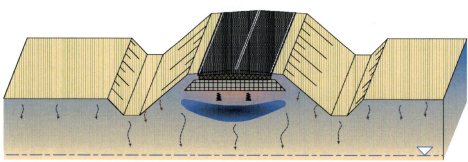

Fig. 39.3 The thawing process in the highway basement during the spring

Fig. 39.4 Laboratory model for
frost heave and thaw weakening
susceptibility

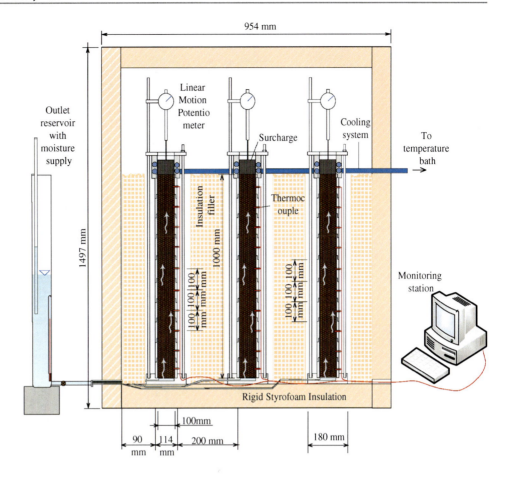

temperature controlled cabinet are connected to a constant
head (Mariotte) supply reservoir, and chilled from the top. A
schematic representation of the test equipment is presented
on Fig. 39.4.

Unlike the ASTM Standard, the height of the soil columns was increased from 15 cm to 1 m, which allows testing
the thermal gradient after each 10 cm in detail. Nine soil
columns can be tested simultaneously, which permits trip-
licate tests.

The plastic mould for the soil column consists of 10
identical acrylic sectional rings which allow easy assembly
and dismantling of the soil column. Each acrylic ring has an
inside diameter of 100 mm, outside diameter of 110 mm, and
height of 100 mm. The soil is separated from the acrylic
rings by a long rubber sleeve.

Each acrylic ring has a mid-height drilled hole for a
thermocouple sensor which takes temperature readings at
regular intervals, with up to 90 thermocouple readings at a
time, collated using a data logger.

The constant head of supplied moisture is provided by a
Marriotte water supply reservoir, which has a vacuum tight
fitting glass tube in the top of the reservoir to provide the
equal water pressure. The reservoir also allows control of the
elevation of the reservoir liquid. The bottom of the reservoir

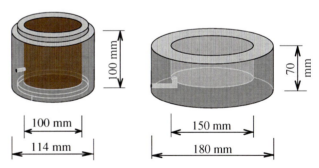

Fig. 39.5 Plastic mould section and base plate for testing a soil sample

is connected to the sample by a flexible plastic tube joined to
the specimen base plate (Fig. 39.5).

Freezing from the top is implemented by chilling collars
and metal surcharges which have good heat conductivity and
cause isotropic chilling on the sample surface. The surcharge
weight by a circular disk with a 2.72 kg of mass is placed on
the top of the soil specimen to simulate natural overloads
from the upward layers of pavement. This cooling construction allows the vertical sliding of the surcharge inside
the chilling collar and at the same time provides the equal
effective vertical stress during the heave frost.

Fig. 39.6 Freezing-thawing cycle tests procedure

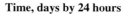

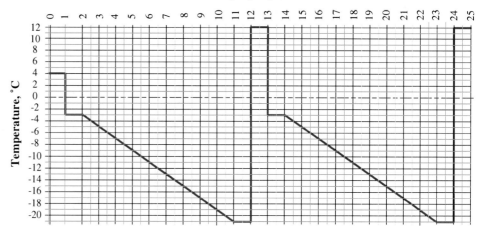

The testing process consists of two freezing-thawing cycles. The bottom temperature remains stable at 4 °C, while the top of the soil column is gradually cooled by 2 °C per day until it reaches −20 °C. After each cycle stage, 3 columns of soil are collected for further tests. The experiment is run using different de-icing agents and with deionized water, to permit a comparison.

The temperature control chamber is to be made from rigid Styrofoam insulation material and adapted to have the capacity to hold 9 samples.

The testing process consists of two freezing-thawing cycles and represented in the Fig. 39.6. The first 24 h is a conditioning period when both the top and bottom are held in 4 °C. In the first freezing period the temperature of the cooling collars is lowered to −3 °C and is held at this grade for 24 h. After this period the cooling temperature decreases steadily at a rate of −2 °C per day. At the same time, the bottom moisture supply temperature remains stable at 4 °C during the all test. When the cooling temperature reaches −21 °C it is again be kept for 24 h. Afterward 3 soil columns are removed for checking moisture and chemical content while the chilling equipment is to be switched off for 24 h to allow the thawing and drain of the remaining soil columns. After this thawing period another 3 soil columns samples are extracted and checked for engineering properties as well as chemical content tests.

The second freezing cycle is identical to the first, with the last 3 soil columns removed and tested at the end.

Testing for mechanical properties in the current research includes: particle size distribution, plasticity limits, index and activity of clays, moisture content after each test stage of testing, dry density and bulk density after each test stage of testing, cohesion C and angle of internal friction φ, CBR charts at each stage after each stage of testing, coefficient of permeability and coefficient of consolidation.

39.3 Expected Outcomes

- Improved understanding of the migration of de-icing solutions in highway sub soils.
- Development of a reliable methodology for the experimental study of the thermal and humidity regime of highway sub soils, including the use of de-icing chemicals.
- Evaluation of the impact of de-icing chemicals' precipitation on the bearing capacity and deformation of sub base soils of roads

39.4 Conclusions

The laboratory investigation will examine the significance of the secondary salinization process and freeze-thaw in highway sub soils. The degree to which the de-icing chemicals migrate upwards towards a freezing front under pavement layers will be assessed. Any change to the soils' mechanical properties due to the presence and behavior of de-icing agents will be evaluated.

References

ASTM D5918-06 Standard (2007) The laboratory method for frost heave and thaw weakening susceptibility. Annual book of ASTM standards, vol 04.08, pp 401–412

Bing H, Ma W (2011) Laboratory investigation of the freezing point of saline soil. Cold Reg Sci Technol 67(1–2):79–88

Blomqvist G, Johansson E (1999) Airborne spreading and deposition of de-icing salt—a case study. Sci Total Environ 235(1–3):161–168

Bronfenbrener L, Bronfenbrener R (2010a) Frost heave and phase front instability in freezing soils. Cold Reg Sci Technol 64(1):19–38

Bronfenbrener L, Bronfenbrener R (2010b) Modeling frost heave in freezing soils. Cold Reg Sci Technol 61(1):43–64

Dreving VP (1954) The phase rule. Moscow State University, Moscow

Giakoumakis SG (1994) A model for predicting coupled heat and mass transfers in unsaturated partially frozen soil. Int J Heat Fluid Flow 15(2):163–171

Hoekstra P (1966) Moisture movement in soils under temperature gradients with the cold-side temperature below freezing. Water Resour Res 2(2):241–250

Kane DL, Hinkel KM, Goering DJ, Hinzman LD, Outcalt SI (2001) Non-conductive heat transfer associated with frozen soils. Glob Planet Change 29(3–4):275–292

Loch JPG (1981) State-of-the-art report—frost action in soils. Eng Geol 18(1–4):213–224

Lundmark A, Jansson P (2008) Estimating the fate of de-icing salt in a roadside environment by combining modelling and field observations. Water Air Soil Pollut 195(1–4):215–232

Lundmark A, Olofsson B (2007) Chloride deposition and distribution in soils along a deiced highway—assessment using different methods of measurement. Water Air Soil Pollut 182(1–4):173–185

Matsumura Shinji, Yamazaki Koji (2012) A longer climate memory carried by soil freeze–thaw processes in Siberia. Environ Res Lett 7(4):045402

Norrström AC, Jacks G (1998) Concentration and fractionation of heavy metals in roadside soils receiving de-icing salts. Sci Total Environ 218(2–3):161–174

O'Neill K (1983) The physics of mathematical frost heave models: A review. Cold Reg Sci Technol 6(3):275–291

Othman MA, Benson CH (1993) Effect of freeze-thaw on the hydraulic conductivity and morphology of compacted clay. Can Geotech J 30(2):236–246

Qi J, Ma W, Song C (2008) Influence of freeze–thaw on engineering properties of a silty soil. Cold Reg Sci Technol 53(3):397–404

Qi J, Vermeer PA, Cheng G (2006) A review of the influence of freeze-thaw cycles on soil geotechnical properties. Permafrost Periglac Process 17(3):245–252

Simonsen E, Isacsson U (1999) Thaw weakening of pavement structures in cold regions. Cold Reg Sci Technol 29(2):135–151

Vidyapin IY, Cheverev VG (2008) The hydraulic of freezing saline soils. Earth Cryosphere 12(4):43–45

West DRF (1982) Ternary equilibrium diagrams. Chapman and Hall, London New York

A Geotechnical and Geochemical Characterisation of Oil Fire Contaminated Soils in Kuwait

40

Humoud Al-Dahanii, Paul Watson, and David Giles

Abstract

As a consequence of the Saddam Hussein 1991 Iraqi led invasion of Kuwait more than 600 oil wells were set fire to as part of a scorched earth policy while retreating from the country. This action created a series of "oil lakes" and hydrocarbon contamination within the desert causing serious environmental damage. Some 23 years later after the fires were extinguished the ground affects of these actions can still be detected. This paper will present the results of a detailed geotechnical and geochemical investigation into the current ground conditions now present in the Burgan Oil Field some 35 km south of Kuwait City. Detailed geotechnical testing together with hydrocarbon analysis using a Gas Chromatograph—Mass Spectrometer (GCMS) have been carried out on samples from varying depths within the Greater Burgan Oil Field. A detailed geological, geotechnical and geochemical ground model has been developed to present the findings of these investigations. The area under study has major development plans for both housing and infrastructure. Subsequent Quantitative Human Health Risk Assessments have been undertaken to determine the potential levels of risk posed to any future urban developments within these affected areas. The paper will report on this assessment detailing the hazards posed and the tools used to assess them. Potential risks will be discussed and mitigation and management scenarios will be highlighted.

Keywords

Oil lakes • Hydrocarbon contamination • Risk assessment • Geotechnical assessment

40.1 Introduction

The state of Kuwait has experienced serious environmental damage as a consequence of the formation of multiple oil lakes and hydrocarbon contamination resulting from the destruction caused in the Gulf War of 1991. Of the 810 active oils wells operating in Kuwait in 1991, 730 were damaged or set ablaze during the conflict (Fig. 40.1). This research set out to assess the hydrocarbon contamination and geotechnical effects of the oil fires still being encountered some 20 years plus after the war.

40.2 Site Description

The area under study was located in the Greater Burgan oilfield (Fig. 40.2) which lies in the Arabian Basin. The Greater Burgan field is the largest clastic oil field in the world covering an area of 838 km^2, located in south eastern Kuwait. The oil field is subdivided into the Burgan, Magwa and Ahmadi sectors based on the underlying geological structure (Kaufman et al. 2000). At the height of the destruction smoke plumes from the Greater Burgan oil fires extended over 50 km from the well sites up to an altitude of 2.5 km. Spillages from ruptured pipelines resulted in

H. Al-Dahanii · P. Watson
School of Civil Engineering and Surveying, Portsmouth, UK

D. Giles (✉)
School of Earth and Environmental Sciences, University of Portsmouth, Portsmouth, UK
e-mail: dave.giles@port.ac.uk

Fig. 40.1 Burning oilfield during Operation Desert Storm, Kuwait (USACE)

Fig. 40.2 Oil fields in Kuwait (Kuwait Oil Company KOC)

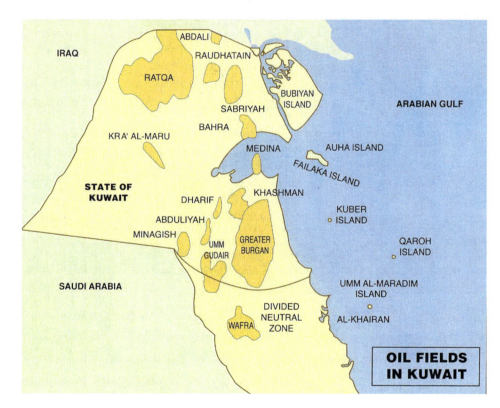

numerous oil lakes which caused extensive contamination and environmental damage.

40.3   Site Geology

Kuwait is dominated by rocks of Tertiary age dating from the Palaeocene to the Eocene (Al-Sulaimi and Mukhopadhyay 2000). Limestones, dolomites and evaporitic sequences (anhydrites) from the Umm Er Radhuma, Rus and Dammam Formations are unconformably overlain by sandstones of the Kuwait Group which include the Fars and Ghar Formations, again overlain by the Dibdibba Formation. The solid geology of the Greater Burgan site is located within the Fars and Ghar Formations with interbedded sands and clays, some sandstones and weak white nodular limestones (Hunting Geology and Geophysics 1981). Superficial deposits are predominantly Aeolian sands, with occasional gravels with sands, muds and calcareous sandstones in the coastal areas. (Table 40.1).

Table 40.1 Generalised stratigraphy of the study area

GENERALISED STRATIGRAPHY				HYDRO GEOLOGICAL UNITS
QUATERNARY SEDIMENTS (< 30m)		Unconsolidated sands and gravels, gypsiferous and calcareous silts and clays		Localised Aquifers
KUWAIT GROUP	DIBDIBBA FORMATION (200 - 200m)	Gravelly sand, sandy gravel, calcareous and gypsiferous sand, calcareous silty sandstone, sandy limestone, marl and shale, locally cherty		Aquifer
	FARS & GHAR FORMATIONS			
	Unconformity	Localised shale, clay and calcareous silty sandstone		Aquitard
HASA GROUP	DAMMAM FORMATION (60 - 200m)	Chalky, marly, Dolomitic and calcarenitic limestone		Aquifer
		Nummulitic limestone with lignites and shales		Aquitard locally
	RUS FORMATION (20 - 200m)	Anhydrite and limestone		Aquiclude
	UMM ER RADHUMA FORMATION (300 - 600m)	Limestone and dolomite (calcarenitic in the middle) with localised anhydrite layers		Aquifer
	Disconformity	Shales and marls		Aquitard
ARUMA GROUP		Limestone and shaly limestone		Aquifer

The Greater Burgan oil fields main producing reservoirs are within the Cretaceous Burgan, Mauddud and Wara Formations, all sandstones (Fig. 40.3).

40.4   Investigation and Testing

A significant number of ground investigation boreholes were commissioned to determine the site specific geology and to collect samples for both geotechnical and geochemical characterization. The Burgan and Magwa sectors were chosen for this more detailed investigation. Both near surface (ground level), shallow (up to 2 m) and deep (up to 6 m) samples were taken to determine the critical zones for hosting potential contamination or problematic ground conditions. Standard in situ and laboratory geotechnical tests were undertaken including Standard Penetration Tests,

Fig. 40.3 Typical site profile showing degraded hydrocarbon contamination

Table 40.2 Example chemical results from Greater Burgan Field Al-Magwa

S.I	T.P. No.	Depth of sample (m)	Water soluble chloride (Cl⁻)		Water soluble sulfate			pH	
			%	PPM	as SO_3 %	PPM	as SO_4 %	PPM	
1		0.00	0.0425	425	0.1240	1240	0.1488	1,488	7.84
2	(0,100 m)	0.25	0.0283	283	0.0919	919	0.1103	1,103	8.94
3		0.50	0.0567	567	0.0746	746	0.0895	895	8.61
4		1.00	0.1645	1645	0.0304	304	0.0365	365	7.83
5		1.50	0.0822	822	0.0902	902	0.1082	1,082	8.29
6		2.00	0.0567	567	0.0554	554	0.0665	665	8.36

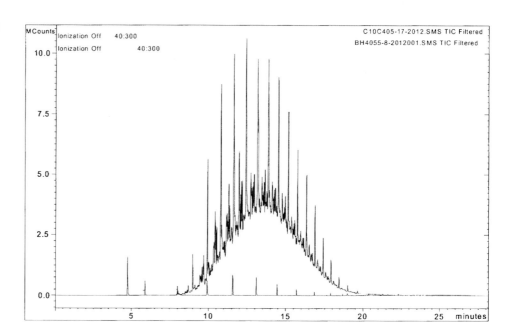

Fig. 40.4 Gas chromatogram of hydrocarbon contaminated soil from Abdally sector

particle size distribution analysis, Atterberg Limits and Direct Shear tests where applicable. Chemical testing was primarily performed using Gas Chromatograph Mass Spectrometry (GCMS) to ascertain the nature of the residue hydrocarbons present together with elemental analysis. Water soluble chlorides and water soluble sulphates were also tested for (Table 40.2). The GCMS enabled the speciation of the hydrocarbons present in order to determine the degradation that had taken place since the original spillages in 1991 (Fig. 40.4).

40.5 Contamination Modeling

Currently a human health exposure assessment utilizing the ground investigation and laboratory test results is being undertaken using the RISC (Risk Integrated Software for Cleanups) software tool for performing human health risk assessments for hydrocarbon contaminated sites using fate and transport models to estimate receptor point concentrations in indoor and outdoor air and groundwater (ESI n.d.). The sites being assessed are potential future housing developments associated with the expansion of the city of Al Ahmadi. Both airborne and ingestions pathways are being assessed for a variety of receptors.

40.6 Summary

The legacy of the Saddam Hussein 1991 invasion of Kuwait and the subsequent destruction of the oil producing facilities is still detectable in the geotechnical and geochemical soil profile. Human Health Risk Assessments are being undertaken and contamination remediation strategies designed to enable future developments on or near these sites. Elevated Total Petroleum Hydrocarbons (TPH) levels have been detected as expected and their potential impact is currently being evaluated.

References

Al-Sulaimi J, Mukhopadhyay A (2000) An overview of the surface and near-surface geology, geomorphology and natural resources of Kuwait. Earth Sci Rev 50(3):227–267. doi:10.1016/S0012-8252 (00)00005-2

ESI (n.d.) RISC—integrated software for cleanups. Retrieved from http://esinternational.com/risc/

Hunting Geology and Geophysics (1981) Photographic survey of the State of Kuwait. Submitted to the Kuwait Oil Company, Kuwait

Kaufman RL, Kabir CS, Abdul-Rahman B, Quttainah R, Dashti H, Pederson JM, Moon MS (2000) Characterizing the Greater Burgan field with geochemical and other field data. SPE Reservoir Eval Eng 3(2):118–126. doi:10.2118/62516-PA

Comparison Between Neural Network and Finite Element Models for the Prediction of Groundwater Temperatures in Heat Pump (GWHP) Systems

41

Glenda Taddia, Stefano Lo Russo, and Vittorio Verda

Abstract

A fundamental aspect in groundwater heat pump (GWHP) plant design is the correct evaluation of the Thermally Affected Zone (TAZ) that develops around the injection well. This is particularly important to avoid interference with previously existing groundwater uses (wells) and underground structures. Temperature anomalies are detected through numerical methods. Computational fluid dynamic (CFD) models are widely used in this field because they offer the opportunity to calculate the time evolution of the thermal plume produced by a heat pump. The drawback of these models is the computational time. This paper aims to propose the use of neural networks to determine the time evolution of the groundwater temperature downstream of an installation as a function of the possible utilization profiles of the heat pump. The main advantage of neural network modeling is the possibility of evaluating a large number of scenarios in a very short time, which is very useful for the preliminary analysis of future multiple installations and optimal planning of urban energy systems. The neural network is trained using the results from a CFD model (FEFLOW) under several operating conditions. The final results appeared to be reliable and the temperature anomalies around the injection well appeared to be predicted well.

Keywords

Groundwater heat pump • Thermally affected zone • FEFLOW • Neural networks • Italy

41.1 Introduction

The market for geothermal heat pumps has grown considerably in the last decade (Lund et al. 2011; Bayer et al. 2012) and is one of the fastest-growing renewable energy technologies. A typical groundwater-based well-doublet scheme for heating or cooling (Banks 2009) typically comprises three elements:

1. An abstraction well, from which water is abstracted at a rate Q and a temperature Te
2. A heat-transfer system (a heat exchanger or a heat pump), which either extracts heat from, or rejects heat to, the groundwater flux;
3. One (or more) re-injection well(s), at a distance L from the abstraction well, where water is reinjected at a rate Q and temperature Tr. For space-cooling schemes, $\Delta T = (Tr - Te) > 0$ and for heating schemes $\Delta T < 0$. Thus, in the case of open-loop heat pumps, the water re-injected into the aquifer has a different temperature from the water of an undisturbed aquifer ($\Delta T \neq 0$). This thermal disturbance (Thermally Affected Zone, TAZ) propagates through the groundwater and may affect the temperature of water withdrawal operated by downstream installations

G. Taddia (✉) · S.L. Russo
Department of Environment, Land, and Infrastructure Engineering (DIATI), Politecnico di Torino, Corso Duca degli Abruzzi 24, 10129 Turin, Italy
e-mail: glenda.taddia@polito.it

S.L. Russo
e-mail: stefano.lorusso@polito.it

V. Verda
Department of Energy (DENERG), Politecnico di Torino, Corso Duca degli Abruzzi 24, 10129 Turin, Italy
e-mail: vittorio.verda@polito.it

(Lund et al. 2011; Chung and Choi 2012). Moreover a thermal plume may pose an internal risk to the sustainability of the well doublet due to the phenomenon of thermal feedback. In fact, if the well separation is insufficient there could be a risk that a proportion of the discharged warm water will flow back to the abstraction well. Computational fluid dynamic (CFD) models are widely used in this field because they offer the opportunity to calculate the time evolution of the thermal plume produced by a heat pump, depending on the characteristics of the subsurface and the heat pump (Yang et al. 2011; Nam and Ooka 2010; Zhou and Zhou 2009). In this case we using the finite element FEFLOW package. Neural networks could represent an alternative to CFD for assessing the TAZ under different scenarios referring to a specific site.

The development of artificial neural networks (ANNs) began approximately 70 years ago (McCulloch and Pitts 1943). This paper aims to propose the use of neural networks to determine the time evolution of the groundwater temperature downstream of an installation, as the function of possible utilization profiles of the heat pump. Due to the large variety of scenarios that can take place, these profiles need to be approximated in order to be easily expressed with a limited number of parameters and simple time functions. Various different simulations covering a wide variability range of the main characteristics of an installation have been conducted on the test case of Politecnico di Torino. These results are used to train the neural network.

41.2 Test Site Hydrogeology and FEFLOW Numerical Modeling

The test site is the Politecnico di Torino and is located in the urban area of Turin, the capital of the Piemonte Region in northwest Italy (geographical coordinates 45°03′45″N, 7°39′43″E, elevation 250 m asl).

The urban area of Turin is mainly situated on the outwash plain of several glaciofluvial coalescing fans connected to the Pleistocene-Holocene expansion phases of the Susa glacier. The plain extends between the external Rivoli-Avigliana Morainic Amphitheatre (RAMA, Susa Glacier) on the west side and Torino Hill on the east (see Fig. 41.1). The site is located in the central part of the urban area between the Dora Riparia River to the north, the Sangone River to the south, and the main draining Po River to the east, which flows northeast along the western border of Torino Hill. Two 47 m-deep wells having the same technical characteristics are present at the site. One is used for groundwater extraction (named P2), the other for injection (named P4). Injection well P4 is located 78 m from extraction well P2, almost directly downflow with respect to the local unperturbed potentiometric gradient. A piezometer (named S2) 47 m deep.

The potentiometric surface 17 m below ground level displays a W-to-E gradient of 0.269 % toward the Po river. The saturated thickness of the unconfined aquifer at the site is approximately 30 m. The buildings connected to the existing GWHP plant, are used for university offices and laboratories. The groundwater levels and temperature are measured in the extraction and injection wells and in the piezometer using through specific monitoring probes submerged 2 m below the groundwater level. The measured average undisturbed groundwater temperature is 15.0 °C along the saturated zone as experimentally determined from multi-temporal temperature logs in the wells and piezometer. This temperature is practically constant throughout the year. The average mesh spacing in the modelling domain is 15 m, and this was refined to 3 m in the area close to the wells to provide enhanced estimation of thermal plumes.

In particular, the time delay required for the thermal plume to reach three points located 20, 30, and 60 m downstream of the injection well has been calculated as a function of the mesh size. Several control points were included downgradient with respect to the injection well in order to check the evolution of the thermal plumes over the space. Control points 19, 21, 24, and 26 are placed along the line that connects the injection well with the piezometer, while control points 20, 22, 23, 25, and 27 are projections of previous control points along the groundwater flow direction. The horizontal angle between the two lines is almost 30°. Control points 19–23, 25, and 27 are located 10 m from the injection well while control points 24 and 26 are 20 m from it (see Fig. 41.2). Rainfall infiltration was not included in the calculations, owing also due to a lack of measured infiltration data. Appropriate FeFlow time-varying functions (TVFs) for Q and ΔT were defined by means of a post-processing phase of the monitoring data obtained from the heat pump plant monitoring system. The TVFs have been discretized considering a time step of one day, while the automatic computational time-step has been used for FeFlow simulations.

41.3 Simulation Results and Neural Network Model Applied to the Test Site

As multiple simulations are performed to train the neural network model, a parameterization of the scenarios is required.

Two additional parameters are associated with the groundwater: the first parameter is the maximum temperature change between water re-injection and withdrawal (from 0 to 12 °C), while the second parameter is associated with the partial load operation. Intermediate values correspond to a combination of these limiting cases. In order to show the main effects of the previous parameters, have been analyzed

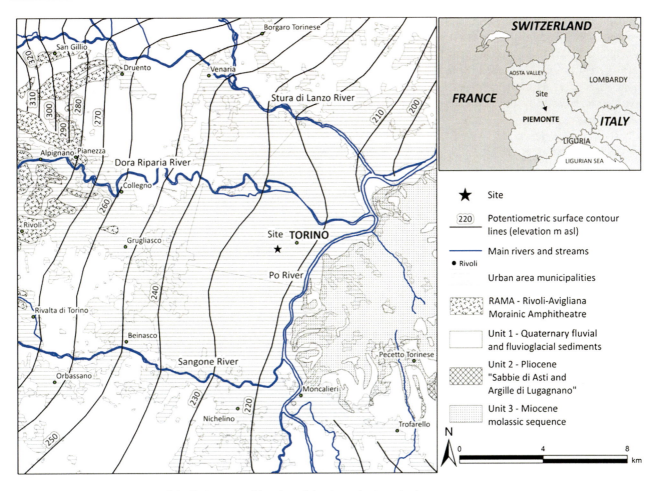

Fig. 41.1 Hydrogeological map of the Turin area and location of the site under investigation

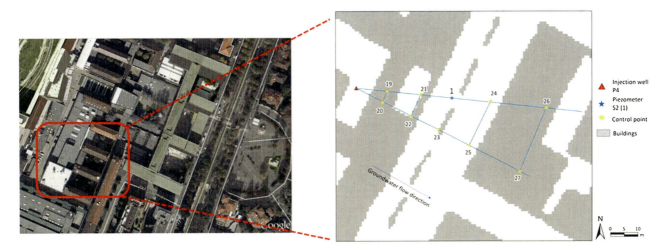

Fig. 41.2 Plan view of the control points on the topographical map

three principal scenarios and it were modelled using FEFLOW and appropriate FEFLOW time-varying functions (TVFs) for Q and ΔT. The first two scenarios are different because of a different maximum value of the re-injection temperature (3.3 °C in the first scenario and 11 °C in the

second scenario). The third scenario is similar to the first one but a small value of the reduction in mass flow rate is considered, which means that when the heating/cooling load decreases, the heat pump is primarily operating by reducing the re-injection temperature change with respect to the extraction temperature. The results obtained by modelling these scenarios are compared by checking the groundwater temperatures at two control points downstream of the injection well: point 27, located 60 m downstream in the groundwater flow direction, and point 26, located 63 m downstream and at 30° with respect to the groundwater flow direction. The maximum temperature in scenario 2 is about 20.5 °C, which is about 5.5 °C less than the injection temperature in this scenario. Temperature reduction is due to mixing effects and heat conduction. In scenarios 1 and 3, the maximum temperature is about 17 °C, which is about 1 °C lower than the injection temperature. This different behaviour is explained by the reduced driving force (smaller temperature gradient) with respect to scenario 2. This also means that groundwater temperatures in the three scenarios tend to converge at longer distances, where temperature gradients are small.

The network is characterized by five inputs: time, maximum heating load (0–600 kW), maximum cooling load (0–600 kW), maximum temperature variation between water re-injection and water withdrawal (0–12 °C), and water mass flow rate withdrawal reduction at partial load (0–1).

Figure 41.3 shows a comparison between the temperature profile at the piezometer calculated using the neural network and the values simulated using FEFLOW. The two curves show similar trends and very close values, which means that the ANN model is quite reliable even in the case of a utilization profile of the heat pump very different from that used for the parameterization. The temperature difference between the peak temperatures obtained with the two models is 0.34 °C, while the root mean square error in a period of 200 days is 0.46 °C. The largest temperature deviation is about 1 °C. This result is considered satisfactory, considering that the simulation with the ANN has been performed treating the case as if the load profile were not known, that is as if the installation of the heat pump were only planned.

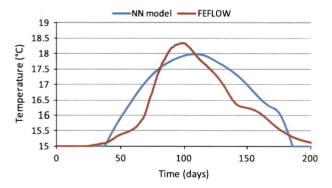

Fig. 41.3 Comparison between temperatures simulated with FEFLOW and calculated with the ANN model

41.4 Conclusions

In this paper, the use of a neural network model to predict groundwater temperature profiles at a specific site is proposed. The network is trained using simulations performed using FEFLOW. In the case of a simulation performed using real profiles of water withdrawal and variable re-injection temperature, the neural network model is still able to predict the groundwater temperature profile as a function of time. The results are quite satisfactory since the average temperature deviation is about 0.46 °C and further improvements are possible, by increasing the number of simulations that are used for the training process. There may not be significant computational advantage in this approach, since the training process requires various simulations, nevertheless the ANN model can be easily implemented into optimization procedures and used by people that is not expert on CFD modelling. Once the model has been trained, it is possible to evaluate a large number of scenarios in a very short time, which is very useful while performing the optimization of operating strategies in order to minimize the overall energy used and environmental impact or for the preliminary analysis of future multiple installations, when the temperature of extracted temperature may be different than the unperturbed value. This is possible provided that the application of the neural network is performed on the same aquifer that has been considered for the model training. If the characteristics of the aquifer changes (in particular, the unperturbed velocity of groundwater), the model must be retrained.

References

Banks D (2009) Thermogeological assessment of open-loop well-doublet schemes: a review and synthesis of analytical approaches. Hydrogeol J 17(5):1149–1155. doi:http://dx.doi.org/10.1007/s10040-008-0427-6

Bayer P, Saner D, Bolay S, Rybach L, Blum P (2012) Greenhouse gas emission savings of ground source heat pump systems in Europe: a review. Renew Sustain Energy Rev 16(2):1256–1267. doi:10.1016/j.rser.2011.09.027

Chung JT, Choi JM (2012) Design and performance study of the ground-coupled heat pump system with an operating parameter. Renew Energy 42:1–2. doi:10.1016/j.renene.2011.08.054

Lund JW, Freeston DH, Boyd TL (2011) Direct utilization of geothermal energy 2010 worldwide review. Geothermics 40(3):159–180. doi:10.1016/j.geothermics.2011.07.004

McCulloch WS, Pitts W (1943) A logic calculus of the ideas immanent in nervous activity. Bull Math Biophys 5:115–133

Nam Y, Ooka R (2010) Numerical simulation of ground heat and water transfer for groundwater heat pump system based on real-scale experiment. Energy Build (42):69–75. doi:http://dx.doi.org/10.1016/j.enbuild.2009.07.012

Yang QC, Liang J, Liu LC (2011) Numerical model for the capacity evaluation of shallow groundwater heat pumps in Beijing plain, China. Procedia Environ Sci 10:881–889

Zhou Y, Zhou Z (2009) Simulation of thermal transport in aquifer: a GWHP system in Chengdu, China. J Hydrodyn 21:647–657

Ground Stiffness Evaluation Using the Soil Stiffness Gauge (SSG)

42

Mário Quinta-Ferreira

Abstract

The stiffness of a ground mass is a useful parameter to understand its mechanical behaviour under service. Despite the advantage provided by the knowledge of the ground stiffness, this parameter was seldom determined in the past, because it required elaborated in situ tests. At present, the stiffness can be easily determined with a soil stiffness gauge (SSG) that is a light and user-friendly equipment. The fast measurement and the good usability of the SSG are great advantages, together with the ability to compute the stiffness modulus of the ground assuming a Poisson's ratio. Based on a case study of a deteriorated pavement of a bus park the contribution of the SSG to the characterization of embankments and pavement layers is presented. The field and laboratory data demonstrated that the low stiffness modulus resulted from low quality embankment materials, reduced pavement thickness and deficient drainage of the pavement foundation.

Keywords

Soil stiffness gauge • Stiffness modulus • Site characterization • Pavement performance

42.1 Introduction to the SSG

The Soil Stiffness Gauge (SSG) allow to easily determine the stiffness of a soil layer and thus to compute the stiffness modulus of the ground, assuming a Poisson's ratio. This equipment can be very useful in quality control of unbound materials, in the determination of the stiffness or deformation modulus related to the relative compaction, in the identification of structural abnormalities, also allowing to quantify the strength increase with time of stabilized materials (Abu-Farsakh et al. 2004; Alshibli et al. 2005; Nazzal 2003; Seyman 2003; Batista 2007; Quinta-Ferreira et al. 2012). The SSG also allows on time corrective action during the construction process, which results in gains in efficiency and cost savings, avoiding subsequent corrective work.

The SSG technology was originally developed by the United States defence industry for the detection of land-mines, having evolved to provide the design of a light equipment with only 10 kg (Fiedler et al. 1998 reported by Nazzal 2003). The stiffness is obtained causing very small vibrations on 25 different frequencies between 100 and 196 Hz, and measuring the resulting deformation ($<1.27 \times 10^{-6}$ m at 125 Hz). The apparatus used is the GeoGauge (Humboldt 2007), that records the stiffness value for each of the 25 frequencies and presents the average value. The soil deforms an amount δ proportional to the outside radius of the base ring (R), the Young's modulus (E), the shear modulus (G) and the Poisson's ratio (υ). The stiffness (K) is obtained dividing the force (P) by the deflection (δ) it produces (K = P/δ). According to Poulos and Davis (1974) the stiffness for a ring load in an elastic half-space is given by: $K = (P/\delta) = (3.54\,GR)/(1 - \upsilon)$, being K the stiffness, G the shear modulus, R the radius of the load ring and υ the Poisson's ratio. Knowing the Poisson's ratio the stiffness modulus is computed using the relation: $Eg = K (1 - \upsilon^2)/1.77R$.

M. Quinta-Ferreira (✉)
Department of Ciências da Terra, Geosciencies Center, University of Coimbra, Largo Marquês de Pombal, 3000-272, Coimbra, Portugal
e-mail: mqf@dct.uc.pt

The depth of measurement with the GeoGauge is between the surface and approximately 31 cm. The range of functionality is between 3 and 70 MN/m for stiffness, and between 26.2 and 610 MPa for the stiffness modulus. The Poisson's ratio may vary between 0.20 and 0.70 in increments of 0.05 (Humboldt 2007).

The tests with the GeoGauge were performed using an internal procedure described by Quinta-Ferreira et al. (2008) based on the recommendation of the GeoGauge Guide (Humboldt 2007) and on the ASTM D6758 (2008).

42.2   Some Applications of the SSG

In the control of embankments using the SSG a reference value of deformability must be used, which should be related to the relative compaction, and the results should be carefully evaluated based on experience. Abu-Farsakh et al. (2004) found that the maximum values of the stiffness modulus and the dry unit weight of two soils, clayey silt and sandy clay, do not occur simultaneously, tending to occur on the dry side of the compaction curves. As the construction procedures tend to require a moisture content within ±2 % of the optimal, they concluded that the SSG modules can vary about 40 % in this range, while the dry unit weight varies only 2.5 %. As a result of this observation, Abu-Farsakh et al. (2004) argue that the change in stiffness in this range is much larger than the dry unit weight, so that the use of stiffness as a criteria for acceptance is difficult to implement due to their sensitivity to the variation of water content.

As indicated by Alshibli et al. (2005) both the SSG and the impact deflectometer (FWD) apply a dynamic force to estimate the elastic modulus of the material, however with FWD the energy applied is much greater than the one applied by the SSG. Both tests assume that it is an elastic half-space in which a load is applied and the measurement of the surface deflection is used to calculate the elastic modulus of the layer.

A comparative evaluation of the in-place stiffness modulus using a van-integrated falling weight deflectometer (FWD) and the GeoGauge on a limestone all-in-aggregate (AIA), used in the base course of a highway pavement (Quinta-Ferreira et al. 2008, 2012) allowed to conclude that the dry unit weight is related with the stiffness modulus obtained with the SSG. The deflections measured in the centre of the FWD plate are related with the modulus obtained with the SSG. Considering as reference the equivalent modulus obtained with the FWD the moduli computed for the 85th percentile with both the FWD and the SSG, show a difference lower than 20 %. The deflections obtained with the FWD and the modulus obtained with the SSG show a similar pattern of dispersion. The results allowed a good coherence between both tests, indicating that

the SSG can be used for in-place modulus evaluation, structural uniformity and pavement design validation.

42.3   Evaluation of Deteriorated Pavements

Shortly after the start of operation of a bus park, huge pavement deterioration were observed, that required to clarify the causes. Based on the visual assessment of the deteriorations four grades were considered (Fig. 42.1): Grade 1—No visible signs of deterioration; Grade 2—Deteriorated Pavement, without loss of functionality; Grade 3—Pavement very Deteriorated, eventually presenting open cracks and losses of bitumen in small areas; Grade 4—Pavement destroyed with complete loss of functionality or even non-existent, occurring in a small areas.

To clarify the causes of bad pavement performance the following works were done: (a) surface geological reconnaissance of the site and surrounding area, (b) measurement of the stiffness of the pavement and foundation ground using the SSG, (c) opening of three test pits in locations presenting the worst behaviour, SSG measurements and collection of samples for laboratory tests; (d) completing laboratory tests (Table 42.1), (e) interpretation of the results of field and laboratory.

Mainly based in the observation of the three test pits, the following units were defined:—Asphalt layer;—aggregate; —fill soil. The thickness of the layers measured in the test pits did not represented the average in the bus park because the prospection was carried out in the more deteriorated locations were the asphalt or aggregates were depleted and mixed with the foundation materials, as in test pit P1. The bedrock of the bus park consists of medium to coarse grained porphyroid biotite granite, having feldspar crystals reaching ten centimetres in length, and presenting a large range of weathering, since slightly weathered to decomposed. The bus park platform was constructed mainly on an embankment of decomposed granite, and in a very small part over an excavated area, at east.

The study with the SSG was done in 18 test locations, evenly distributed through the bus park (Fig. 42.1). Three readings in each location were made, considering the average result. In each one of the test pits two measurements with the GeoGauge were done: one in the pavement and the other in the soil underneath. The value used for the Poisson's ratio was 0.35. The average values obtained for the deformability modulus are shown in Fig. 42.1. From the analysis of Fig. 42.1 it can be stated that the lowest Eg values are located in the worst areas classified as grade 4 and grade 3. In areas without pavement deteriorated (grade 1) the deformability values were higher, which is consistent with the best material performance.

Fig. 42.1 Zoning of the pavement degradation in the bus park and test locations

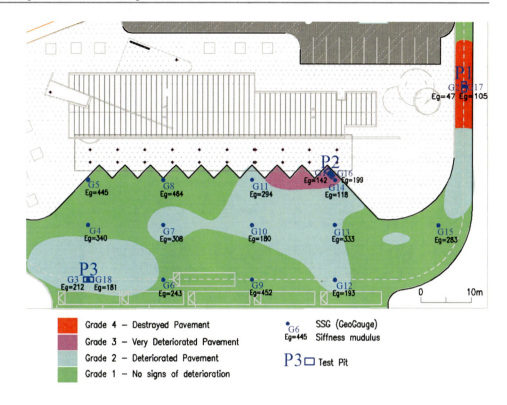

Table 42.1 Properties of soils tested

Location	P1	P2	P3
Depth (m)	0.1–1.3	0.1–0.6	0.15–1.0
Stiffness (MN/m)	4.9–11.1	15.0–30.0	19.1–22.3
Stiffness modulus (MPa)	46.7–105.0	142.1–198.7	181.1–211.8
LL (%)	36	33	33
IP (%)	11	5	6
<0.074 (%)	24.3	21.3	18.7
Sand equivalent	17	16	23
$\gamma_{dmáx}$ (kN/m^3)	18.0	19.0	–
W_{opt} (%)	13.5	11.2	–
CBR (%)	3	5	–
ASTM	A-2-6	A-1-B	A-1-B
USCS	SM	SM	SM

Based on the tests performed it is possible to conclude that soils P1 and P2 have worse characteristics than soil P3 (Table 42.1). These poor results are in accordance with the higher grade of degradation of the pavement observed around test pit P1 (Fig. 42.1) with a total loss of asphalt and even aggregate that was pushed by the passage of the buses.

In test pit P2 it was observed that the pavement waved during the passage of the buses, due to the saturation of the soils in the foundation. It was also observed that the deteriorated asphalt was pushed by the wells of the buses, overriding the sidewalk.

The observation of the pavement and the laboratory tests allowed to identify, characterize and understand that the pavement structure presented low stiffness modulus in significant areas, which was aggravated by the reduced foundation drainage. The soils in the pavement foundation had too much fines, high liquid limits and plasticity index as well as low CBR values. The pavement rehabilitation was based on an appropriate redesign, together with an efficient drainage, both at surface and in deep.

42.4   Final Remarks

Both our results and the ones that it was possible to find in the bibliography, showed that the SSG (the GeoGauge) is a device with great practical applicability, that is possible to relate with other geotechnical tests. The ease of use is a great advantage of the SSG. In the case study presented the field and laboratory data demonstrated that the low stiffness modulus resulted from low quality embankment materials, reduced pavement thickness and deficient drainage of the pavement foundation.

Acknowledgments This work was funded by the Portuguese Government through FCT—Fundação para a Ciência e a Tecnologia under the research project PEst-OE/CTE/UI0073/2011 of the Geosciences Center. The support of Instituto Pedro Nunes during the field work is acknowledge. The permission to publish the data is acknowledge to the company Encobarra, Engenharia e Construções, S.A.

References

Abu-Farsakh MY, Alshibli K, Nazzal M, Seyman E (2004) Assessment of in situ test technology for construction control of base courses and embankments. Technical report n° FHWA/LA.04/385. Louisiana Transportation Research Center, Baton Rouge, LA, USA, p 126

Alshibli KA, Abu-Farsakh M, Seyman E (2005) Laboratory evaluation of the geogauge and light falling weight deflectometer as construction control tools. J Mater Civil Eng 17(5):560–569

ASTM D6758 (2008) Standard test method for measuring stiffness and apparent modulus of soil and soil aggregate in-place by an electro-mechanical method. American Society for Testing and Materials, ASTM International, West Conshohocken, PA

Batista LCM (2007) Determinação de parâmetros de deformabilidade de camadas de um pavimento a partir de ensaios de campo. Dissertação de Mestrado, G.DM-159/07, Dep. Engenharia Civil e Ambiental, Universidade de Brasília, Brasília, DF, p 182 (in Portuguese)

Fiedler S, Nelson C, Berkman F, DiMillio A (1998) Soil stiffness gauge for soil compaction control. Public Road, FHWA 61(5):5–11

Humboldt (2007) GeoGaugeTM user guide. Model H-4140. Soil stiffness/modulus gauge. Version 4.0. Humboldt Mfg. Co. Norridge, Illinois, USA

Nazzal MD (2003) Field evaluation of in situ test technology for Qc/Qa during construction of pavement layers and embankments. Louisiana State University, Master of Science thesis in civil engineering

Poulos HG, Davis EH (1974) Elastic solutions for soil and rock mechanics. Wiley, Hoboken, pp 167–168

Quinta-Ferreira M, Santarém Andrade P, Castelo Branco F (2008) Alguns resultados do GeoGauge e do deflectómetro de impacto (FWD) na caracterização de pavimentos. XI Cong. Nacional de Geotecnia/IV Cong. Luso Brasileiro de Geotecnia, Coimbra, p 8 (in Portuguese)

Quinta-Ferreira M, Fung E, Andrade PS, Branco FC (2012) In-place evaluation of a limestone base course modulus, using a van integrated falling weight deflectometer (FWD) and the GeoGauge (SSG). Road Mater Pavement Design, p 15. doi:10.1080/14680629.2012.735794

Seyman E (2003) Laboratory evaluation of in situ tests as potential quality control/quality assurance tools. Louisiana State University, Master of Science thesis in Civil Engineering

Influence of Fracture Systems and Weathering on the Sustainability of Rock Excavation Made for the Purpose of Infrastructure Construction

43

Radoslav Varbanov, Miroslav Krastanov, and Rosen Nankin

Abstract

The problems related to rock slope stability of excavations realized in differing in genesis rock complexes are the object of the present paper. The stability of these slopes depends on the natural fracture system in the massif, the additional fracture systems occurring in the rock masses during the excavation works as well as on the resistance against weathering of the rock massif. The investigation of the above problems was carried out in the excavations of the Sofia-Mezdra railway line, which crosses the Stara Planina Mountain chain. The mountain ridges are built of rock complexes with various composition and genesis. The railway construction was started in the middle of the 19th century. The commissioning of the line was on November 8, 1899. Rock slopes with different height and inclination were formed during the railway construction. In the course of a period exceeding 110 years the rock slopes have been subjected to weathering processes that have inevitably exerted effect on their stability too. The modern safety requirements for the railway line operation impose the performance of different stabilization activities. The design of the safety measures is connected with the assessment of the rock slope stability as of the present moment. The diverse geotechnical characteristics and geological conditions of rock massifs, built of sedimentary rocks (argillites and sandstones), metamorphic rocks (gneisses and quartzites) and volcanic rocks (various diabase types), have been considered.

Keywords

Fracture system • Rock slope • Weathering • Rock mass rating

43.1 Introduction

A spatial model of the fracture systems is composed on the base of field measurements. Qualitative assessment (rock mass rating) of the massifs is made on the base of this model. The results show that in some zones of the rock massifs "rock wedges" have been formed, which are unfavorably inclined towards the excavation slope. The origin of slope deformations has been already observed along some of these weak zones. Shear strength assessment is made along the fractures of the "rock wedges", formed both in the natural terrain and in the railway line excavations. Data are presented for the quantitative assessment of the risk of rock massif slope destruction in the natural state, as well as in the zone of the artificial rock slopes.

43.2 Location and Geomorphology

The studied region is situated in West Bulgaria in the Iskar River valley, at a distance of 40 km to the north of the capital Sofia (Fig. 43.1). The relief in the region is mountainous, with an average altitude above 700 m. The lowest parts of

R. Varbanov · R. Nankin
University of Architecture, Civil Engineering and Geodesy, Sofia, Bulgaria

M. □rastanov (✉)
Geological Institute "Strashimir Dimitrov", Bulgarian Academy of Sciences, Sofia, Bulgaria
e-mail: miroivanov@mail.bg

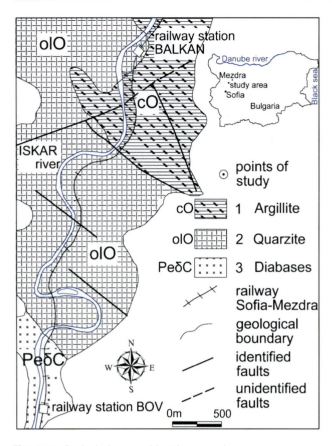

Fig. 43.1 Geological map and location on study area

the relief are in the Iskar River bed—about 450 m. In geomorphologic respect the region is a part of the Stara Planina Mountain chain system. Its contemporary relief was formed under the influence of diverse geological and erosion denudation processes.

43.3 Geological-Tectonic Structure

In geological respect, rocks of Carboniferous and Ordovician age are outcropped in the direction from north to south. The Carboniferous rocks are built of the diabases (PeδC2) of the Petrohan pluton, occupying a narrow strip in the initial zones of the studied railway section. The rest part of the railway alignment is built almost entirely of the Olistostrome unit (olO), represented by quartzites, aleurites and phylitized argillites. These rocks are formed as stacks, separating several olistostromes with relatively large sizes. The thickness of the olistostrome unit exceeds 800 m. The end zones fall within a region built mainly of the argillites and aleurites of the Clayey Metagroup (cO), which are metamorphized in a green schist facies. The lower boundary represents an abrupt lithological contact with the underlying aleurite-quartzite metagroup. It is discordantly covered on top by the rocks of

the Petrohan Terrigenic Group. The thickness of the metagroup exceeds 300 m. The considered region falls within the Berkovitsa tectonic unit, building the southern most highly uplifted segment of the West Balkan zone. Numerous longitudinal, transverse and oblique fault disturbances are established in this region. Some of them are with proved older embedding and repeated tectonic activation. The oblique faults, affecting the investigated railway section, are grouped in two systems with directions 140–170° and 65–80°. The great diversity of the tectonic structures influencing the explored region determines the intense tectonic activity, shaping the relief as well as providing prerequisites for the development of hazardous and risk phenomena and processes.

43.4 Methods of Investigation

Rock massif assessment (Rock Mass Rating—RMR) has been performed for the purposes of the design of stabilization activities in three of the most widespread rock complexes along the railway alignment. The two most widely utilized classifications of rock masses are applied—of Bieniawski (1976, 1989) and Barton et al. (1974). The following six criteria are used in the RMR evaluation: (1) Uniaxial compressive strength of the natural massif in its natural state and after about 100 years of weathering; (2) Determination of rock massif quality (Rock Quality Designation—RQD); (3) Measurement of the distances between the fracture systems for "a natural rock massif and a fractured rock massif due to construction activities and weathering processes"; (4) State of the fracture systems; (5) Hydrogeological conditions in the zone of the fracture systems; (6) Spatial orientation of the fractures and type of fracture surface.

43.5 Engineering Geological Characteristics

The engineering geological conditions are characterized on the base of the mapping and in situ investigations carried out. Most generally, one qualitative assessment is made for the present state of the rock varieties, subjected to the influence of the weathering processes in the course of 113 years, which build the slopes of the studied section (Bov station and Balkan stop) of the Sofia-Mezdra railway line. Rocks of the three genetic groups are outcropped in this section of the railway, from south to north, as follows: magmatic (diabases), metamorphic (quartzites) and sedimentary (argillites), as shown in Fig. 43.1. The qualitative assessment is made on the base of the field investigations mainly by means of two parameters: RMR (Rock Mass Rating—Bieniawski) and GSI (Geologycal Strength Index—Hoek and Marinos), as well as on one probabilistic

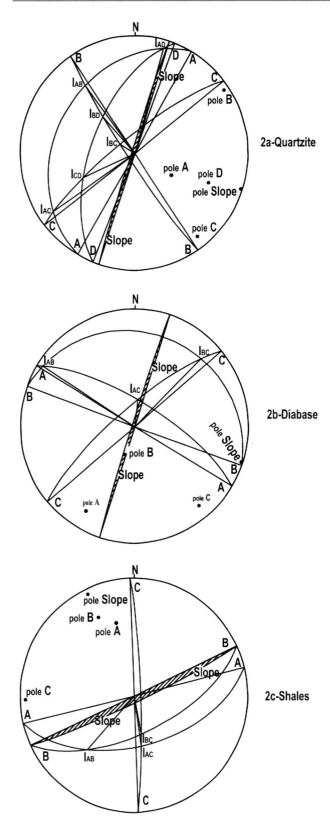

2a-Quartzite

2b-Diabase

2c-Shales

Fig. 43.2 Stereographic projections on system cracks and slope. *A–A* —projection of the system cracks *A*; pole *A*—pole cracks of the system *A*; *I_{AB}*—projection of the intersection of the systems cracks *A* and *B*

Table 43.1 Values of the elements of the fracture systems and the slope

Rock masses	System of cracks/ slope	Elements of occurrence crack systems and slopes	
		Dip direction $\alpha(\ldots°)$	Dip $\psi(\ldots°)$
Quartzite	System A	300	31
	System B	236	85
	System C	322	79
	System D	291	59
	Slope	288	88
Diabases	System A	31	73
	System B	20	21
	System C	320	77
	Slope	288	88
Argillite	System A	166	54
	System B	155	63
	System C	88	85
	Slope	155	85

assessment concerning the possibility of wedge formation, along which there is a hazard of rock mass caving. The probabilistic assessment was realized by field measurements of the elements of embedding of the basic fracture systems and the slope, exerting impact on the rock massif stability. The investigations were realized by means of stereographic projections of the fracture systems and the slope (Fig. 43.2) of the excavation. The studied elements of embedding of the fracture systems and the slope are given in Table 43.1. The rest part of the investigated section is almost entirely built of quartzite metamorphic rock, for which it is obvious (Fig. 43.2a) that the projections of the intersection lines (I_{AB} and I_{AC}) of the fracture systems lie and cross the slope projection, which provides the possibility of caving wedge formation in the rock massif, while the projection of the intersection line I_{BC} does not lie on the slope projection and hence there is no hazard of rock massif caving. The zone around the Bov station is built of diabase magmatic rock and it is seen in Fig. 43.2b that the projections of the intersection lines (I_{AB}; I_{AC}; I_{AD}; I_{BC}; I_{BD} and I_{CD}) of the fracture systems lie on and cross the slope projection, providing the possibility of caving wedge formation in the rock massif. The slopes in the end zones of the considered section are built by argillite sedimentary rock and it is seen in Fig. 43.2c that the projections of the intersection lines (I_{AB}; I_{AC} and I_{BC}) of the fracture systems lie on and cross the slope projection, which gives the possibility of forming hazardous caving wedges in the rock massif. The assessment of the state and the classification of the different types of rock massifs according to RMR and GSI are shown in Table 43.2.

Table 43.2 Values of RMR and GSI for different rocks, revealing along the study route

Classification parameter	Exploration of various geological rock masses		
	Diabases	Quartzite	Argillite
Uniaxial compressive strength (MPa)	39	48	24
Rating	4	4	2
Rock quality designation (%)	87.5	87	72.5
Rating	17	17	13
Spacing of discontinuities (m)	0.19	0.25	0.14
Rating	8	10	8
Condition of discontinuities	Slightly rough surface, separation <1 mm, slightly weathered walls	Rough surface, separation <1 mm, slightly weathered walls	Slickenside surfaces or gouge <5 mm thick or separation 1–5 mm continuous
Rating	25	25	10
Ground water conditions	Completely dry	Completely dry	Dripping
Rating	15	15	4
Orientation of discontinuities	Fair	Fair	Favorable
Rating (slopes)	−25	−25	−5
Rock mass rating (RMR)	44	46	32
Rock mass classes determined from total ratings			
Class number	III	III	IV
Description	Fair rock	Fair rock	Poor rock
Geological strength index (GSI)	39	41	27
Structure of rock masses	Blocky/disturbed	Blocky/disturbed	Disintegrated
Surface conditions	Good	Good	Poor

43.6 Conclusion

The rock massifs built of diabases and quartzites, which occupy definite zones of the considered alignment, are classified according to RMR as class III, with geotechnical state of medium strong rock massif. While these built of argillites are classified as class IV, with geotechnical state of weak rock massif. In accordance with the GSI classification of rock massifs, the diabases and quartzites have a small-block structure with disintegrated zones and good state of the fractures, while the argillites are with fragmentary structure and poor state of the fractures. The argilites are with the lowest value of RMR, which explains the problems that occured during exploitation of the railway line in the zones, where there are some.

The above mentioned characteristics of the rock massifs, established as a result of the researches done, are a sign of weathering processes that create conditions for formation of fracture systems. The latter, in turn, lead to the formation of rock wedges of different sizes in the massif. To protect the railway line are fulfilled protective measures, such as caving of rocks, removal of vegetation on the slopes, construction of retaining walls and safety nets, anchored in the bedrock.

References

Angelov V et al (2008) Explanatory note to map sheet K-34-35-G (Lakatnik), Sofia

Barton NR, Lien R, Lunde J (1974) Engineering classification of rock masses for the design of tunnel support. Rock Mech 6(4):189–239

Bieniawski ZT (1973) Engineering classification of jointed rock masses. Trans S Afr Inst Civ Eng 15:335–344

Hoek E (2006) Practical rock engineering

Hoek E, Dieberichs M (2006) Empirical estimation of rock mass modulus. Int J Rock Mech Min Sci Geomech Abstr 43:203–215

Ilov G (2009) Applied mechanics rocks: rocks and building foundations in rock

Route Alignment and Optimization of Railway Based on Geological Condition

44

Weihua Zhao, Nengpan Ju, and Jianjun Zhao

Abstract

Characterized by high seismic intensity and high geological disaster risk, the geological condition of Longmenshan is complex and unique. How to carry out route alignment in such complicated geological background and avoid geohazards is one of key issues. Based on the geological background, including landform, lithology, Geological tectonic, geohazards distribution, acquired by remote sensing and field investigation, this paper firstly compared route schemes of the Chengdu-Lanzhou railway in Longmenshan mountain area, and regarded the Jushui river basin as recommendation for relatively less geohazards and no gobs. Then this paper secondly evaluated route schemes related to geohazards distribution and geohazards features in Jushui river basin, and regarded the D2K scheme as preferred option, which avoid large-scale and typical geohazards in the form of "early into tunnel and later out of tunnels" by longer tunnels.

Keywords

Geological condition • Geohazards • Route alignment • Route optimization

44.1 Introduction

The Chengdu-Lanzhou railway is a key infrastructure of The Western Development Strategy of China. The general trend of the project determined the railway must across the Longmenshan faults zone (Zhu et al. 2009). The geological condition in this zone is complex and unique with the feather called "four extremely, three high". The "four extremely" are extremely strong topological incision, extremely complex and active tectonic, extremely weak rock mass and extremely significant Wenchuan earthquake effect. The "three high" are high crustal stress, high seismic intensity and high risk of geohazards (Huang 2011; Du et al. 2012). How to carry out route alignment in such complicated geological background and maximally avoid geohazards is one of key issues (Wu et al. 2010; Yang et al. 2010; Huang et al. 2013).

44.2 Geological Conditions of Study Area

The geomorphology of the study area is characterized by high mountains and deep valleys. The gradient of slopes along river is generally larger than 35°. Lithologies of stratums expose completely, comprising limestone, dolomite, sandy conglomerate, shale and magmatic rock. The proposed schemes in Longmenshan zone are located between the Front Fracture of Longmenshan and the Central Fracture of Longmenshan. Suffered by the Wenchuan earthquake, a large number of slopes failed and developed abundant geohazards with high density (Huang and Li 2009). These geohazards distribution is controlled by river system and fault lines (Huang and Li 2009).

44.3 Route Comparison and Selection Based on Geological Conditions

The geohazards are acquired by RS interpretation and checked by investigation. By comparing the landform, geological tectonic, hydrology, gobs and related aspects, this paper analyzed advantage and disadvantage of each scheme

W. Zhao (✉) · N. Ju · J. Zhao
State Key Laboratory of Geohazard Prevention and Geoenvironment Protection, Chengdu University of Technology, Chengdu 610059, Sichuan, China
e-mail: weihuageo@gmai.com

Table 44.1 Comparison of route schemes in three river basin in Longmenshan region

Schemes	Shitingjiang	Mianyuanhe	Jushuihe
Route	A6K et al.	A16K, C2K et al.	CK, C1K, DK, D1K, D2K et al.
Landform	Steep mountains; narrow and deep valley	Narrow valley except Qingping; steep mountains	Open terrain, wide valley along the lines in Jushui town and Erlangmiao et al.
Lithology	Magmatic rock; diorite; limestone; dolomite	Limestone; dolomite; siliceous rock	Thick-layer limestone; dolomite; sandy slate
Geological tectonic	• Route perpendicular to the Longmenshan faults	• Route perpendicular to the Longmenshan faults	• Route perpendicular to the Longmenshan faults
	• The front Longmenshan fracture has no obvious movement	• The surface ruptured by front Longmenshan fracture	• The front Longmenshan fracture has no obvious movement
	• The central fracture dislocated the Guanghan-Yuejiashan railway	• The ground surface at the central fracture has severe motion and lifted to 1.3 m	• The ground surface at the central fracture has small rupture
Geohazards (landslides, rock falls and debris flow)	• Large density of geohazards, the maximal value up to $12/km^2$	• Large density of geohazards, the maximal value up to $11/km^2$	• The maximal value of density of geohazards is $7/km^2$
	• More large-scale landslides, such as Ganhekou landslide	• More large-scale landslides, such as Tianchixiang landslide, yibadao landslide and Xiaogangjian dam lake	• Less large-scale landslides, the large-scale landslides such as Daguangbao, Laoyingyan and Dazhuping far away from railway
	• Landslides and rockfalls provided abundant resources for debris flow	• Landslides and rockfalls provided abundant resources for debris flow, such as the outbreak of the Qingping debris flow in 2010, against construction of tunnels	• Abundant resources for debris flow in Jinxigou valley
Gobs	More gobs along the route within 1 km with large range, such as Jinhe phosphorite, Songlin coal mine et al., have obvious deformation	There are more gobs along the route within 2 km with three surface subsidence, and Tianchi first well, Qingping phosphorite et al. are nearby the route	There are no gobs along the route within 1 km. The straight-line distance from closest coal mine is 1.3 km. There's no deformation

in the three river basins and suggested the best one. The comparison listed in Table 44.1. Compared with Shitingjian river basin and Mianyuanhe river basin, the Jushui river basin has relatively open terrain, the Longmenshan faults in this basin has no obvious movement and rupture, density of geohazards relatively small, resources and scale of debris flow relatively less and small. Taking all geological conditions in account, the Jushui river basin was recommended as the best region.

44.4   Route Optimization in Jushui River Basin Based on Geohazards

The Jushui river basin was recommended as best region, but specific to this basin, development degree of geohazards in different parts is inhomogeneous. Based on investigation and according to the principle that as far as

possible to avoid geohazards, this paper next compare and optimize routes in the Jushui river basin listed in Table 44.2. Route Schemes and Geohazards' Distribution that Infecting Route Alignment are shown in Fig. 44.1. The proposed schemes in Longmenshan are located between the Front Fracture of Longmenshan (F2 faults system) and the Central Fracture of Longmenshan (F3 faults system). With F2 and F3, the region was divided to Jushui, Jushui to Gaochuan and Gaochuan to Maoxian three parts. In Table 44.2, compared with other schemes, the D2K scheme pass through this basin by adopting long tunnels, and successfully avoid typical large-scale geohazards triggered by Wenchuan earthquake, such as the Guantan landslide (Fig. 44.1) (Zhao et al. 2010), Yongjiashan unstable slope, Shiziyuan rockfall, Ganmofang landslide (Fig. 44.1). Slopes of tunnel entrances and exits have lesser hazards with small-scale. Thus, the D2K scheme has some advantages and was recommended as optimization.

Table 44.2 Comparison of route schemes in Jushui river basin

Segment	Route Scheme	Tunnels		Number of geohazards near exits and entrances of tunnels		Geologic instruction
				Entrance	Exit	
Jushui (Piedmont of Longmenshan)	CK	Kuzhuan tunnel		8	35	The Guantan landslide is a large-scale anti-inclined landslide induced by strong earthquake and follow-up rain, which slid from top to toe and formed dammed lake. The route scheme C1K go through Yudongshan–Guanxian–Lianghekou, and infected by this landslide, and the route DK, D1K and D2K exit at Yongjiashan and avoid the Guantan landslide
	C1K	Number one tunnel		0	1	
	DK	Dapingshan tunnel		1	10	
	D1K	Yudongshan tunnel		0	8	
	D2K	Anxian tunnel		0	8	
Jushui to Gaochuan (between the front fracture and the central fracture)	CK	Yongjiashan	Shibanlou	59/57	50/24	• The Yongjiashan unstable slope is close to the front fracture of Longmenshan, which has many dangerous rocks that threaten CK
	C1K	Number 2	Number 3	3/14	22/5	• Slopes along Yuejinqiao–Shiziyuan–Dengjiaping are steep and not suitable to build railway bed or bridge. The D2K avoid the hazards
	DK	Shiziyuan	Dengjiaping	10/3	7/1	• The Shibanlou tunnel was located on unstable slope with landslides and rockfalls. The geohazards developed in groups against entrance and exit of tunnels and should be avoided
	D1K	Shiziyuan tunnel		0	7	• Exit of Shiziyuan tunnel is located at Erlangmiao, which slope has low altitude difference and gentle slope. The river valley is broad and low hazard density
	D2K	Shiziyuan tunnel		0	4	
Gaochuan to Maoxian (behind the central fracture)	CK	Longmenshan tunnel		44	37	• The exit of Longmenshan tunnel of CK and C1K is located at upper reaches of Lengjingou valley. Abundant loose accumulation triggered by earthquake is located in the valley and easily form debris flow under storm
	C1K	Longmenshan tunnel		3	1	
	DK	Longmenshan tunnel		2	1	
	D1K	Longmenshan tunnel		0	1	
	D2K	Longmenshan tunnel		0	1	• The geological conditions of D1K and D2K in Gaochuan are better

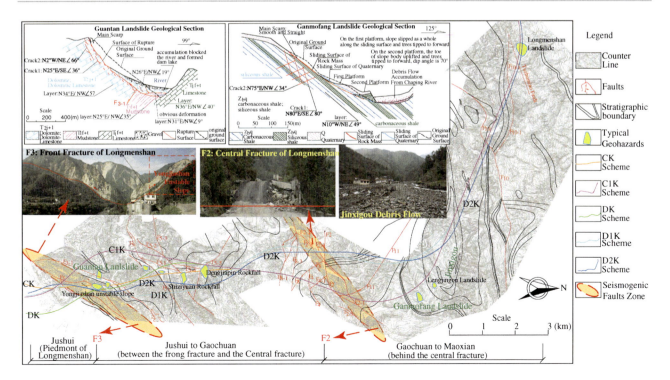

Fig. 44.1 Route schemes and Geohazards' distribution that infecting route alignment

44.5 Conclusion

Based on geological background acquired by RS and field investigation, this paper firstly compared routes schemes of the Chengdu-Lanzhou railway in Longmenshan mountain area, and regarded the Jushui river basin as recommendation with relatively less geohazards and no gobs. Then this paper secondly compared routes schemes in Jushui river basin, and regarded the D2K scheme adopting long tunnels, as preferred option, which avoid large-scale and typical geohazards. The work of this paper provided geological basis for the Chengdu-Lanzhou railway alignment.

Acknowledgments This study was financially supported by the Nature Science Foundation of China (No. 41372306)

References

Du YB, Yuan CB, Wang YD, Tao Y (2012) Major geological hazard and geological alignment of Chengdu-Lanzhou railway. J Railway Eng Soc 8:11–15 (in Chinese)

Huang R, Li W (2008) Research on development and distribution rules of geohazards induced by Wenchuan earthquake on 12th May, 2008. Chin J Rock Mech Eng 27(12):2585–2592

Huang R, Li W (2009) Analysis on the number and density of landslides triggered by the 2008 Wenchuan earthquake. China. J Geol Hazards Environ Preserv 20(3):1–7 (in Chines)

Huang R, Li Y, Qu K, Wang K (2013) Engineering geological assessment for route selection of railway line in geologically active area: a case study in China. J Mt Sci 10(4):495–508

Wu G, Xiao D, Jiang L et al (2010) Problems about engineering geology of high-grade railway route selection in complicated mountainous areas. J Southwest Jiaotong Univ 45(4):527–532 (in Chinese)

Yang C (2012) Analysis of the controlling factors of Chengdu-Lanzhou railway location in 5.12 Wenchuan earthquake-stricken areas. In: Proceedings of the 1st international workshop on high-speed and intercity railways, vol 147. Lecture notes in electrical engineering, pp 385–394

Zhao J, Ju N, Li G, Huang R (2010) Failure mechanism analysis of Guantan landslide induced by Wenchuan earthquake. J Geol Hazards Environ Preserv 21(2):92–96

Zhu Y, Wei Y, Zhong X (2009) Route selection and overall design of railway in high seismic intensity mountain area: a case study of the Chengdu-Lanzhou railway. Analysis and investigation on seismic damages of projects subjected to Wenchuan earthquake, pp 808–816 (in Chinese)

Engineering Properties of Permian Clay Tuffs

45

John Johnston, Stephen Fityus, Olivier Buzzi, Chris Rodgers, and Robert Kingsland

Abstract

The economic Permian coal deposits of the Newcastle Coal Measures of Eastern Australia are characterized by frequent tuffs and tonsteins, of predominantly clay character. Compared with their associated sedimentary rocks, these have undesirable engineering properties. This paper describes the results from a series of studies undertaken to characterize these materials in their context as engineering materials. The large number of tuff units throughout the coal measures are found to vary greatly in their composition and texture, with many being dominated by high plasticity, expansive clays. This makes them difficult to compact, with low dry densities and high optimum water contents. Their treatment as earthworks materials is reviewed and data is presented which quantifies variability in their engineering properties, as determined from a major earthworks project. Methods of successfully incorporating them into earthworks designs are discussed.

Keywords

Tuff • Clay tuff • Tuffaceous clay • Swelling clay

45.1 Introduction

The Newcastle coal measures (NCM) are a late Permian aged geological sequence, located in the north east corner of the Sydney Basin. This sequence is well known for its economic coal deposits and unique geology, and for the large proportion of tuff and tuffaceous strata found within it, when compared to other geological sequences of the Sydney Basin (Ives 1995).

Tuffs are sedimentary rocks of volcaniclastic origin. They may be crystalline or vitric in nature. Tuffs derived from thin ashfalls, that have undergone devitrification to form kaolinitic or bentonitic claystones, are described as tonsteins (Diessel 1985). Air-fall from volcanic eruptions often deposit over large areas, and so tuff units display high lateral continuity. In the NCM they are commonly associated with coal and carbonaceous rocks, as shown in Fig. 45.1, making them useful as stratigraphic markers (Kramer et al. 2001). Estimates on the amount of tuffaceous material within the Newcastle coal measures vary from 19 and 20 % (Ives 1995 and Diessel 1985, respectively) to 25 % (Brakel 1989).

Tuffs in the NCM are highly variable in chemical composition and mineralogical makeup, and this makes their engineering properties highly variable. Different ash sources, together with varying depositional environments, have resulted in different clay mineral assemblages being

J. Johnston
RCA Australia, 92 Hill Street, Carrington, 2294, Australia
e-mail: johnj@rca.com.au

J. Johnston · S. Fityus (✉) · O. Buzzi
The University of Newcastle, University Drive, Callaghan, 2308, Australia
e-mail: stephen.fityus@newcastle.edu.au

C. Rodgers
Roads and Maritime Services, Sydney, NSW, Australia

C. Rodgers · R. Kingsland
Hunter Expressway Alliance, Sydney, NSW, Australia

R. Kingsland
Parsons Brinckerhoff, Sydney, NSW, Australia

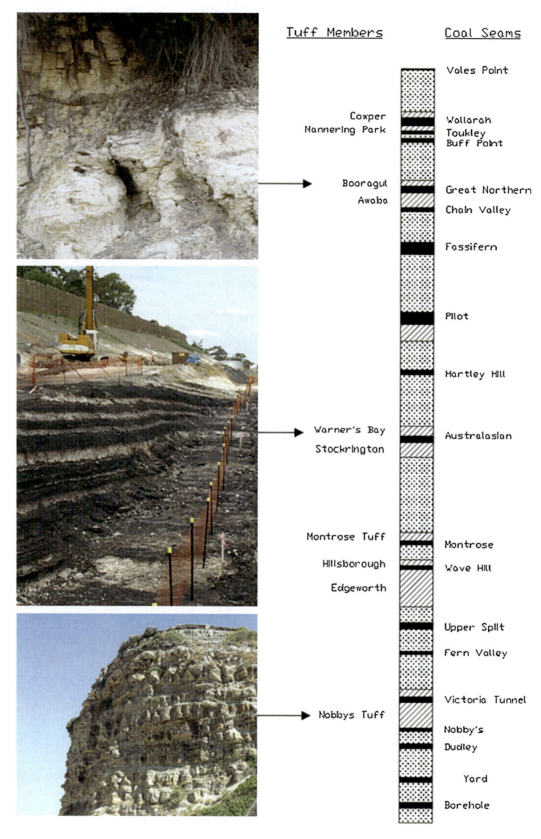

Fig. 45.1 Stratigraphic column of the Newcastle coal measures showing named and unnamed tuff units, with photographs of some well-known units

Fig. 45.2 Tuffs encountered while constructing the Hunter expressway. **a** Several tuff bands in a cutting. **b** Reworked material in compacted fill

produced, with the wet, humid swamps forming highly montmorillonitic clays, and locally acidic environments forming kaolinite clay minerals (Seedsman 1989).

The high proportion of clay minerals in the Permian tuffs and tonsteins makes them problematic when used in engineering works. Problems relate primarily to their typically low strength and their tendency for significant volume change upon water absorption. This makes them prone to slake when exposed and to heave when unloaded (Seedsman 1989). When disturbed and remolded, they readily form highly plastic clays with low dry densities and high optimum water contents which are undesirable for earthworks applications (Fityus et al. 2005).

Due to the large amount of tuffaceous materials contained within the NCM, these materials are frequently encountered in earthworks and excavations. This was particularly so for the Hunter Expressway Project, where numerous tuffaceous strata were encountered. In its 13 km long eastern part through the NCM, the 2 million cubic metres of earthworks included around 70 % of intermixed tuffaceous and carbonaceous materials. Examples are shown in Fig. 45.2. Under normal circumstances, these materials would be spoiled or relegated to non-structural uses. However, this was not an option for this project, and instead, they were utilized in encapsulated fill in several large embankments. Use of these materials in structural embankments, however, has afforded an excellent opportunity to comprehensively characterize their earthworks properties. This paper provides a summary of this.

45.2 Engineering Properties of Tuffs on the Hunter Expressway Project

45.2.1 Compaction Properties

When remolded and compacted, the tuffs used in this project display highly variable standard Proctor compaction behaviour, as shown the 3 curves in Fig. 45.3a, for tuffs coming from adjacent geological sequences. Figure 45.3b shows that across the entire project, there is a general trend of lower OMC values correlating to higher MDD values, reflecting the variability in clay content and its systematic effect on engineering properties. Figure 45.3c shows that the MDD values span the range from very low values of 1.3 t/m³ to more normal values of 2.0 t/m³, but that the distribution is skewed toward lower values with a mean of around 1.6 t/m³.

Figure 45.3d shows that the optimum water contents are generally normally distributed, with all samples falling between 10 % and the relatively high value of 40 %, with a mean of around 22 %.

45.2.2 Plasticity and Swelling Properties

Figure 45.4a shows that more than half of the CBR values were found to be between 0 and 5 %, with 80 % of all values between 0 and 20 %. Values as low as 1 % were measured. Figure 45.4b shows the values found for plasticity index

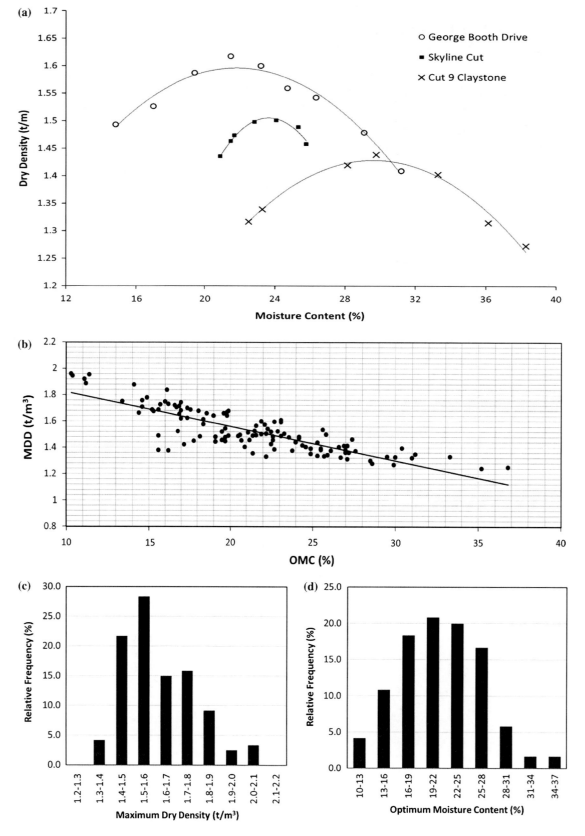

Fig. 45.3 Compaction data for NCM tuffs. **a** Standard proctor compaction curves for three typical NCM tuffs. **b** Maximum dry density versus optimum moisture content for 120 samples. **c** Relative frequency of MDD. **d** Relative frequency of OMC

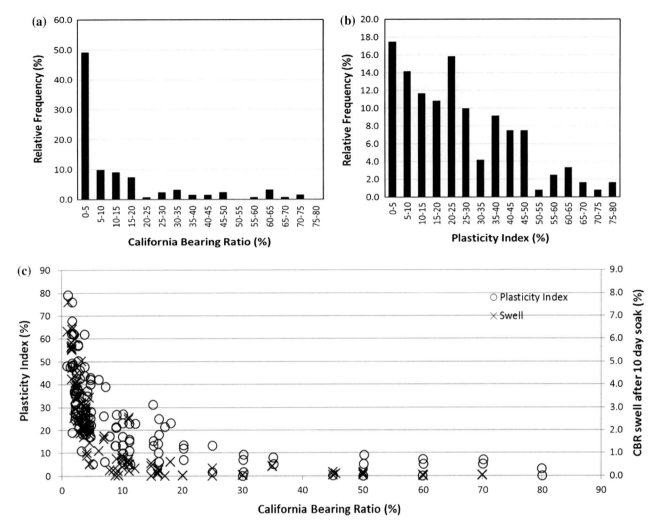

Fig. 45.4 California bearing ratio and plasticity index data. **a** Relative frequency of CBR. **b** Relative frequency of PI. **c** Correlation between 10 day soaked CBR and plasticity index and swell

were far more variable, with 90 % of values falling between 0 and 50 %, but with some extreme values as great as 80 %. Figure 45.4c shows that for samples with lower CBR values, the swell after 10 days of soaking was as great as 8 %, with all materials with CBR values less than 10 % recording some swell, and the many samples with CBR less than 5 % swelling from 3 to 8 %.

45.3 Conclusion

The results presented show that the engineering properties of tuffs and tuffaceous claystones from the Newcastle Coal measures are highly variable and generally unfavorable. Swell is a major issue for dimensional control of earthworks. On this project, surcharged encapsulation was used to control swelling of remolded tuffs in embankments up to 13 m high. Swelling is a complicated problem, influenced by compacted dry density, initial water content, confinement

and moisture change (Buzzi et al. 2011). In designing embankments from compacted remolded tuffs, it is necessary to place them at a water content and dry density, such that under their confining stress, they will neither expand nor collapse as their water content changes. This was a key aspect in the design phase of the Hunter Expressway.

References

Brakel AT (1989) Correlation of the Permian coal measure sequences of eastern Australia. Bur Min Res Bull 231:193–203

Buzzi O, Giacomini A, Fityus S (2011) Towards a dimensionless description of soil swelling behaviour. Geotechnique 61:271–277

Diessel CFK (1985) Tuffs and tonsteins in the coal measures of New South Wales, Australia. Dixieme Congres International de Stratigraphie et de Geologie du Carbonifere, Madrid, 1983, pp 197–210

Fityus S, Hawkins G, Delaney M, Morton S (2005) An overview of engineering geology and geotechnical challenges in the Newcastle region. Aust Geomech 40(1):5–28

Ives MJ (1995) Stratigraphy and engineering properties of the Newcastle and Tomago coal measures. Engineering geology of the Newcastle-Gosford region. Sydney, Australian Geomechanics Society, pp 223–240

Kramer W, Weatherall G, Offler R (2001) Origin and correlation of tuffs in the Permian Newcastle and Wollombi coal measures, Australia, using chemical fingerprinting. Int J Coal Geol 47:115–135

Seedsman RW (1989) Claystones of the Newcastle coal measures, NERDDC project C0902, report number 749. ISSN: 0811-9570

Vertical Harbour Quay Rehabilitation Using Ground Anchors

46

Liliana Ribeiro and Alexandre Santos-Ferreira

Abstract

Issues related to structure design are a danger to future user's safety, as well as leading to higher economic costs, resulting in overruns of the initial budget. Sometimes, during works, unexpected situations could arise, requiring constant monitoring and verification. In recent years, ground anchors usage in soil and rock, has been increasing. It's efficiency makes them widely applied. Regarding harbour constructions, the use of permanent ground anchors is restricted, due to constraints related to surroundings and corrosion, being usually reserved to special cases where other solutions are unavailable or inappropriate. This paper describes the problems derived from ground anchors rupture, used in a vertical quay wall, located in Pinhão (Portugal) fluvial harbour, and the necessary studies for a remedial project, including a new design for the ground anchors, performed as rehabilitation and strengthening of the quay structure. Finally, some comparisons and conclusions are made between the pre-design, which supported construction and lead to its collapse, and the results reached during rehabilitation process.

Keywords

Ground anchors • Rehabilitation • Quay • Harbour

46.1 Introduction

In order to improve docking conditions in Pinhão fluvial harbour (Portugal), a vertical quay wall with 81 m length was built, with enough capacity for tourism boats (hotel boats) to use. Those boats usually require 80 m long docking posts.

Pinhão vertical quay wall was developed in upstream junction of Douro and Pinhão rivers.

L. Ribeiro (✉) · A. Santos-Ferreira
IPTM, I. P, Edf. Vasco da Gama, Rua General Gomes Araújo, 1399-005, Lisbon, Portugal
e-mail: liliananinhosribeiro@gmail.com

A. Santos-Ferreira
e-mail: asf1954@netcabo.pt

A. Santos-Ferreira
CICEGe—Faculty of Sciences and Technology (FCT), University Nova of Lisbon, Quinta da Torre, 2829-516, Caparica, Portugal

The location of this quay wall, on the right bank of Douro River, is shown in Fig. 46.1.

The structural solution adopted consists in an anchored Larssen sheet pile wall.

46.2 Pre-design

The pre design considerations, for both the initial and rehabilitation design, will be briefly described.

46.2.1 Local Geology

In order to access local geology a set of Dynamic Probing Light (DPL) tests, along the alignment of the future structure, were performed. The results showed formations of weathered shale, presented in Fig. 46.2.

When those formations were not emerged, a soil layer overtops them with usually less then 2 m thickness. The

Fig. 46.1 Location of Pinhão quay harbour

Fig. 46.2 Pinhão geological profile

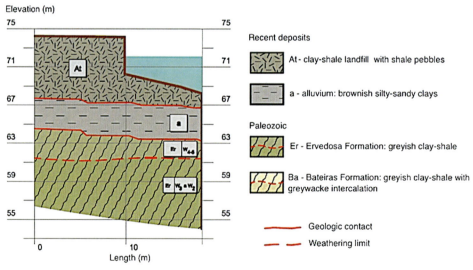

geotechnical characteristics obtained through the DPL test are presented on Table 46.1.

46.2.2 Pre-design Considerations

The ship type considered in the design is a passenger ship intended for river tourism with 1,500 deadweight tonnage (dwt), 80 m overall length, 10 m breadth and 3.6 m draught.

Table 46.1 Geotechnical characteristics of site material

Material specific weight (kN/m³)	Emerged landfill	Material geo-technical characteristics	Landfill material
	$\gamma = 18$		$\phi' = 30^{\circ}$
			$c' = 0,0$
	Submerged landfill		**Rock fill material**
	$\gamma' = 10$		$\phi' = 45^{\circ}$
	$\gamma = 20$		$c' = 0,0$
	Emerged rock fill		**Weathered shale**
	$\gamma = 18$		$\phi' = 25^{\circ}$
			$c' = 0,0$
	Submerged rock fill		**Shale**
	$\gamma' = 10$		$\phi' = 40^{\circ}$
	$\gamma = 20$		$c' = 60 \ kN/m^2$
	Water		
	$\gamma = 10$		

Table 46.2 Obtained results in pre design

Service phase					Ground anchors
Considered water levels	Embeddedness (m)	Sheet pile (m)	Momentum (kN · m/m)	Reaction on tendon (kN/m)	Total length
No water	4.05	9.40	339.5	158.9	16 m
Minimum level	4.25	9.65	286.1	154.9	Bond length
Maximum level	3.89	9.29	199.7	116.9	4 m
Full	3.73	9.13	169.4	87.7	Inclination 25°

Table 46.3 Active pressures over the structure, in pre-design phase

Active pressures	
Pre-design (first project) (kN)	Pre-design (rehabilitation project) (kN)
134.7	159.83

Table 46.4 Results obtained to ground anchors bond length

Ground anchors bond length		Effectively used (m)
Pre-design (first project) (m)	Pre-design (rehabilitation project) (m)	
4.0	3.0	6.0

Table 46.5 Results obtained to ground anchors free length

Ground anchors free length	
Pre-design (first project) (m)	Pre-design (rehabilitation project) (m)
12.0	15.0

Table 46.6 Results obtained to stability of potential slip surfaces analysis

Stability of potential slip surfaces analysis			
Pre-design (rehabilitation project)			Pre-design (first project)
Safety factor	Static analysis	Pseudo-static analysis	
7.562	2.604		3.02

Fig. 46.3 Global safety analysis and shear stress analysis

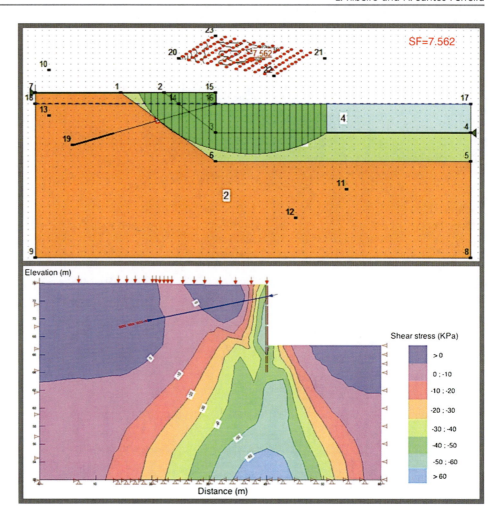

The active static impulses acting on the structure were defined by the Coulomb theory. Seismic impulses were calculated by Mononobe-Okabe method.

For the first pre-design phase Larix Cubus software was used and the results are presented in Table 46.2.

Short after finishing the vertical quay harbour construction, the ground anchors tendons suffered rupture.

In this paper a review is presented to access the collapse causes. For this attempt, a comparison of obtained results in the initial project, and those obtained in the rehabilitation work phase, will be made.

46.3   Rehabilitation Phase

Although the quay wall was not subject to normal use, some ground anchors have broken. The sheet pile wall, and overtopping beam did present neither deformations, nor excess stresses.

To rehabilitate the structure two options were available:

- First option was demolish the entire quay wall. This option was step aside due to timing and economic constrains;
- Second option was, re-design new ground anchors. Those new ground anchors would be placed at the mid point between the first stage anchors.

46.3.1 New Ground Anchors Pre-design

In the rehabilitation phase pre-design, the empirical diagrams of Terzaghi and Peck (1948) were used to assess the active impulses over the sheet pile wall, and the (Bustamante and Doix 1985) method was used to ground anchors design (Ribeiro 2012).

A 2D stress-strain analysis, as well as a limit equilibrium analysis was also performed to assess the overall stability.

The results are presented in Tables 46.3, 46.4, 46.5 and 46.6 and Fig. 46.3.

46.4 Final Remarks

The obtained results of rehabilitation phase pre-design allows us to state that:

- In the initial design (pre-collapse), the ground active pressures over the structure were underestimated. This may be a fact to justify the ground anchors bad behavior (rupture), however this is may not be the only reason;
- In the initial design (pre-collapse), the ground anchors free-length was too short. So, it's possible that fixed anchor zone lies within a ground mass prone to failure.

The study of global stability can be said that:

- The global security coefficients, evaluated in the study compared with those obtained by the designer, both are within required values for this type of work. So, apparently this is not the reason for the disruption of anchors of the initial project.

The stress-strain analysis developed in rehabilitation pre-design phase show that the rehabilitated structure should present a good behavior; as the rehabilitated quay is in service since November 2011, the observation of its behavior shows a good fit within pre-design phase.

References

Bustamante M, Doix B (1985) Une méthode pour le calcul des tirants et des micropieux injectés. Bull. Liaison Ponts et Chaussées. Paris, Laboratoire Central des Ponts et Chaussées. 140:75–91

Ribeiro L (2012) Ancoragens em estruturas portuárias. Análise de um caso de obra. Master's thesis, FCT-UNL, Quinta da Torre, Almada

Terzaghi K, Peck R (1948) Soil mechanics in engineering practice, 1st edn. Wiley, New York

The Importance of the Existing Engineering Geological Conditions During the Building Construction on the Terrain Affected by Sliding

47

Dragoslav Rakić, Zoran Berisavljević, Irena Basarić, and Uroš Đurić

Abstract

Performing of the deep excavations leads to an imbalance in the terrain and consequences are frequent occurrences of local sliding, collapse, settlement and even destruction of the adjacent facilities. Therefore, knowledge of the existing engineering geological data is very important, because in urban areas problems of interaction of the new facility and geological environment are not the only ones that need to be solved, but also the influences of the other structures located in vicinity. Aggravating circumstances are terrains with complex engineering geological conditions affected by active, dormant but also stabilized landslides. In this paper the importance of knowing engineering geological conditions and history of landslide processes is highlighted with the example of the construction of a shopping mall in a densely populated area of the Serbian capital-Belgrade. The results of engineering geological and geotechnical researches are chronologically presented starting from the first landslide activation in 1970 and its reactivations in 1981 and 1992. By the latest research results, the existence of "fossil" landslide is registered for the first time in this part of the terrain. Based on that, the project for the protection of surrounding terrain was done due to the deep foundation pit excavation.

Keywords

Engineering geological conditions • Fossil landslide • Residual strength

47.1 Introduction

In developed urban areas, the lack of available space on the surface is a problem and it is solved by the use of underground space. This often includes the performance of deep excavations with the protection of existing facilities. However, the choice of inadequate protection measures often leads to negative consequences with significant material costs and in some cases human lives are endangered. Due to the specific conditions of performing a deep excavation such as: variation in different soil types, limited space, difficult and demanding work conditions, excavation speed etc., conditions of the natural geological environment are especially important. Therefore, the base for the deep excavations performances is the engineering geological maps which should be practical, concise, clear and with adequate graphical and numerical representation of the terrain.

47.2 Geological Terrain Composition and Characteristics of the Sliding Process

The terrain where the shopping mall was built is a densely populated, hilly area of the capital of Serbia—Belgrade. The terrain basis, within the exploration area, consists of marine basin sediments which are the oldest sediment layers of Paratetis near Belgrade. They are presented with marls (Lg),

D. Rakić (✉) · I. Basarić · U. Đurić
Faculty of Mining and Geology, University of Belgrade, Djušina 7, 11000 Belgrade, Serbia
e-mail: rgfraka@rgf.bg.ac.rs

Z. Berisavljević
Koridori Srbije d.o.o., Kralja Petra 21, 11000 Belgrade, Serbia

G. Lollino et al. (eds.), *Engineering Geology for Society and Territory – Volume 6*,
DOI: 10.1007/978-3-319-09060-3_47, © Springer International Publishing Switzerland 2015

which are intensively modified in the surface layer, and they formed a surface weathering zone of degraded marl clays (Lg*). Quaternary cover is formed over this complex and it is made of different lithogenetic sediments in which diluvial sediments dominate on the slope ground part and alluvial-proluvial sediments on the flat part of the terrain next to a stream. Loess diluvium (dl-l) and a thin layer of diluvial clays (dl-gl) are separated within the diluvial sediments. Alluvial-proluvial sediments are characterized by polycyclic sedimentation with material gradation in vertical direction, so that within these sediments three areas are distinguished: clays (al-gl), clayed sands (al-gp) and clayed gravels (al-gš). Greater part of the terrain is covered by uncontrolled fill mainly made of construction waste, and it is divided into two parts: the so-called old fill (n_s) which was formed before 1970 and the fill which was formed after that (n). The wider zone of the terrain belongs to the region that includes an area of active, dormant and stabilized landslides (Fig. 47.1). The shopping mall is located on the landslide that was activated for the first time in 1970. Its activation started due to the trench excavation for the installation of sewer pipes along local streets. Then the first engineering geological terrain explorations were performed and based on these results its remediation was carried out by the system of drainage

trenches filled with sand and gravel. This drainage system was not sufficient to perform a permanent repair, because the landslide was reactivated in 1981. The researches that were performed in 1981, had an aim to protect the street that propagates through the frontal area of the landslide scar, but the remediation of the entire slope towards the local stream was not considered. For those reasons, in the frontal part of the landslide the retaining structure of reinforced concrete piles was built, with the average length of 12 m. Afterwards, the terrain surface affected by sliding was arranged. The slope got a natural layout that did not indicate the existence of the active sliding process. However, in 1992, the sliding process was reactivated and expanded to the southwest of the site, and the landslide re-affected central parts of the slope below the retaining structure. For these reasons, the retaining structure was extended along the street, and the gravel embankment (n_k) was built out in the central part, at the bottom of the slope which beside the function of ballast provided mitigation of the slope inclination (Fig. 47.2). It can be concluded that the sliding process on this site was periodically active and with uneven temporal frequency. It is exactly due to these frequent sliding processes, that the terrain was avoided for construction even though the site is in the narrow city core.

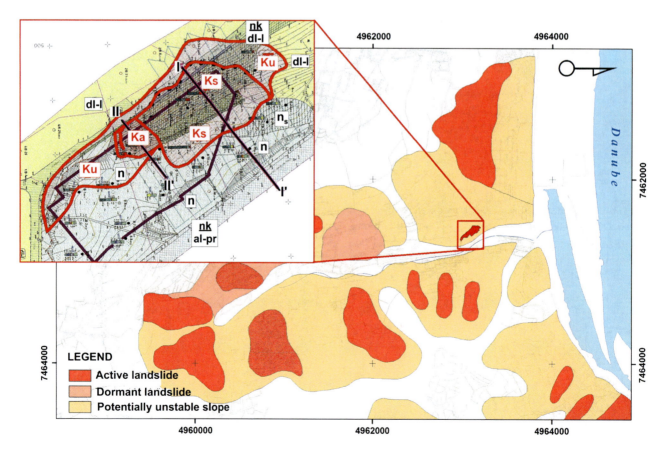

Fig. 47.1 Map of registered landslides along the local stream with engineering geological map of microlocation

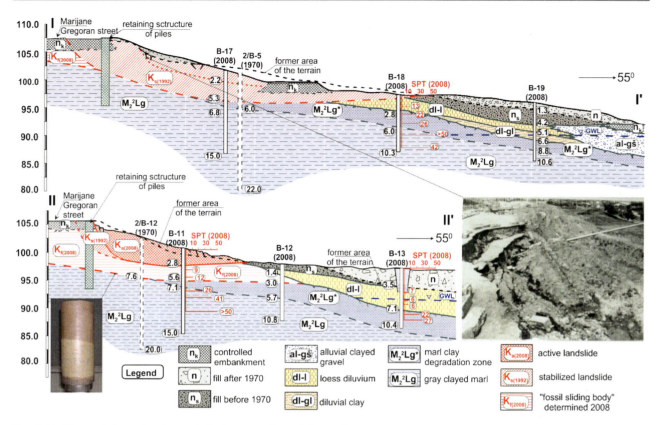

Fig. 47.2 Characteristic engineering geological cross sections of the terrain

47.3   Defining Engineering Geological Models

The latest engineering geological explorations at this site were performed for the construction of the shopping mall "Merkur", which was built in the meantime. For this occasion, the terrain zoning according to degree of stability was performed (Fig. 47.1), where the terrain was separated into: terrain affected by an active sliding (Ka), terrain on which the remediation measures were carried out (Ks) and terrain on which the landslide was dormant (Ku). Based on these researches, the active slip surface at a depth of 5.0 m (Fig. 47.2) was determined. In the repaired part of the sliding body, the groundwater table was determined at a depth of 3.5 m up to 8.0 m, while in the landslide foot part in the active zone of the sliding body, the groundwater table was measured at a depth of 1.6 m. Apart from the fact that the causes of relatively recent slides formation were determined, the latest engineering geological researches, helped discover for the first time the existence of "fossil landslide" in this part of Belgrade's terrain (Fig. 47.2).

This confirms the known fact that in the wider Belgrade area along the right bank of the Danube, the sliding processes occurred in the past and in most cases were stopped

by the formation of the loess cover (Rakić et al. 2009). This periodic activity of terrain sliding had caused a chaotic mixture of several lithological members within the colluvial mass where degraded marl clays (gL*) and diluvial clays (dl-gl) dominate. Lithological heterogeneity also affected the parameter values of shear strength that varied in a wide range. Due to the prevailing primary brittle, crystallizing and cementation bonds, cohesion of immovable degraded marl clays was $c' = 42$–60 kPa, while in the predominantly saturated, cracked and softened weathering zone it is minimized to the so-called apparent i.e. temporary cohesion of the weak and unstable hydrocolloid relations and equalled $c' = 5$–22 kPa (Rakić et al. 2000). In 1970 and later in 1981 laboratory tests were performed and gave the results of the residual internal friction angle of saturated colluvial soil samples. Also, the back-analyses were performed giving the mobilized shear strength parameters at failure along the slip surface. The latest laboratory tests have mostly yielded lower residual values for the internal friction angle $\varphi'_r = 11$–$12°$ (Fig. 47.3).

In the process of the stability analysis, several engineering geological cross sections of the terrain were considered taking into account the groundwater level. Considering that the newly formed sliding body affected surface parts of "fossil

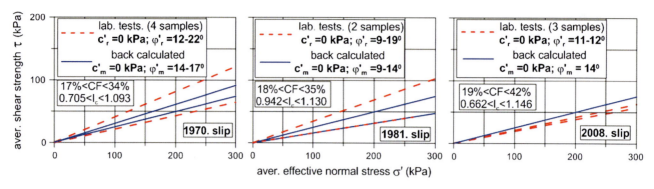

Fig. 47.3 Residual shear strength depending on the time activity of the landslide

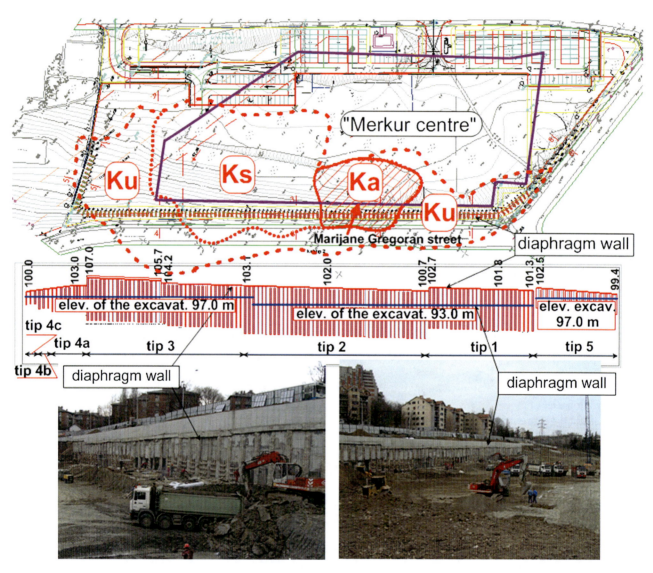

Fig. 47.4 Protection of foundation excavation

landslide" in one part of the terrain, as a corresponding slip surface for the securing of the foundation pit, a contact of the zone of degraded marl clays and grey marls was proposed. Conditional internal friction angle was determined from the limit equilibrium condition by back-analysis i.e. for the adopted safety factor $F_s = 1.0$, and assuming that along the slip surface $c'_r = 0$ kPa (Popescu 2002). The value of the conditional internal friction angle $\varphi'_m = 14°$ (Fig. 47.3) was obtained by the back-analysis method, which was later used to determine the force of a potential sliding body on the retaining structure.

47.4 Remediation Measures

Depending on the morphology of the terrain, the depth of the foundation pit excavation ranged from 8.3 to 14.0 m. This implied removing the entire colluvial part in the facility domain, whether it is the active, stabilized, dormant or "fossil" part. It also required taking into account the possibility of cutting the existing drainage trenches which were built after sliding in 1970.

Thus formed excavation had a lower elevation related to the pile base elevation of the existing retaining structure along the street, questioning its stability and the stability of the major road i.e. the slope above it on which there are residential buildings (Fig. 47.4). As the basic remediation measure and the foundation pit protection measure, the reinforced concrete diaphragms were designed. Along the street of Marijana Gregoran they were placed in the so-called "comb arrangement" perpendicularly to the reinforced concrete structure of piles, and connected with the overhead slab at a certain depth (Anagnosti and Stambolić 2008). Groundwaters from the slope are collected by the drainage curtain, which was placed between the diaphragms, and transferred in a controlled process through drainage pipes and shafts system to the city sewage. The reinforced concrete diaphragms were designed according to deepest determined

slip surfaces ("fossil" or dormant), which were defined in the contact of degraded marl clay zone and grey marl zone.

47.5 Conclusion

Due to the inaccessibility of locations, as well as the economic factors and scarce resources for research purposes, we are not always able to perform the necessary amount of exploration works which are objectively necessary to obtain reliable data. This especially refers to urban areas, where not only the problems of interaction between the new structure and geological environment need to be solved but also the influence of other structures located in the vicinity. In this regard, systematization and reinterpretation of the existing engineering geological and geotechnical data are very useful, because in these areas the unfavourable engineering geological conditions are not the only problem, but also insufficient knowledge of them. Therefore, we should not forget that the absence of sliding traces on a surface does not always prove its past stability.

References

Anagnosti P, Stambolić S (2008) The main project for securing the foundation pit of the "Merkur" shopping mall. Karaburma, Belgrade

Popescu ME (2002) Landslide causal factors and landslide remedial options. In: Proceedings of 3rd international conference on landslides, slope stability and safety of infra-structures. Keynote Lecture, Singapore, pp 61–81

Rakić D, Čaki L, Dragaš J (2000) Progressive softening of the belgrade clayey deposits owing to underground construction. In: An international conference on geotechnical and geological engineering. GeoEng2000, Melbourne, Australia

Rakić D, Nikolić A, Kostić S, Berisavljević Z (2009) Geotechnical investigations terrain for building the trade center "Merkur". In: Karaburma, 3th conference geotechnics in civil engineering, Association of Civil Engineers of Serbia (ACES), Zlatibor, pp 123–128

GIS Based, Heuristic Approach for Pipeline Route Corridor Selection

48

Ludwig Schwarz, Klaus Robl, Walter Wakolbinger, Harry Mühling, and Pawel Zaradkiewicz

Abstract

Large diameter pipeline systems play an increasingly important role in the supply of population and economy with natural resources like oil, gas and water all over the world. Due to the fact that the sources are often located far off from the consumer, pipeline systems consequently have to cross large distances, facing various social, environmental and geological constraints along their route. Therefore pipeline route selection is a constraint based corridor selection and narrowing process. With the first definition of the project corridor and the successive narrowing of this corridor down to the corridor of interest decisive milestones of the final pipeline alignment are set, which, if at all possible, can only be modified with extensive timely and economic effort during later project stages. On the other hand pipeline routing teams are increasingly confronted with an extensive amount of data and ever tighter project schedules specified by the pipeline owner. In order to cope with the enormous amount of data in short time and to define a safe and economic route corridor in a transparent and traceable way, not only for the client but also for lenders, authorities, stakeholders, NGOs, etc. a GIS based, heuristic pipeline corridor selection approach was developed for the Trans Anatolian Natural Gas Pipeline Project (in the following referred to as TANAP) in close cooperation of engineering geologists, routing engineers and GIS experts. The process is a desktop based procedure including data collection, classification and above all their spatial evaluation utilizing GIS technology.

Keywords

Infrastructure • Pipeline route selection • Constraint mapping • GIS • Heuristic method

48.1 Introduction

The TANAP project is an approx. 1,800 km long, 56 in. pipeline system that intends to transport natural gas to be produced in Caspian region via Turkey to Europe. It traverses the whole of Turkey in east-west direction between the Turkish-Georgian border and the Turkish-Greek, respectively the Turkish-Bulgarian border. These three border crossings constituted the benchmarks for pipeline route selection. Additionally it has to be mentioned that an offshore route via the Black Sea was excluded by the client. However in the western Turkey a short offshore section had

L. Schwarz (✉) · K. Robl
ILF Beratende Ingenieure ZT GmbH, Feldkreuzstraße 3, 6063, Rum, Austria
e-mail: ludwig.schwarz@ilf.com

W. Wakolbinger
LandMark GmbH&Co.KG, Peilsteinstraße 3a, 83435, Bad Reichenhall, Germany

H. Mühling
ILF Beratende Ingenieure GmbH, Werner-Eckert-Straße 7, 81829, Munich, Germany

P. Zaradkiewicz
ILF Consulting Engineers Polska SP.zo.o., Osmanska 12, 02-823, Warsaw, Poland

Fig. 48.1 Overview of the project area (framed area)

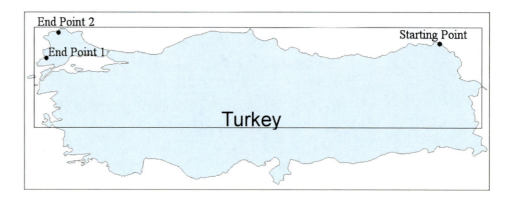

to be considered some place along the Bosporus, Marmara Sea or Dardanelles (Fig. 48.1).

The basic principal of pipeline route selection is to find the shortest connection between the start and the end point. The degree of deviation from this principal is governed by the constraints encountered within the corridor.

Along the TANAP project area various constraints are encountered that cover aspects of social and environmental impact, public safety, constructability, structural integrity of the pipeline and land use. Therefore pipeline route selection is a constraint based corridor selection and narrowing process. The following corridors are typically established during the narrowing process:

1. Project Corridor (width typically 20–40 % of the pipeline length)
2. Corridor of Interest (width typically 2–4 % of the pipeline length)
3. Preferred Corridor (typically 1–2 km wide)
4. Specified Corridor (typically 200–500 m wide)
5. Pipeline Centerline and Construction Corridor

The constraints utilized in each narrowing step depend on the size of the constraint with respect to the respective corridor width. Thus small sized constraints that can easily be bypassed within the corridor of concern are not considered in the respective narrowing step (e.g. landslide prone areas are considered in the definition of the corridor of interest rather than individual landslides, archeological sites are typically not considered before defining the preferred corridor). Consequently the investigation detail and the route accuracy increase inversely proportional to the width of the corridor under investigation in the course of the narrowing process. Early stages of corridor narrowing are purely based on desktop work while field work and ground truthing become increasingly important with the reduction of the corridor width. The subject paper describes the first two steps of the narrowing process from the definition of the project corridor to the selection of the corridor of interest.

48.2 Methodology

Modern information society provides extensive amount of data in electronically processible form available on short term including geological and terrain data (lithology, landslides, karst, seismicity, active fault lines, mining areas, digital terrain model, etc.), environmental data (protected areas, land use, etc.) and social data (population density, settlement areas, development areas, existing and planned infrastructure, military areas, etc.). Handling the enormous amount of data in short time has turned into a major challenge especially for large infrastructure projects. Therefore GIS based data management has become state of the art for several decades now (e.g. Avtar et al. 2011; Blais-Stevens et al. 2012; Sydelko and Wilkey 1994). GIS software also provides sophisticated tools for data processing, evaluation and presentation.

For the TANAP Route corridor selection a process was adopted utilizing GIS Cost Distance analysis tools. After the various datasets are fed into the system and organized in different layers, the individual datasets are further classified with each class assigned a cost factor (please note that costs in GIS language is a synonym for a function of time, distance, or any other factor that incurs difficulty or an outlay of resources). Within the described procedure the assignment of the cost factor is the most critical process with respect to the final result. For constraints related directly to construction cost factors are fairly easy to estimate since they are based on actual costs and thus can be derived from previous projects (e.g. influence of terrain on pipeline construction). For constraints not directly related to construction (e.g. protected areas, land use etc.) cost factors are much more difficult to quantify since they depend on "soft" factors like the political situation, social acceptance, etc. The process of cost factor selection is further complicated by the fact that not only the cost factors within one dataset have to match each other but also the cost factors between all datasets. Over or under prediction of the cost factor of one dataset can have a significant impact on the result. Thus the selection of cost factors is to a large extent a heuristic process that considers experience from previous projects but is also based on assumptions to account for missing information and uncertainties. The selection process is furthermore an iterative process where the individual cost factors have to be calibrated comparing the outcome of the evaluation in specific route sections to experience from previous projects.

GIS cost analysis tools are usually raster based. Therefore the area of concern is split into a raster of cells where the size of the individual cells is selected based on the size of the area of concern and the level of detail of the considered constraints. In a next step the total cost factor of each cell is determined as a cumulative cost factor of the individual constraints encountered within one cell. For this purpose the software also offers the possibility to weight the different constraint raster that make up the cost raster. Thereby the "Weighted Sum" tool overlays several raster, multiplying each by their given weight and summing them together.

The result of this calculation already provides first clues about suitability of the area for pipeline construction (see Fig. 48.2).

The informative value of these cumulative cost factors is still limited as they provide spot data only and not data over length. This means that considering the entire pipeline length it may be more reasonable to cross critical areas exhibiting high costs if a reduction of pipeline length can be achieved. To overcome this limitation a Cost Distance raster is calculated for the entire project corridor. The cost distance tools determine the shortest weighted distance (or accumulated travel cost) from each cell to the nearest source location. When moving from a cell to one of its four directly connected neighbors (vertical or horizontal movement), the costs to move across the links to the neighboring node is 1 times the cost factor of cell 1, plus the cost factor of cell 2, divided by 2. If the movement is diagonal, the costs to travel over the link is 1.41 (or the square root of 2) times the cost factor of cell 1 plus the cost factor of cell 2, divided by 2 (ESRI 2012) (Fig. 48.3).

The graphical output of the result shows a raster map with increasing costs from the start point to the end point (see Fig. 48.4).

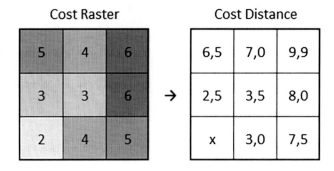

Fig. 48.3 Example for "cost distance" calculation: lowest accumulated cost distance to move from start point x to upper-right cell = (3 + 6) *SQRT(2)/2 + 3,5 = 9,9

To receive a corridor map the calculation is run twice, (1) from the start point and (2) from the end point. Finally the results of both calculations are summarized. Classifying the result from low to high values enabled us to determine the project corridor (see Fig. 48.5).

48.3 Results and Discussion

The Cost Corridor map shown in Fig. 48.5 provides a very distinct picture of the project corridor. The dashed area was thus selected for the next assessment step the definition of the corridor of interest. For this step additional and more detailed constraints were added and the classification of the constraints used in the previous assessment partly refined. As expected the result of the project corridor assessment enabled further narrowing of the corridor. But the result also showed a braided net of possible corridors rather than just one distinct corridor. This outcome is only an apparent limitation of the procedure since it actually accommodates the requirement of the authorities to investigate route alternative during environmental impact assessment. So based on the cost corridor map of the corridor of interest three alternative route corridors, each approx. 2 km wide, were defined and investigated in detail, both on desktop but also by extensive field works. These works concluded in the definition of the 2 km wide preferred route corridor which was then subject to more detailed investigation works including

Fig. 48.2 Combined constraint map, costs increase from white to black, *dashed line* indicates derived project corridor

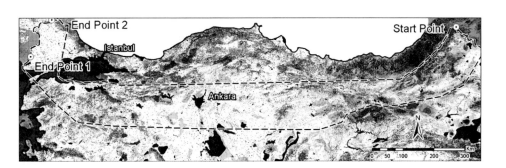

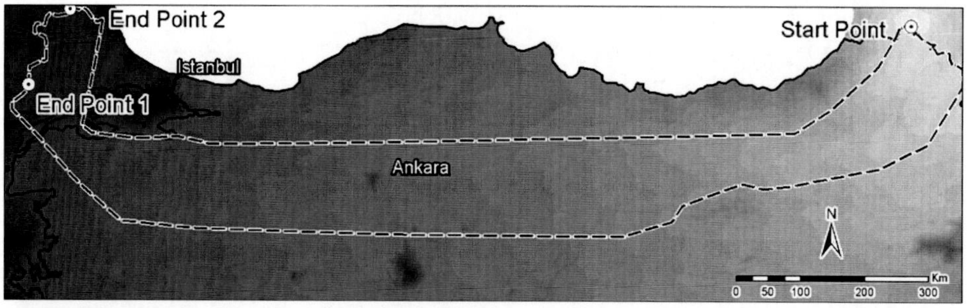

Fig. 48.4 Cost distance map, calculated from start point, costs increase from white to black, *dashed line* indicates derived project corridor

Fig. 48.5 Cost corridor map, costs increase from white to black, *dashed line* indicates derived project corridor

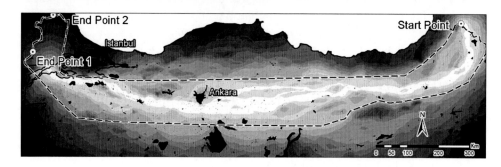

remote sensing techniques, geohazard mapping, environmental and social impact assessment, etc.

The assignment of cost factors proofed to be the most sensitive part of the entire procedure and thus has to be conducted with care. For this certain sections have to be identified within the corridor which allows an empirical calibration of the cost factors. Furthermore it has to be considered that cost factors are dependent on numerous project specific factors. Therefore there is no general set of cost factors to be utilized on similar projects. In fact the selection has to be carried out individually for each project in form of an iterative process considering experience, both of the involved geologists/engineers and from existing projects in the region, as well as local and project specific conditions.

One big advantage of GIS based corridor assessment is that changes like reclassification of the constraints or adjustment of the cost factors can be conducted within a small time frame, allowing covering large areas within short time.

48.4   Conclusions

Due to the fact that constraint classification and cost factor allocation requires experience of the involved disciplines (route engineers, engineering geologists) the described methodology is not suitable for novices in the field of infrastructure corridor selection. Despite this limitation GIS based pipeline corridor assessment proved to be a powerful tool for fast and reproducible early corridor selection phases. Its use is not confined to pipeline projects but can equally be utilized for different types of above- and underground infrastructure projects whereat the classification of the constraints and quantification of the allocating cost factors has to take the type of infrastructure into account.

Acknowledgement   The authors wish to thank TANAP in general and Mr. Sinan Elaslan in particular for the permission and support to publish this article.

References

Avtar R et al (2011) Landslide susceptibility zonation study using remote sensing and GIS technology in the Ken-Betwa River Link area, India. Bull Eng Geol Environ 70:595–606

Blais-Stevens A et al (2012) Landslide susceptibility mapping of the sea to sky transportation corridor, British Columbia, Canada: comparison of two methods. Bull Eng Geol Environ 71:447–466

Esri (2012) ArcGIS 10.1

Sydelko PJ, Wilkey PL (1994) GIS least-cost analysis approach for sitting gas pipeline ROWs. Technical Information Center, Oak Ridge, Tennessee

A Study of Ground Natural Temperature Along Tabriz Metro Line 2, Iran

49

Ebrahim Asghari-Kaljahi, Karim Yousefi-bavil, and Mahyar Babazadeh

Abstract

The cities extension, request for developing and increasing underground transportation system. One of this ways in big cities is urban tunnels or metro. Tabriz often located on alluvial deposits, so these tunnels repose in these deposits. In this study we investigate engineering geology aspects of Tabriz Metro line 2, approximately 22 km long. Then it is described about natural temperature measuring of ground layers, and effect of this temperature in metro construction and using. Ground natural temperature is depended on atmosphere temperature and mineralogy of soil and groundwater condition. Some of clayey minerals are radioactive and increase ground temperature. Also, groundwater flowing may affect the ground natural temperature. In metro construction for tunnel boring machine (TBM) and exploitation of tunnel, knowing nature temperature is necessary. In this study, ground natural temperature is measured between 15 and 25 m depth. These measuring shown that the ground natural temperature is between 13 and 18 °C.

Keywords

Ground temperature • Thermometer • Tabriz metro

49.1 Introduction

Today, heavy traffic and city transportation are the big cities problems which can be reduce by subway transportation. Tabriz, one of the Iran's crowded metropolitan, faced the traffic problem, can be managed by subway/Metro construction. In line with Tabriz Metro extension, Tabriz Metro Line 2 (TML2) construction is carrying out along E-W with approximate length of 22 km. More than 120 boreholes with

E. Asghari-Kaljahi (✉)
Department of Geology, University of Tabriz, Tabriz, Iran
e-mail: e-asghari@tabrizu.ac.ir

K. Yousefi-bavil
Department of Geological Engineering, Middle East Technical University, Ankara, Turkey
e-mail: yousefibavil.karim@metu.edu.tr

M. Babazadeh
Young Researcher Club, Science and Research Branch, Islamic Azad University, Tabriz, Iran

depth of 25–45 m were drilled for Geotechnical study of the project (P.O. Rahvar 2008). To study ground strata and their strength, Standard Penetration Tests (SPT), Pressure meter and Permeability tests were accomplished. During boreholes drilling desired disturbed and undisturbed samples were obtained and necessary physical, mechanical and chemical tests were fulfilled.

Knowing natural ground temperature in tunneling is essential especially for its ventilation in utilization phase (Department of Justice and Attorney-General 2012). In order to earth layers thermometry, thermometer installed in some boreholes and natural ground temperature in tunnel (depth of 15–25 m) was measured.

49.2 Tabriz Geology

Tabriz is surrounded by Oun-Ebne Ali Mountains with trend of E-W in north and Sahand volcanic with low heights in south. Tabriz plain stretch out E-W due to mentioned mountains. General slope of plain is due to the west and result in general drainage of surface and sub-ground water toward west. Plain mostly covered with alluvial sediments (Geological Survey of Iran 1993). TML2 route from Qaramalek in west to Baghmisheh has rather smoother slope and toward east they become hills and associated with numerous domes. Along the TML2, elevation difference between lowest and highest point reaches almost 280 m. In east part of the TML2 between Baghmisheh and Marzdaran it can be seen small folds and faults that result in layers rotation, fracturing and displacements (Hooshmand et al. 2012).

49.3 Engineering Geology

TML2 route from Garamalek area in west to Baghmisheh area is covered by alluvial sediments, and in continue to east Marlstone and Claystone and Siltstone layers have outcrops and/or with a thin cover of sediment in ground surface. Under this alluvium sediments near Abbasi Street toward east, layers of Marl, Sandstone and Conglomerate located in depth of less than 10 m. According to investigations, length of TML2 can be categorized into 5 general zones:

Zone 1: distance between Sanat Square in Garamalek to Jahad Square Conjunction. In this zone, west of Tabriz, up to studied depth (almost 30 m) alluvial deposits consist of fine grain and sand alternation. Groundwater table varies between 5 and 18 m.

Zone 2: distance between Jahad Square Conjunction to Selab-Aghzi in Abbasi Street. In this zone, subsurface layers mainly consist of coarse grain alluvial sediments (Gravel and Sand) whit rock fragments (Boulder and cobble) floated in them. Also, interlayer fine grain alluvial exist among this coarse grain sediments, however tunnel routes in this area mainly passes through coarse grain deposits.

Zone 3: distance between Selab-Aghzi in Abbasi Street to Shahid-Fahmideh Square. In this zone along Abbasi Street, week rock layers of claystone, mudstone, sandstone and marlstone underlain with surface alluvial layer with thickness of 5–15 m. In this part metro tunnel settles inside rock layers. Surface alluvial layers constitute of fine and coarse grain alluvial deposits that mainly were classified as SM, ML and CL. Groundwater table reached between 5 and 25 m in this part and inside rock layers occasionally gaseous artesian water exist.

Zone 4: distance between Shahid-Fahmideh Square to Eastern of Baghmisheh.

In this zone, subsurface layers mainly formed from coarse grain (Sand-Gravel) and fine grain (Clay-Silt) alluvial. Tunnel routes in this segment passes through the fine and coarse grain sediments. This zone groundwater depth varies between 5 and 20 m and gaseous artesian water has been observed.

Zone 5: distance between Baghmisheh to Tabriz International Exhibition. Mostly, this zone constitute of weak rock layers of marl, claystone, mudstone and sandstone.

49.4 Groundwater Condition

Groundwater table varies greatly along TML2. Although, groundwater flows out as artesian in Fahmideh square's boreholes, in some boreholes groundwater did not reach to the great depth. Overly, groundwater depth varies between 2 up to over 30 m. Water table levels decline from east to west which representative of groundwater flows from east to west, and this condition somehow coincides with Tabriz Plain slopes.

During drilling in BH-2, West of Fahmideh Square, groundwater reached approximately in 5 m but after end of drilling groundwater table raised and overflowed with slight flow. The noticeable point regarding groundwater in this borehole exists of CO_2 solution in the water. The amount of gas inside the groundwater is remarkable and based on conducted tests it is approximately 1 %.

Studies have shown that aquifer within alluvial sediments is separated from aquifer within rock bed in east part. In other words, within impermeable rock layers, fossil water and occasionally gaseous under-pressure water exist as lenses with no extension. The reason of existence of under-pressure water within rock strata is alternation of impermeable layers (Claystone and Marlstone) and permeable layers (weak Conglomerate) which water was constrained within permeable layers.

49.5 Ground Natural Temperature

Ground natural temperature has been studied in located depth of tunnel. To do so, along TML2 route with distance of 2–2.5 km from each other, electrical thermometers were installed in depth of 15–25 m and were read in different times. Figure 49.1 shows the location of boreholes where thermometers were installed. In installing of thermometers it should be ensured that there is a complete contact between instrument and surrounding layers. After installation in subjected depth, inside boreholes completely filled with site soil. This process depends on the accuracy required in testing. Main aim in filling the boreholes is minimizing the impact on thermal regime of the earth which is not possible without free groundwater flow prevention. After installation of thermometers, borehole will be filled with bored soil of boreholes through two ways:

(a) Layers of material to be dumped and compacted
(b) Appropriate ratio of material mixed with water and dumped inside borehole by teremi pipe.

Fig. 49.1 Thermometer boreholes location on Tabriz map

Fig. 49.2 Electrical thermometer tool

Fig. 49.3 Natural ground temperature measurements

Also, where installed thermometers levels were below the groundwater level, borehole can be filled with bentonite-cement slurry, the bentonite pellets and/or any similar waterproof material which satisfies the complete filling of the borehole.

A picture of thermometer tool and reading set has shown in Fig. 49.2 and a picture of thermometer and measuring devise has shown in Fig. 49.3.

Boreholes Location which thermometers installed and their special position has shown in Table 49.1. In all boreholes, thermometry instrument set below the water table.

Immediately after installation of tools, thermometer readings carried out in different times, and their results have shown in Fig. 49.4. Several days after installation, temperature values stabled and natural thermal of earth gained. It is between 13 and 18 °C. Great primary temperature (immediately after installation) is due to

dehydration of bentonite-cement grout that poured for sealing the borehole. It was observed relatively lower temperature in east part boreholes than the west part, which had high values. Of course ground type affects the results that constitute of soft rocks in east part rather than centre and west, consist of alluvial, for its low temperature. The temperature changes diagram in different boreholes were illustrated in Fig. 49.4.

49.6 Conclusion

Ground natural temperature of TML2 Tunnel route (depth of 15–25 m) studied by installation of some electrical thermometers. Studies show that ground natural temperature varies between 12.8 and 18.1 °C. Almost higher temperature observed in fine alluvial deposits, medium in coarse grain

Table 49.1 The location of thermometers measured temperatures

Borehole	Borehole location	Borehole location coordinates			Temperature (°C) after 120 days	Ins. depth (m)	GWD (m)
		X	Y	Z (m)			
A_2B_1	Khane Sazi	46° 14′ 00.0″	38° 05′ 20.1″	1,372	17.1	18	15.9
D_2B_1	Qare Aghaj St.	46° 15′ 53.2″	38° 04′ 52.6″	1,392	17.6	20	8.2
BH-16	Qajil area	46° 16′ 59.1″	38° 04′ 48.3″	1,404	17.1	24	8.8
J_2B_2	Abbasi Sq.	46° 19′ 33.9″	38° 04′ 41.4″	1,458	15.3	21	9.0
BH-4	Mikhak Park	46° 21′ 02.6″	38° 04′ 31.5″	1,485	14.8	19.5	6.1
N_2B_1	Fahmideh Sq.	46° 21′ 50.1″	38° 04′ 36.2″	1,508	15.2	23	16.1
L_2E_6	Baghmishe town	46° 22′ 43.4″	38° 04′ 19.0″	1,533	15.3	16	15.1
P_2B_1	Marzdaran town	46° 23′ 22.9″	38° 03′ 26.2″	1,565	14.1	20	10.3
S_2B_1	Tabriz int. exh.	46° 24′ 02.6″	38° 02′ 10.4″	1,614	12.9	23.3	10.6

Fig. 49.4 Temperature variations with elapsing time in various boreholes

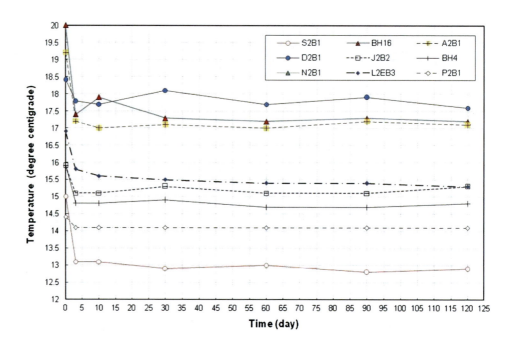

alluvial sediments and lower temperature in Marly layers. Where groundwater table is in shallow depth, relatively higher temperature and where it is deeper, the lowest temperature observed.

Acknowledgement The writers would like to acknowledge Iman Sazan Consulting Engineers Institute (Project manager of TML2) and P.O. Rahvar Consulting Engineers (Geotechnical consultant of TML2) for their co-operation.

References

Department of Justice and Attorney-General (2012) Tunneling code of practice 2007, Workplace Health and Safety Queensland

Geological Survey of Iran (1993) Tabriz geology map, scale 1/100,000

Hooshmand A, Aminfar M, Asghari E, Ahmadi H (2012) Mechanical and physical characterization of Tabriz marls, Iran. J Geotech Geol Eng 30(1):219–232

P. O Rahvar Consulting Engineers (2008) Geotechnical investigation report of TML2 project

The Challenges of Site Investigations, Dredging, and Land Reclamation: A Port Hedland (Western Australia) Project Perspective

50

P. Baker, J. Woods, M. Page, and F. Schlack

Abstract

Port Hedland, Western Australia is the largest bulk commodities export port in the world. The construction of wharves and stockyards requires significant dredging and land reclamation. Significant planning, logistical and technical challenges must be overcome for nearshore geotechnical site investigations to be successful. The sub-surface geology of Port Hedland harbour is often considered relatively straightforward, yet geotechnical investigations have not always provided sufficient or appropriate data, with some projects culminating in spectacular disputes between contractors and developers. The dredged materials are deposited in offshore spoil dumps, as well as onshore in dredge material management areas (DMMAs). Successful land-reclamation is achieved by separating the fines fraction leaving a sand-gravel soil (known as grits). The fines are typically pumped to designated fines settlement areas.

Keywords

Port hedland • Nearshore • Investigations • Dredging • Reclamation

50.1 Introduction

Port Hedland, Western Australia is the largest bulk commodities export port in the world. In the year ending 30 June 2013, 286 Mt (million tonnes) of cargo left Port Hedland, 280 Mt of which was iron ore bound for Asian steel mills, the remainder consisting mostly of salt, manganese, chromite, and copper.

Prior to the first major development of the port in the 1960s, the maximum natural depth of the harbour was about 9 m. With ongoing development and decreasing availability of deep water parts of the port, recent expansions and proposed facilities are increasingly situated in shallower areas. Hence the significance of dredging and land reclamation components of port expansion projects continues to increase, as does the onus on engineering geologists to provide timely, accurate and appropriate input of data from nearshore site investigations into large multidisciplinary project engineering teams. The current extent of dredged areas is clearly visible at low tide (Fig. 50.1).

50.2 Preparation, Logistical and Technical Challenges

50.2.1 Health Safety and Environmental Management

Prior to any site investigation a hazard identification workshop (HAZID) is conducted with all relevant parties involved in the investigation, including the engineering

P. Baker · M. Page
WorleyParsons, Perth, 6000, Australia
e-mail: paul.baker@worleyparsons.com

M. Page
e-mail: michael.page@worleyparsons.com

J. Woods (✉)
WorleyParsons, Calgary, SP09-19, Canada
e-mail: jonathan.woods@worleyparsons.com

F. Schlack
Port Hedland Port Authority, Port Hedland, 6721, Australia
e-mail: frans.schlack@phpa.com.au

G. Lollino et al. (eds.), *Engineering Geology for Society and Territory – Volume 6*,
DOI: 10.1007/978-3-319-09060-3_50, © Springer International Publishing Switzerland 2015

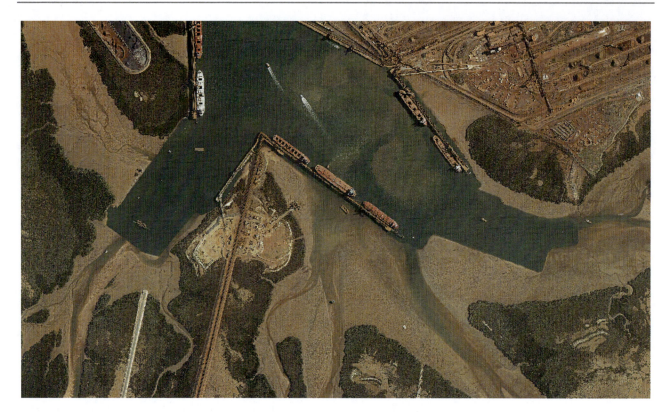

Fig. 50.1 Aerial photograph of Port Hedland Harbour at low tide (courtesy of Port Hedland Port Authority 2013)

geologists. The purpose of the HAZID is to identify potential hazards from activities that are conducted on site, assign a risk (hazard and consequence) to these activities, and identify ways to mitigate risk.

All environmental issues for the nearshore investigation are assessed and these range from oil spills from hydraulic equipment through to the risk of a cyclone.

A key health and safety focus for nearshore investigations during the Australian summer is the need for detailed cyclone management plans. An average of about five tropical cyclones form off the Pilbara coast each year, with generally one or two systems resulting in the port being placed on alert or ultimately evacuated and closed. The systematic and efficient approach to cyclone management is essential for any work conducted in the port.

50.2.2 Operational and Logistical Challenges

The initial challenge is in securing a window of time where there is a berth and suitably-sized crane available for assembly of the jack-up on which the drilling rig and auxiliary equipment are placed.

Prior to any borehole moves or tows the port authority has to be contacted and permission granted. The towing of a fully loaded jack-up barge occurs at a slow pace (3–4 knots) and the amount of time required for towing has to allow for

sea state, shipping movements, mechanical failure, and potential for grounding.

Port Hedland experiences a large tidal range, with a maximum astronomical variation of 7.5 m above Chart Datum (CD) and a maximum flood tide rate of 1.5 knots. The large tidal range and powerful currents typically dictate when operations such as barge movements are viable and when shallow areas of the port can be accessed. The large tidal range necessitates the need for flexibility in timing of work shifts and durations, placing unusual demands on personnel as well as equipment. At times, the low tide restricts access to the jack-up or prevents the support/emergency evacuation vessel from standing nearby.

50.3 Geology and Dredging

50.3.1 Nearshore Geology

Basement geology comprises Archaean granitic rocks of the Pilbara Craton. Overlying the basement rocks is an accumulation of mostly Pleistocene sediments with a relatively thin surficial cover of Holocene sediment (Fig. 50.2).

Surficial Holocene deposits include shallow marine, beach and dune sediments comprised of calcareous shells, reworked alluvium and calcareous rock fragments, which accumulate along the coastline, and fine grained deposits

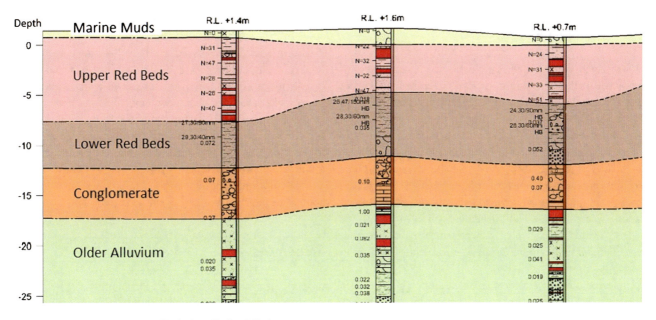

Fig. 50.2 Typical geological profile in Port Hedland Harbour

comprising estuarine muds which accumulate in tidal creeks and flats. The majority of the underlying Pleistocene deposits comprise siliciclastic terrigenous sediment, including quartz, feldspar, lithic fragments and clays, eroded from basement rocks in the hinterland and transported to the coastal plain. Characteristically, much of the alluvium is 'red' due to grain coating and staining of the clayey matrix by iron oxides and is known locally as the 'Red Beds'.

Changes in sea level during the Quaternary have resulted in periodic exposure and submergence of the coastal plain sediments, as well as influencing the elevation of the groundwater table. This has contributed to post-depositional alteration/diagenesis of sediments in the Port Hedland area. Dissolution and precipitation has resulted in the induration of sediments into weakly to well-cemented rock. The presence of calcrete cementation in the Lower Red Beds differentiates it from the Upper Red Beds (Fig. 50.2). The underlying Conglomerate is variably cemented and is high strength rock in parts. Induration may also be imparted by the in situ alteration of clays (e.g. kaolin clays) into cementing forms such as palygorskite.

50.3.2 Implication of Geology on Dredging

The purpose of a nearshore geotechnical investigation is to determine the nature of the material to be dredged, used and disposed of (Bray et al. 2001). Geotechnical aspects that are fundamental to the understanding of the dredging process and the evaluation of dredging projects include the in situ characteristics of the material to be dredged, the change in strength and volume of the material during the dredging and

relocation process, and the potential for change in material grading and the behaviour of the material during and after placement in the reclamation area.

The sub-surface geology of Port Hedland harbour is often considered relatively straightforward, yet the geological interpretation has not always accurately assessed or adequately communicated the geological risks for dredging and reclamation to project teams or contractors. Consequently, expensive claims have arisen as a consequence of extreme abrasion and excessive cutter, pump and pipeline wear caused by the combination of angular quartz and plastic fines forming armoured clay balls, with associated lost time and schedule delays.

The characteristics of the material within Port Hedland harbour have an effect on the suitable dredging plant to be used for a dredging campaign. The Marine Muds can be dredged by small suction dredges or backacter dredges. This material is often deposited in offshore spoil grounds as opposed to on land because the Marine Muds are a known acid sulfate soil. The dredging of berth pockets and turning circles penetrates the Red Beds and Conglomerate layer. These units contain quartz and lithic fragments; the hardness, shape, size and angularity of which has a significant effect on the abrasiveness, particularly in pumping operations from a cutter suction dredge. Of particular effect in Port Hedland is the formation of clay balls which contain abrasive quartz gravel; the quartz tends to protrude from the clay balls, making them highly abrasive.

The amount of fines being dredged will also have to be monitored in terms of the extent of sediment plumes (environmental impact) and its management in land reclamation, as discussed below.

50.4 Materials—Land Reclamation and Fines Management

Once dredged, some materials are deposited in offshore spoil dumps, as well as onshore in dredge material management areas (DMMAs). Reclamation management during the construction phase has to be tightly controlled. Selection of pipeline routes has to be executed according to a predetermined Reclamation Management Plan. The point at which dredge spoil is discharged into the reclamation area may change during the course of reclamation and survey control is required to monitor the rapidly advancing reclamation. Excess water and suspended silt also has to be guided towards a distal pond where pumps extract it from the reclamation area, to name just a few aspects that have to be strictly enforced.

Understanding how particle size distribution changes as a consequence of the action of cutter-suction dredging, proves invaluable in developing strategies for land reclamation. Successful land-reclamation is achieved by separating the fines fraction leaving a gravelly sand soil (known as grits). Dozers working the dredge spoil as it is discharged into the reclamation area contour and traffic compact the material. As the dozers work the dredge spoil, excess water carries the fines across the carefully contoured surface to a holding pond, from which the fines are typically pumped to designated fines settlement areas. With good management, spoil from dredging the Red Beds and Conglomerate (discussed in Sect. 3) becomes gravelly well-graded sand with a fines content between 5 and 10 %. This material is an excellent construction material for use in civil construction of roads, stacker and reclaimer embankments and general industrial land. When compacted to 98 % maximum modified dry density, the material can have an internal angle of friction (phi') of 40–44° (derived from direct shear tests) and a unit weight around 20.5 kN/m^2.

To size a fines settlement pond, a bulking factor of 5 is applied to the fines content of the Red Beds.

References

Bray RN, Bates AD, Land JM (2001) Dredging, a handbook for engineers, 2nd edn. Butterworth-Heineman, Oxford

Bureau of Meteorology (2013) Climatology of Tropical Cyclones in Western Australia. Available from http://www.bom.gov.au/cyclone/climatology/wa.shtml. 19 Aug 2013

Fire & Emergency Service Authority of Western Australia (2013) Prepare cyclone smart. Government of Western Australia. Available from http://www.dfes.wa.gov.au/safetyinformation/cyclone/Pages/publications.aspx. 19 Aug 2013

Port Hedland Port Authority (2013) Cargo statistics and port information leaflet

The Role of Geological Analysis in the Design of Interventions for the Safety of the Road Asset. Some Examples

51

Serena Scarano, Roberto Laureti, and Stefano Serangeli

Abstract

The Design Management—Geotechnical Unit of ANAS (National Public Road Company) is often involved in study and monitoring activities regarding some instability events affecting the road asset; generally, those events tend to compromise the functionality of the infrastructure. These circumstances are generally caused by geomorphological, hydrogeological or stratigraphic arrangements, and triggered by specific rainfall conditions that tends to modify the whole stability of the road body and the slope complex as a whole. In particular, as examples of ANAS experiences, two case histories are illustrated. Both situations concern reinforced embankments that, following rainy periods, have shown strong evidence of instability. In order to size the stabilization measures, necessary for the road restoration, a thorough study of the phenomenon, in both cases, was developed. It was realized by different stages of investigation and monitoring of roadway displacement. In addition to a thorough geological and geomorphological survey, useful in the identification of particular instability surface forms, specific site-investigation campaigns were prepared and completed by the installation of topographic, geotechnical and interferometric monitoring instruments. At the same time, the geometry of the instability and the evolution mechanisms of the movements were quantified, determining the relationships between the movements and the external conditions, especially the meteorological and hydraulic ones. This analysis is aimed to obtain all the information useful to define the lines of action that ensure the final safety of the road asset.

Keywords

Road embankment • Instability • Strains • Stabilization

51.1 Introduction

Road infrastructures are designed keeping into account the stresses deriving from the modifications of the environmental context that includes them, in order to guarantee their functionality during their whole life. In recent times there have been examples where stretches of road embankments, built in geomorphological contexts and hydrogeological conditions of particular sensitivity, have shown over time, as a result of particular climatic conditions, internal deformations greater with respect of those provided by the project and tolerable from the structure.

The case-histories illustrated as follows relate to earth-reinforced embankments that, following rainy periods, have shown strong evidence of instability, which have partially compromised their functionality, leading to the temporary closure to traffic of the road sections. The Geotechnical Unit of the Design Management of ANAS S.p.A., frequently interested in the study of this kind of events, has been involved in the study, monitoring, site-investigation activities, in order to identify the causes and the mechanisms of instability evolution in act and, finally, to propose and design the stabilization solutions.

S. Scarano (✉) · R. Laureti · S. Serangeli
Design Management, Anas S.p.A., Rome, Italy
e-mail: se.scarano@stradeanas.it

G. Lollino et al. (eds.), *Engineering Geology for Society and Territory – Volume 6*,
DOI: 10.1007/978-3-319-09060-3_51, © Springer International Publishing Switzerland 2015

Subsequently, in order to acquire all the knowledge necessary to understand the geological and geotechnical reference context, as well as to establish the project interventions of final consolidation of the embankments, specific site investigation campaigns have been prepared, completed by instrumental and monitoring survey. These campaigns consisted of the realization of boreholes pushed deep inside the laying surface of the embankment, with the aim to reconstruct the geological reference model and the thickness of the embankment involved in the movement. The monitoring consisting, however, in the implementation of several independent topographical, instrumental and interferometric systems of reading, as well as instrumented sections such as inclinometers, assestimeter, piezometers, mechanical fessurimeter, optical targets, in order to evaluate the areal extent of the instability phenomenon and to record the different rates and trends of the displacement.

51.2 Case Histories

Two examples of damages occurred inside hearth-reinforced embankments, part of the italian road network, are after described. In that cases it was necessary to analyze the events and, therefore, to design the consolidation works and safety settings, in order to restore the circulation of traffic. Despite the two cases are concerning two different environmental and climate context, in both the events the main factor was represented by the rainfall seepage inside the slope of the road body. In fact, the main instability happened after long rainy periods. The consequent growth of the pore pressure, with the exceeding of the shear resistance of the embankment-slope system, therefore, triggered the strain events.

51.2.1 Rome Hinterland

In December 2010, along an important road, following a prolonged rainy period, there has been a deformation phenomenon with significant proportions, that affected a stretch of the reinforced embankment, causing the partial closure to traffic of the road (Fig. 51.1).

In the first emergency phase a provisional safety intervention was prepared by means of a drainage system and installation of metal gabions, in order to lower the pore pressure and, at same time, to create an overload to the foot of the slope, with a stabilizing function. A campaign of site-investigation, instrumental and topographical monitoring was subsequently implemented. It included the execution of 13 boreholes, the use of three piezometers, one inclinometer and one assestimeter, with cadenced readings.

The acquired data have allowed to define a very detailed geological reference model, centred on the instability area. It is represented by geological formations belonging to the prevulcanic sedimentary sequence of the Roman area (De Rita et al. 1995), consisting, for the most part, by sands and gravels, with clay and silty levels, oxidized horizons and peaty levels, referred to beach and infralittoral and, furthermore, to fluvial and brackish depositional environments

Fig. 51.1 Damages and tension cracks along the roadway

(Faccenna et al. 1995). These deposits are followed by pyroclastic materials, belonging to the Sabatini Mounts volcano, with cineritic matrix with pumice. They contains, sometimes, slag and lithic lava and volcaniclastic reworked levels (Fig. 51.2) (Ventriglia 2002).

The geological and hydrogeological models, so defined, showed that the sand deposits underlying the road embankment contain an appreciable water circulation that influenced the equilibrium conditions of the roadway. This aquifer is of semiconfined kind, because it's enclosed between the pliocenic clayey substrate (to the bottom) and a layer of silty clay (to the top). Because of a prolonged infiltration due to the rainfall, the sandy aquifer has developed a growth of the pore pressure and the rising of the piezometric surface. It caused filtration phenomena within the body of the overlying embankment, and the creation of a sliding circular surface, placed immediately behind the reinforced-heart body.

The definitive safety interventions consisted of the realization of a bulkhead of large diameter piles, interventions of consolidation of the soil and, furthermore, of civil works of water gathering (Fig. 51.3) (Facchini and Nart 2006).

51.2.2 Liguria (Northern Italy)

In the Imperia province (Liguria—Northern Italy) a section of road interchange, consisting of a series of ramps located along a slope and supported by reinforced embankments (Comendini and Rimoldi 2013), suffered the first signs of instability in early 2011, as a result of high rainfall intensity events, that have affected the whole Region.

Following the deformational events, preliminary works were made. They were represented by draining trenches above the road and, later, by local consolidation works of the enbamkment.

Between the months of October and November 2012, because of the repetition of high intensity and long-term rainy events, a recovery of deformation was recorded, causing the appearance of large tension cracks along the road surface, which led ANAS to define a complete consolidation design of the body of the embankment (Fig. 51.4).

At this point, another intervention has been made, represented by the realization of sub-horizontal drains in the body of the embankment. A monitoring plan has also set up, in parallel with a deep geotechnical site-investigations campaign, represented by 69 topographic control points, 10 inclinometers, 4 assestimeters, 10 piezometers, 7 mechanics fessurimeter and the implementation of an interferometrical monitoring, after described.

The complex of the acquired data have allowed to identify a reference geological model (Lanteaume 1968), represented by a calcareous-marly substrate, belonging to the "Borghetto d'Arroscia-Alassio" and "Moglio-Testico" Units, which with "S. Remo-M. Saccarello" Units constitutes the "Flysch with Elminthoides" Formation of Ligurian-Piedmont Domain (Boni and Vanossi 1960). The substrate is covered, in surface, by a layer of eluvial-colluvial deposits, with significant clay content. The site-investigations have also shown the presence of a main sliding surface inside the cohesive materials that covers the bedrock, immediately below the reinforced embankment affected by the deformation, together with evidence of waterflow inside the fractured part of the rock mass.

So, works of definitive stabilization were realized. They consist of two lines of bulkheads with large diameter piles, with the aim to support the actions of pushing from upstream and to anchor the embankment to the rock substrate. The

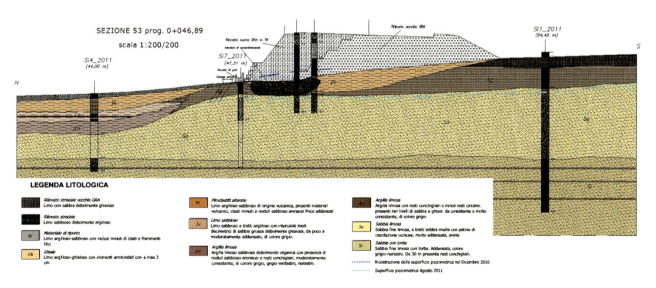

Fig. 51.2 Cross geological section in the damaged site

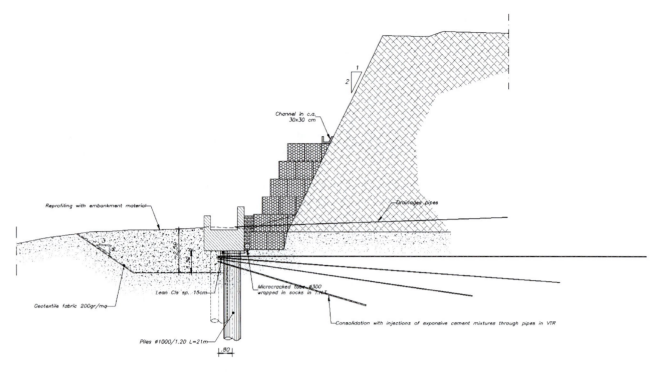

Fig. 51.3 Cross section showing the stabilization works

second bulkhead has been placed on the foot of the embankment and it also reach the bedrock. The works are completed with drainage works along the slope upstream of the embankment (Fig. 51.5) (Bianco 2001).

51.2.2.1 Interferometric Monitoring (Nhazca Data): Comparison Before and After the Interventions

For the evaluation of the areal extension of the instability phenomenon and for a determination of the rates and trends

Fig. 51.4 Overview, from the opposite slope, of the road interchange

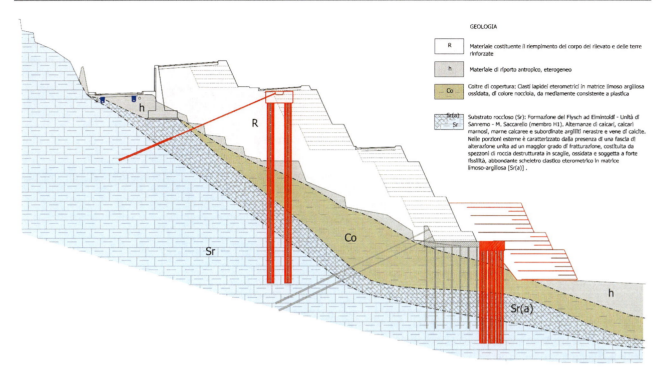

Fig. 51.5 Cross geological section, completed with the designed stabilization works

of the displacements, a monitoring station with SAR interferometry Terrestrial (TInSAR) has been installed, by Nhazca (spin-off of La Sapienza University) on the opposite slope. Through this technique, chromatic maps of the survey area were obtained, calculated by comparing the phase value of each pixel of images acquired at different times, that represent the bidirectional movements along the line of sight of the instrument (line of sight) (Fig. 51.6).

The magnitude of displacement in the following time, from the monitoring network installation, were between −1.5 and −3 mm/day, on approach to the sensor along the line of sight. The embankment has undergone major shifts at the base and in the most western part, while further upstream and eastern portions the movements were more contained. The terrains beyond the foot of the embankment, retaining walls at the entrances of tunnels and the wall on the side of

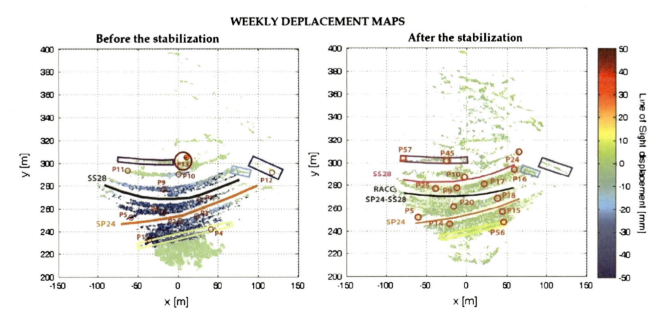

Fig. 51.6 Shift TInSAR map before and after the stabilization works

the embankment, however, have undergone no appreciable movement. The rates recorded over the entire displacement have undergone a gradual deceleration in subsequent periods from the initial movement. Just after the completion of the first measures stabilization (anchored bulkheads) the movements were reduced abruptly to zero.

51.3   Conclusions

In the field of road engineering, as well as the design of new road, the occurrence of damages or slides affecting the road body, causes to acting in the maintenance and safety of existing works that, over time, have been damaged.

In the first emergency, usually, preliminary work are carried out in order to temporary safe the road asset and to assure the traffic along it. In a second phase it is essential to operate with definitivesolutions.

To achieve this, the most information about the soils on which the road will be located and surrounding area are essential. Such information can help to formulate hypotheses about kinematics and mechanism of landslides, in order to propose appropiate design solutions for the last safety.

For this reason it is important to have an adequate campaign of geological site investigations and instrumental and topographic monitoring of the area. So the knowledge of the geological reference model is fundamental in the choice of design solutions to be adopted in the road planning.

The case histories show the fundamental role, as a cause of the damages, played by the response of the road body related to the particular hydrogeological features.

References

Bianco PM (2001) Strutture in terra rinforzata contro il dissesto idrogeologico. Maggioli Editore

Boni A, Vanossi M (1960) Ricerche e considerazioni sul Flysch della Liguria occidentale. Atti Istituto Geol Univ Pavia, vol XI

Comendini M, Rimoldi P (2013) Terre rinforzate. Flaccovio Editore, Palermo

De Rita D, Faccenna C, Funiciello R, Rosa C (1995) Stratigraphy and volcano-tectonics. In: Trigilla (ed) The Volcano of the Alban Hills, Tipografia SGS Roma, 33–71

Faccenna C, Funiciello R, Marra F (1995) Inquadramento geologico strutturale dell'area romana. Mem Descr Carta Geol D'It L:31–47

Lanteaume M (1968) Contribution à l'étude géologique del Alpes Maritimes franco—italiennes. Mèm Carte Géol Fr:405

Facchini Massimo, Nart Massimiliano (2006). Ripristino di sedi stradali con tecniche a basso impatto ambientale. Le strade, vol 6

Ventriglia U. (2002). Geologia del territorio del comune di Roma, Amministrazione provinciale di Roma, Roma

Groundwater Level Variation and Deformation in Clays Characteristic to the Helsinki Metropolitan Area

52

Tiina-Liisa Toivanen and Jussi Leveinen

Abstract

The aim of this study was to use various modelling approaches to examine the groundwater level changes and other factors influencing subsidence and soil compaction in clay deposits characteristic to the Helsinki metropolitan area in Finland. The research area is Perkkaa in Espoo, located in the Helsinki metropolitan area in the coastal area of Southern Finland. The surface of the studied area is mostly covered by clay, with a thickness varying between 5 and 15 m. The urban development of the Perkkaa area began in the beginning of the 1970s. The hydrogeological environment in the area has been considerably altered by construction, and it has also led to lowering of the groundwater levels. As a result, there are large deformations, especially on the streets and the yards. Using an available laser scan data, GPS-measurements of the levels of ditches and minor streams, and drilling and geological maps provided by the city of Espoo and Finnish Geological Survey of Finland, it has been possible to create a 3d model of the geological main units in the research area and estimate the depth of the clay deposits. The 3d model was used further to model the changes in the groundwater flow in the research area. On the basis of this study, significant reduction in the groundwater levels are induced by the drainage systems and particularly the underground pipeline systems rather than changes in the recharge in the area due to construction.

Keywords

Engineering geology • Groundwater level • Clay deposits • Subsidence

52.1 Introduction

The factors influencing deformation and subsidence in clay deposits include manmade loadings and variation of groundwater level. In addition to natural seasonal variation, groundwater levels are influenced by changes in land use, underground construction, changes in climate, and geological uplift. The aim of this study was to use various modelling approaches to examine the groundwater level changes and other factors influencing subsidence and soil compaction in clay deposits characteristic to the Helsinki metropolitan area in Finland.

The index properties and stress deformation characteristics of Finnish clay deposits have been the subject of a number of studies since the 1960s and geotechnical investigation results are documented in databases particularly in the Helsinki metropolitan area. Therefore, somewhat surprisingly, geotechnical designs rely typically at most on few site specific measurements of index properties. A part of the study was to examine how reliably the existing data could be used in determining the risks of excessive consolidation and soil deformation by estimating the deformation characteristics based on the available statistical data.

T.-L. Toivanen (✉) · J. Leveinen
Department of Civil and Environmental Engineering, Aalto University, Espoo, Finland
e-mail: tiina-liisa.toivanen@aalto.fi

J. Leveinen
e-mail: jussi.leveinen@aalto.fi

52.2   Engineering Geological Setting

The research area is Perkkaa in Espoo, Finland, located in the Helsinki metropolitan area in the coastal area of Southern Finland. The size of the research area is approximately 2.7 km^2. The development of the Perkkaa area began in the beginning of the 1970s. Since then, the hydrogeological environment has been considerably altered by construction, and it has also led to lowering of the groundwater levels. As a result, there are large vertical deformations, especially on the streets and yards. The city of Espoo has been observing the deformations on the two main streets in the area. According to the measurements, the maximum soil subsidence on the two streets is roughly 0.5 m compared to the original surface levels. Considering the constant re-paving and repair works on the streets, it is likely that the actual maximum subsidence is on the order of 0.7–0.8 m.

In general, the geology and topography of the area is characteristic to Southern Finland. The surface of the studied area is mostly covered by clay, with a thickness varying between 5 and 15 m. In the southern and middle parts of the area, the soft soil layer can be over 20 m thick. The water content of the clay varies considerably, mostly being between 70 and 140 %. A silt and sand layer a couple of meters thick can be commonly found beneath the clay deposits, on top of a 1- to 5-m-thick layer of till that overlies the bedrock. The studied area is bordered by a fill area to the east, a silt/sand moraine area to the west, and a moraine/bedrock area to the north and south. Most of the area is elevated less than 2 m above the Baltic Sea level. In the northern part of the area, at the moraine and exposed bedrock areas, the elevation varies, being 10 m at its highest (Ojala 2011). The research area is a part of well-defined catchment area, which provides good boundaries for modelling the flow of the groundwater and the changes in its level.

The investigations carried out by State Technical Research Centre 40–50 years ago provide still the most comprehensive assessment of soil engineering geological properties in Finland. The results summarized in Gardemeister (1975) involved 48 drill cores of fine grained soil samples collected from different parts of the country. The data published by Gardemeister include estimates of water content, and void ratio, which can be used to assess consolidation deformation of fine sediments, based on the method introduced by Helenelund (1951). Since this approach relies on relatively easily obtainable index properties, the method is widely utilized in prediction of soil deformations in Finland. Therefore, in the following, the distributions of the relevant index properties for the method are used to assess stochastically the soil consolidation and subsidence due to changes in water table and man-made loadings.

52.3   Land Use of the Research Area

52.3.1   The Urban Development of the Area

The development of the Perkkaa area started in the 1970s, which was a period of rapid urbanization in the Helsinki metropolitan area. The first buildings in Perkkaa were residential high-rise apartment buildings, and a bit later, first few office blocks were constructed. After the 70s the area remained mostly static, until the 90s, which was the period of gradual built up with mostly residential projects. Today, the Perkkaa area is home to about 4,000 inhabitants. During the past few years some office blocks has been built in the east side of the area. The development of the area will continue within next few years.

52.3.2   Foundations and Drainage

In the southern and central parts of the area the depth of the soft soil can be up to 20 m and the buildings have been built on piles. The western parts of the area have less or no soft soil layers on them, and the buildings have been constructed on shallow foundations. In most of the buildings the subgrade surface reaches the depth of a few meters below the ground level. In each property groundwater is drained a couple of meters below the original ground surface.

All urban infrastructure e.g. plumbing and drinking water systems, heating pipelines that are susceptible to frost damage are buried subsurface to a sufficient depth. If no insulation is used this depth typically exceeds 1.8 m.

The yards and alleys have not had originally any specific foundations, and no ground improvement methods have been used.

52.4   Data and Modeling Approaches

Using an available laser scan data, GPS-measurements of the levels of ditches and minor streams, and geotechnical soil investigations and geological maps provided by the city of Espoo and Finnish Geological Survey of Finland, it has been possible to create a 3d model of the geological main units in the research area and estimate the depth of the clay deposits. The geological 3d-model and available groundwater level estimates made in the geotechnical investigations were used to compile a groundwater flow model using USGS Modflow in GMS—software utilities. The objective of the flow modeling was not to estimate actual elevation of the groundwater level but assess the drawdown induced by the drainage systems and pipeline trenches (Astm 1999).

Subsequently, soil subsidence/consolidation was estimated stochastically after the following assumptions:

- The water content method (Helenelund 1951) can be estimated to assess the soil deformation (due to consolidation).
- The unsaturated part of the fine sediments, "the dry crust", is 1.0 m thick and remains undeformed.
- The input soil properties are water content and dry specific weight for the dry crust and the saturated zone and are assumed to follow distributions obtained by Gardemaister (1975) for Litorina sediments.
- Groundwater drawdown that has been resulting mainly from draining associated with various subsurface infrastructures, is local and within the range of 0–2 m.
- Aggregate material layers applied in building sites and road construction are 1–1.5 and consequently induce a load of 26–39 kPa.

52.5   Results and Discussion

The results of the stochastic calculations indicate that similar construction practices as applied in the study area will likely yield substantial consolidation or subsidence in Southern Finland. The thicker the clay-formation, the more substantial the subsidence is. Based on the sensitivity analysis, the water content appears as the key parameter. Even if the site specific water contents would be close the minimum end of the distribution obtained by Gardemaister (1975), the subsidence will exceed 0.1 m with 95 % probability if the soil thickness reaches 3.0–5.0 m. Both the predictions of subsidence and the uncertainties involved will increase with the thickness of the clay deposits. For over 5 m thick clay deposits obtaining a 0.2 m subsidence or less appears a mission impossible.

In the study area rain and snow-melt is collected from the roofs of the flat roofed buildings to the municipal waste water system. Also at the yards of the buildings the surface run-off is directed to the drainage systems. Also substantial part of the area is today asphalt paved areas, which could also be expected to reduce the overall recharge. However, in quite opposite way, the optimized i.e. automatically calibrated recharges in outcrop slope areas suggest about 10 % higher recharge when calibration is done to the groundwater observations taken during the late 80s and the 90s compared to observations taken in the late 60s. The observations represent sporadic measurements of the "undisturbed" groundwater levels related to the geotechnical investigation activities during preceding construction, not systematic monitoring and reliable time series. Taking into account the uncertainty of the groundwater data, results suggest at most that not significant changes in the catchment scale recharge appear to have happened. Since significant part of the development has taken place in the areas of fine sediments, which originally low to insignificant recharge, the overall

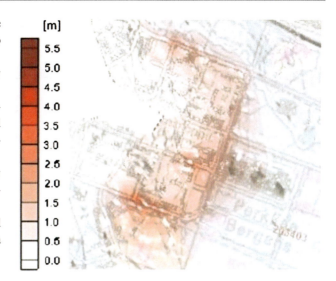

Fig. 52.1 Drawdown in groundwater levels in the Perkkaa area based on the numerical groundwater flow model

impacts of the recharge have been small. As suggested by the numerical groundwater flow model, significant reduction in the groundwater levels are induced by the drainage systems and particularly the underground pipeline systems rather than changes in the recharge. Results based on the numerical groundwater flow model are shown in Fig. 52.1.

52.6   Conclusions

Based on this study, lowering of the ground water levels due to construction can have a significant effect on subsidence in clay deposits even just a few meters deep. Lowering of the ground water levels cause deformations and in that way major costs of repairs. In addition to the direct costs it can also have an effect on the prestige of the area and the values of apartments in the district. The condition of the street and yards has an obvious influence on the satisfaction of the area residents, and thereby on their experience of personal safety (Kyttä et al. 2013).

In contrast with the current planning practices, the effects of possible lowered ground water levels should thus be taken into account proactively in the zoning and planning processes, to avoid structural damage and expensive recurring repair works on the streets and yards.

References

Astm (1999) Standard guide for subsurface flow and transport modelling, D5880–95
Gardemaister R (1975) On engineering-geological properties of fine-grained sediments in Finland. Dissertation, Technical Research Centre of Finland, Building Technology and Community Development, Publication, vol 9, 91 p

Helenelund KV (1951) Om konsolidering och sättning av belastade marklager (On consolidation and settlement of loaded soil layers). Dissertation, Finland Technical Institute, Helsinki

Kyttä M, Broberg A, Tzoulas T, Snabb K (2013) Towards contextually sensitive urban densification: location-based softGIS knowledge

revealing perceived residential environmental quality. Landscape Urban Plan 113:30–46

Ojala AEK (2011) Construction suitability and 3D architecture of the fine-grained deposits in Southern Finland—examples from Espoo. Geol Surv Finland Spec Pap 49:126–136

Determine of Tunnel Face Stability Pressure in EPB Machine with Use Analytical Methods (Case Study: Mashhad Metro Line2)

53

Mehdi Abbasi and Mohsen Abbasi

Abstract

This article discusses the face stability of the soil in front of the TBM cutterhead and includes calculations of face stability for 27 cross sections for Mashhad Urban Railway Line 2 with use analytical methods. In the case of the MURL2 Project a TBM will be applied with an Earth Pressure Balance shield. The objective of this study is to advise on a support pressure which has to be applied during the different phases of the tunnel boring process of MURL2. The 27 cross sections have been selected for their spreading over the sections between the stations and the locally present soil overburden. During the shield tunnelling process, subsoil is cut loose by the cutting wheel. The main failure mechanism, which may occur, is inward collapse or cave in. To prevent the cutting face from collapsing, a supporting pressure can be applied by the TBM.

Keywords

Face stability • Analytical methods • EPB machine • MURL2

53.1 Introduction

MURL2 with total length of about 14 km is extended from North-East of Mashhad, to South-East and contain 12 stations along the route. The line 2 shall be able to prepare services for about 10,000 passengers per hour per direction. The tunnel was bored by two earth pressure balance (EPB) shield. The characters of this EPB machine presented in Table 53.1.

The stability of the face is one of the most important factors in selecting the adequate method of excavation of a tunnel. This is particularly true for mechanized tunneling and specific boring machines (TBM).

53.2 Local Properties

Calculations for the face stability have been performed for 27 cross sections of the MURL 2. This part discusses the soil properties at the relevant cross sections. The 27 cross sections have been selected for their spreading over the sections between the stations and the locally present soil overburden.

53.2.1 Geotechnical Soil Parameters

The geotechnical soil parameters which have been derived from the geotechnical reports (IMN consultant engineers Co. 2010) and are used for the calculations of the face stability, are summarized in the following Table 53.2.

53.3 Face Stability Theory

In the case of the MURL2, a TBM will be applied with an EPB shield. In the case of an EPB shield, the soil that is excavated is collected in the chamber directly behind the

M. Abbasi (✉)
IMN Consulting Engineers Co., Tehran, Iran
e-mail: Mehdi_abbasi1980@yahoo.com

M. Abbasi
IMN Consulting Engineers Co., Mashhad, Iran
e-mail: Abbasi.mohsen1982@gmail.com

G. Lollino et al. (eds.), *Engineering Geology for Society and Territory – Volume 6*,
DOI: 10.1007/978-3-319-09060-3_53, © Springer International Publishing Switzerland 2015

Table 53.1 TBM's technical parameters

Parameter	Value	Unity
Diameter of cutterhead	9.43	m
Diameter of front shield	9.38	m
Diameter of rear shield	9.35	m
Length of TBM shield	10	m
Weight of TBM	7,000	kN
External diameter of tunnel lining	9.1	m

cutting wheel. The excavation chamber needs to be completely filled for the support pressure that is delivered by the hydraulic jacks to be conveyed to the cutting face. The supporting pressure on the cutting face can be regulated by varying the pressure delivered by the hydraulic jacks. A second way of regulating the supporting pressure can be achieved by varying the speed at which the soil that has been excavated is transported from the excavation chamber by the auger.

53.3.1 Analytical Calculation Models

A first conservative estimation of the upper and lower bound of the required supporting pressure can be found relatively easy by assuming the soil mass acts according to Mohr-Coulomb's failure criterion and by applying Rankine's theory of earth pressure (Eq. 53.1). At inward collapse, horizontal effective stresses will be in the active state (Rankine) and thus a lower bound for the required supporting pressure can be found.

$$q_{min} = \sigma'_h + p_w = K_a \cdot \sigma_v'' - 2 \cdot c \cdot \sqrt{(K_a)} + p_w \quad (53.1)$$

$$K_a = (1 - \sin \emptyset)/(1 + \sin \emptyset) \quad \text{and} \quad p_w = \gamma_w \cdot h$$

The formula mentioned above is a relatively simple approach to the problem and is only applicable for 2D situations. In practice, the occurring failure mechanisms in shield tunnelling will be 3D. It can be expected that the minimum support pressure in a 3D situation will be lower than in a 2D situation due to the higher strength of the soil, caused by cohesion and friction forces along the sliding planes of a 3D failure mechanism (e.g. arching effects). A number of analytical face stability models are available from literature, which take into account both the 2D situations as

well as the 3D situations. Three method use in this paper consist: Anagnostou and Kovári (1994), Jancsecz and Steiner (1994) and Leca and Dormieux (1990).

53.3.2 Comparison of Calculation Models for MURL2

Each model for the analysis of face stability as described in Sect. 53.3.1 has its own strengths and weaknesses. To assess which model is the most suitable for the MURL2 project, a comparison has been made for two cross sections, 2a and 9b (Table 53.4). Cross section 2a is representative for the clayey soils of the first part of the trajectory and section 9b for the more sandy and gravelly soils of the latter part. The results of the calculations of the minimum required face support pressures, applying the methods described in Sect. 53.3.1, are given (Table 53.3). The crown of the tunnel has been taken as a reference level for comparison purposes.

53.4 Calculation of Face Stability

53.4.1 Calculation Method Minimum Soil Pressure

The minimum soil pressure is the pressure in the lower boundary of the soil pressure at which no instability occurs of the soil in front and above the excavation face. The minimum support pressures are calculated according to the Jancsecz and Steiner method. Jancsecz and Steiner have designed a 3D face stability model based on a soil wedge in front of the excavation face. This method considers the forces on a possible and probable failure mechanism, with the use of limit equilibrium analysis to determine the limit earth pressure acting on the tunnel face. The model is designed to calculate minimum support pressures only. The maximum support pressure is discussed in Sect. 53.3.2. The Jancsecz and Steiner failure model consists of two parts:
- soil wedge in front of the face (lower part);
- soil silo above the wedge (upper part)

The method is based on the analysis of the force equilibrium on a soil wedge in front of the tunnel face. With the different values of overburden and angle of internal friction, as present in the cross sections, the three dimensional earth pressure coefficient KA3 is found. With this parameter and

Table 53.2 Averages of geotechnical parameters all sections

Geotech. unit no.	Geotechnical unit description	Unit bulk weight dry (kN/m³)	Unit dry weight γ dry (kN/m³)	Ø°	c' (kPa)
I	Silty clays	18	16.5	30	14.5
II	Clayey sands	18.5	17	32.2	4.9
III	Clayey gravel (with sand)	20.5	18.5	37.5	0

Table 53.3 Face support pressures calculated applying different methods

	Section 9b-gravelly soil	Section 2a-clayey soil
Ka	60	62
Jancsecz and Steiner	62	67
Anagnostou and Kovari	65	54
Leca and Dormieiux-lower bound	12	5
Leca and Dormieiux-upper bound	110	107

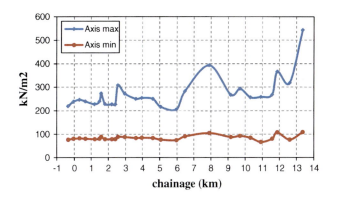

Fig. 53.1 Min and max support pressures at tunnel axis

the vertical earth pressure, the horizontal soil pressures at different TBM levels can be found. The minimum earth pressures at the relevant soil pressures are assessed at the crown, the axis and the heel of the tunnel. The model of Jancsecz and Steiner takes the ratio of C/D into account. If this C/D ratio is smaller than 2, the model assumes that no

arching in the soil above the excavation face will occur. For the calculation of the minimum soil face pressure, the surface loads are taken into account. For those situations where the tunnel is below a road, traffic loads (20 kN/m^2) are

Table 53.4 Min and max support pressures at the tunnel crown, axis and heel

Cross section (km)	Overburden height (m)	Geo-techn. unit at tunnel axis	Crown (kN/m^2) max.	min.	Axis (kN/m^2) max.	min.	Heel (kN/m^2) max.	min.
-1a	-0.35	8.7	I	137	59	219	77	301	95
-1b	-0.05	9.8	I	156	63	238	81	320	99
-1c	0.3	10.3	I	165	65	246	83	328	101
-1d	0.65	9.8	I	157	63	239	81	321	99
-1e	1.2	9.2	I	145	61	227	79	335	123
-1f	1.48	9.8	I	155	63	239	81	354	130
0a	1.58	10.3	I	188	70	273	89	388	138
0b	1.8	9.2	I	144	61	228	79	325	111
0c	2.2	9.2	I	145	61	227	79	309	97
1a	2.4	8.9	I	144	61	226	79	307	97
1b	2.55	13.6	I	226	72	308	89	390	105
1c	2.95	11.6	I	190	70	272	88	354	106
2a	3.6	10.4	I	168	66	250	84	332	102
2b	3.95	10.6	I	172	67	254	85	336	103
3a	4.6	10.5	I	168	66	250	84	332	102
3b	5.05	8.6	I	134	59	216	77	298	95
3c	5.95	7.9	I	123	56	205	75	287	93
4a	6.45	12.3	I	201	73	282	91	364	109
5a	7.9	18	II	309	88	393	105	477	121
6a	9.15	11.1	I	185	69	267	88	349	106
7a	9.69	12.9	I	212	75	293	93	375	111
7b	10.3	10	I	174	67	256	85	338	103
8a	10.91	10.1	III	165	52	258	67	352	82
8b	11.57	11.1	II	183	64	267	81	351	98
9a	11.87	16.8	I	284	90	366	108	448	126
9b	12.6	12.1	III	225	62	318	77	411	91
10a	13.35	25.4	III	449	95	542	109	635	123

applied. If the cross section is below a building, the building loads are applied in the model.

53.4.2 Calculation Method Maximum Soil Pressure

The maximum soil pressure is the pressure at which the soil in front of the excavation face does not exceed the vertical pressure. The maximum soil pressure is determined at the different levels of the TBM. Surface loads are not taken into account for this upper pressure boundary, because they are not permanently present.

53.4.3 Minimum and Maximum Support Pressure

In tunneling by TBM, it is good practice to ascertain a safety buffer between the minimum and maximum soil pressure, usually with a value of 20 kN/m^2. This pressure buffer will also be applied for the MURL2 tunnel. Applying the theory of Jancsecz and Steiner as described in Sect. 53.3.1 and the method of determination of the maximum soil pressure of Sect. 53.3.2, the lower and upper boundaries for the minimum and maximum soil pressure are found. To these values the buffer pressure of 20 KN/m^2 is added to come to the minimum and maximum support pressures. In the Fig. 53.1 a schematization of the upper and lower boundary of the support pressures are given. The support pressure of the TBM should remain in the hatched area. The calculations have been performed for the supporting pressure at the tunnel crown, axis and heel (Table 53.4).

53.5 Conclusions

The evaluation of the tunnel face-support pressure is a critical component in both the design and construction phases of TBM. In this article, lower bound minimum support pressures and upper bound maximum support pressures have been determined for 27 cross sections. It is tried to assess the lower and upper bound at critical locations of the tunnel route. In between the cross sections, the support pressures of the TBM should be maintained near the most feasible levels of the support pressures, based on the local conditions and geology. It is strongly recommended to obey the limits of the lower and upper support pressures, in order to avoid problems with the stability of the excavation face.

References

Anagnostou G, Kovari K (1994) The face stability of slurry-shield-driven tunnels. Tunn Undergr Space Technol 9(2):165–174

IMN consultant engineers Co (2010) Geotechnical reports of MURL2

Jancsecz S, Steiner W (1994) Face support for a large mix-shield in heterogeneous ground condition. In: Tunneling '94, Institution of Mining and Metallurgy, London

Leca E, Dormieux L (1990) Upper and lower bound solutions for the face stability of shallow circular tunnels in frictional material. Geotechnique 40(4):581–606

Numerical Modeling of Interrelationships Between Linear Transportation Infrastructures and Hydro-geological Hazard in Floodplains

54

Rosamaria Trizzino

Abstract

The development and efficiency of transportation infrastructures has always been a central element in the planning and management of territory, not only with regard to the social and economic aspects but also for the management of environmental emergencies, in particular those of Civil Protection. In the valleys most usual design solution is the road embankment. In the presence of a surface water table the realization of a road embankment causes more or less marked alterations of the piezometric levels, with serious problems to vehicular traffic. Typically in such situations to avoid the risk of flooding of the roadway engineers tend to increase the height of the embankment, thus increasing the overload in the foundation soil and thereby creating new risk situations. In this paper it is proposed a modeling study aimed at determining the interactions between the geometric and geomechanical properties of foundation—road body—roadway and surface and deep water table levels, in order to identify risk scenarios. The numerical analysis has been carried out by a finite element calculation code taking into account different combinations of water level depth and embankment height for different lithological and geotechnical properties of the foundation soils. The obtained results show that there are some critical combinations of the above parameters that can cause the rise of the water level well above the ground surface not only at the road embankment but also in the surrounding areas up to tens of meters from the road.

Keywords

Hydro-geological hazard • Embankments • Roads • Soil stresses • Flooding

54.1 Introduction

The development and efficiency of transportation infrastructure has always been a central element in the planning and management of territory, not only with regard to the social and economic aspects but also for the management of environmental emergencies, in particular those of Civil Protection. In this context, the efficiency of road and railroad infrastructures appears to be a fundamental issue or even a discriminating element to the effectiveness of specific interventions in the presence of natural and human made environmental "disasters".

It is well understood then the importance of properly assessing the interactions between transportation infrastructure and the natural and built environment and in particular the influence of the hydrogeological, geological and geomorphologic conditions on the serviceability of the roads and railroads. But it is only in recent years, with the beginning of the great highway construction, that it has been highlighted the complexity of the territory—road—vehicle relationship and thus the need for extensive research in the design of the geometry of the road both in relation to the physical characteristics of the territory (in particular geomorphologic and hydro-geological) and the conditions of vehicular traffic.

R. Trizzino (✉)
CNR, IRPI, via Amendola, 122/i, 70125, Bari, Italy
e-mail: r.trizzino@ba.irpi.cnr.it

As a matter of fact, water levels play a fundamental role both in relation to the triggering and evolution of landslides (Cotecchia et al. 1996; Polemio and Trizzino 1999; Spilotro et al. 2000) and for interference with the operation of transportation infrastructure (see for example the case of the landslide of Montaguto, Irpinia, Southern Italy, where the earthflow covered the SS90 road and the railway line) (Giordan et al. 2013) (Fig. 54.1). Many of the issues relating to the roads in floodplains are in fact due to the presence of surface water table and/or runoff of meteoric water. In valley floor areas most usual design solution is the road embankment. In the presence of a surface water table the realization of a road embankment can cause more or less marked alterations of the piezometric levels, which can create serious problems to vehicular traffic. Typically, in such situations, to avoid the risk of flooding of the roadway engineers tend to increase the height of the embankment, thus increasing the overload in the foundation soil and thereby creating new risk situations.

In this regard, Eurocode 7 (Geotechnical design of Civil Engineering works) establishes that the design shall ensure that the deformation of the embankment will not cause a serviceability limit state in the embankment or in structures, roads or services sited on, in or near the embankment. Actions in which ground- and free-water forces predominate shall be identified and the possibility of deformations due to changes in the ground-water conditions should be taken into account.

When deriving the actions that embankments impose on adjacent infrastructures or any reinforced parts of the ground, the differences in the stiffnesses should be considered (EN 1997-1:2004, Sect. 12.3). In the same way, in deriving design distributions of pore-water pressure, account shall be taken of the possible range of anisotropy and heterogeneity of the soil.

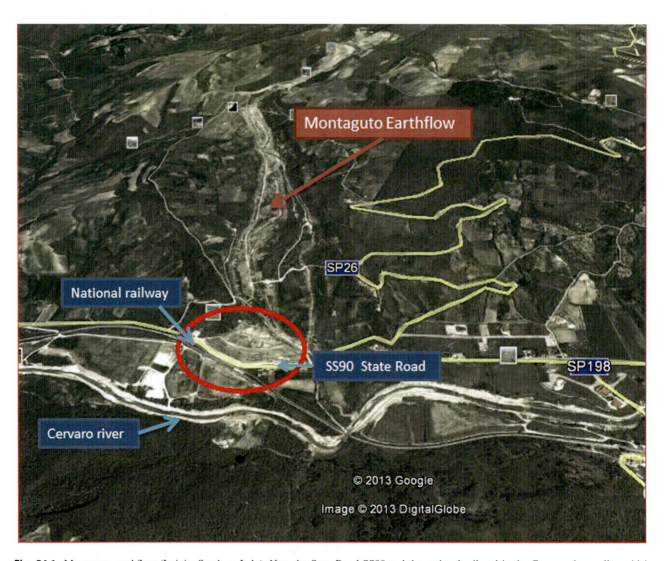

Fig. 54.1 Montaguto earthflow (Irpinia, Southern Italy). Note the State Road SS90 and the national railroad in the Cervaro river valley which were covered by the landslide debris

Starting from these issues, in this paper it is proposed a modeling study aimed at determining the interactions between the geometric and geomechanical properties of the foundation soil- road body—roadway system and surface and deep water table levels. The analysis has been carried out in a parametric form by a finite element calculation code taking into account different combinations of water level depth and embankment height for different lithological and geotechnical properties of the foundation soils.

54.2 Principles of the Method

As stated in the Eurocode 7 (Cap. 12.7) the stability of an embankment acting as a dam to a large degree depends on the pore-water pressure distribution in and beneath the embankment.

As well known, the construction of an embankment generates an increase of the soil stresses beneath the infrastructure; this causes the modification of the void ratio and then of the hydraulic conductivity of the subgrade soil. In the presence of a surface water table the fulfilment of a man made embankment can create an excess of pore-water pressure that leads to the rise of the piezometric surface up to the ground level. When the water table reaches the ground level downstream we can have the flooding of more or less large areas even to several meters (15–20 m and more) from the transportation infrastructure (Maltinti et al. 2000).

Moreover, the rise of the piezometric level upstream can cause the instability of the up-slope scarp of the road embankment creating failure surfaces and more or less large soil slips that lead to the global instability of the infrastructure with the loss of serviceability of the road (Serviceability Limit State).

To determine the stress-strain behavior of the soil beneath the embankment and to compute the consequent excess pore-water pressures a Finite Element Model has been developed together with a coupled consolidation analysis. The first step has been to select a particular constitutive model that was consistent with the soil conditions and the objective of the analysis. It has been assumed an effective stress elastic-plastic model with pore-water pressure changes.

In the presence of a road embankment it is fundamental to take into account the capillarity phenomena that can lead to plasticize the foundation levels. In this analysis unsaturated flow conditions have been considered by defining a hydraulic conductivity function $k = k(s)$, where s is the negative pore-water pressure (suction) of the sub-soil (Cafaro et al. 2008). As well known, this function will show

a different trend for different material properties, so the response of the foundation soil to the construction of an embankment will depend on the sub-soil strata succession (Hoffman and Tarantino 2008; Kawai et al. 2000; IGWMC 1999; Schnellmann et al. 2010).

To take into account the decrease of the hydraulic conductivity when the soil grain structure becomes more compact, a K-modifier function has been defined modifying the Ksat by a factor depending on the vertical effective stress state. It has been assumed that K diminishes by a factor of 10 as the effective stress increases from 10 to 100 kPa.

Finally, the soil consolidation problems have been solved using a coupled stress/pore-pressure analysis to determine the excess pore-water pressures dissipation with time by solving simultaneously both equilibrium and flow equations.

54.3 Results and Conclusions

The conceptual model has been applied to a hypothetic road embankment made of gravelly-sandy soil classified as A1 in the Italian CNR-UNI 1006/63 Code for the road body materials. Different embankment heights have been considered from 2 to 20 m. The analysis has been carried out in a parametric form considering different properties of the sub-soil strata and various depths of the water table. The stress-strain modifications beneath the embankment and the Pore-Water Pressures have been calculated by means of the Finite Element Computation Code GEOSTUDIO 2007, SIGMA/W and SEEP modules. Main results are shown in Figs. 54.2, 54.3 and 54.4. In Fig. 54.2 are reported the vertical stress modifications and the piezometric level rise after the placement of an embankment of 4 m for an initial water table depth of 5 m beneath the ground level. The piezometric height increase resulting in these conditions is of 1.50 m, whereas for an embankment of 10 m and a water table depth of 5 m the Δhw is equal to 2.60 m (Fig. 54.3), dangerously closer to the road foundation. The situation is ever more dangerous for an embankment of 10 m and an initial water table depth of 1 m (Fig. 54.4): in this case, as shown in the figure by the shadows areas, the piezometric level rises up to and above the ground level. But the unexpected result is that this water rising takes place at a distance of about 15 m from the road and involves an area up to 100 m up and down stream that is, in particular hydro-geological conditions, flooding risk.

In conclusion, the results obtained from this preliminary study showed that there is a significant relationship between the geo-mechanical soil properties and the road safety and serviceability that induce to carry out further investigations and computations on this topic.

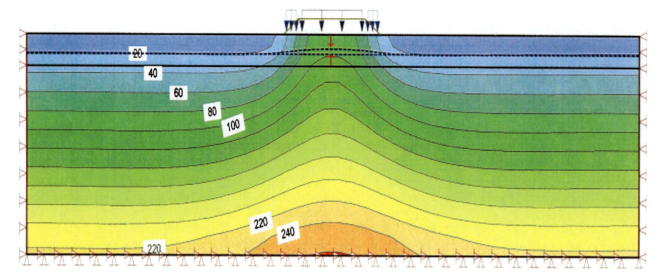

Fig. 54.2 SIGMA/W solve analysis results with isolines of vertical stresses distribution (kPa) for an hypothetical embankment. Height $H = 4$ m; water table depth $h_w = -5$ m; thickness of compressible stratum $S1 = 8$ m; piezometric level rise (*red arrows*) $\Delta h_w = 1.50$ m

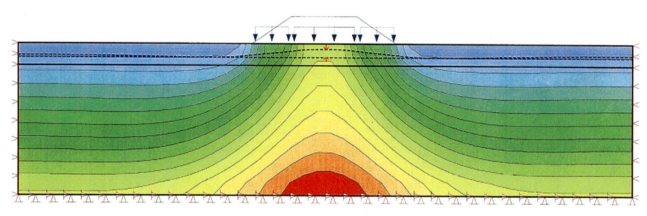

Fig. 54.3 SIGMA/W solve analysis results with isolines of vertical stresses distribution (kPa) for an hypothetical embankment. Height $H = 10$ m; water table depth: $h_w = -5$ m; thickness of compressible stratum: $S1 = 8$ m; piezometric level rise (*red arrows*) $\Delta h_w = 2.60$ m

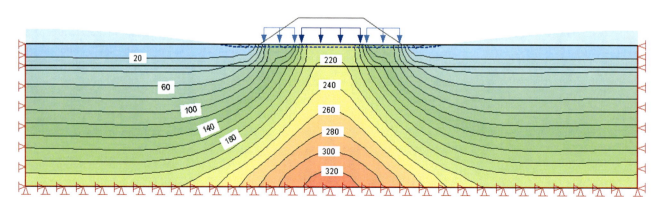

Fig. 54.4 SIGMA/W solve analysis results with isolines of vertical stresses distribution (kPa) for an hypothetical embankment. Height $H = 10$ m; water table depth: $h_w = -1$ m; thickness of compressible stratum: $S1 = 8$ m. Note the shadow light blue areas (water surface) above the ground level

References

Cafaro F, Hoffman C, Cotecchia F, Buscemi A, Bottiglieri O, Tarantino A (2008) Modellazione del comportamento idraulico di terreni parzialmente saturi a grana media e grossa. Rivista Italiana di Geotecnica, no. 3/2008, pp 54–72

Cotecchia V, Parise M, Polemio M, Trizzino R, Wasowski J (1996) Primi risultati della ricerca sulla frana di Acquara-Vadoncello (Senerchia,AV) eseguita nell'ambito del progetto CEE –Environment Landslide evolution controlled by climatic factors in seismic areas—Prediction methods and warning criteria. Accademia Nazionale dei Lincei, Conv. "La stabilità del suolo in Italia: zonazione della sismicità-frane", Roma

Giordan D, Allasia P, Manconi A, Baldo M, Santangelo M, Cardinali M, Corazza A, Albanese V, Lollino G, Guzzetti F (2013) Morphological and kinematic evolution of a large earthflow: the montaguto landslide, southern Italy. Geomorphology, 187, pp 61–79

Hoffman C, Tarantino A (2008) Effect of grain size distribution on water retention behaviour of well graded coarse material. In: 1st European conference on unsaturated soils, 2–4th July 2008, Durham, UK

Kawai K, Kato S, Karube D (2000): The model of water retention curve considering effects of void ratio. In: Rahardjo H, Toll DG, Leong EC (eds) Unsaturated soil for Asia, pp 329–334

IGWMC (1999) Hydrus 2D: simulating water flow, heat and solute transport in variably saturated media .In: U. S. Salinity Lab. Agriculture. Research Service and U. S. Dept of Agriculture Riverside California

Maltinti F, Portas S, Annunziata F (2000) Soluzioni progettuali per il ripristino delle condizioni idrogeologiche preesistenti alla costruzione di un rilevato stradale. X Convegno S.I.I.V, Catania, pp 1–11

Polemio M, Trizzino R (1999) Hydrogeological, kinematic and stability characterization of the 1993 Senerchia landslide(Southern Italy). Landslide News, no. 12:12–16

Schnellmann R, Busslinger M, Schneider HR, Rahardjo H (2010) Effect of rising water table in an unsaturated slope. Eng Geol 114:71–83

Spilotro G, Coviello L, Trizzino R (2000) Post failure behaviour of landslide bodies. In: Proceedings of 8th ISL, Cardiff, Wales, pp 26–30

Convergence Predictions and Primary Support Optimization of the Tunnel Progon

55

Zoran Berisavljevic, Svetozar Milenkovic, Dusan Berisavljevic, and Nenad Susic

Abstract

The tunnel Progon is being constructed as a part of Corridor 10 highway project about 5 km north from the border crossing with Bulgaria in southern Serbia. During the design process 2D finite element analysis was utilized with the objective to investigate an influence of different support systems on the convergence across an excavation. Two models were used in calculations. The first model is elastic perfectly-plastic with constant stiffness independent of the stress level. The second model assumes hyperbolic relationship between the deviatoric stress and axial strain and stress dependent stiffness. This model also accounts for small strain stiffness of soil. In order to allow for a certain convergence across an excavation the load reduction method was utilized. Convergence monitoring is currently ongoing for the excavated portion of approx. 700 m of the tunnel exit portal. The measurements show cumulative displacements in the range of 1–3 cm. Results obtained by the analyses prior to the tunnel construction are in good agreement with the measurements. Convergence is approx. 5 cm when MC model is used and around 1 cm in the case of HS-small model, thus limiting the measurement results from the lower and upper bound.

Keywords

Mohr-Coulomb • HS-small • Plastic points • Convergence • β-method

55.1 General Settings

The tunnel "Progon" is located in the southern part of the Republic of Serbia on the highway E-80 (Dimitrovgrad bypass). According to design (Highway Institute 2012) two paralel tunnel tubes are to be constructed on the axial distance of 27.5–30.0 m. Length of the left and right tubes are L = 1,084.44 m and L = 1,065.74 m, respectively.

Approximately 100 m of tunnel entrance portal and 60 m of tunel exit portal will be constructed in an open excavation (cut and cover method). The tunnel will be excavated by means of Sprayed Concrete Lining (SCL) method. The tunnel structure shall be closed and arched with an equivalent diameter of D ≈ 12 m.

Results of geotechnical field investigations showed that the tunnel tubes will be excavated in sediments of quaternary age consisting of clay with gravel and cobbels interbedded with parts consisting predominantly of clay. Height of overburden varies and an average height of 20 m is adopted for the analyses.

The objective of this paper is to present results of the primary support optimization done during the preparation of the main design. Support types are a combination of different structural elements: shotcrete, anchors, micropiles, forepole umbrella, etc. Support consisting of shotcrete and forepoles will be more closely examined, Fig. 55.1.

Z. Berisavljevic (✉)
Koridori Srbije Ltd, 21 Kralja Petra St, 11000 Belgrade, Serbia
e-mail: berisavljevic_zoran@yahoo.com

S. Milenkovic
Highway Institute, 257 Kumodraska St, 11000 Belgrade, Serbia

D. Berisavljevic · N. Susic
Institute for Material Testing, Blvd. vojvode Stepe 43,
11000 Belgrade, Serbia

G. Lollino et al. (eds.), *Engineering Geology for Society and Territory – Volume 6*,
DOI: 10.1007/978-3-319-09060-3_55, © Springer International Publishing Switzerland 2015

Fig. 55.1 View on borehole core samples, tunnel tubes and forepoles as a face protection

55.2 Model Assumptions and Analysis of Results

All analyses are performed by assuming plane strain conditions in software packadge Plaxis that is based on the finite element method (FEM). The model consists of 2,476 15-noded triangular finite elements (average size of 1.782 m). The mesh is refined in the zone of tunnel excavations. It is assumed that material cannot sustain tensile stresses. Influence of pore water pressure is not considered as the groundwater level was not observed during the site investigations.

Firstly, the analysis is performed with an assumption of unsupported excavation of tunnel tubes. The material is described with an elastic perfectly-plastic model assuming linear Coulomb-Mohr failure criterion. The model consists of six parameters, namely: γ, E, c, φ, υ i ψ, i.e. unit weight of soil, modulus of elasticity, cohesion, angle of shearing resistance, Poisson's ratio and angle of dilatancy,

respectively. Adopted parameters correspond to the B category of the main design with values of: $\gamma = 20$ kN/m^3, E = 70 MPa, c = 0.04 MPa, $\varphi = 24°$, $\upsilon = 0.3$ i $\psi = 0°$.

In the first calculation phase the initial stress state is generated based on the K_0 procedure. This assumption is valid due to tunnel construction in relatively young, normally consolidated and tectonically undamaged sediments. After the initial stress generation the excavation phase is performed.

Results showed that without an adequate support system excavations would collapse (chimney type failure).

This can be seen on a plot of plastic points, Fig. 55.2. Abovementioned indicates that support of excavations is needed.

Subsequent analyses are performed with an assumption of immediate installation of tunnel lining. The case considered is lining made of shotcrete and previously installed forepoles. Shotcrete lining is modeled as a plate element (Brienkgreve and Broere 2011), described with two parameters, namely axial (EA) and flexural (EI) rigidity. According to

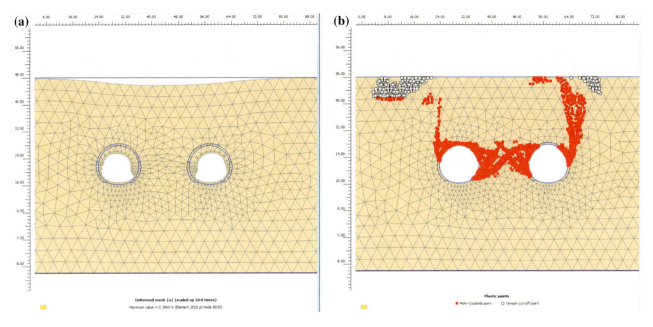

Fig. 55.2 **a** Deformed mesh showing ground surface settlement and deformations of tunnel contours, **b** development of plastic tension and shear zones

design the thickness of shotcrete is d = 35 cm. Elastic modulus of shotcrete is reasonably assumed to be E = 20,000 MPa (Hoek and Bawden 1993). It is worth mentioning that elastic modulus of shotcrete is time dependent, but this property is not included in the analyses.

Forepoles are modeled as proposed by Hoek (2004), where the zone of influence is taken to have properties obtained in a process of weighted averages of properties of

steel, grout and surrounding soil. This material with predefined thickness of ≈0.60 m has following values of parameters: E = 1,200 MPa, c = 0.15 MPa (other parameters are the same as for surrounding soil).

Figure 55.3a shows analysis results with an immediate installation of tunnel lining. The heave of an invert and ground surface can be observed. The heave is a consequence of applied Coulomb-Mohr elastic perfectly-plastic model.

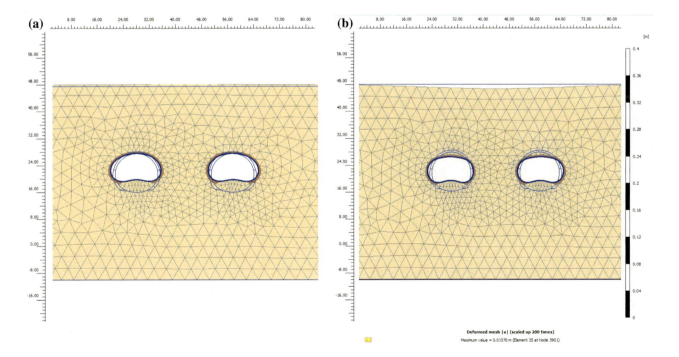

Fig. 55.3 Ground displacements in case of **a** MC model, **b** HSsmall model

Table 55.1 Parameters of the HSsmall model

γ kN/m^3	E_{oed}^{ref} MPa	E_{50}^{ref} MPa	E_{ur}^{ref} MPa	c MPa	φ°	ψ°	G_0^{ref} MPa	$\gamma_{0.7}$ –	m –	R_f –
20	10.90	21.40	85.60	0.04	24	0	120	0.0001	0.5	0.9

where: E_{oed}^{ref}—tangent stiffness for primary oedometer loading at the reference level of normal stress of 0.1 MPa, E_{50}^{ref}—secant stiffness for 50 % of ultimate load in standard drained triaxial test, where $\sigma_{3\,=}$ 0.1 MPa, E_{ur}^{ref}—unloading/reloading stiffness in standard drained triaxial test, where $\sigma_{3\,=}$ 0.1 MPa, G_0^{ref}—reference shear modulus at very small strains, $\gamma_{0.7}$—shear strain at which $G_s = 0.722\ G_0$, m—power for stress-level dependency of stiffness, R_f—failure ratio

The convergence of an invert, with shotcrete support, is 3.87 cm and if considering both shotcrete and forepoles convergences are insignificantly smaller, i.e. 3.78 cm.

In order to show an influence of the higher order model on convergence predictions, analysis is repeated by applying HSsmall model. This model assumes stress dependent stiffness taking into account stiffness at small strains. Parameters used to describe the model are presented in Table 55.1.

Certain parameters are derived from laboratory tests, and others are adopted as proposed in Brienkgreve and Broere (2011). Advantage of this model compared to others is his relative insensitivity to chosen domain size.

Figure 55.3b shows results of a n analysis with HSsmall model. Convergence of invert is 1.60 cm without forepoles and 1.57 cm with forepoles. Ground surface settlements are 0.79 and 0.64 cm, respectively. The differences between the two models are obvious if Fig. 55.3a, b are compared.

Above mentioned analyses are not realistic due to some time needed for support to be installed, thus it is necessary to allow for an initial convergence of the excavation prior to support installation. This is achieved by applying β-method. In this way it is possible to consider 3D arching effect and deformations of the tunnel face. The idea is that the initial stresses acting around the location where the tunnel is to be constructed are divided into a part $(1-\beta)p_k$ that is applied to the unsupported excavation and a part βp_k that is applied to the supported excavation. The β coefficient $(0 < \beta < 1)$ is an experience parameter depending on the tunnel round length and equivalent diameter. Some proposals for its determination could be found in literature (Moller 2006; Moller and Vermeer 2004). Baudendistel (1979) proposed values for parameter β after considering vertical crown displacements of tunnels from 3D linear-elastic analysis. Table 55.2 shows values of parameter β for a tunnel with a horse shoe profile.

The design proposed that the round length d, of the tunnel, is to be not more than 1 m (Milenković et al. 2009). If this length is compared to equivalent diameter of D ≈ 12 m, the β coefficient is found to be 0.5.

Results of analyses performed by using MC model are presented in Fig. 55.4. In Fig. 55.4a contours of unsupported excavations (for β = 0.83) are deformed over 30 cm which would eventually lead to a collapse as shown in Fig. 55.2.

Table 55.2 β coefficient for different ratios of round length d, and excavation diameter D (after Baudendistel 1979, reproduced in Moller 2006)

d	1.5 D	D	0.5 D	0.25 D	0.125 D	0
β	0.0	0.02	0.11	0.23	0.41	0.72

The mode of deformation indicates that the largest displacements are in the tunnel crown, so this zone needs to be strengthened. Second analysis, Fig. 55.4b, is performed assuming excavation under forepoles (according to design), for β = 0.5. In this case the displacements are largest in the tunnel crown and heave of an invert is observed. If forepoles are extended around an excavation the largest displacements are found in the zone of invert, Fig. 55.4c.

The following analyses include the installation of lining by allowing it to accept the rest of the stresses acting around the tunnel excavation. Figure 55.4d shows the displacements of the tunnel excavations after applying shotcrete support. Fig. 55.4e shows the displacements of combined shotcrete support and forepoles installed according to design. The displacements induced by installation of shotcrete lining can be found as a difference between the displacements shown in Fig. 55.4e, b. Similar conclusions can be made if Fig. 55.4d, f are compared, when forepoles are extended around excavation contours. Results show that the differences in displacements are largest in the zone of an invert. The reason for this is installation of forepoles which transferred the displacements from the tunnel crown towards the invert, causing its heave. Performed analyses imply that besides the shotcrete support with forepoles, prevention from invert heave needs special consideration.

55.3 Comments and Conclusions

Convergence monitoring is currently ongoing for the excavated portion of approx. 700 m of the tunnel exit portal (reaching steady state plane strain conditions). The measurements show cumulative displacements in the range of 1–3 cm, Fig. 55.5.

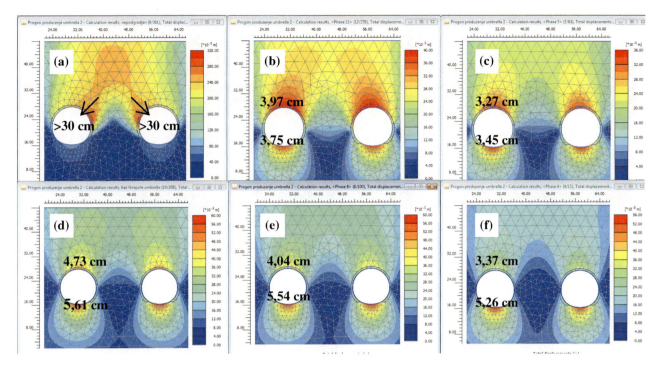

Fig. 55.4 Shadings of total displacements around tunnel excavations **a** Unsupported tunnel excavations, β = 0.83, **b** Tunnel excavation with forepoles according to design, β = 0.5, **c** Forepoles extended towards the invert, β = 0.5, **d** Supported excavation with the shotcrete lining, β = 1, **e** Shotcrete and forepoles according to design, β = 1, **f** Shotcrete and forepoles extended towards the invert, β = 1

Fig. 55.5 Typical convergence profile of one cross-section in plane strain conditions

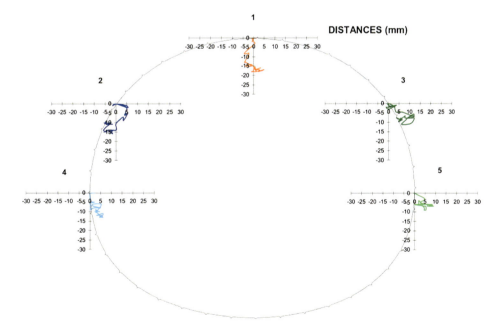

If HSsmall model is used for convergence predictions (accounting for arching effect and deformations of the tunnel face) values of approx. 1 cm are obtained.

In this way convergence is approx. 5 cm when MC model is used and around 1 cm in the case of HSsmall model, thus limiting the measurement results from the lower and upper bound.

Obtained results are highly dependent on the constitutive model used to represent material behavior and the value of β parameter.

References

Baudendistel M (1979) Zum Entwurf von Tunneln mit großen Ausbruchsquerschnitt. In: Berechnung, Erkundung und Entwurf von Tunneln und Felsbauwerken / Computation, Exploration and Design of Tunnels and Rock Structures. Rock Mechanics / Felsmechanik / Mécanique des Roches, vol 8. pp 75–100. doi:10.1007/978-3-7091-8564-3_6

Brienkgreve RBJ, Broere W (2011) Plaxis 2DV10: finite element code for soil and rock analyses. Delft University of Technology & Plaxis b.v. Delft

Hoek E (2004) Numerical modelling for shallow tunnels in weak rock. http://geotecnica.dicea.unifi.it/hoek_shallow.pdf

Hoek E, Kaiser PK, Bawden WF (1993) Support of underground excavations in hard rock. Mining Research Directorate and Universities Research Incentive Fund, Ontario, p 225

Milenković S et al (2011) In geotechnics of hard soils—weak rocks: geotechnical conditions for the construction of tunnel "progon" on the dimitrovgrad bypass in Serbia. In: Anagnostopoulos A, Pachakis M, Tsatsanifos Ch. (eds) 15th European conference on soil mechanics and geotechnical engineering, Athens, September 2011. IOS Press BV, Amsterdam, p 2064

Moller SC (2006) Tunnel induced settlements and structural forces in linings (Dissertation). Institute of Geotechnical Engineering, Stuttgart

Moller SC, Vermeer PA (2004) In underground space use. Analysis of the past and lessons for the future: on design analyses of natm-tunnels. In: Solak E (ed) Proceedings of the international world tunnel Congress and the 31st ITA general assembly, Istanbul, May 2005. Taylor and Francis Group, London, p 1384

Quarry Site Selection and Geotechnical Characterization of Ballast Aggregate for Ambo-Ijaji Railway Project in Central Ethiopia: An Integrated GIS and Geotechnical Approach

56

Regessa Bayisa, Raghuvanshi Tarun Kumar, and Kebede Seifu

Abstract

The main objective of the present study was to select potential quarry sites and to characterize the rock for ballast aggregate for proposed Ambo-Ijaji railway project in central Ethiopia. The study area is located in the Oromia Regional state which is around 120 km from Addis Ababa on way to Ambo town. The quarry site selection criteria was formulated and applied in GIS environment by incorporating factors such as; slope of the quarry rock face, distance of the quarry site, land use/land cover, accessibility, overburden thickness, rock type and degree of weathering. Thus, six quarry sites were evaluated for their suitability. Further, representative rock samples from quarry sites were collected and tested for physical, mechanical and chemical properties to ensure that the source rock is suitable for ballast aggregate. In addition, petrographic analysis was also made to understand the correlation of engineering properties with petrographic parameters. The results revealed that all six quarry sites satisfies the selection criteria and are suitable to provide ballast aggregate. The engineering laboratory tests indicates that the test values for unit weight, Los Angeles Abrasion value, soundness test by sodium sulfate, specific gravity, and unconfined compressive strength are within the specified limits except for three quarry sites which have water absorption values higher than the standard specifications. From the correlation of petrographic test with different engineering test results it has been found that the petrographic parameters has an important role in controlling engineering properties of rocks.

Keywords

Potential quarry • Ballast aggregate • Los-Angeles abrasion • Aggregate soundness • Water absorption

56.1 Introduction

For railway, ballast is the most important sub-structure. The selection of suitable quarry site for ballast is based on pre-defined criteria such as; type of rock and its weathering grade, distance from the proposed rail alignment, thickness of overburden at proposed quarry site, landuse and land-cover, accessibility to proposed quarry site etc. After the selection of suitable quarry site the next component of evaluation is to determine the quality of ballast rock based on different physical and mechanical properties of the source ballast rock (Raisanen et al. 2006).

R. Bayisa (✉) · R.T. Kumar · K. Seifu
School of Earth Sciences, College of Natural Sciences, Addis Ababa University, Po Box 1176, Addis Ababa, Ethiopia
e-mail: bayyee.2001@gmail.com

R.T. Kumar
e-mail: tkraghuvanshi@gmail.com

K. Seifu
e-mail: seifukebede@yahoo.com

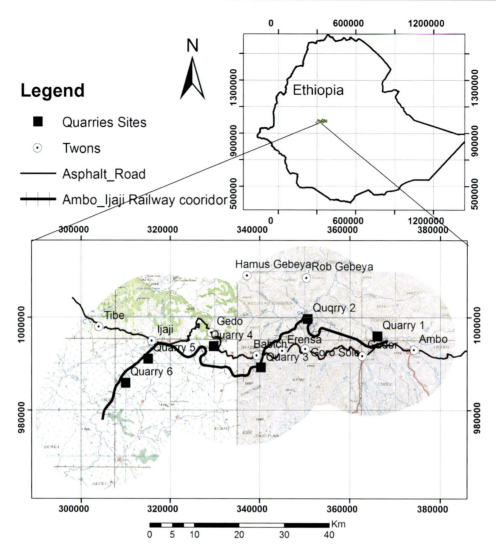

Fig. 56.1 The location map of the study area

The present study was conducted for Ambo-Ijaji Railway Project in Central Ethiopia. The corridor of the study area has the total length of 94 km which starts from Ambo and extend upto Ijaji. Geographically, the area is bounded by the UTM coordinates of 304809-368349E and 977872-994953N (Fig. 56.1). The regional geology of the present study area is characterized by three main geological formations; (i) The Precambrian basement rocks, biotite gneiss with minor intercalation of fine to medium grained undifferentiated schist are exposed in the North of the study area (Abebe et al. 1998; Kidane 2010). (ii) The Mesozoic sedimentary rocks; sandstone, Limestone, gypsum and shales, mainly exposed around northwest of the study area. (iii) The Cenozoic volcanic rock covers large parts of the study area and mainly consists of lower and upper volcanic sequences. In addition, the western part of Ijaji and Tibe area is covered by Quaternary Sediments (Fig. 56.2). The main objective of the present study was to identify the suitable quarry sites for

the proposed Ambo-Ijaji railway project and to assess the geotechnical characteristics of the ballast material.

56.2 Suitability Evaluation of Quarry Sites as per the Selection Criteria

During the present study the selection criteria was formulated by considering parameters such as; slope, distance of the site, overburden soil thickness, landuse/landcover, general accessibility and rock type with its degree of weathering. In total six quarry sites were initially identified by considering the suitable rock type and optimum distance from the proposed rail line alignment (Fig. 56.1). Later, the general selection criterion was formulated (Table 56.1) and was applied in GIS environment to identify the suitable quarry sites. The evaluated suitability based on selection criteria is presented in Table 56.2.

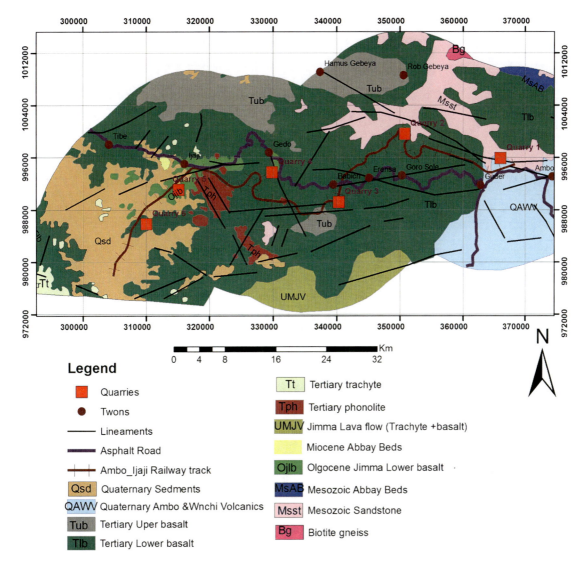

Fig. 56.2 Geological map of the study area

Slope factor was considered and evaluated in terms of quarry's hazard. It is imperative to mention that more steeper the slope more instability problems may be encountered. Thus, the surrounding environment will be degraded due to such slope instability problems.

Quarry site distance and their accessibility condition may affect the economy, increase in transportation time and may result into environmental pollution. Transporting the rock aggregate to a long distance can consume more fuel and much time which can ultimately affect the economy of the project.

Overburden soil thickness is also an important parameter. Removal of thick soil cover impart additional cost to the quarrying activity, thus directly affects the project economy.

The *land use/land cover* is important parameter in evaluating the impact of quarry on the surrounding environment. Thus, during the selection of suitable quarry site, type of land use/land cover that has less susceptibility to the impact of quarrying activity has been considered.

Rock quality is an important parameter as different rock types with varied weathering grades can demonstrate different engineering properties.

In general, to determine the overall suitability of each quarry site based on the six evaluated parameter, equal weights were given to the parameters and later their average value was calculated. According to the average evaluation results (Table 56.2) all quarries are identified as suitable except quarry 1 which is identified as highly suitable.

56.3   Ballast Rock Characterization

To ensure the suitability of the source rock from the selected quarry sites sampling and laboratory analyses were performed. To assess the suitability of ballast material, sodium sulfate soundness test, compressive strength test, specific gravity test, Los Angeles Abrasion value test, Water

Table 56.1 Formulated potential quarry site selection criteria

Types of parameters	Conditions of parameter	Suitability condition	Value
Quarry site slope angle parameter in degree	>45	Unsuitable	1
	30–45	Slightly suitable	2
	15–30	Suitable	3
	0–15	Highly suitable	4
Quarry site distance from each other along the corridor	<10 and >50 km	Unsuitable	1
	10–15 and 40–50 km	Slightly suitable	2
	15–20 and 30–40 km	Suitable	3
	20–30 km	Highly suitable	4
Quarry site distance from the proposed track corridor (offset)	<30 m and >5 km	Unsuitable	1
	30–50 m and 4–5 km	Slightly suitable	2
	50–100 m and 3–4 km	Suitable	3
	100 m to 3 km	Highly suitable	4
Estimated overburden soil thickness of the quarry sites	>4 m	Unsuitable	1
	2.5–4 m	Slightly suitable	2
	1.5–2.5 m	Suitable	3
	0–1.5 m	Highly suitable	4
Landuse/Landcover of the quarry sites	Dense settlement and cultural heritage area	Unsuitable	1
	Moderately Settlement and forested area	Slightly suitable	2
	Rarely populated, agricultural land and less vegetated grass land	Suitable	3
	Bare land and grass land	Highly suitable	4
Type of rock in the proposed quarry sites (if it is not weathered)	Shale, slate and pumice	Unsuitable	1
	Limestone, Sandstone and Siltstone	Slightly suitable	2
	Granite, Diorite, Gabbro, quartzite and hard mineral and well-cemented sedimentary rock	Suitable	3
	Basalt, rhyolite, andesite	Highly suitable	4
Weathering degree of the proposed quarry material	Highly weathered	Unsuitable	1
	Moderately weathered	Slightly suitable	2
	Slightly weathered	Suitable	3
	No weathering	Highly suitable	4

absorption test, unit weight test and petrographic thin section analysis were performed. These laboratory tests were conducted following "American Railway Engineering Maintenance-of-way Association Manual" (AREMA 2009). The laboratory test results with recommended specifications for representative samples from six quarry sites are presented in Table 56.3.

Sodium sulfate soundness test is used to estimate the aggregate soundness when they are subjected to the weathering action. The test conducted on representative samples

Table 56.2 General suitability conditions of the proposed quarry sites as per formulated selection criteria

Quarry sites	Selection criteria parameters	Condition of quarry sites as per given parameter	Suitability conditions	Value
Quarry 1	Slope angle	Slopes fall within 0°–15° class	Highly suitable	4
	Distance and accessibility	Distance from the corridor 1.4 km and distance from the existing road is 1.8 km	Highly suitable	4
	Overburden thickness	Overburden soil thickness ranges from 0.5 to 1 m	Highly suitable	4
	Land use/Land cover	It is located partly on the small grass land and partly on the farm land	Suitable	3
	Rock type and weathering degree	It is the slightly weathered porphyritic olivine basaltic material	Suitable	3
Quarry 2	Slope angle	The dominant slope class is 30°–45°	Suitable	3
	Distance and accessibility	It is 0.6 km from the corridor and 21 km from the quarry 1	Highly suitable	4
	Overburden thickness	Overburden soil thickness range from 0.5 to 1 m	Highly suitable	4
	Land use/Land cover	It is located on less vegetated land and surrounded by small farm lands	Suitable	3
	Rock type and weathering degree	It is moderately weathered porphyritic basaltic material	Slightly suitable	2
Quarry 3	Slope angle	In general slope is within 15°–30° slope class	Suitable	3
	Distance and accessibility	It has 0.5 km offset and 17.5 km distance from quarry 2. Earth road exists	Suitable	3
	Overburden thickness	Overburden soil thickness ranges from 0.5 to 1.5 m	Highly suitable	4
	Land use/Land cover	Mainly it is covered by less vegetated land.	Suitable	3
	Rock type and weathering degree	Slightly weathered trachy-basalt is present	Suitable	3
Quarry 4	Slope angle	Majority of the slopes fall into 30°–45°	Slightly suitable	2
	Distance and accessibility	Its distance from quarry 3 is 19.3 km and it needs 1.8 km access road	Suitable	3
	Overburden thickness	Overburden soil thickness range from 0.5 to 1 m	Highly suitable	4
	Land use/Land cover	It is covered by less vegetated land	Suitable	3
	Rock type and weathering degree	It is a slightly weathered porphritic olive basalt	Suitable	3
Quarry 5	Slope angle	Almost entire slopes fall within 15°–30° slope class.	Suitable	3
	Distance and accessibility	Earth road exists and its distance from preceding quarry 4 is 16.5 km	Suitable	3
	Overburden thickness	Overburden soil thickness range from 0.5 to 1.5 m	Highly suitable	4
	Land use/Land cover	It is mainly covered by less vegetated grass land and is surrounded by farm land	Suitable	3
	Rock type and weathering degree	It is the slightly weathered aphinitic basalt	Suitable	3
Quarry 6	Slope angle	Almost entire slopes fall within 0°–15° slope class	Highly suitable	4
	Distance and accessibility	It is 8.2 km from quarry 5	Slightly suitable	2
	Overburden thickness	Overburden soil thickness ranges from 0.5 to 1.5 m	Highly suitable	4
	Land use/Land cover	It is covered by less vegetated grass land and surrounded by farm land	Suitable	3
	Rock type and weathering degree	Slightly weathered trachy basalt rock is present	Suitable	3

demonstrated values within the range of 0.9–1.4 % which is well within the permissible limits (≤5.0 %).

Uniaxial Compressive strength (UCS) test is performed to determine the strength of the intact rock. According to Instruction manual on concrete test hammer, as cited in

Tawake (2007), the rock with UCS value > 20 MPa can be accepted as suitable aggregate source rock. The test results showed that the UCS values for all samples falls in the range of 34–45 MPa which are well within the permissible limits (Table 56.3).

Table 56.3 Geotechnical laboratory test results

Test types		Quarry 1	Quarry 2	Quarry 3	Quarry 4	Quarry 5	Quarry 6	Permissible Limit*
Soundness loss % by sodium sulfate		1.3	1.4	1.1	0.98	0.90	0.92	≤5.0 %
Compressive strength (Mpa)		36	34	38	42	45	44	>20 Mpa
Specific gravity	Apparent (Bulk) specific gravity	2.938	2.918	2.958	2.988	2.999	2.990	≥2.60
	Saturated surface dry (SSD) specific gravity	2.804	2.801	2.814	2.913	2.926	2.916	
	Oven dry (OD) specific gravity	2.801	2.799	2.806	2.899	2.906	2.900	
Unit weight	Unit weight in dry condition (kg/m^3)	2,238	2,218	2,298	2,301	2,305	2,303	≥1,400 (kg/m^3)
	Unit weight in SSD condition (kg/m^3)	2269.3	2249.5	2329.7	2322.9	2325.8	2324.2	
Percentage of void content (%)		19.94	20.60	17.94	20.47	20.36	20.43	
Los Angeles abrasion value (%)		14.8	14.9	14.5	14.3	14.0	14.1	≤25 %
Water absorption test values in %		1.4	1.42	1.38	0.95	0.90	0.92	≤1 %

*AREMA (2009)

Specific gravity is related to the density of the rock and can control both vertical and horizontal holding capacity of ballast aggregate. According to Indraratna et al. (2006), the higher specific gravity of parent rock ensures greater holding capacity and lower degradation of the ballast aggregate. The bulk specific gravity for all representative samples satisfies the standard specification (Table 56.3), as all the values are above 2.60.

The *Bulk density* (unit weight) of an aggregate is the weight (mass) of an aggregate per its unit volume. The results showed that the Unit weight for all representative rock samples satisfies the standard specification (Table 56.3), as all values are above 1,400 (kg/m^3).

The *Los Angeles Abrasion test* is to measure the load resistance of ballast material. This test measure the toughness of ballast aggregate and is used to determine the ability of ballast aggregate to survive the contact force (Raymond and Bathurst 1994). The results shows that all the values for representative rock samples satisfies the standard specification (Table 56.3), as the values are less than 25 %.

Water absorption test is conducted to measure the ability of an aggregate rock to absorb water and to know the sensitivity to degradation. The results shows that the rocks from quarry 1, 2 and 3 have water absorption values higher than the specification while for quarry 4, 5 and 6 the values are within the specification (Table 56.3). However, even if the water absorption value of samples from quarry 1, 2 and 3 is more, it is expected that more fresh suitable rock would be available during the quarrying operation, as the samples during the present study were collected near to the surface.

56.4 Correlation of the Engineering Properties with Petrography

The engineering properties of ballast aggregate source rock mainly depends on its mineralogical composition, the size and shape of its minerals, presence or absence of micro cracks in it and the degree of weathering. During the present study the thin sections were prepared for the representative rock samples from each quarry site and later these were analyzed under petrographic microscopy.

In the present study the rock sample from *quarry* 5 is the 1st most suitable in comparison to samples from other quarry sites in terms of its engineering test results. The better engineering quality of the rock from quarry 5 has resulted from abundance of compacted opaque minerals that were surrounded with very fine grained ground mass. Both fine grained texture of ground mass and the abundance of highly compacted opaque minerals in this rock has played a significant role in providing better engineering quality to the ballast aggregate material. However the opaque minerals have high potential for alteration, therefore if the ballast aggregate has more opaque minerals such as; iron oxide it will readily be affected by weathering.

The rock sample from *quarry* 3 is acidic and has high silica content in comparison to samples from other quarry sites. However, laboratory test results revealed that the quality of this rock falls into 4th order in its suitability. Further, the rock sample from quarry 3 has more than 70 % feldspar minerals that are elongated and lath shaped with trachitic texture. This lath shape and trachitic texture may affect grain to grain interlocking capacity of the minerals in the rock, thus resulting into poor engineering quality of the rock.

Further, existence of micro cracks in the rock material can affect its engineering properties. Particularly, when the compressive stress is applied on the rock aggregate, the preexisting inter-granular micro cracks and the grain boundary micro cracks extends and transform into the trans-granular micro cracks. These micro cracks were observed in the rock samples from *quarry* 1, *quarry* 2 and *quarry* 6.

The rock sample from *quarry* 1 has more volcanic glass in comparison to other samples. Besides, presence of calcite phenocrysts was also observed. Based on the laboratory test result quarry 1 was classified as 5th order in its suitability, as compared to other quarry sites. Relatively low suitability of quarry 1 has resulted probably from the presence of abundant volcanic glass, existence of micro cracks and from the presence of soft secondary mineral calcite. Based on laboratory test results the rock sample from *quarry* 2 is placed into 6th order of suitability as ballast source material. From the hand spacemen observation it has been observed that the rock sample from quarry 6 was highly weathered as compared to other samples. Thus, presence of micro cracks and high degree of alteration (weathering) has resulted into reduction of engineering quality of this rock.

56.5 Conclusion

The present study was conducted for Ambo-Ijaji Railway Project in Central Ethiopia. The main objective of the study was to identify the suitable quarry sites for the proposed project and to assess the geotechnical characteristics of the ballast material. To meet out the objective of the study, selection criteria was formulated by considering parameters such as; slope, distance of the site, overburden soil thickness, landuse/landcover, general accessibility and rock type with its degree of weathering. The general selection criterion was applied in GIS environment to identify the suitable quarry sites. The results revealed that all quarries are suitable as per the selection criteria. Further, to assess the suitability of ballast material, sodium sulfate soundness test, compressive test, specific gravity test, Los Angeles Abrasion value test, Water absorption test, unit weight test and petrographic thin

section analysis were carried out on the representative samples. The test results revealed that all the samples possess values within the specified limits, except rocks from quarry 1, 2 and 3 which have water absorption values higher than the standard specifications. However, it is expected that more suitable rock would be available during the quarrying operation, as the samples during the present study were collected near to the surface. Further, based on the petrographic examination results it is deduced that the high engineering quality rock from quarry 5 has probably resulted from the abundance of opaque minerals and its fine grained texture, whereas low quality rock from quarry 2 was from its high degree of alteration and existence of micro cracks in it. Similarly, based on the laboratory test result relative low suitability of quarry 1 has resulted probably from presence of abundance volcanic glass, existence of micro cracks and presence of soft secondary mineral calcite. In general, from the correlation of petrography with different engineering test results it may be conclude that the petrographic parameters (mineralogical composition, grain texture, micro crack and secondary minerals) play an important role in controlling engineering properties of ballast aggregate rocks.

References

Abebe T, Mazzarini F, Innocenti F, Manetti P (1998) Yerer-Tullu Wellel volcanotectonic lineament: a transtensional structure in central Ethiopia and the associated magmatic activity. J Afr Earth Sci 26(1):135–150

American Railway Engineering and Maintenance of Way Association (AREMA) (2009) Manual for railway engineering, vol. 1, Roadway and Ballast, USA, 1292 pp

Indraratna B, Khabbaz H, Salim W, Christie D (2006) Geotechnical properties of ballast and the role of geosynthetics in rail track stabilization. Univ Wollongong Res Online 3:91–101

Kidane T (2010) Final Report on Structural and Geological Mapping for the Wolkite Ambo Ground Water Potential Assessment Project. Water Works Design and Supervision Enterprise (unpublished), Addis Ababa Ethiopia

Raisanen M, Mertamo M, Stenlid L, Viitanen J (2006) Laboratory crushing of rock aggregates. Paper number 411, Geological Survey of Finland

Raymond GP, Bathurst RJ (1994) Repeated load response of aggregates in relation to track quality index, vol. 31. Queen University, Canada, pp 547–554

Tawake AK (2007) Assessment of potential terrestrial aggregate sources on Ghizo Island, Solomon Islands. EU EDF8-SOPAC project Report No 107

Radon Emanation Techniques as an Added Dimension in Site Investigation of Water Storage Facilities

Gary Neil Davis and Mannie Levin

Abstract

For dams and reservoirs, the assessment of the basin watertightness relies on identification of structural weaknesses. Conventional site investigations can readily be complimented by carrying out radon emanation studies, relying on unique properties of this radioactive noble gas, and easier transmission via fractured, faulted and porous rock zones. Recent studies for an abstraction weir supplemented the conventional ground investigations with radon emanation studies to confirm the locality of a suspected fault zone which traverses the founding bedrock. The successful location of the structural weakness, hidden beneath 20–30 m of alluvium, allowed optimal location of monitoring boreholes; proving the value of this technique in such studies.

Keywords

Site investigation • Radon emanation • Structural weaknesses

57.1 Introduction

Dam and reservoir studies must include assessment of the basin watertightness. Conventional studies would typically comprise investigation of the regional and structural geology, and would include engineering geological mapping, as well as utilising satellite imagery; with follow-up studies potentially including drilling and test pitting, as well as laboratory testing.

Although well known in groundwater studies (Levin 2000), radon emanation techniques are finding increasing application in other areas, such as for the investigation of dams and reservoirs.

The radon emanation technique enables identification of structural weaknesses, particularly when obscured by the drift geology, and allows the subsequent investigations to be more focused.

The technique rests on the emission of the α-particle by the radioactive isotope Radium (Ra226) which produces the gas Radon (Ra222). The radium in turn is naturally produced during the decay of Uranium (U238); where uranium is present in most rock types, in minerals such as zircon, mica, apatite etc.

Radon, being a noble gas, has the ability to migrate from its source without chemical interference, and emanates from the mineral surfaces into the rock pores or dissolves in the water phase where present. Radon (Ra222) is radioactive with a half-life of 3.8 days. Due to its half-life it is present in groundwater for up to 15 days. This unique combination of features underlies the interest in radon gas as a geophysical tracer.

Various researchers have shown that radon is released mainly through rapid diffusion along imperfections to the particle surface and that crystal imperfections therefore play a vital role. This means that the fraction of the radon released to ground water and soil air will depend largely on the physical nature of the rock and associated fluid phases, rather than on the uranium concentration in the rock. It therefore follows that increased concentration can be expected from fractured, faulted and porous rock zones, as illustrated in (Fig. 57.1).

G.N. Davis (✉) · M. Levin
Aurecon South Africa (Pty) Ltd, Pretoria, South Africa
e-mail: gary.davis@aurecongroup.com

G. Lollino et al. (eds.), *Engineering Geology for Society and Territory – Volume 6*,
DOI: 10.1007/978-3-319-09060-3_57, © Springer International Publishing Switzerland 2015

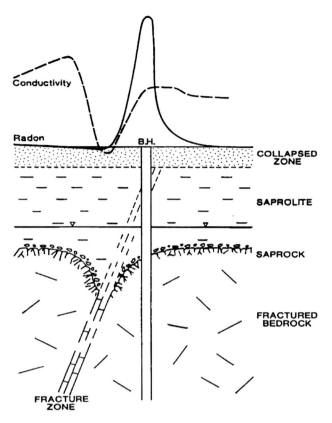

Fig. 57.1 Radon technique illustrated, after Wright (1992)

57.2 Methodology

The radioactivity makes radon easy to measure in small concentrations. The most widely used method of detection in aerial surface radon surveys is the nuclear tracks ("track etch") method, due to its simplicity. The method is applied mainly in soils covering the rock formations from which the radon is assumed to migrate. The distribution (diffusion) pattern depends on the structures in the underlying rock and therefore the areas of higher radon concentration correspond with fractures, faults or other more permeable zones. Knowledge of the underlying geology is essential for interpretation of results.

The radon emanation technique can therefore be employed to confirm the presence of discontinuities such as mentioned, at a particular site. This knowledge is crucial to ensure appropriate measures are not considered in the design and construction phases; otherwise these features could be key in later problems linked to uncontrolled seepage or leakage.

This method has been employed with some success in South Africa; a case in point being an abstraction weir planned as an integral part of a water transfer scheme.

57.3 Case History—Vlieëpoort Abstraction Weir

The abstraction weir is a key component in an interbasin water transfer scheme. The envisaged structure will be roughly 350 m in length. The structure will only be up to 15 m above river level, but the substantial alluvial deposits will see innovative founding solutions. The need for a detailed understanding of the foundation seepage potential gave impetus to utilizing these radon emanation studies to supplement the site investigations.

57.3.1 Geological Setting and Investigation History

The weir site is located at a narrowing of a valley where the Crocodile River cuts through the Vlieëpoort Mountains, and is characterised by a substantial thickness of alluvial material, overlying bedrock comprising Banded Ironstone Formation (BIF). Initial concerns that the site might have been underlain by dolomitic rocks—which would have added a further dimension to the study—proved unfounded.

The alluvium thickness generally varies between 20 and 30 m, but is shallower towards the flanks where alluvium thicknesses between 11 and 13 m were recorded. A maximum alluvium thickness of 39.5 m was recorded.

Limited exploratory drilling was conducted at the weir site during initial feasibility-level investigations (DWA 2008). The follow-up design investigations (TCTA 2010) commenced with geophysical surveys, comprising gravimetric and resistivity surveys, with the aim of identifying any linear features which might be present, as well as defining variations in bedrock elevation. These anomalies were then targeted for further exploratory drilling of rotary core boreholes as well as percussion boreholes.

The gravity and resistivity surveys revealed the presence of possible fault zones, and extrapolation between these geophysical anomalies indicated possible alignments of these geological features, which were subsequently targeted for follow-up drilling.

57.3.2 Radon Emanation Studies

The significant thickness of the alluvium, coupled with indications of possible fault zones, contributed to concerns relating to potential seepage via the founding horizons. At the same time the downstream farming community also had concerns regarding the potential foundation cut-off, and the possible effects on the aquifer represented by the

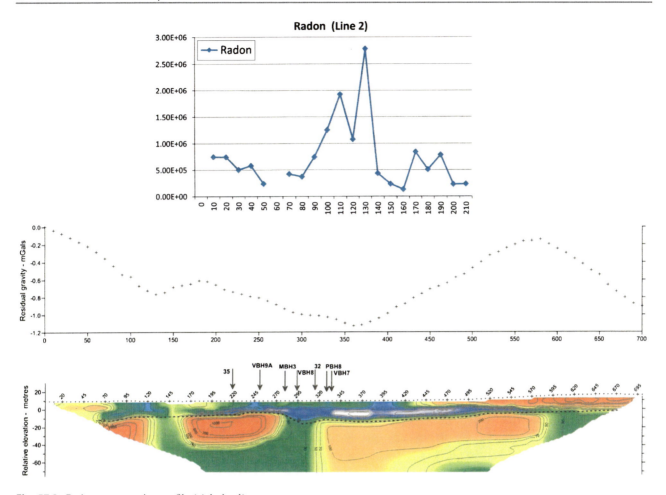

Fig. 57.2 Radon concentration profile (*right bank*)

downstream alluvium. It was subsequently decided to conduct a radon survey; both to confirm the presence and orientation of the potential fault traversing the site and further to use these results to optimally locate a series of monitoring boreholes.

Two radon profile lines were selected; one each on the opposite banks, running parallel to the river, centered on the anomalies recognized in the geophysical survey and the drilling results. The passive Radon Gas Monitors (RGMs) were placed in shallow auger holes about 500 mm deep and 10 m apart. The RGMs were left for 2 weeks and were then removed for processing and recording. The analysed radon values were plotted against the traverse distance for each profile (Fig 57.2).

57.3.3 Study Findings

The anomalous peak radon values were considered to indicate vents in the geological structure, with high radon emanating at these points. Furthermore, these localities were found to correspond to the geophysical resistivity anomalies (Fig. 57.2).

The corresponding anomalous radon survey results from the respective traverses were considered to indicate the location of the fault immediately downstream of the planned weir structure, and slightly oblique to the river. This information was used to determine optimal positions of a number of monitoring boreholes; these boreholes were subsequently drilled and continue to be monitored.

57.4   Conclusions

The radon emanation studies proved invaluable in confirming the position of a structural weakness, i.e. the presumed fault, located downstream of the abstraction weir. This was particularly notable when considering the thickness of the alluvium; typically between 20 and 30 m.

In this example the locality of the presumed fault was not the primary concern in terms of founding conditions, as the substantial alluvium thickness presents its own unique challenge in terms of a founding solution. In any event it had been assumed that faults would intersect the founding bedrock.

In terms of a quick and reliable method of locating structural weaknesses within the founding bedrock, however, the radon studies were invaluable; and allowed optimal locations for groundwater monitoring boreholes to be determined.

References

Department of Water Affairs, South Africa (2008) Mokolo crocodile (West) water augmentation project (MCWAP) feasibility study. Report 8, detail geotechnical investigations, phase 2. Prepared by Africon in association with Kwezi V3 Engineers, Vela VKE, WRP Consulting Engineers and specialists

Levin M (2000) The Radon emanation technique as a tool in ground water exploration. Borehole Water J 46:22–25

TCTA (2010) Mokolo crocodile water augmentation project (MCWAP). Contract No TCTA 07-041. Consulting services for MCWAP. Phase 1: geotechnical investigations, vol 3. Geotechnical interpretive report, July 2010. Prepared by Mokolo Crocodile Consultants

Wright EP (1992) The hydrogeology of crystalline basement aquifers in Africa. In: Wright EP, Burgess WG (eds) The hydrogeology of crystalline basement aquifers. The Geological Society, London

Part V

Capturing and Communicating Geologic Variability and Uncertainty

Convener Dr. Jeffrey Keaton—*Co-conveners* Helen Reeves, William Haneberg, Steve Parry, Rosario Basurto

Geologists tend to talk to each other using jargon that is not useful to non-geologists who could benefit from the geologic information. Dashed and queried lines geologists use to represent variability and uncertainty are not adequately documented or quantified for communication with non-geologists. Engineers need good site characterization for risk assessment and reliability-based design projects which require rigorous expression of uncertainty. Geologists need to participate in this process or be marginalized by engineers treating geology as a completely random variable or expressed only in terms of quantitative geotechnical data. This session will focus on capturing variability and uncertainty in geology and communicating it for broad use.

Improving Geotechnical Uncertainty Evaluation in Reliability-Based Design

58

Fred H. Kulhawy

Abstract

Modern geotechnical engineering often embraces uncertainty directly. For foundations, load-and-resistance-factor-design (LRFD) and multiple-load-and-resistance-factor-design (MRFD) formats are used. For "basic LRFD", the ground is simply treated as another structural component. In the more advanced design formats, the resistance factors attempt to capture some key geologic and geotechnical issues. These factors are a function of the general soil/ground strength or stiffness, as well as the quality of the key design data (strength, modulus, etc.) as obtained from different methods of site investigation, expressed as a coefficient of variation (COV) of the parameter related to the mean or trend line in-situ. These issues are addressed herein, and it is shown that the advanced formats deal with actual ground conditions more realistically. As ground professionals, we need to push code developments to reflect the reality that ground conditions need to be addressed explicitly. Recommendations are made for foundations in both soil and rock.

Keywords

Reliability-based design • Foundations • Uncertainty • Ground variability • Coefficient of variation

58.1 Background

The basic principles of reliability-based design (RBD) were introduced into civil engineering well over 60 years ago, and their origin and basic development, for all practical purposes, were driven by the structural engineering community. As would be expected, the design equations were set up to reflect structural practice and convenience. For simplified RBD, the basic design equation is:

$$\eta \, F_n \le \psi \, Q_n \qquad (58.1)$$

in which F_n = nominal (unfactored) load, Q_n = nominal capacity, η = load factor (≥ 1), and ψ = resistance factor (≤ 1),

resulting in the name "load-and-resistance-factor-design" (LRFD). Much effort was placed on refining the loading, and it is common to see ηF_n combinations for dead and live loads and, in advanced design scenarios, additional terms for other load combinations or mechanisms. By contrast, there was relatively little effort made in resistance evaluation, because the properties of structural materials, such as steel or concrete, do not vary greatly. In fact, the structural designer specifies the desired material and its properties. In this development, the load and resistance factors are specified in increments of 0.05, with no finer "grading". These apparently are sufficient for structural design.

The geotechnical engineering community did not embrace RBD as readily, likely because of the quantification of the capacity and the resistance factor. In fact, when the early structural RBD codes were being developed, Lumb and others (Kulhawy 2010) were just in the early phases of developing the basic statistics of geotechnical properties. To illustrate, the First International Conference on Applications of Statistics and Probability to Soil and Structural

F.H. Kulhawy (✉)
School of Civil and Environmental Engineering, Cornell University, Hollister Hall, Ithaca, NY 14853-3501, USA
e-mail: fhk1@cornell.edu

Engineering was held in Hong Kong in 1971. There were 33 papers in the proceedings (15 structural), and 11 of them focused on soil statistical properties. There were 26 attendees (soil and structural). Clearly, although there were some pioneers working on some aspects of geotechnical RBD 40–50 years ago, they certainly were the exception.

Only during the past 20–30 years has geotechnical RBD been embraced more widely, although there remains a segment of the community that still questions its merits. The geotechnical research has shown that the resistance factor must be examined more carefully, because it can be highly variable. Also studies have shown that it is more efficient and accurate to formulate the basic design equation as:

$$\eta \, F_n \leq \psi_s Q_{sn} + \psi_t Q_{tn} + \psi_w W \qquad (58.2)$$

in which the ψ values are calibrated for each distinctive term in the geotechnical capacity equation (side resistance, tip resistance, weight). This format is known as multiple-load-and-resistance-factor-design (MRFD). And the protocol of 0.05 increments for the resistance factor should be dropped.

58.2 Foundations in Soil

There has been much research done for RBD of foundations in soil. Some has been rather simplified, while others have been sophisticated to varying degrees. Clearly, the status to date is varied. An overview is given by Kulhawy et al. (2012).

In its most simplified form, there is "basic LRFD", which employs Eq. 58.1 and makes numerous simplifying assumptions. For example, consider the case given in Table 58.1 by the AASHTO code, which basically governs the design of transportation structures (bridges) in the U.S. In this code, Eq. 58.1 is used, a recommendation is given for use of a design equation to calculate Q_n (but is not mandatory), there is no guidance on how to use this design equation, and there are no requirements for the specific site investigation and testing to be done to characterize the site and evaluate the soil properties to be used to calculate Q_n. To be fair, there are general guidelines for site investigation and testing in the code. But once the site soil type has been broadly characterized, a table would be entered for a particular foundation type, loading mode, and ground response.

For illustration, Table 58.1 would be used for drilled shafts in clay during undrained uplift loading. The calibration used to assess the resistance factor for this code is simplified, based on simple distributions, judgment, and fitting. As noted in the table, it is a "nominal" value that is to cover a range of target reliability index values.

This approach implicitly assumes that a good quality site investigation and appropriate soil testing is to be done. The reality is that approaches differ between states, designers, and certainly between bridge types. Clearly, a major long-span crossing will require more attention than a small rural crossing.

Geotechnical engineers who understand reliability well are fully aware of the shortcomings of this "basic LRFD" approach and have developed and promoted more rigorous and accurate alternatives. Perhaps the first "complete" study of this type was done by Phoon et al. (1995). This work focused on the need for proper and thorough site characterization that will be sufficient to delineate clearly the site stratigraphy. Once the stratigraphy is defined, the pertinent design properties must be evaluated. This process requires quality site investigation and testing that can establish the property mean or trend line, by layer with depth. The testing should be sufficient to quantify, at least simply, the property variability via its standard deviation (SD). The coefficient of variation (COV) is then the SD/mean.

Detailed studies have shown that the COV varies greatly as a function of field or laboratory test (e.g., Phoon et al. 1995; Phoon and Kulhawy 1999a, b). The precise COV would only be used directly in RBD if detailed numerical simulations were to be employed. For most designs, only a good approximation is necessary to select meaningful resistance factors. Table 58.2 shows the nominal ranges of

Table 58.1 Undrained ultimate uplift resistance factor for drilled shafts designed using AASHTO (2010)

Soil	Ψ_u
Clay	0.35[a]

Note Target reliability index (β_T) = 2.5–3.5 (nominal 3.0)
[a] reduce by 20 % if a single shaft (equivalent to β_T = 3.5)

Table 58.2 Ranges of soil property variability for reliability calibration (Phoon et al. 1995, updated Phoon and Kulhawy 2008)

Geotechnical parameter	Property variability	COV (%)
Undrained shear strength, s_u	Low[a]	10–30
	Medium[b]	30–50
	High[c]	50–70
Effective stress friction angle, ϕ'	Low[a]	5–10
	Medium[b]	10–15
	High[c]	15–20
Horizontal stress coefficient, K_o	Low[a]	30–50
	Medium[b]	50–70
	High[c]	70–90

[a] typical of good quality direct lab or field measurements
[b] typical of indirect correlations with good field data, except for the standard penetration test (SPT)
[c] typical of indirect correlations with SPT field data and with strictly empirical correlations

Table 58.3 Undrained ultimate uplift resistance factors for drilled shafts designed by $F_{50} = \Psi_u Q_{un}$ or $F_{50} = \Psi_{su} Q_{sun} + \Psi_{tu} Q_{tun} + \Psi_w W$ (Phoon et al. 1995)

Clay	COV of s_u (%)	Ψ_u	Ψ_{su}	Ψ_{tu}	Ψ_w
Medium	10–30	0.44	0.44	0.28	0.50
(mean s_u = 25–50 kN/m^2)	30–50	0.43	0.41	0.31	0.52
	50–70	0.42	0.38	0.33	0.53
Stiff	10–30	0.43	0.40	0.35	0.56
(mean s_u = 50–100 kN/m^2)	30–50	0.41	0.36	0.37	0.59
	50–70	0.39	0.32	0.40	0.62
Very stiff	10–30	0.40	0.35	0.42	0.66
(mean s_u = 100–200 kN/m^2)	30–50	0.37	0.31	0.48	0.68
	50–70	0.34	0.26	0.51	0.72

Note Target reliability index = 3.2

property variability based on the results of extensive studies. As shown, the COV groupings are minimal and represent ranges that are easily implemented in practice. Note that the deformation modulus would have similar values as K_o. Akbas and Kulhawy (2010) illustrate well how site-specific data can improve the variability estimates.

Based on the actual or nominal COV for the soils at the site, more realistic resistance factors can be selected that are representative of the actual site conditions. Table 58.3 illustrates resistance factors for the same class of undrained problem as Table 58.1, but taking into account ground conditions more realistically. The first column defines the overall clay stiffness/strength, the second defines the property variability as a function of exploration and testing general quality, and the third is the rigorously calibrated resistance factor for the defined conditions. Note that better quality investigation/testing allows for a higher resistance factor. This approach can be defined as "extended LRFD". Columns 4–6 present the resistance factors for the MRFD approach given by Eq. 58.2, in which each capacity term has its own calibrated resistance factor. MRFD allows for the most rigorous and accurate design matching the calculated and target reliability indices. The calibration for MRFD is a bit more complicated, and different terms dominate nonlinearly in the process, but its application in practice is straightforward. Note that all factors change gradually, without the 0.05 abruptness.

58.3 Foundations in Rock

There has been relatively less research done for RBD of foundations in rock. Some of this work is summarized in Kulhawy and Prakoso (2007). Most codes have adopted the "basic LRFD" approach described previously, with all of its limitations and with correlations to the intact rock uniaxial compression strength (q_u). Prakoso and Kulhawy (2011) describe a better "extended LRFD" approach, in which there is no needed differentiation for general rock stiffness/strength as in soil, but there are still three categories of COV for the strength (q_u), with correlated resistance factors. The same range of COV variability (10–30, 30–50, 50–70) is appropriate for q_u. Low corresponds to good quality drilling, sampling, and testing; medium corresponds to lower grade field efforts or indirect correlations such as the point load test; and high corresponds to minimal efforts, index correlations such as the Schmidt hammer, or field estimates with geologic hammer and such.

Although RBD evaluations for foundations in rock have been correlated with the intact rock q_u, there will be inevitable misuses of this work. While doing a project review this past year, I encountered a foundation RBD based on the rock mass q_u computed from the geological strength index (GSI), with settlements computed from the GSI as well. This RBD is simply wrong.

The GSI has become an important tool for various rock engineering problems, but it must be remembered that it is based upon a careful engineering geologic description of the rock mass. It is a qualitative index and empirical and, as Evert Hoek has stated numerous times in his papers on the GSI, "do not try to be too precise" and "quoting a range is more realistic than citing a single value". Before using the GSI, be sure to read the following papers that provide solid guidance on its proper use (Marinos and Hoek 2000; Marinos et al. 2005).

It should also be noted that, being a qualitative index, the GSI does not have any quantitative statistics. Assigning any means or COVs to the GSI are purely ad-hoc and may be misleading. However, the deformation modulus correlations with GSI are somewhat more quantitative, but they still are strictly empirical. If they are to be used, a high variability should be expected, with COV = 70–90%, based on Table 58.2.

58.4 Concluding Comments

Modern geotechnical engineering has been moving toward embracing uncertainty directly. Design equations for foundations can be expressed in basic LRFD, extended LRFD, or MRFD formats. In basic LRFD, the ground is simply treated as another structural engineering entity that is characterized by a single resistance factor defined solely by broad material type. Hopefully, proper ground investigation and testing issues are addressed in the analysis/design evaluation, but design codes are particularly deficient and lax in this regard.

The other, more advanced, formats used in soil attempt to capture some key geologic and geotechnical issues. The resistance factors are a function of the general soil/ground strength or stiffness, as well as the quality of the key design data (strength, modulus, etc.) as obtained from different methods of site investigation, expressed as a coefficient of variation (COV) of the parameter related to the mean or trend line in-situ. Generic guidelines are suggested to assist in this overall evaluation of ground condition and its variability. MRFD formats provide the most accurate assessments. Thorough local site evaluations can improve on these generic guidelines.

For foundations in rock, the basic LRFD format is comparable to that for soil. The extended LRFD format again attempts to capture some key geologic and geotechnical issues, but this format is still relatively simple. MRFD is not available. Cautions must be exercised in rock property evaluations, especially if the GSI is used. It is qualitative and empirical, and it must be used as it was intended.

To advance our geo-profession and capture geologic and geotechnical issues properly, we need to move away from "basic LRFD" and into at least "extended LRFD" or preferably MRFD.

References

AASHTO (2010) LRFD bridge design specifications, 5th edn. American Association of State Highway and Transportation Officials, Washington, DC

Akbas SO, Kulhawy FH (2010) Characterization of geotechnical Variability in Ankara clay: a case history. Geotech Geol Eng 28 (5):619–631

Kulhawy FH (2010) Uncertainty, reliability and foundation engineering: the 5th Peter Lumb lecture. Trans Hong Kong Inst Eng 17 (3):19–24

Kulhawy FH, Phoon K-K, Wang Y (2012) RBD of foundations—a modern view. Geotechnical state of art and practice (GSP 226). ASCE, Reston, pp 102–121

Kulhawy FH, Prakoso WA (2007) Issues in evaluating capacity of rock sockets. In: Proceedings of the 16th SE Asian geotechnical conference, Kuala Lumpur, pp 51–61

Marinos P, Hoek E (2000) GSI: a geologically friendly tool for rock mass strength estimation. In: Proceedings of the GeoEng2000, Melbourne, pp 1422–1442

Marinos V, Marinos P, Hoek E (2005) The geological strength index: applications and limitations. Bull Eng Geol Environ 64:55–65

Phoon K-K, Kulhawy FH (1999a) Characterization of geotechnical reliability. Can Geotech J 36(4):612–639

Phoon K-K, Kulhawy FH (1999b) Evaluation of property variability. Can Geotech J 36(4):625–639

Phoon K-K, Kulhawy FH (2008) Serviceability limit state RBD. Chapter 9 in RBD in geotechnical Engineering. Taylor & Francis, London, pp 344–384

Phoon K-K, Kulhawy FH, Grigoriu MD (1995) RBD of foundations for transmission line structures. Report TR-105000. Electric Power Research Institute, Palo Alto

Prakoso WA, Kulhawy FH (2011) Some observations on RBD of rock footings, GeoRisk 2011(GSP 224). ASCE, Reston, VA, pp 600–607

Communicating Geological Uncertainty: The Use of the Conceptual Engineering Geological Model

59

Christopher Jack and Steve Parry

Abstract

Rock engineering requires an in-depth understanding of the rock mass. However, geological data is often superficially evaluated and interpreted in isolation, without reference to an overall model. Given geological variability and complexity, the lack of an interpretative framework can result in misleading or incorrect interpretations. The existence of a conceptual engineering geological model is particularly useful at the feasibility stage of a project. This paper illustrates a simple conceptual engineering geological model for a proposed cavern site in Hong Kong. The model allows the identification and evaluation of geological uncertainty and has applications including the optimisation of the site investigation, facilitating early risk assessment and decision making and allowing cost estimates to be made.

Keywords

Conceptual model • Caverns • Hong Kong

59.1 Introduction

This paper presents a conceptual engineering geological model for the rock engineering aspects of a proposed cavern site at Sha Tin, Hong Kong. The purpose of the caverns will be to allow the relocation of a nearby sewage treatment works underground. The model was developed for tender evaluation purposes. Key uncertainties and risks identified by the model are also discussed, along with some initial findings and applications. It has been assumed that the caverns will be situated at >30 m depth and will be of similar dimensions to some existing caverns in Hong Kong (120 m long, 17 m high, 15 m span).

59.2 Conceptual Engineering Geological Model

The conceptual model approach follows the recommendations of IAEG commission C25 (Parry et al. in press) in that it is '*based on understanding the hypothetical relationships between engineering geological units, their likely geometry, and anticipated distribution. This approach, and the models formed, are based on concepts formulated from knowledge and experience and are not necessarily related to real three-dimensional space or time*'. The models in this paper are based on the published geological maps and memoirs and an interpretation of a 1:5,000-scale digital elevation model (DEM). The focus is on the evidence for structural geological control and the models incorporate the authors' experience and knowledge. No site specific ground investigation was available.

Figure 59.1 is a conceptual model of the faults and lineaments in the region of the cavern interpreted from these data sets. The main trends, in decreasing order of magnitude, are NE-SW, NW-SE, ENE-WSW and WNW-ESE. Regional dyke swarms also follow these trends, with NE being the dominant trend. As a result of the tectonic and igneous

C. Jack (✉)
Christopher Jack Rock Engineering Limited, 20/8 Bruntsfield Avenue, Edinburgh, EH10 4EW, UK
e-mail: chris.jack@cjrockeng.com

S. Parry
Parry Engineering Geological Services, 12a Riverside Court, Calver Mill, Calver, Derbyshire, UK

G. Lollino et al. (eds.), *Engineering Geology for Society and Territory – Volume 6*,
DOI: 10.1007/978-3-319-09060-3_59, © Springer International Publishing Switzerland 2015

Fig. 59.1 Proposed cavern site —location and regional structural geological model (image © 2011 Google, © GeoEye, ©2011 DigitalGlobe)

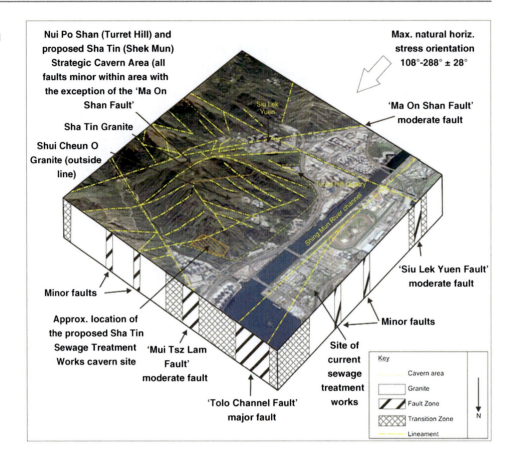

history, it is considered that structures with similar orientations are likely to exist at all scales from microstructures, through shear fractures and joints, to faults. The geology of the cavern site comprises the Shui Chuen O Granite of Cretaceous age. This intrusion was strongly controlled by the dominant NE structural trend and is elongated in this direction. Tectonic activity continued after the granite intrusion, with subsequent faulting and intrusion by dykes, again mainly following the NE orientation. Whilst some of the data shown is observational (i.e. from the published geological map), this has significant uncertainties associated with it, with most faults denoted as 'uncertain'. In addition, the lineaments extracted from the DEM have no ground control.

Once the geological history of the cavern area was established, an evaluation of likely discontinuity types and their potential engineering implications was undertaken and synthesized (Jack et al. 2012). The discontinuities are discussed below and are shown schematically in Fig. 59.2. Note that while the model in Fig. 59.2 shows spatial relationships, these are only indicative. The discontinuity data is extrapolated from the regional conceptual model and combined with published data; the depth of weathering and weathering grades are based on local experience.

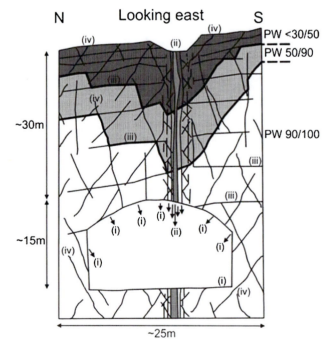

Fig. 59.2 Conceptual engineering geological cross section of the cavern site, notation: (i) potential block fall, (ii) minor fault with potential for cave-in, (iii) sheeting joints, (iv) subvertical discontinuities, PW—partial weathering (GEO 1988)

Faults. No major or moderate faults appear to cross the cavern site, although some possible minor faults do, particularly at the southern end of the site. Possible materials along the minor faults include breccia and fault gouge. Adjacent to the faults, the frequency of shear fractures may increase markedly and the rock may be shattered. The faults may also be partially or fully silicified and/or intruded by dykes. Unless unweathered or re-cemented by high strength secondary mineralisation, the fault material will be "weak". The minor faults may result in increased overbreak, block falls, cave-in and high groundwater inflow in the caverns which can be difficult to control. Assessed feasibility parameters: GSI 10–30 (20). Q, RMR and RMi have not been assessed as it is better to assess faults individually and in detail.

Sheeting joints: Such joints are commonly developed in Hong Kong in coarse grained granites and coarse ash tuff. Orientations will likely vary with slope aspect at the cavern site, possibly towards the NE and E. Sheeting joints are unlikely to be encountered in the caverns as they are rarely developed at depths >30 m. If present, they are likely to be widely spaced and weakly developed (Hencher et al. 2011) and they may form surfaces from which blocks could fall, or release planes for sidewall blocks. For feasibility purposes the sheeting joints are considered to have an effective friction angle of 42° (Hencher and Richards 1982). However, the angle of friction could be much less if some of the sheeting joints are dilated, weathered and/or have significant kaolin or other infill. Joint Roughness Coefficient (JRC, Barton 1973) values might range between JRC 10 and 20.

Steeply dipping discontinuities: It is assumed that the principal discontinuity sets will have broadly the same orientation as the faults. This assumption is reasonable for feasibility purposes, as the main discontinuity sets in Hong Kong granites are typically found to follow the main fault trends. These will typically control the stability in the proposed caverns. In the absence of other information, the discontinuities are assumed to be rough planar with an angle of friction of 40° (Hencher and Richards 1982), although this could be much less where discontinuities are slickensided or weathered. JRC values between 5 and 10 have been assumed.

Rock mass: The cavern site is assumed to be situated in slightly decomposed Shui Chuen O Granite. Based on Palmstrom and Stille (2010), the granite is classed as '*jointed rocks or blocky materials—Class B—rocks intersected by joints and partings—jointed homogenous rocks*' and '*jointed rocks or block materials—Class C—jointed rocks intersected by seams or weak layers—prominent weathering along joints*'. Consequently, it is assessed that the main issues associated with the typical rock mass will be block falls and areas of water inflow. The following ranges of parameters are considered to be appropriate for feasibility purposes (averages in brackets). Strength: 100–200 MPa (150 MPa); block volume: 0.01–15 m^3 (0.5 m^3); GSI: 50–80 (70); Q: 0.7 [RQD 75, J_n 9, J_r 1, J_a 3, J_w 0.66, SRF 2.5] to 38 [RQD 100, J_n 4, J_r 1.5, J_a 1, J_w 1, SRF 1] (11 [RQD 90, J_n 6, J_r 1.5, J_a 2, J_w 1, SRF 1]); RMR: 40–90 (70).

In situ *stress*: The maximum horizontal component of the natural stress field is assumed to be orientated at $108° - 288° \pm 28°$ (Free et al. 2000) as site-specific data was not available for incorporation into the model.

59.3 Key Uncertainties and Risks

The following key uncertainties have been identified based on the model:

- The possible presence, nature and engineering properties of any faults.
- The nature and properties of the shear fractures and joints.
- The groundwater regime.
- The state of in situ stress.

The nature and properties of the rock material, including weathering, are of secondary importance providing that the cavern can be located in good quality rock with one diameter or more of rock above the crown. In this case it is the discontinuities that will control the overall stability of the cavern. The key risks highlighted by the model are associated with large uncertainties because of the current lack of site-specific engineering geological data and include block fall, cave-in and excessive groundwater inflow.

These uncertainties and risks should be managed through the use of an uncertainty register and a risk register. The uncertainty register will recommend actions to be taken to address the uncertainties and the risk register will recommend actions to mitigate the risks. Key uncertainties that cannot be satisfactorily resolved during later stages of the project will become risks and will be added to the risk register. However, such registers are of little use if their contents are not communicated effectively to the project team and this must be one of the engineering geologist's key priorities throughout the project.

Where possible uncertainties should be quantified. This makes it easier to communicate the importance of the uncertainties to other specialists including planners and engineers. It also makes uncertainties more tractable to analysis and incorporation within economic, risk, reliability and probabilistic analyses. Examples of how uncertainties can be quantified include assigning monetary values to uncertainties or probabilities of occurrence. However, some uncertainties cannot be sensibly quantified and judgment will always rule in geological considerations.

59.4 Findings and Applications

The paper demonstrates the usefulness of the conceptual model approach in allowing the visualisation of the engineering geological setting and the derivation of potential uncertainties and risks. Despite its simplicity and the significant uncertainties, it is possible to derive the following from the model.

- It appears that the geology of the site is suitable for cavern development.
- Future site investigations should focus on the discontinuities (at all scales), the groundwater regime and the in situ stress.
- The key risks are block fall, cave-in and excessive groundwater inflow.
- The depth of the caverns should be greater than 30 m, eliminating most or all of the concern for sheeting joints, highly weathered rock and undesirable effects of low in situ vertical stress.
- For feasibility stage considerations of a cavern at shallow or intermediate depth, the longitudinal axis should ideally be orientated along the bisection line of the largest intersection angle of the strikes of the two dominant sets of discontinuities. This should reduce instability and overbreak. With reference to Fig. 59.1, the ideal direction is N–S.
- In addition, at the feasibility stage, it is a common consideration that the major in situ horizontal stress should be parallel to the longitudinal axis of the cavern to reduce instability, which is approximately E–W.
- Therefore on balance the model suggests the best orientation is E–W.

- The assessment of the ground conditions, together with the rock mass classifications, allows for an initial estimation of construction costs.

References

Barton N (1973) Review of a new shear strength criterion for rock joints. Eng Geol 7:287–322

Barton N, Lien R, Lunde J (1974) Engineering classification of rock masses for the design of tunnel support. Rock Mech 6(4):189–236

Free MW, Haley J, Klee G, Rummel F (2000) Determination of in situ stress in jointed rock in Hong Kong using hydraulic fracturing and over-coring methods. In: Proceedings of the conference on engineering geology HK 2000, Institution of Mining and Metallurgy, Hong Kong Branch, pp 31–45

GEO (1988) Geoguide 3, Guide to soil and rock description. Geotechnical Engineering Office, Civil Engineering Department, Hong Kong

Hencher SR, Richards LR (1982) The basic frictional resistance of sheeting joints in Hong Kong granite. Hong Kong Eng 11(2):21–25

Hencher SR, Lee SG, Carter TG, Richards LR (2011) Sheeting joints: characterisation, shear strength and engineering. Rock Mech Rock Eng 44(1):1–22

Jack CD, Parry S, Hart JR (2012) Structural geological input for a potential cavern project in Hong Kong. In: Proceedings of the 32nd annual seminar, geotechnical division, the Hong Kong Institution of Engineers

Marinos P, Hoek E (2000) GSI—a geologically friendly tool for roc mass strength estimation. In: Proceedings of GeoEng 2000 conference, Melbourne, pp 1422–1442

Palmstrom A, Stille H (2010) Rock engineering. Thomas Telford, London

Parry S, Baynes FG, Culshaw MG, Eggers M, Keaton JF, Lentfer K, Novotny J, Paul D (in press) Engineering geological models—an introduction: IAEG Commission 25. Bull Eng Geol Environ

Evaluating the Effects of Input Cost Surface Uncertainty on Deep-Water Petroleum Pipeline Route Optimization

60

William C. Haneberg

Abstract

A resampling-based stochastic simulation approach was used to evaluate the uncertainty that may be associated with geologically constrained least-cost path pipeline route optimization. A smoothed version of a composite geocost surface from a deep-water pipeline routing project was resampled and the results used to generate a series of equally probable cost surface realizations, which were in turn used as the basis for the same number of route optimizations. Eighty percent of the simulated routes followed a 500–2,000 m wide corridor nearly parallel to the baseline route (based upon complete information) between two hypothetical pipeline termini located about 25 km apart. Twenty percent followed an alternate corridor of approximately the same width. These results suggest that, while the general method of geologically constrained pipeline route optimization is a relatively robust one, uncertainties in geological input will at the least create a least-cost route corridor rather than a single least-cost route and may suggest realistic alternatives that must be critically evaluated in light of the available geological information.

Keywords

Pipelines • Route optimization • Cost surface • Marine geohazards • Stochastic simulation

60.1 Introduction

Proper assessment of geologic hazards for deep-water oil and gas developments in which billions of dollars of capital may be at risk requires both a reproducible logical framework and an understanding of the uncertainty and natural variability inherent in geological information. In pipeline route selection, the overriding objective is to find the shortest route that satisfies both primary requirements such as terminus locations and secondary constraints such as areas of geological, biological, or cultural concern (Tootill et al. 2004).

One logical framework that has proven useful on a number of pipeline route evaluations, including deep-water petroleum pipeline routes crossing tens of kilometers of potentially hazardous seafloor, is geologically constrained least-cost path optimization (e.g., Feldman et al. 1995; Haneberg et al. 2013; Price 2010; Yildirim et al. 2007; Luettinger and Clark 2005). In least-cost path optimization, a composite cost surface is developed from a variety of qualitative and quantitative geologic and bathymetric information. Each hazard or attribute is mapped separately, weighted, normalized, and then added to the others to create the composite cost map. Then, steepest path algorithms are used to determine the least expensive route in terms of the spatially variable cost integrated over the length of the route. Implicit in the analysis is the assumption that the composite cost used in the optimization is proportional to the actual cost of pipeline route characterization, design, construction, maintenance, operation, and, if appropriate, decommissioning.

W.C. Haneberg (✉)
Fugro GeoConsulting Inc., 6100 Hillcroft, Houston, TX 77081, USA
e-mail: whaneberg@fugro.com

G. Lollino et al. (eds.), *Engineering Geology for Society and Territory – Volume 6*,
DOI: 10.1007/978-3-319-09060-3_60, © Springer International Publishing Switzerland 2015

(a)   (b)   (c)

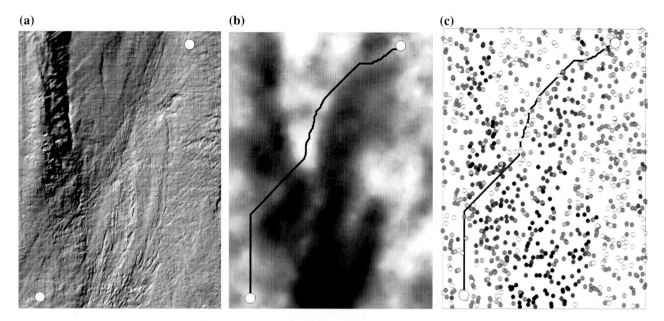

Fig. 60.1 **a** Shaded relief seafloor rendering of the study area showing the two hypothetical pipeline termini used in this paper. Illumination is from the northwest. **b** Smoothed representation of the cost surface used in the real pipeline route selection along with the baseline least-cost optimized route. Dark areas are high cost and light areas are low cost. **c** Locations and relative costs of 1,000 randomly sampled points based on the kriged cost surface. Images cover an area of approximately 18 km by 24 km

One of the complications of least cost path optimization is that considerable professional discretion is required to select and evaluate the relative importance of the variables used to construct the composite cost map. Moreover, the variables may differ from project to project and region to region, making a completely standardized or codified approach unrealistic.

Some kinds of information used to guide pipeline route selection, for example the mapped extents of past slope failures, can be uncertain because they are inherently subjective. That is to say, different geologists are almost certain to produce slightly different maps from the same data (Haneberg and Keaton 2012; Keaton and Haneberg 2013). Other kinds of information, for example the results of regional process-based slope stability simulations, are uncertain because they incorporate model uncertainty, parameter uncertainty, and spatial variability of geotechnical properties. The latter can be partially accommodated using probabilistic formulations that treat input values as random variables with distributions that may differ among soil units or engineering geologic facies (Haneberg et al. 2009, 2013; Haneberg 2012). Similarly, some kinds of information, for example the location of a past slope failure, are categorical (either it exists at a point or it does not, although the exact extent may be subjective, as discussed above, and if only two states are possible then the variable becomes binary) whereas others, for example, slope angle or seafloor radius of curvature, are continuous and can take on a range of values.

As a consequence of these complexities, the composite geocost surface will itself be continuously variable and uncertain to a degree that is difficult to analyze because each of its components will have its own kind and degree of uncertainty. One practical approach to begin understanding the effect of composite cost surface uncertainties, which is described in this paper, is to sample a sufficient number of random points from a cost surface, which is known in statistics as bootstrapping (Efron 1979), and then use those points to seed conditional simulations of equally probable alternative cost surfaces (known as realizations). Haneberg (2006) used a similar approach to generate realizations of digital elevation model error fields and evaluate their effects on subsequent slope stability calculations.

60.2 Method

In this work, the effects of cost surface uncertainty were evaluated using a resampling-based (bootstrap) geostatistical conditional simulation approach. The cost surface map for an actual deep-water pipeline routing project was first sampled at 1,000 randomly located points. Composite cost values from the sampled points were then kriged and used to generate 10 equally probable realizations of the cost surface using conditional Gaussian simulation with normal score transformation as implemented in the Geostatistical Analyst

Fig. 60.2 Four of the ten equally probable cost surface realizations and associated least-cost optimized pipeline routes. Each realization is generally similar but different in detail from the others, simulating the effects of input uncertainty on the routing algorithm

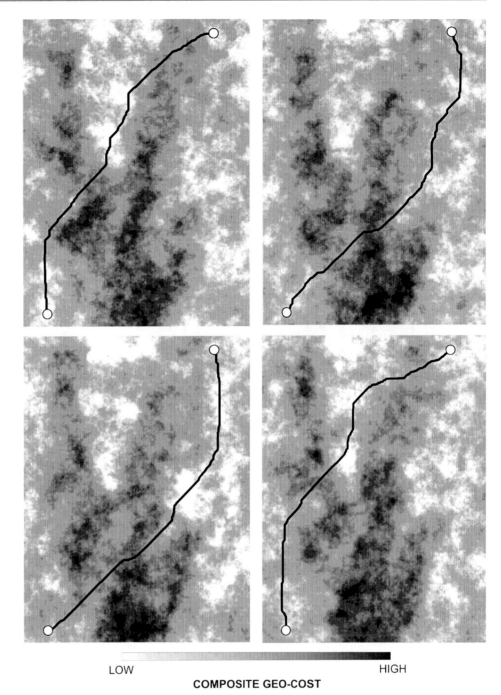

LOW   HIGH

COMPOSITE GEO-COST

extension to ArcMap 10.1. The term "conditional" means that values were fixed at the 1,000 randomly chosen points, in essence to anchor the results to the real map upon which they are based, and simulated at all other points. The process is easily automated, and in a real project one would typically generate many more equally probable realizations, perhaps 100 or more, for consideration. The much smaller number of 10 used in this paper was chosen to simplify the illustrations while still demonstrating the principles of the method.

Two hypothetical pipeline termini about 25 km apart and separated by a variety of seafloor conditions were selected. Then, a least-cost route between the two termini was calculated for each of the 10 cost surface realizations. Neither the actual pipeline termini nor the specific geohazard layers used to create the actual cost surface, which are confidential, are described here. Only a smoothed version of the actual cost surface is shown and then in terms of relative, not absolute, values.

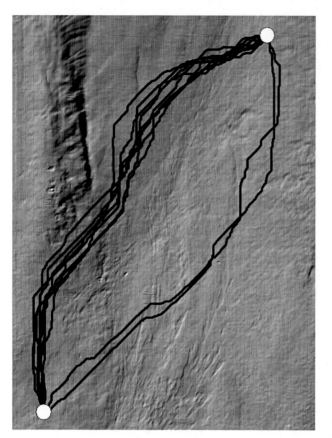

considered the baseline case. Note how the route follows a path of relatively light shaded (i.e., low cost) areas between the two termini.

Figure 60.1c shows the 1,000 randomly sampled points, as above shaded according to their relative cost (dark for high and light for low) along with the baseline route for reference.

A total of 10 cost surface realizations were generated for this study and 4 of those are shown in Fig. 60.2. Note that while the general distribution of high and low costs is similar in each of the 4 realizations, details such as the continuity of low-cost corridors differ slightly from realization to realization. As in Fig. 60.1, dark shades represent high costs and light shades represent low costs. Each of the 4 cost surface images in Fig. 60.2 also shows the least-cost path route calculated for that surface.

Figure 60.3 shows all 10 route realizations superimposed upon a seafloor rendering to illustrate the range of variability. Out of the 10 realizations, 8 follow a fairly well defined corridor ranging in width from about 500–2,000 m. Two of the simulated routes, as were shown in Fig. 60.2, depart from that corridor and follow a second corridor of similar width. Comparison of the 4 cost surfaces in Fig. 60.2 shows that in those two realizations the simulated routes departed significantly from the baseline case because of the existence of a relatively low cost alternative pathway.

Fig. 60.3 All 10 simulated routes superimposed on a shaded seafloor rendering. Actual pipeline routing projects use many realizations, perhaps 100 or more, to generate a stochastic cloud of equally probable pipeline routes and highlight areas most sensitive to input uncertainty

60.3   Results

The example area for this study covers an area of approximately 18 km by 24 km as illustrated in Fig. 60.1a, which is a shaded relief rendering generated from the seafloor return in a 3-D seismic data volume. It was created by digitally picking the seafloor return using seismic interpretation software and then converting it from two-way travel time to depth using a standard polynomial equation (Advocate and Hood 1993). Water depths range from about 200 m in the south to more than 2,000 m in north. The white circles in the SW and NE corners of the image are the two hypothetical pipeline termini used in this analysis.

Figure 60.1b shows a smoothed representation of the actual composite cost surface used for a pipeline routing project in this area (the actual project used different pipeline termini than those shown in Fig. 60.1a). Dark shades represent areas of high cost whereas light shades represent areas of low cost. No scale is shown because the costs are relative and normalized to each other; hence, their absolute magnitudes are inconsequential. The black line in Fig. 60.1b is the least-cost route obtained using the kriged cost surface and

References

Advocate DM, Hood KC (1993) An empirical time-depth model for calculating water depth, northwest Gulf of Mexico. Geo-Mar Lett 13:207–211

Efron G (1979) Bootstrap methods: another look at the jackknife. Ann Stat 7:1–26

Feldman SC, Pelletier RE, Walser E, Smoot JC, Ahl D (1995) A prototype for pipeline routing using remotely sensed data and geographic information system analysis. Remote Sens Environ 53:123–131

Haneberg WC (2012) Spatially distributed probabilistic assessment of submarine slope stability. In: Offshore site investigation and geotechnics: proceedings of 7th international conference, London, 12–14 Sept 2012, pp 551–556

Haneberg WC (2006) Effects of digital elevation model errors on spatially distributed seismic slope stability calculations: an example from Seattle, Washington. Environ Eng Geosci 12:247–260

Haneberg WC, Bruce B, Drazba MC (2013) Using qualitative slope hazard maps and probabilistic slope stability models to constrain least-cost pipeline route optimization. In: Proceedings of offshore technology conference, 5–9 May 2013, Houston, Texas, Paper OTC 23980

Haneberg WC, Keaton JR (2012) Ground truth: an obstacle to landslide hazard assessment? Geol Soc Am Abst Programs 44(7):345

Haneberg WC, Cole WF, Kasali G (2009) High-resolution LiDAR-based landslide hazard mapping and modeling. Bul Eng Geol Environ 68:273–286

Keaton JR, Haneberg WC (2013) Landslide inventories and uncertainty associated with ground truth. In: Wu F, Qi S (eds) Global view of engineering geology and the environment. Taylor & Francis, London, pp 105–110

Luettinger J, Clark T (2005) Geographic information system-based pipeline route selection process. J Water Resour Plan Manage 131:193–200

Price G (2010) Geographic information systems pipeline route optimization (GISPRO). In: Proceedings of 10th ESRI petroleum user group conference, Houston, 22–24 Feb 2010, paper 1237

Tootill NP, Vandenbossche MP, Morrison ML (2004) Advances in deepwater pipeline route selection—a Gulf of Mexico case study. In: 2004 offshore technology conference, Paper OTC 16633

Yildrim V, Aydinoglu AC, Yomralioglu T (2007) GIS based pipeline route selection by ArcGIS in Turkey. In: Proceedings of ESRI international user conference, 18–22 June 2007. http://proceedings.esri.com/library/userconf/proc07/papers/papers/pap_2015.pdf (accessed 6 Jan 2013)

Scanline Sampling Techniques for Rock Engineering Surveys: Insights from Intrinsic Geologic Variability and Uncertainty

61

Helder I. Chaminé, Maria José Afonso, Luís Ramos, and Rogério Pinheiro

Abstract

Discontinuity surveys are based on collecting rock data from fieldwork and are an essential component of rock-mass quality estimation in rock engineering. Strength, deformability and permeability characteristics of a rock-mass are strongly influenced by its discontinuities. Scanline surveys are a reliably technique in which a line is drawn over an outcropped rock surface and all the discontinuities intersecting it are measured and described. The discontinuity geometry for a rock-mass is characterised by the number of discontinuity sets, mean density and the distributions for location, orientation, size and spacing/fracture intercept. Rock site investigation deals with several key elements that need to be addressed, namely the information required to characterise the rock system and the intrinsic uncertainty associated with this information. This way, quantifying the information content of the on-site measurements and creation a database is vital to be used for decision making processes and risk assessment on rock engineering design projects. In addition, a clear geology framework plays a key-role to support the investigation of all rock engineering projects. Nevertheless, the intrinsic variability of geological, petrophysical and geotechnical properties must be quantified for reliability-based design and to decrease the geological uncertainty. All geologists and engineers' practitioners must have the aim to contribute to the correct study of the ground behaviour of soil and rock, their applications in sustainable design with nature and environment and to satisfy the society's needs.

Keywords

Scanline techniques • Rock-mass • Engineering geosciences • Rock engineering • Uncertainty

H.I. Chaminé (✉) · M.J. Afonso · L. Ramos · R. Pinheiro
Laboratory of Cartography and Applied Geology (Labcarga),
Department of Geotechnical Engineering, School of Engineering
(ISEP), Polytechnic of Porto, Porto, Portugal
e-mail: hic@isep.ipp.pt

M.J. Afonso
e-mail: mja@isep.ipp.pt

L. Ramos
e-mail: lcr@isep.ipp.pt

R. Pinheiro
e-mail: rfsp@isep.ipp.pt

H.I. Chaminé · M.J. Afonso
Centre GeoBioTec, University of Aveiro, Aveiro, Portugal

61.1 Introduction

Barton (2012) argues the *"Discontinuous behaviour provides rich experiences for those who value reality, even when reality has to be simplified by some empiricism"*. This impressive quotation describes the general framework of the complexity of the heterogeneous rock-mass behaviour. The lessons learned on several geoengineering projects stress the importance of the accuracy of the basic geological and geotechnical data information related to the rock masses characterization and assessment.

Linear or circular sampling or sampling within windows along a scanline are accurate approaches to the systematic record of discontinuities (joints, fractures, faults, veins, etc.).

In several geologic and geotechnical frameworks this is, moreover, the easiest and fastest way to collect discontinuities data (e.g., Priest and Hudson 1981; Hudson and Priest 1983; Priest 1993; Mauldon et al. 2001; Rohrbaugh et al. 2002; Priest 2004; Peacock 2006; Chaminé et al. 2010, 2013; Pinheiro et al. 2014). Scanline surveys will provide an amount of reliable information concerning structural geology, petrophysical and geotechnical features of rock masses, either in boreholes or exposed rock surfaces (Fig. 61.1). However, some procedures must be fulfilled to avoid systematic or random errors (Terzaghi 1965; ISRM 1981; CFCFF 1996; Hudson and Cosgrove 1997). Collecting data for the basic geotechnical description of rock masses is of considerable importance for the prediction of scale effects in rock mechanical behaviour (Cunha and Muralha 1990).

The characteristics of discontinuities can be estimated using scanline sampling techniques (Fig. 61.2), but the accuracy is subject to bias (e.g., Priest and Hudson 1981; Priest 1993; Park and West 2002; Rohrbaugh et al. 2002). According to Mauldon et al. (2001) the circular sampling tools and estimators (such as fracture trace intensity, trace density and mean trace length) eliminate most sampling biases, due to orientation and also correct many errors owing to censoring and length bias. Conversely, Wu et al. (2011) argue the predictions based on the rectangular window methods were found to be more accurate than that based on the circular window methods.

In this work, we highlight the importance of an integrative approach for geoengineering purposes of field surveys performed with scanline techniques on free rock-mass faces in diverse contexts, such as quarrying, underground excavations and hard-rock hydrogeotechnical studies. All studies should be developed in a GIS platform by using the following tools: field mapping, morphotectonic analysis, structural geology, rock geotechnics and hydrogeomechanics. This approach led us to a better understanding of the relevance of rock masses heterogeneity for geoengineering purposes at different scales and to reduce the intrinsic variability and uncertainty in collecting geologic and geotechnical data.

61.2 Rock Scanline Surveys: A Reliable Tool to Unbiased Sampling

Discontinuity features play a major role in controlling the mechanical behaviour of a rock-mass (Priest and Hudson 1981). Discontinuities are generally characterised in terms of the following properties (e.g., ISRM 1981; Priest 1993, 2004): orientation, frequency or spacing, size and shape, aperture, conductivity, surface geometry, strength and stiffness. Describing only the discontinuities which seem to be important can be considered as a subjective method of fracturing surveying. From a statistical perspective, it is important to set up a rigorous unbiased sampling regime at the rock face such as (Priest 2004): sampling all traces of discontinuities within a defined area (window sampling), all that intersect a circle (circle sampling) or all that intersect a straight line (scanline sampling). ISRM (1981) stated that a scanline survey is an objective method for recording and describing rock fracturing on a rock-mass exposure (Fig. 61.3).

Fig. 61.1 The main scientific and technical fields of applications of scanline sampling technique surveys related to engineering geosciences, rock engineering and geotechnical engineering

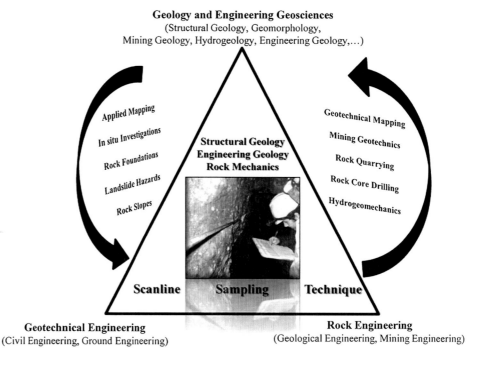

Geology and Engineering Geosciences
(Structural Geology, Geomorphology, Mining Geology, Hydrogeology, Engineering Geology,...)

Applied Mapping
In situ Investigations
Rock Foundations
Landslide Hazards
Rock Slopes

Structural Geology Engineering Geology Rock Mechanics

Geotechnical Mapping
Mining Geotechnics
Rock Quarrying
Rock Core Drilling
Hydrogeomechanics

Scanline   Sampling   Technique

Geotechnical Engineering
(Civil Engineering, Ground Engineering)

Rock Engineering
(Geological Engineering, Mining Engineering)

Fig. 61.2 Rock scanline surveys framework to rock design (slope, tunnel, quarry and cavern): a reliable tool to reduce the intrinsic geologic variability and uncertainty

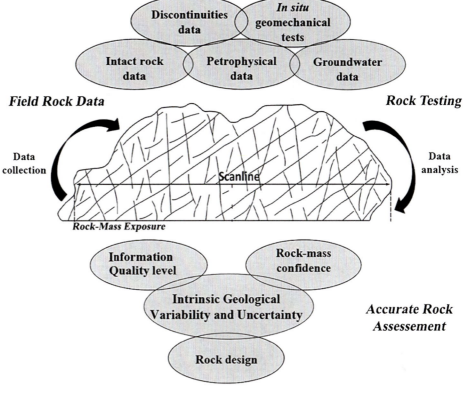

Fig. 61.3 The description/ classification/behaviour *versus* assessment/design/modelling of heterogeneous and fractured rock masses

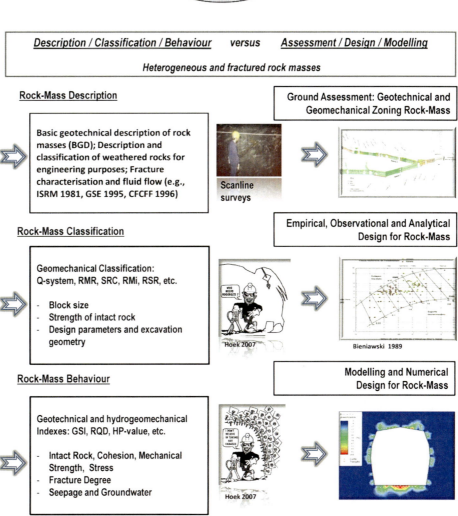

61.3 Concluding Remarks

A clear geology and structural geology framework plays a key-role to support the investigation of all rock engineering projects (Hudson and Cosgrove 1997; Hoek 2007; De Freitas 2009; Chaminé et al. 2013; Shipley et al. 2013). The heterogeneity of the geological properties of rock masses is very significant in geoengineering issues (Hudson and Cosgrove 1997). Particularly, the assessment of in situ block size plays a key-role in rock engineering design projects, such as mining, quarrying and highway cutting operations (e.g., Lu and Latham 1999; Haneberg 2009; Chaminé et al. 2013). In addition, the evaluation based on engineering geosciences, geohydraulic and geotechnical features of rock masses involve combining parameters to derive quantitative geomechanical classifications for geoengineering design (e.g., Bieniawski 1989; Gates 1997; Smith 2004; Barton 2006, 2012; Hoek et al. 2013). In short, good rock engineering must be based in good engineering geosciences, and the big issue raised by Pells (2008), *"what happened to the mechanics in rock mechanics and the geology in engineering geology"*, is still valid. However, the intrinsic variability of geological, petrophysical and geotechnical properties must be quantified for reliability-based design and to decrease the uncertainty (e.g., Mazzoccola et al. 1997; Hoek 1999; Keaton 2013). In addition, Mazzoccola et al. (1997) stated an important issue: *"Is there enough information available for design?"*.

Acknowledgements This work is under the framework of the Lab-carga-IPP-ISEP|PADInv'2007/08 and Centre GeoBioTec|UA (PEst-C/CTE/UI4035/2014). RP was supported partially by LGMC|ISEP Centre. We acknowledge the anonymous reviewers for the constructive comments that helped to improve the clarity of the manuscript.

References

Barton N (2006) Rock quality, seismic velocity, attenuation and anisotropy. Taylor & Francis, London, 729 p

Barton N (2012) From empiricism, through theory, to problem solving in rock engineering: a shortened version of the 6th Müller Lecture. ISRM News J 14:60–66

Bieniawski ZT (1989) Engineering rock mass classifications: a complete manual for engineers and geologists in mining, civil, and petroleum engineering. Interscience, Wiley, New York, 272 p

CFCFF [Committee on Fracture Characterization and Fluid Flow] (1996). Rock fractures and fluid flow: contemporary understanding and applications. National Research Council, National Academy Press, Washington DC, 568 p

Chaminé HI, Afonso MJ, Silva RS, Moreira PF, Teixeira J, Trigo JF, Monteiro R, Fernandes P, Pizarro S (2010) Geotechnical factors affecting rock slope stability in Gaia riverside (NW Portugal). In: Williams A et al (eds) Proceedings 11th Congress, IAEG. CRC Press, Auckland, pp 2729–2736

Chaminé HI, Afonso MJ, Teixeira J, Ramos L, Fonseca L, Pinheiro R, Galiza AC (2013) Using engineering geosciences mapping and GIS-based tools on georesources management: lessons learned from rock quarrying. Eur Geol Mag J Eur Fed Geol 36:27–33

Cunha AP, Muralha J (1990) Scale effects in the mechanical behaviour of joints and rock masses. Memória 763. LNEC, Lisbon 44 p

De Freitas MH (2009) Geology: its principles, practice and potential for geotechnics. Q J Eng Geol Hydrogeol 42:397–441

Gates W (1997) The hydro-potential (HP) value: a rock classification technique for evaluation of the groundwater potential in bedrock. Environ Eng Geosci 3(2):251–267

Haneberg WC (2009) Improved optimization and visualization of drilling directions for rock mass discontinuity characterization. Environ Eng Geosci 15(2):107–113

Hoek E (1999) Putting numbers to geology: an engineer's viewpoint. Q J Eng Geol Hydrogeol 32(1):1–19

Hoek E (2007) Practical rock engineering. Hoek's Corner, RocScience, p 342

Hoek E, Carter TG, Diederichs MS (2013) Quantification of the geological strength index chart. In: Proceedings geomechanics symposium 47th US rock mechanics, San Francisco, CA, ARMA 13-672, pp 1–8

Hudson JA, Cosgrove JW (1997) Integrated structural geology and engineering rock mechanics approach to site characterization. Int J Rock Mech Min Sci Geomech Abs 34(3/4):136.1–136.15

Hudson JA, Priest SD (1983) Discontinuity frequency in rock masses. Int J Rock Mech Min Sci Geomech Abs 20:73–89

ISRM [International Society for Rock Mechanics] (1981) Basic geotechnical description of rock masses. Int J Rock Mech Min Sci Geomech Abs 18:85–110

Keaton J (2013) Engineering geology: fundamental input or random variable? In: Withiam JL, Phoon K-K Hussein M (eds) Foundation Engineering in the face of uncertainty: honoring Fred H. Kulhawy. ASCE. GSP 229. pp 232–253

Lu P, Latham JP (1999) Developments in the assessment of in situ block size distributions of rock masses. Rock Mech Rock Eng 32 (1):29–49

Mauldon M, Dunne WM, Rohrbaugh MB Jr (2001) Circular scanlines and circular windows: new tools for characterizing the geometry of fracture traces. J Struct Geol 23(2–3):247–258

Mazzoccola DF, Millar DL, Hudson JA (1997) Information, uncertainty and decision making in site investigation for rock engineering. Geotech Geol Eng 15:145–180

Park HJ, West TR (2002) Sampling bias of discontinuity orientation caused by linear sampling technique. Eng Geol 66(1–2):99–110

Peacock DC (2006) Predicting variability in joint frequencies from boreholes. J Struct Geol 28(2):353–361

Pells PJN (2008) What happened to the mechanics in rock mechanics and the geology in engineering geology. J South Afr Inst Min Metall 108:309–323

Pinheiro R, Ramos L, Teixeira J, Afonso MJ, Chaminé HI (2014) MGC–RocDesign|CALC: a geomechanical calculator tool for rock design. In: Alejano LR, Perucho A, Olalla C, Jiménez R. (eds) Proceedings of Eurock2014, Rock Engineering and Rock Mechanics: Structures in and on Rock Masses (ISRM European Regional Symposium, Vigo, Spain). CRC Press/Balkema, Taylor & Francis Group, London. pp. 655–660. (on pen-drive insert, ISRM Paper CH100)

Priest SD (1993) Discontinuity analysis for rock engineering. Chapman and Hall, London 473 p

Priest SD (2004) Determination of discontinuity size distributions from scanline data. Rock Mech Rock Eng 37(5):347–368

Priest SD, Hudson JA (1981) Estimation of discontinuity spacing and trace length using scanline surveys. Int J Rock Mech Min Sci Geomech Abs 18(3):183–197

Rohrbaugh MB Jr, Dunne WM, Mauldon M (2002) Estimating fracture trace intensity, density, and mean length using circular scan lines and windows. AAPG Bull 86(12):2089–2104

Shipley TF, Tikoff B, Ormand C, Manduca C (2013) Structural geology practice and learning, from the perspective of cognitive science. J Struct Geol 54:72–84

Smith JV (2004) Determining the size and shape of blocks from linear sampling for geotechnical rock mass classification and assessment. J Struct Geol 26(6–7):1317–1339

Terzaghi RD (1965) Sources of errors in joint surveys. Géotechnique 15:287–304

Wu Q, Kulatilake PHSW, Tang H-M (2011) Comparison of rock discontinuity mean trace length and density estimation methods using discontinuity data from an outcrop in Wenchuan area, China. Comput Geotech 38(2):258–268

A Suggested Geologic Model Complexity Rating System

62

Jeffrey R. Keaton

Abstract

A geologic model captures selected geologic qualities in a region and at a site. An engineering geologic model is relevant to a project and includes specific project aspects. Geology expressed quantitatively, describing variability and uncertainty, is needed in projects employing reliability-based design. Geologists use line style to convey confidence in interpretation of location and nature of lithologic contacts and faults. Mapped lines have error related to scale, terrain, and methods, and uncertainty related to interpretation and allotted field time. A geologic model complexity rating system suggested herein has nine components, four of which are related to regional-scale geologic complexity and five of which pertain to site-scale complexity, terrain characteristics, information quality, geologist competency, and time allotted to prepare the model. Rating criteria and scores are organized into four levels. Simple, uniform, predictable conditions are assigned scores of 3, whereas complex, unpredictable conditions have scores of 81; intermediate conditions have scores of 9 or 27. Possible cumulative scores range from 27 to 729. Cumulative scores can be converted into a form suitable for reliability-based design by defining the highest possible score as the mean value for the geologic model and the actual score as the model standard deviation. Conditions with a maximum score would have coefficient of variations COV = 1.0. The lowest possible COV would be 0.037 (=27/729). Geologic model COVs could be added to geotechnical COVs. This paper seeks to encourage geologists to translate their interpretive understanding into geotechnical parameters needed by engineers for use in Ground Models.

Keywords

Geologic variability • Geologic uncertainty • Reliability-based design • Coefficient of variation

62.1 Introduction

A geologic map or section is a type of geologic model. It is an artistic representation of one interpretation of geologic features and relationships inferred from limited observations of the distribution of rock types, surficial deposits, and geologic structures often with little or no subsurface data or laboratory test results (Keaton 2013). An engineering geologic model is a geologic model relevant to a proposed project that includes pertinent engineering aspects. Geology must be expressed in quantitative terms that describe complexity and uncertainty explicitly to have value in engineering projects, particularly those that employ reliability-based design (RBD). Geologists convey confidence in interpretation with solid, dashed, dotted, and queried lines on geologic maps and sections. The mapped lines have location

J.R. Keaton (✉)
AMEC Americas, Inc, 6001 Rickenbacker Road, Los Angeles 90040, USA
e-mail: jeff.keaton@amec.com

error related to map scale, terrain, vegetation, and field methods, and uncertainty related to the competency level of the geologist and the time allotted for field mapping (FGDC 2006). The FGDC does not specify target values for zones of confidence for digital geologic maps because they are made to satisfy widely varied purposes and needs.

Geologic complexity refers to the qualities and details that are the focus of geologists seeking to understand the history of geologic processes that have occurred to produce the formations as they appear in the field today (Morgenstern and Cruden 1977). Geotechnical complexity relates to variability in strength, stiffness, and hydraulic conductivity of soil and rock masses as these properties might affect the performance of engineered works. Morgenstern and Cruden (1977) state "The most important contribution to increased reliability of site characterization of complex conditions comes from an extra effort associated with geological mapping with the interpretation of the nature of the geotechnical complexity."

Hoek (1999), in the introduction to the second Glossop Lecture, recognizes that geologists tend to be uncomfortable in putting numbers to qualitative observations. He discusses aspects of rock mass rating systems and then comments on the rockfall hazard rating system by Pierson and van Vickle (1993). Hoek (1999) comments that this rating system is based on a set of simple observations that can be made from a slow-moving vehicle and that it contains all the components required for an engineering evaluation of public risks. Hoek (1999) also comments that no direct instructions are given as to how the rating score should be used.

Reliability-based design in geotechnical engineering is challenged by a need for improved calibration of factors used in the process and specifically for layered soil profiles (Kulhawy et al. 2012). Quantitative tools for subsurface investigation, particularly the cone penetration test (CPT), have been used in geotechnical engineering for many years as a means for developing site soil profiles. The quantitative nature of CPT results and the need for geotechnical parameters make geologic methods of interpreting site stratigraphy undesirable, particularly if the geologist is unable to express uncertainty and variability in useable terms.

Burland's Soil Mechanics Triangle (Burland 1987; Keaton 2013) has Geologic Model, Ground Model, and Geotechnical Model in the corners (Fig. 62.1). The Geologic Model represents the site geologic conditions relevant to the proposed project, whereas the Ground Model is the Geologic Model expressed in terms of engineering parameters. The Geotechnical Model is predictive based on project loads and performance requirements. It is the goal of this paper to aid in taking one step towards translating the geologists' interpretive understanding into geotechnical parameters needed by engineers for the Ground Model.

62.2 Geologic Model Complexity Rating System

A geologic model complexity rating system proposed herein has nine components (Fig. 62.2), four of which are related to the regional-scale complexity of the geology (genetic

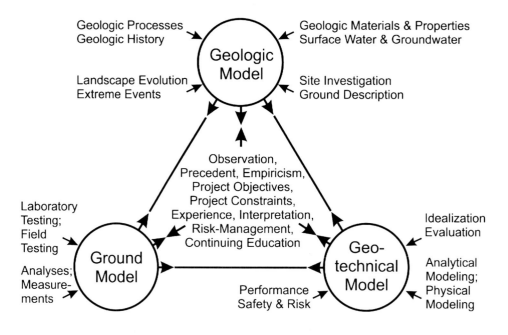

Fig. 62.1 Burland's (1987) soil mechanics triangle as modified by Keaton (2013)

Category / Component		Rating Criteria and Score			
	Points	3	9	27	81
Regional-scale geologic complexity	Genetic - deposition or emplacement	Simple, uniform conditions	Generally simple, predictable conditions	Somewhat complex, generally predictable conditions	Highly complex and variable conditions
	Epigenetic - structural or deformational	No faulting or folding observed or expected	One episode of limited faulting and folding expected	Two episodes of limited faulting and folding expected	Multiple episodes of major faulting and folding expected
	Epigenetic - alteration or dissolution	Unlikely because of geologic setting	Possible because of geologic setting	Likely because of geologic setting	Known to exist
	Epigenetic - weathering and erosion	Uniform weathering profile; minor erosion	Generally regular weathering profile; some erosion	Irregular weathering profile; moderate erosion	Highly irregular weathering; extensive erosion, buried valleys
Site-scale geologic complexity		Vertically and laterally uniform over project site	Generally regular over project site	Irregular over project site	Highly irregular over project site
Terrain features		Some relief; many good exposures	Some relief; some good exposures	Strong relief; poor exposures	Heavy vegetation; few or very poor exposures
Information quality		Extensive data from multiple sources	Limited data from few sources	Reconnaissance level information only	Existing information only; desktop study
Geologist competency level		Professional Geologist with local field experience	Professional Geologist with field experience in non-similar geology	Geology degree or training with some field experience	No geology training or field experience
Alotted time or level of effort		Ample time; well-developed interpretation	Adequate time; thoughtful interpretation	Brief time; thoughtful interpretation	Brief time; rushed interpretation

Fig. 62.2 Geologic model complexity rating system, rating criteria, and scores

deposition or emplacement, epigenetic deformation, epigenetic alteration, and epigenetic weathering and erosion). The other five components pertain to site-scale complexity of the geology, terrain characteristics, information quality, geologist competency, and time allotted to prepare the model. These components were mentioned in Morgenstern and Cruden (1977; geologic and geotechnical complexity), FGDC (2006; geologic complexity, terrain, information quality, geologist competency, allotted field time), FHWA (2011; information quality levels), NRCS (2002; outcrop confidence), and Pierson and van Vickle (1993; rating levels and scoring system).

Rating criteria and scores are organized into four levels (Fig. 62.2) similar to rockfall hazard rating systems (Pierson and van Vickle 1993). Simple and uniform conditions can be predicted with confidence and are assigned scores of 3, whereas complex and nonuniform conditions cannot be predicted and are assigned scores of 81; intermediate conditions have scores of 9 or 27. Actual scores will have values associated with multiple conditions for different components.

Thus, possible cumulative scores for the nine categories range from 27 to 729. A possible way to convert cumulative scores into a form suitable for reliability-based design would be to define the coefficient of variation (COV, ratio of standard deviation to the mean value) of the highest possible score as 1.0 and use 729 as the designated mean value for the geologic model. The actual score for the conditions of the site being evaluated would be accepted as the standard deviation since COV is an expression of variability. The COV of the lowest possible score would be 0.037 (=27/729). The COV of the interpreted geologic model should not be used alone in geotechnical analyses; instead, it should be applied to the geotechnical parameters that are used to convert the geologic model into a ground model. For example, if the CPT data and laboratory test results produced strength parameters that the geotechnical engineer believed have $COV_{geotech} = 0.2$ and the geology were simple, predictable, and well documented (i.e., $COV_{geology} = 0.037$), then the combined $COV_{ground} = 0.2 + 0.037 = 0.237$. The engineer may choose to round the COV_{ground} to 0.2, but at least the geology would be considered explicitly and have its own score.

Consider a site condition with tectonic activity, deformation, alteration, complicated surficial deposits (landslides), heavy vegetation, poor topographic maps, and an entry-level geologist who is given less than a week to develop an interpretation. The score might be $COV_{geology} = 1.0$. If the engineer's CPT data and laboratory test results were taken to have $COV_{geotech} = 0.25$, then the combined $COV_{ground} = 1.25$, which would attract the attention of the design team members in their application of reliability-based design methods. In reality, the COV can greatly exceed 1.0; therefore, quantifying geologic complexity is an important topic that deserves further consideration and discussion.

62.3 Conclusions

The geologic model complexity rating system proposed in this paper seems to cover a spectrum of components that contribute to complexity and variability in geologic conditions. It appears to be simple and scoring for a project should be straightforward. The project management aspects could be brought into focus as a source of risk (Baynes 2010) if inadequate time had been allocated for the site geologic characterization task. At a minimum, application of the geologic model complexity rating system would provide a reason for meaningful dialog between engineers and geologists and for geologists to begin to work harder at "putting numbers to geology" (Hoek 1999). Hopefully, this process will help geologists better understand what is needed from a geotechnical perspective for engineers to deliver reliability-based design projects.

This paper is part of the broad range of considerations under the "actuarial" initiative being given by Commission No. 1 (Engineering Geological Characterisation and Visualisation) of the International Association for Engineering Geology and the Environment (IAEG).

References

Baynes FJ (2010) Sources of geotechnical risk. Q J Eng GeolHydrogeol 43:321–331

Burland JB (1987) Nash lecture: the teaching of soil mechanics—a personal view. In: Groundwater effects in geotechnical engineering, vol 3. Proceedings of 9th European conference on soil mechanics and foundation engineering, Balkema, Rotterdam/Boston, pp 1427–1441

FGDC (2006) FGDC digital cartographic standard for geologic map symbolization. US Geological Survey Geologic Data Subcommittee, Federal Geographic Data Committee Document Number FGDC-STD-013-2006, 33 (plus 250 pages of appendices)

FHWA (2011) Subsurface utility engineering. Federal Highway Administration, US Department of Transportation, http://www.fhwa.dot.gov/programadmin/sueindex.cfm (13 Jun 2012)

Hoek E (1999) Putting numbers to geology—an engineer's viewpoint (second glossop lecture). Q J Eng Geol 32:1–19

Keaton JR (2013) Engineering geology: fundamental input or random variable? In: Withiam JL, Phoon KK, Hussein MH (eds) Foundation engineering in the face of uncertainty. ASCE, Reston, pp 232–253 (Geotechnical Special Publication 229)

Kulhawy FH, Phoon KK, Wang Y (2012) Reliability-based design of foundations—a modern view. Geotechnical Engineering State of the Art & Practice: Keynote Lectures from GeoCongress 2012 Geotechnical Special Publication 226. ASCE, Reston, pp 102–121

Morgenstern NR, Cruden DM (1977) Description and classification of geotechnical complexities. In: International symposium on the geotechnics of structurally complex formations, vol 2, Italian Geotechnical Society, pp 195–204

NRCS (2002) Rock material field classification system. US Department of Agriculture Natural Resources Conservation Service, National Engineering Handbook Part 631 Geology, Chapter 12, http://directives.sc.egov.usda.gov/viewerFS.aspx?hid=21423 (13 Jun 2012)

Pierson LA, van Vickle R (1993) Rockfall hazard rating system—participants' manual. Federal Highway Administration Publication No. FHWA-SA-93-057, Washington, DC

Managing Uncertainty in Geological Engineering Models for Open-Pit Feasibility

Rosalind Munro and Jeffrey R. Keaton

Abstract

Open-pit mines rely on geology for economic feasibility based on mineral value and geotechnical parameters. Pit slope angles may control economics; therefore, open-pit mine design utilizes a reliability approach that specifies confidence interval, precision index, and practical strength values. The geological model for economic evaluation is developed before pit slope stability analyses are undertaken and is based on lithology and mineralization, both of which are essential for the geological engineering model. The preliminary mine plan includes a shell for the ultimate pit based on assumed pit slope configuration that includes haul-road benches. Geological engineering characterization utilizes the geological model to identify locations for geotechnical bore holes and detailed mapping of rock structure to document aspects critical for slope stability. Uncertainty is common; rock structure variability is represented from direct observation of selected outcrops and from detailed logging of rock core. Lithologies that form the ultimate pit walls arc sampled for unconfined compression testing. Samples of a single lithology from different bore holes reflect formation variability. Typically, a few (3–15) samples of each lithology are tested. The desired precision index and reliability may not be met for a specified confidence interval with a limited program; these parameters cannot be determined until test-result variability is known. Geological uncertainty may be managed for open-pit mines by using practical strength values, drilling a few additional bore holes, and performing additional unconfined compression tests.

Keywords

Confidence interval • Precision index • Reliability • Practical strength value

63.1 Introduction

Open-pit mine feasibility involves geology in two different but related ways. The initial use of geology is for general economic feasibility based on mineral value, concentration, and distribution. The preliminary mine layout provides access to the ore using realistic overall pit slope angles and haul-road grades based on assumed geotechnical parameters. Subsequent use of geology is for evaluating the assumed geotechnical parameters to define the steepest stable pit slope angles needed to access the ore body. Small slope angle changes have immense effect on economics of an open-pit mine that is hundreds of meters deep because of the volume of rock and overburden that must be excavated, hauled, and stored during mine development.

The mining industry is accustomed to performing statistical analyses of mineral values that start with relatively few assay tests and are supplemented incrementally with additional assay results. It stands to reason that the mining industry would use a reliability-based approach for selecting strength parameters for designing open-pit mine slopes

R. Munro (✉) · J.R. Keaton
AMEC Americas, Inc, 6001 Rickenbacker Road,
Los Angeles, 90040 USA
e-mail: rosalind.munro@amec.com

initially based on few data points that are supplemented over time. Even the "final" geological model of a mine is tested continuously with blast-hole samples assayed for ore control. A reliability-based approach specifies confidence interval (typically 95 %) and precision index (initially 1.5 for general use), as well as a reliability factor (i.e., non-failure likelihood) appropriate for the project or component being analyzed. The reliability-based approach supports selection of practical strength values for use in engineering stability analyses.

The geological model for economic evaluation of mineral resources usually is developed before the geological engineering for pit slope stability is undertaken. The geological model of the mine is based on lithology, mineralization, and alteration (Fig. 63.1a), all of which also are essential for the geological engineering model. Mine geologists put much effort into developing an accurate three-dimensional (3D) geological model which tends to be accepted in total as the basis for the engineering geological model. The preliminary mine plan will include a shell for the ultimate pit based on assumed pit slope configuration that includes haul road ramps and inter-ramp benches (Fig. 63.1b). The geological engineering characterization program utilizes the geological model of the mine to identify a few locations, often no more than 10, for geotechnical bore holes and locations based on exposed lithology for detailed mapping of rock structure to document aspects that are critical for slope stability. Uncertainty is common in geological engineering characterization. A scale dependency for pit slope stability also exists with bench face angles being steeper than inter-ramp angles, and inter-ramp angles being steeper than the overall pit slope angle (Fig. 63.1b). Rock structure variability is based on direct observation of selected outcrops and from detailed logging of rock core and treated statistically separate from rock strength.

Samples of rock core selected from lithologies in which the pit will be excavated are tested for unconfined compression. Samples of the same lithology obtained from different bore holes are considered collectively in an attempt to represent the variability of the lithologic unit. Typically, a small number (3–15) of unconfined compression tests will be performed for each lithology. Drawing defensible conclusions from analyses of small sets of data requires a statistical approach described by Gill et al. (2005).

63.2 Statistics of Small Data Sets

Statistical analyses of data such as unconfined compression test results utilize the degrees of freedom associated with the number of tests performed, the t-distribution related to the degrees of freedom, the calculated confidence interval based on test results, the probability associated with the t-distribution, and the Chi squared distribution based on the t-distribution probability (Gill et al. 2005). Therefore, the desired confidence interval, precision index, and reliability may not be met with a limited geological engineering characterization program and the parameters cannot be determined until the variability in test results is calculated. The results of the initial characterization lead to a practical value of rock strength for use in stability analyses and a recommendation for managing uncertainty with a minimum number of additional bore holes and unconfined compression tests intended to attain the desired confidence interval, precision index, and reliability factor for the open-pit mine.

Laboratory test results are a subset of the possible tests that could be performed on the rock controlling the stability of open-pit mine slopes (Gill et al. 2005). The reliability of the statistics calculated from small data sets (e.g., mean and standard deviation) depends on the number of samples tested. The number of tests needed to attain a desired confidence interval is calculated with Eq. 63.1 (Appendix A). Gill et al. (2005) recommend that values of precision index Pi be based on project importance; $Pi \leq 1.5$ is suggested for routine mining projects, $Pi \leq 1.35$ is suggested for mining

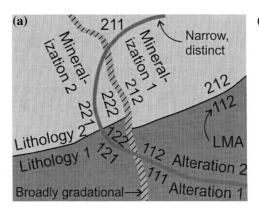
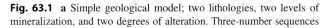
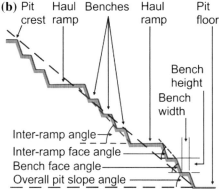

Fig. 63.1 **a** Simple geological model; two lithologies, two levels of mineralization, and two degrees of alteration. Three-number sequences denotes lithology-mineralization-alteration (*LMA*). **b** Pit slope geometry and terms. Both parts are modified from Read and Stacey (2012)

facilities with higher importance, such as shafts, and for routine civil engineering projects, whereas $Pi \leq 1.2$ is suggested for projects in which public safety may be an issue. Gill et al. (2005) also recommend that 95 % be used for Ic for all projects. They note that $Pi = 1.0$ cannot be reached because doing so would require that an infinite number of specimens be tested. Furthermore, the mean and standard deviation of test results would have to equal the true mean and standard deviation of the rock formation, and Ic would be 100 %.

If Ic is designated to be 95 %, then $\beta = 0.975$ from Eq. 63.2 (Appendix A) and t_β can be determined based on the number of samples tested and the appropriate degrees of freedom for the specific analysis. The ratio of true mean of the rock formation and the mean of the laboratory tests is given by Eq. 63.3 (Appendix A). At this point, all parameters in Eq. 63.1 (Appendix A) are known, with Pi having been specified for the project importance. Therefore, Eq. 63.1 can be solved for Pi (Eq. 63.4, Appendix A) so that its actual value can be determined. Equation 63.1 also can be solved for t_β (Eq. 63.5, Appendix A) using Pi calculated from Eq. 63.4, which will enable calculation of β and Ic for the actual test results. t_β can be obtained by interpolating from tables in statistics textbooks or using utilities in electronic spreadsheets or mathematics applications. The ratio of true standard deviation of the rock formation and the standard deviation of the laboratory tests is given by Eq. 63.6 (Appendix A).

A practical strength value associated with a desired or target stability probability (i.e., non-failure probability) is given in Eq. 63.7 (Appendix A) as the lower bound of the true value of mean strength minus the upper bound of the true standard deviation of strength. Target values of Ic, Pi, and Ps not being met is an indication that additional testing is needed to increase the degrees of freedom. Additional testing may or may not result in a smaller coefficient of variation (COV, the ratio of standard deviation to mean), but it will improve the t-distribution parameters.

A feasibility assessment of an open-pit mine pit shell used samples from five borings in two rock types; the unconfined compressive strength (UCS) data are from a real project, but no details are needed for this discussion. Twelve UCS tests were performed on samples of Rock Type 1, whereas Rock Type 2 had four tests (Table 63.1). Initial values used in the feasibility-level statistical analysis were $Ic = 95$ % and $Pi = 1.5$. Selected parameters for the two rock types are summarized in Table 63.2 and Fig. 63.2. Target values of Ic and Pi were exceeded for Rock Type 1, but were not met for

Table 63.1 Unconfined compressive strength (UCS) test results for two rock types

Rock type	UCS (MPa)	Ns	$\bar{X}$	sd	COV
1	60, 65, 66, 77, 80, 89, 90, 93, 106, 113, 122, 141	12	91.83	24.77	0.270
2	27, 41, 53, 68	4	47.25	17.44	0.369

Table 63.2 Statistical values for UCS data in Table 63.1

Rock type	Ic (%)	Pi	Lower μ (MPa)	Upper σ (MPa)	Ps (%)	PSV(Ps) (MPa)
1	96.83	1.44	75.39	46.47	85	21.29
					92	0.03
2	58.26	5.22	37.80	34.19	39	0.02

Target $Pi \leq 1.5$, $Ic \geq 95$ %

Fig. 63.2 Summary of unconfined compression strength of two rock types

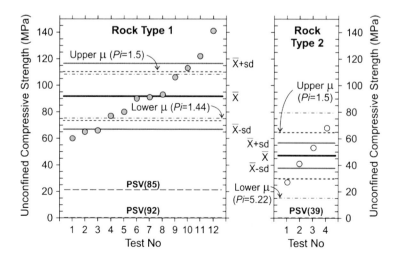

Rock Type 2, indicating that additional tests are needed to attain target Ic and Pi values. Gill et al. (2005) show that $Ps = 85\%$ results in a true stability probability of approximately 99% for $Ic = 95\%$, whereas the stability probability is approximately 90% for $Ps = 65\%$. The highest Ps value for Rock Type 2 that produces $\mathrm{PSV}(Ps) \approx 0$ is 39% for $Ic = 58\%$. Rock Type 1 tests support $Ic = 97\%$ so the PSV (85) strength would be appropriate for stability analyses, and a higher PSV might be justified by using a lower Ps value because $Ic > 95\%$.

63.3   Conclusions

Uncertainty in geological engineering models at the mine feasibility level can be managed by relying on geological models of lithology, mineralization, and alteration, along with UCS tests on a small number of samples of the rock that will support the pit slopes. Statistical analyses of UCS test results are essential to select strength values that have suitable stability probabilities and are consistent with desired confidence limits and precision indexes. The geological engineering model at the mine feasibility level provides the foundation for model refinements and improved precision in later phases of mine development.

Appendix A: Statistics Equations for Sect. 63.2

Equations for Sect. 63.2 are in this appendix (modified from Gill et al. 2005).

$$Ns = \left[\left(\frac{Pi+1}{Pi-1} \right) t_\beta \frac{sd}{\bar{X}} \right]^2 + 1 \tag{63.1}$$

$$Ic = 100(2\beta - 1); \text{ consequently, } \beta = 0.5(1.0 + Ic/100) \tag{63.2}$$

$$1 - \left(\frac{Pi-1}{Pi+1} \right) = 1 - \varepsilon_m \le \frac{\mu}{\bar{X}} \le 1 + \left(\frac{Pi-1}{Pi+1} \right) = 1 + \varepsilon_m \tag{63.3}$$

$$Pi = \frac{\bar{X}\sqrt{Ns-1} + sd\, t_\beta}{\bar{X}\sqrt{Ns-1} - sd\, t_\beta} \tag{63.4}$$

$$t_\beta = \frac{\bar{X}}{sd}\sqrt{Ns-1}\,\frac{Pi-1}{Pi+1} \tag{63.5}$$

$$\sqrt{\frac{Ns}{\chi_\beta^2}} \le \frac{\sigma}{sd} \le \sqrt{\frac{Ns}{\chi_{(1-\beta)}^2}} \tag{63.6}$$

$$\mathrm{PSV}(Ps) = \frac{2}{Pi+1}\bar{X} - n_\alpha \sqrt{\frac{Ns}{\chi_{(1-\beta)}^2}}sd \tag{63.7}$$

Ns = the number of test results, Pi = precision index, t_β = the t-distribution, β = parameter related to the desired confidence interval, Ic, $\bar{X}$ = mean of the test results, sd = standard deviation of the test results, μ, = true mean of rock formation, σ = true standard deviation of rock formation, ε_m = maximum relative error, χ^2 = Chi squared distribution, $\mathrm{PSV}(Ps)$ = practical strength value associated with a designated or target stability probability (i.e., non-failure probability), n_α = probability coefficient obtained from the normal density function, and $\alpha = Ps/Ic$ with the restriction that $Ps \ll Ic$ for conditions with small mean and large standard deviation to avoid PSV $(Ps) < 0$. The t-distribution requires $Ns - 1$ degrees of freedom.

References

Gill DE, Corthesy R, Leite MH (2005) A statistical approach for determining practical rock strength and deformability values from laboratory tests. Eng Geol 78:53–67

Read J, Stacey P (2012) Guidelines for open pit slope design. CRC Press, Balkema, 496 p

Construction in Complex Geological Settings - The Problematic of Predicting the Nature of the Ground

Convener Prof. Ana Paula F. da Silva—*Co-conveners* Lazaro Zuquette, Ricardo Oliveira, Joaquim Pombo

Nowadays, development has been pushing the occupation of brownfields or any other type of ground previously set aside due to its poor geotechnical characteristics. Additionally, civil engineering projects have been growing in complexity and therefore require higher geotechnical knowledge. In this scope, the role played by the specialist in engineering geology during design has gained even more relevance, since its expertise is fundamental for maximizing the data gathered from site investigation and, afterwards, for building an approximate geological model of the ground and defining its geotechnical properties. The aim of this session is the presentation and discussion of case studies that illustrate the way engineering geology deals with geological complexity and their modelling, since no matter how sophisticated tools the design team might use, basic knowledge of the ground sis still the cornerstone of everything.

Engineering Geological and Geotechnical Cartographic Modeling as a Methodological Basis for Engineering Surveys and Design in Complex Geological Environment

64

Felix Rivkin, I. Kuznetsova, A. Popova, I. Parmuzin, and I. Chehina

Abstract

The methodological principles and methods of the creation of new geotechnical cartographic models are considered on the basis of engineering-geological and permafrost cartographic models. Creation of the geotechnical cartographic models is considered as a method of optimizing the interaction between project designers and geologists working in complex natural environment and presenting the results of research in the form adapted for the design. The results of application of the geotechnical models are examined for the implementation of the pipeline projects in complex environment.

Keywords

Engineering geocryological cartographic model • Geotechnical cartographic model • Complex geoenvironment

64.1 Introduction

The main purpose of the investigations of pipeline routes is to obtain information for the design and construction. The more complex the natural conditions of the study area are the greater demands are placed on the results of investigations.

The engineering-geological and permafrost conditions predetermine the choice of technical design solutions at a large extent. There is a feedback on the other hand, i.e., the pre-selected design solutions greatly affect the scope and methods of the natural environment studies. Therefore, the establishment of functional connections between the engineering-geological conditions and design solutions allows optimizing of both site investigations and the design itself. It provides the opportunity to focus the study of natural conditions and ground properties at the most challenging areas. Further studies, therefore, are directed towards detailing of engineering-geological conditions and the specific technical solutions.

GIS-based research is one of the methods of site investigations, design and construction optimization in complex natural environment. Development of the multi-scale GIS cartographic models as an information support method for engineering research and design is often used in Russia. Normally, these methodological solutions are applied for the challenging projects, for example, during the development of oil and gas fields in the Arctic, the construction of trans-regional systems of main pipelines, etc. (Rivkin et al. 2004, 2006).

It should be noted that the creation of multi-scale and multi-purpose (Rivkin et al. 2010; Rivkin 2012) cartographic models (with their further consistent detailing) is most effective at the early stages of the project, prior to the major investments in the field engineering investigation and design.

64.2 Methods

Contemporary engineering-geological and geocryological conditions in the permafrost area result from a complex interaction of various environmental factors. Therefore, for

F. Rivkin (✉) · I. Kuznetsova · A. Popova · I. Parmuzin ·
I. Chehina
Fundamentproject OJSC, Moscow, Russia
e-mail: f-rivkin@narod.ru

the development of engineering and geocryological carto-graphic models it is pertinent to analyze the interaction of the main factors that determine geotechnical conditions that, in turn, specify permafrost conditions. It means that it is necessary to combine two disparate groups of factors—the engineering-geological (mainly regional) and the permafrost (mainly zonal) data in a general analytic scheme. This analysis is performed in a synthetic matrix form, since it is virtually impossible to construct a hierarchical system to analyze groups of heterogeneous factors.

Geostructural and tectonic conditions, geomorphology and lithology are traditionally regarded as the regional factors for engineering-geological and permafrost conditions assessment. At first, geostructural and major tectonic elements of the study area are singled out, than the main geomorphic features such as watersheds, slopes, valleys, rivers, sea, river and glacial terraces are distinguished.

At the end, differentiation of regional factors is completed by identifying the main types of lithological or cryo-litho-logical sections.

The zonal factors depend on the latitudinal location of the study area: the climate, vegetation, landscape, the depth of seasonal freezing and thawing—that is, the factors that determine the current permafrost conditions in upper part of the ground cross-section.

These two groups of factors are grouped in a hierarchical manner and are arranged along the axes of the matrix. Thus, the matrix of the regional and zonal factors multivariate analysis presents not only a form for the analysis of the interaction of environmental factors, but, in fact, the legend to the GIS-based cartographic models of engineering-geocryological conditions (GIS-EG). In such a methodological statement GIS-EG illustrates a variety of engineering and permafrost conditions and identifies most complicated areas.

Fig. 64.1 Geotechnical cartographic models. **a** The schematic map of the pipeline route with complex natural conditions. **b** The schematic geotechnical cartographic model of the pipeline main design solutions

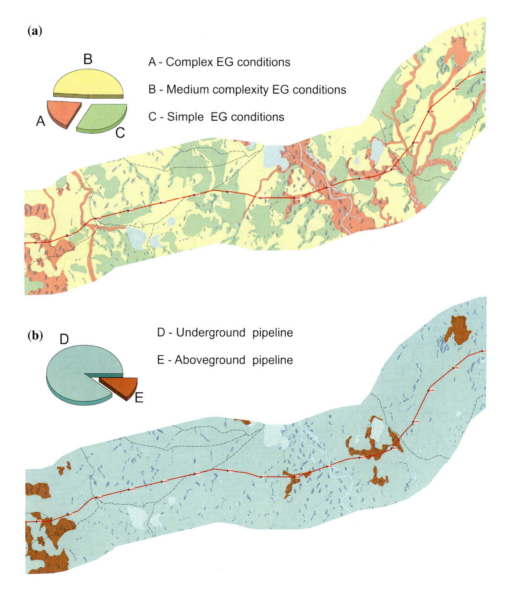

(a)

A - Complex EG conditions

B - Medium complexity EG conditions

C - Simple EG conditions

(b)

D - Underground pipeline

E - Aboveground pipeline

So, the first important step of the pipeline route investigations is the development of the GIS-based cartographic models of engineering geological conditions in the permafrost areas and consistent specification of those models and their characteristics.

A large number of standard technical solutions for foundation structures is used in the Arctic. Each of them was designed for certain engineering-geological and permafrost conditions. Good practice has been developed for application of standard technical solutions in well-defined natural conditions. In other words, the correspondence between the natural environment and the types of appropriate foundations has been formed.

The applicable range of technical solutions includes only those of them which were justified for technical, engineering-geological, environmental, and economic characteristics. Thus, the second step involves the methodical development of a set of possible technical solutions in the areas with complex engineering-geological environment. These areas

have been identified at the first step, and their spatial position was presented at GIS-based cartographic model of the engineering-geocryological conditions (GIS-EG). So, the second step is the development of a wide range of geotechnical solutions for foundation structures for the different types of the geocryological conditions.

The third step involves the methodical integration of the results of the first and second phases—the cartographic expression of the location of various technical solutions along the pipeline route in the form of a spatial geotechnical model. It starts from the transformation of a synthetic matrix legend for map of engineering-geological conditions into the matrix legend for the geotechnical cartographic model. On the basis of the transformed matrix that comprises the legend for the map, geotechnical cartographic model (GIS-GCM) is developed. All selected engineering-geological plan units are shown on the geotechnical models as the areas where the recommended technical solutions or construction methods should be applied.

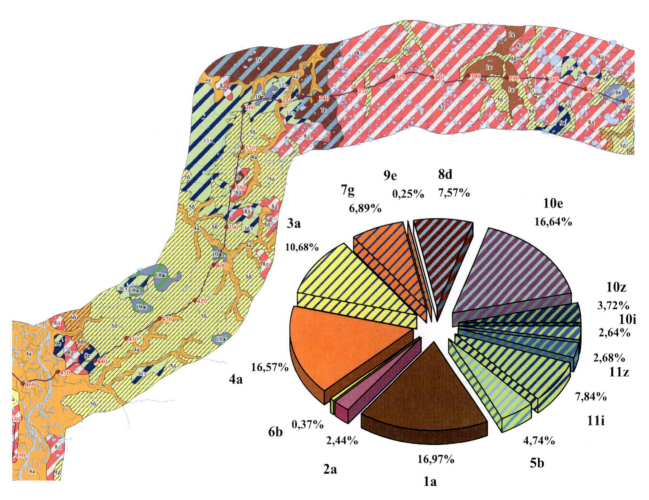

Fig. 64.2 The schematic geotechnical cartographic model (fragment of the design solutions map of pipeline route). Diagram presents percentage of the total length of all selected areas with different design solutions. Indexes (1a, 2a, …, 11z) show types of engineering-geocryological areas with various pipeline foundation solutions are recommended (e.g. 1a—10 m long pile foundation without thermal stabilization of soils; 11z—12 m long pile foundation for the pre-thawing of frozen grounds, with a thermal shield)

64.3 Results and Discussion

Geotechnical cartographic model is actually a methodological bridge between the results of engineering-geological investigations and design. This model explicitly shows the connection between the design and the geotechnical and geocryological conditions. Such methodological solutions come out from the necessity to optimize the site/route investigations and design according to the particular natural environment. They are usually applied for large-scale trans-regional pipeline projects laid in complicated geological conditions. They are also used for the infrastructure development, oil and gas fields in the permafrost area (Arctic and sub-Arctic).

Analysis of this information provided recommendations for certain design solutions. The above-ground laying was recommended at the geologically-complex sites, whether at the other sites the simpler and less expensive underground laying was applied. The GIS-based cartographic model allows interpreting this information in a quantitative way. Figure 64.1 shows the quantitative ratio of above and underground pipeline laying and exemplifies the geotechnical cartographic model developed at the preliminary stage of the investigation.

The complex areas on the Fig. 64.1 are highlighted by red: propagation of icy ground and hazardous geological processes. This information allowed to recommend certain design solutions. Within the complex areas above-ground pipeline was recommended, whereas at other sites—underground installations, which are simpler and less expensive.

Further development of this method is associated with a detailed GIS-EG and GIS-GCM. It allows not only to differentiate areas with above-ground and underground laying, but also highlights areas with various pipeline-supporting piles design (Fig. 64.2). Analysis of the pipeline design solutions map allows to perform a quantitative analysis of the results of route investigations for geotechnical conditions assessment and design. The diagram on Fig. 64.2 shows the percentage of sites with complex conditions, and a performed quantitative assessment of the application of certain

technical solutions that forms the ground for the construction cost estimate.

Therefore, the informational support of the site investigations and design can justify the choice of design solutions based not only on the complexity of natural conditions and the length of such sites, but also taking into account the cost estimate of construction.

64.4 Conclusion

The establishment of specialized GIS-EM & GCM is a modern methodological interconnection basis between geotechnical and permafrost studies and construction design. It allows optimizing both the engineering-geological investigations and design in complex environment, and creating of an information base to justify the selection of design solutions and construction methods. On the other hand, this method makes it possible to concentrate investigations at the most challenging areas, increasing their information value, and therefore their efficiency.

References

Rivkin FM (2012) Geotechnical maps in the structure of information support for geotechnical investigations. In: Proceedings of tenth international conference on permafrost. Extended Abstracts, The Northern Publisher, Salekhard, pp 471–472

Rivkin FM, Ivanova NV, Kuznetsova IV (2006) Geoinformation modeling of engineering geocryological conditions for risk assessment during survey of the Yamal-Western Europe Pipeline (Yamal Segment). In: Proceedings of XIth international congress for mathematical geology, Liège, Belgium, 3–8 Sept, S09-23, p 4

Rivkin FM, Kuznetsova IL, Ivanova NV, Popova AA, Parmuzin IS (2010) Multi-purpose engineering geological cartographic for information support for engineering surveys, designing, and monitoring. In: Proceedings of 11th congress of the IAEG, Auckland, 5–10 Sept 2010, pp 2415–2419

Rivkin FM, Kuznetsova IL, Ivanova NV, Suhodolsky SE (2004) Engineering-geocryological cartographic for construction purposes in permafrost regions. In: Engineering Geology for Infrastructure Planning in Europe. Springer, Berlin, pp 172–178

Experiences Learned from Engineering Geological Investigation of Headrace Tunnel on Sedimentary Rock—Xekaman3 Hydropower Project—Lao PDR

65

Nguyen Song Thanh and Dao Dang Minh

Abstract

Xekaman3 hydropower project is located in Sekong province, Southern of Lao PDR. Site investigation for the project was carried out from January 2003 to May 2010 at damsite, water intake, headrace tunnel (7,400 m) and powerhouse area. Unfortunately, for the tunnel section from vertical well No 1 (GD1) to vertical well No 2 (GD2) with the length of 710 m, many complex geological phenomena happened and it influences significantly the whole construction scheduled and total investment of the project. This paper presents the causes of those phenomena, ways to solve problems and methods of improvement for the tunnel section mentioned above. The authors also mentioned some experiences learned through geological investigation process for headrace tunnel of Xekaman3 hydropower project and possibility of applying these experiences for the projects with similar geological conditions.

Keywords

Site investigation • Headrace tunnel • Complex geology • Xekaman3

65.1 Introduction

The headrace tunnel of Xekaman3 hydropower project with total length of 7,400 m runs through some different formations such as: Long Dai formation (O_3-S_1 ld), Song Bung formation ($T_{1-2}sb$) and Quaternary system. In the tunnel section from vertical well No 1 (GD1) to vertical well No 2 (GD2—Km6 + 830) with the length of 710 m (see Fig. 65.1), the whole section belongs to Song Bung formation with main composition as follows :

- Upper member: Grey–reddish grey thick bedded siltstone, sandstone with fine grain; sandstone, gritstone with coarse grain in upper part.
- Middle Member: Mainly reddish brown thick bedded siltstone; intercalated with lenses of blackish grey sandstone, siltstone, shale.

- Bottom Member: Mainly red thick bedded sandstone with fine-medium grain intercalated with lenses of red siltstone.

Two alternatives of tunnel in this section were considered. Initially, this tunnel section was designed on the ground with the depth of pillar is about 30–40 m from the ground. And then, due to complex geology, it is changed to underground alternative and the depth of tunnel is about 145 m below ground surface.

This paper presents some experiences learned during investigation and construction of this tunnel section, especially for changing of alternative from On the ground to Underground and treatment experiences for tunneling section in complex geology.

65.2 Site Investigation Phases and Results

In the pre-Feasibility Stage (FS) stage, in the section from surge tank to the powerhouse, the tunnel was initially designed on the ground. So, base on this design, boreholes

N.S. Thanh (✉) · D.D. Minh
FECON Foundation Engineering and Underground Construction Company, 15th Floor CEO Building, Pham Hung St, Hanoi, Vietnam
e-mail: thanhns@fecon.com.vn

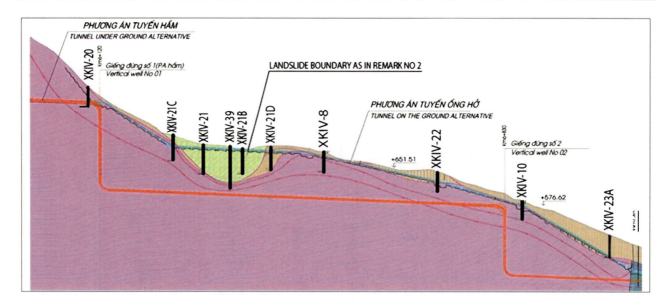

Fig. 65.1 Landslide boundary at penstock based on Remarks No 1 and No 2

XKIV-08, XKIV-09, XKIV-10, XKIV-20, XKIV21, XKIV-22, XKIV-23, XKIV-28 were located at pillars with the depth ranges from 30 to 40 m (see Fig. 65.1). During FS stage, many kinds of geotechnical investigation were implemented. As the result of investigation, geotechnical engineers had delimitated the complex geology area for this tunnel section (Fig. 65.1) and figured out the engineering geological parameters for designers, as mentioned hereafter.

65.2.1 Remark No 1

- Engineering geological map, large scale for the studied area and the distribution of different soils and rocks, thickness of weathering layers in the studied area to the depth of 40 m was established.
- Preliminary definition of the boundary of old landslide along the axis of On the ground tunnel section surrounding borehole XKIV-21 to the depth of about 50 m and confirmed that the section from XKIV-09 to XKIV-22 on the engineering geological profile does not belong to the old landslide.
- In the borehole logs of XKIV-09, XKIV-22, from depth of 16.4 m (XKIV-09), 23.7 m (XKIV-22) and deeper, rock belongs to IIA zone (fractured rock zone) or IIB zone (fresh rock zone).
- Determination of the physico-mechanical properties of weathering rock zones.
- Determination of the hydrogeological conditions of study area.

As listed above and concerning complex geology, designers decided that additional site investigation must be carried out in FS stage to:

- Determine exactly the boundary of old landslide on the ground and underground along the tunnel axis.
- Determine exactly depth of weathered rock zones IB, IIA and IIB.

Then, the final decision to change this tunnel section from On the ground to Underground was made by the designers and approved by the Investor—Viet Lao Power Company Ltd. As requested, on August 2004, 6 additional boreholes, XKIV-21A, XKIV-21B, XKIV-21C, XKIV-21D, XKIV-23A, XKIV-23B, were located around borehole XKIV-21 with depths of 30–50 m (Fig. 65.1). The results of these boreholes lead to Remark No 2.

65.2.2 Remark No 2

- The boundaries of old landslide was redefined: 209 m long, starts from left of XKIV-21C to XKIV-21D borehole along the axis of tunnel; its deepest point rest 74 m below the surface (Fig. 65.1).

Base on Remark No 2, designers decided to change this tunnel section from on the ground to underground and this design will be applied for investigation of technical design (TD) stage. The main purposes of the investigation for this stage were:

- Assess the exact boundaries of old landslide underground. It is noted that their limits up to this stage is not a concern anymore!
- Determined the exact engineering geological conditions from surface to a depth below the tunnel.

For those purposes above, in TD, only one borehole XKIV-39, 90 m deep, was carried out. The results of XKIV-39 lead to the Remark No 3.

65.2.3 Remark No 3

The maximum depth for the bottom of old landslide was 70 m.

Based on this finding, designers decided that the depth of tunnel's bottom of the new design section will be located at elevation of +590 m, 20–25 m lower the bottom of the old landslide and, with this elevation, the tunnel section would run completely in fractured rock zone IIA or fresh rock zone IIB. With this design and since March 2009, the tunneling from GD2 towards the GD1 was undertaken. Unfortunately, during construction of this tunnel section, many complex geological phenomena happened.

Confronted with situation of very poor and complicated geological conditions as mentioned, the Investors requests designers and geological engineers to verify and assess again the geological conditions of tunnel section from vertical No 1 (GD1) to vertical No 2 (GD2), based on the detected geological conditions, geological data records of constructed tunnel sections and recommends to carry out additional geotechnical drilling for the remaining unconstructed tunnel section.

65.3   Changes During Tunneling

65.3.1 Reassessment of Results of Previous Stage and Data Records During Tunneling

Based on the detected very poor and complicated geological conditions of the tunnel section as mentioned above, engineering geological engineers had carefully reassessed the results of geological investigation of FS and TD stages and analyzed the geological description report for constructed tunnel section. Unfortunately, some mistakes were found, namely:

- As in Remark No 1, in borehole logs of XKIV-09 and XKIV-22, the estimation of bedrock zone encountered is not correct.
- The extension of old landslide is not 209 m as assessed in Remark No 2. In fact, along the axis of tunnel, the landslide is 530 m (Fig. 65.2).
- Bottom of old landslide as indicated in Remark No 3 with the depth of 70 m is not correct. It leads that the selection of depth of tunnel 90 m is not enough to avoid the landslide. One of other mistake is that, during drilling additional borehole XKIV-39, the required depth is 90 m, but the final depth of this borehole is only 81 m, not reaching the bottom of tunnel! In TD stage, this is unacceptable.
- The geological description document during tunneling is too sketchy.

As analyzed and due to some incorrections mentioned above, according to the requirement of Investor, a survey unit carried out 3 additional boreholes, XKIV-BS1, XNKIV-BS2 and XKIV-BS3, for the remaining unconstructed tunnel section (Locations in the Fig. 65.2). Main purpose of these additional drilling is to determine exactly the extension and depth of old landslide and evaluate the geological conditions in the unconstructed tunnel section from GD2 to GD1.

In case of whole tunnel section locates on the body of old landslide, then possibility of adjustment for this section from underground to on ground would be considered. The result of these additional boreholes lead to the Remark No 4.

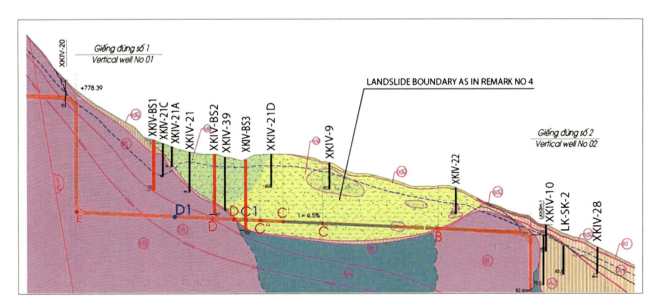

Fig. 65.2 Landslide along the axis of tunnel section after adjustments according to Remark No 4

65.3.2 Remark No 4

- The maximum depth of old landslide is 125 m, not 70 m as stated in Remark No 3 (Fig. 65.2)
- The remaining old landslide body along the axis of tunnel from current tunnel face (Km6 + 392) to the original bedrock boundary is 20 m and there is no need to adjust the tunnel from underground to on the ground. To ensure safety for tunneling in this 20 m remaining, reinforcement solutions must be applied.

As mentioned above, reinforcement grouting for the 20 m remaining old landslide body is selected by the Investor to be performed.

65.3.3 Reinforcement Grouting for Remaining Tunnel Sections

Base on the existing and actual geological conditions, the reinforcement grouting while drilling parallel with the axis at the tunnel face is impossible because it is not safe for workers and equipments; therefore, reinforcement grouting using vertical drilling along the axis of tunnel was selected. To ensure the flows of liquid cement mortar would not go into tunnel, the section near the final tunnel face (km6 + 392) must be blocked by concrete during grouting.

At present, the whole tunnel section construction from tunnel face km6 + 392 to the vertical well GD1 has finished successfully.

65.4 Conclusion and Recommendations

1. In this particular project, at the beginning of FS-stage investigation, by studying topography-geomorphology,

the existing old landslide is confirmed and then, it could have been recommended adjusting the location of tunnel to avoid running through it.

2. After an engineering geological object which could be harmful to the safety of project is defined (as the old landslide mentioned above), it is necessary to perform some additional investigation methods and, for each method, specific criterion should be listed in advanced. For any reasons, if any one of specific criterion is not defined clearly then additional site investigation must be implemented. In this case, some investigation experiences should be noted :

- The main purpose of borehole XKIV-39 is to define the depth of old landslide but it is stopped at 81 m while design depth is 90 m. This is not acceptable because the depth of borehole did not underpass the depth of tunnel. In this case, the borehole XKIV-39 should have been redrilled.
- The geological description of tunnel while tunneling should be detailed enough so that reinforcement solution for specific tunnel section could be applied successfully. In dealing with geological problems, the collaboration between designers and constructor is very important.

References

SDCC Song Da Consulting JS Company (2005) Report on engineering geological investigation of main works of Xekaman3 hydropower project—FS (Dec-2003) and technical design stage (Mar 2005)

SDCC Song Da Consulting JS Company (2010) Additional report on engineering geological investigation of tunnel (section from vertical well No1-GD1 to vertical well No2-GD2) of Xekaman3 hydropower project, detailed design stage—Lao PDR

Engineering Properties of Badlands in the Canadian Prairies

Khan Fawad and Azam Shahid

Abstract

The main objective of this paper was to investigate the engineering properties of badlands in the Canadian Prairies. Under the prevalent semi-arid climate, soils in the area undergo extensive variations due to alternate wet-dry cycles. The soil profile has three distinct sediments: fissured sandstone with a steep slope of 60°; popcorn-textured mudrock with a mild slope of 30° and; eroded pediment with a flat slope of 3°. The fines content increased from dry to wet state with 17–33 % for sandstone, 4–98 % for mudrock, and 21–42 % for pediment. The consistency limits indicated that the water adsorption capacity is highest for mudrock followed by sandstone and then by pediment. The water retention curve of sandstone showed bimodal distribution with a low air entry value (6 kPa) pertaining to drainage through cracks and a high air entry value (150 kPa) associated with flow through the soil matrix. The mudrock and pediment followed a unimodal water retention curve with a single matrix air entry value of 5 kPa. These results explain the observed surface erosion and internal piping through the sandstone and the genetic relationship between the sandstone and the pediment.

Keywords

Engineering properties • Badland sediments • Canadian prairies

66.1 Introduction

Badlands are rugged landscapes comprising of loose materials and are commonly found in arid and semi-arid regions of the globe. The engineering properties of such sediments are derived from geologic history and climatic conditions. This is particularly true for the Canadian Prairies where badlands originated from Cretaceous rocks, comprise of sands through clays, and undergo cyclic saturation and desaturation due to seasonal variations in meteorological parameters. The surface lithology continuously evolves due to alternate fluvial erosion (rain splash, sheet wash, overland flow, concentrated flow, and pipe flow) and water deficiency (evaporation, transpiration, cracking, sapping, and mass movement). The effect of these processes is experienced differently by the various badland sediments at a given site.

A typical example of an incessantly transforming terrain is the Avonlea badland (latitude 50.0367 and longitude 105.0667) in southern Saskatchewan, as given in Fig. 66.1. The region has experienced eight glacial advances and retreats during the Quaternary (Christiansen 1979). The last glacier, the Laurentide ice sheet, reached its maximum extent about 18,000 years B.P. This 1,000 m thick ice gradually retreated in the north-eastward direction and eventually disappeared around 8,000 years B.P. The emerging surface rocks were extensively eroded by the preceding scouring actions of the advancing glaciers. The melting ice cut the exposed materials and created steep-sided channels and deeply incised rills. With increasing floods, the less resistant Cretaceous rocks of the Eastend Formation were washed

K. Fawad · A. Shahid (✉)
University of Regina, 3737 Wascana Parkway, Regina S4S 0A2, Canada
e-mail: shahid.azam@uregina.ca

Fig. 66.1 Map of southern
Saskatchewan showing the
investigated area

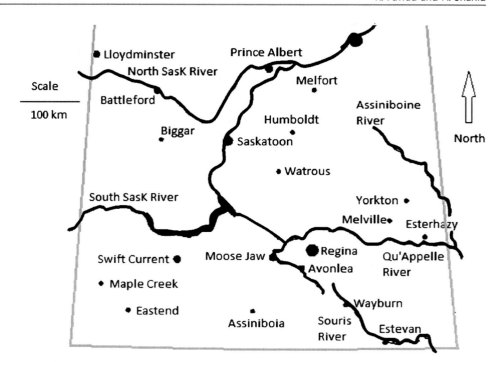

away and deposited on the plains (Byers 1959). The present-
day sandstone and mudrock originated from the Maastrich-
tian age (66–72 Ma).

The prevalent seasonal weather variations dictate the
engineering properties of the deposited materials. Overall,
the area falls under a semi-arid (BSk) climate according to
the Köppen climate classification system. The average
monthly temperature varies between −15 °C in January to
19.6 °C in July with an annual mean of 3.2 °C. Likewise, the
average annual precipitation is 366 mm with a minimum of
10 mm in June and a maximum of 64 mm in February.
Precipitation occurs as winter snowfall (November to
March) that freezes the soil and as summer rainfall (April to
October) that results in high surface runoff. Further, temperature variations between day and night or successive
rainfall events during the summer result in cyclic saturation-
desaturation of the exposed materials. A clear understanding
of the water movement through these surface sediments is
required from an engineering perspective.

The main objective of this paper was to determine the
engineering properties of Avonlea badlands. Based on site
investigations, three representative soils (sandstone, mudrock, and pediment) were selected for laboratory characterization. Index properties were determined for preliminary
soil assessment. The water retention curve was determined to
understand the water holding behavior of the three sediments
and their interaction through erosion.

66.2   Materials and Methods

Field investigations were carried out through several 1-day
visits during periods of no rainfall. Representative soil
samples of the three distinct sediments (sandstone, mudrock,
and pediment) were collected from the top 300 mm layer for
detailed material characterization. The samples were
obtained in 20 L containers, sealed to conserve the field
water content, transported to the Advanced Geotechnical
Engineering laboratory at the University of Regina, and
stored at 24 °C.

The geotechnical index properties were determined for
preliminary soil assessment according to the following
ASTM test methods: specific gravity (Gs) by the Standard
Test Method for Specific Gravity of Soil Solids by Water
Pycnometer (D854-06); grain size distribution (GSD) under
both dry and wet conditions using the Standard Test Methods for Particle-Size Distribution (Gradation) of Soils Using
Sieve Analysis (D6913-04(2009)); and consistency limits by
the Standard Test Methods for Liquid Limit, Plastic Limit,
and Plasticity Index of Soils (D4318-10).

The water retention curve was determined in accordance
with the ASTM Standard Test Methods for Determination of
the Soil Water Characteristic Curve for Desorption Using a
Hanging Column, Pressure Extractor, Chilled Mirror
Hygrometer, and/or Centrifuge (D6836-02(2008)e2). To

Table 66.1 Summary of index properties

Soil property	Sandstone	Mudrock	Pediment
Specific gravity, G_S	2.7	2.8	2.7
<0.075 mm (%)	33 (17)	98 (4)	42 (21)
<0.002 mm (%)[3]	15	67	17
Liquid limit, w_l (%)	39	96	31
Plastic limit, w_p (%)	31	47	23
USCS symbol	SM	CH	SC

Note Numbers in parentheses pertain to dry condition

develop a clear understanding of the entire curve, data over a wide range were generated using a pressure extractor for high water content samples and a dew point potentiometer (WP4-T) for low water content samples.

66.3 Index Properties

Table 66.1 summarizes the geotechnical index properties of the investigated sediments. The measured G_s correlated well with material type, namely; 2.7 for sandstone containing iron-based constituents, 2.8 for mudrock possessing clay minerals, and 2.7 for pediment receiving washed materials from the above (Imumorin and Azam 2011). The fines content (<0.075 mm) increased from dry to wet states for all materials: sandstone, from 17 to 33 %; mudrock, from 4 to 98 %; and pediment, from 21 to 42 %. This is attributed to the removal of particle coating from the larger grains due to physical detachment of ultrafine particles, chemical dissolution of soluble materials, and breakdown of larger aggregates. Further, the corresponding clay size fraction (<0.002 mm)

due to wetting measured 15 % for sandstone, 67 % for mudrock, and 17 % for pediment. These data suggest that grain size thinning in sandstone (classified as silty sand, SM) and pediment (classified as clayey sand, SC) was due to coating removal from sand size grains thereby resulting in erosion whereas the phenomenon was related to breakdown of clay aggregates in mudrock (classified as a fat clay, CH).

66.4 Water Retention Curve

Figure 66.2 gives the water retention curves in the form of volumetric water content (θ) versus suction for the investigated materials. The measured data for the sandstone fitted well to a bimodal distribution with two air entry values (AEV): a lower value (6 kPa at θ = 65 %) corresponding to drainage through cracks followed by a higher value (150 kPa at θ = 40 %) associated with flow through soil matrix. When the field samples were progressively desaturated, air first entered into the discontinuities at low suction (Fredlund et al. 2010). The fissures originate from geologic overburden

Fig. 66.2 Water retention *curves* for the investigated sediments

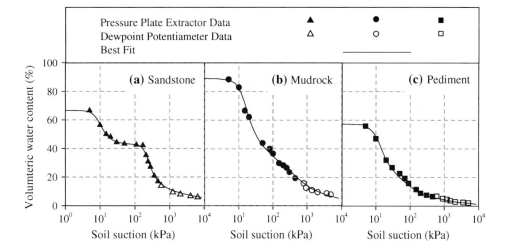

removal and grow over time due to material erosion and dissolution during water flow. According to Azam and Khan (2014), seasonal variations in water availability (snow melt in spring and rainfall in summer) and water deficiency (low rainfall and freezing in fall and winter) result in physical and chemical weathering at Avonlea. Because of the associated increase in fines, these particles got trapped in the relatively bigger pore spaces around the coarser particles. Water flow through the recently developed smaller pores resulted in the observed matrix AEV that, in turn, correlated well with the dense nature of the material. Finally, the residual suction was found to be 500 kPa (at $\theta = 15$ %) and is attributed to the low clay content of the sandstone.

The water retention curve of the mudrock exhibited a unimodal trend with an AEV of 5 kPa (at $\theta = 83$ %) due to drainage through large pores. Such a low AEV for a fat clay is attributed to desiccation cracks in the material from an initially saturated condition. Despite some healing due to expansive clay minerals, numerous swell-shrink cycles over geologic time render these discontinuities to have much lower tensile strengths than the soil aggregates thereby leading to a quick drainage through these paths of least resistance. Subsequent application of suction affected the aggregated soil structure and eventually forced air to enter into the pore system of the popcorn-like motif. Further desaturation resulted in driving water through the individual aggregates and eventually resulted in a residual suction of 1,500 kPa (at $\theta = 12$ %). Overall, the water retention curve correlated well with the high clay content and the high water adsorption capacity of the mudrock.

A unimodal water retention curve was obtained for the pediment. The AEV for this material was found to be 5 kPa (at $\theta = 53$ %) and the residual suction was 80 kPa (at $\theta = 15$ %). These values corroborated well with the granular and loose nature of the pediment, as observed in the field and measured in index properties.

The volumetric water content values at saturation indicated that water storage was highest for mudrock ($\theta = 83$ %) followed by sandstone ($\theta = 65$ %) and then by pediment ($\theta = 53$ %). These saturated conditions during a rainfall together with the water retention curve (unsaturated conditions during dry weather) mean that the eroded and dissolved materials from the sandstone are washed away and get deposited in the pediment because pores in the intermediate mudrock are sealed due to clay swelling. This confirms the genetic relationship between the sandstone and the pediment, as postulated by Raghunandan and Azam (2012).

66.5 Conclusions

Knowledge of the engineering properties of soils is vital for civil infrastructure construction in surface sediments that are directly affected by seasonal weather variations. Three distinct sediments (fissured sandstone with a steep slope of 60°; popcorn-textured mudrock with a mild slope of 30° and; eroded pediment with a flat slope of 3°) found at the Avonlea badland site were characterized. Based on laboratory investigations, the main conclusions of this work can be summarized as follows. The fines content increased from dry to wet state with 17–33 % for sandstone, 4–98 % for mudrock, and 21–42 % for pediment. The consistency limits indicated that the water adsorption capacity is highest for mudrock followed by sandstone and then by pediment. The water retention curve of sandstone showed bimodal distribution with a low AEV of 6 kPa pertaining to drainage through cracks and a high AEV of 150 kPa associated with flow through the soil matrix. The mudrock and pediment followed a unimodal curve with a single AEV of 5 kPa. These results mean that the eroded and dissolved materials from the sandstone are washed away and get deposited in the pediment because pores in the intermediate mudrock are sealed due to clay swelling. This confirms the genetic relationship between the sandstone and the pediment.

Acknowledgments The authors acknowledge the University of Regina for providing laboratory space and computing facilities. Thanks to the Natural Sciences and Engineering Research Council of Canada for providing financial support.

References

Azam S, Khan F (2014) Geohydrological properties of selected badland sediments in Saskatchewan, Canada. Bull Eng Geol Environ 73:389–399

Byers AR (1959) Deformation of the Whitemud and Eastend Formations near Clay bank, Saskatchewan. Trans R Soc Can 53:1–11

Christiansen EA (1979) The Wisconsinan deglaciation of Southern Saskatchewan and adjacent areas. Can J Earth Sci 116:913–938

Fredlund DG, Houston SL, Nguyen Q, Fredlund MD (2010) Moisture movement through cracked clay soil profiles. Geotech Geol Eng 28:865–888

Imumorin P, Azam S (2011) Effect of precipitation on the geological development of badlands in arid regions. Bull Eng Geol Environ 70:223–229

Raghunandan ME, Azam S (2012) Effect of dry-wet cycles on material composition of badlands. In: Proceedings, 65th Canadian geotechnical conference, vol 338. Winnipeg, Canada, pp 1–6

Case Studies of Post Investigation Geological Assessments: Hunter Expressway

67

David J. Och, Robert Kingsland, Sudar Aryal, Henry Zhang, and Geoff Russell

Abstract

This paper presents two case studies where detailed site specific engineering geological assessment during construction justified major changes to the approved design of road infrastructure elements to suit actual site conditions which resulted in a better engineering outcome and substantial cost savings. The road project was the Hunter Expressway, located in the Hunter Valley, some 120 km north of Sydney, which is a 40 km long four-lane dual carriageway motorway currently at the final stage of construction. Two locations (Bridge Viaduct 3 and Retaining Wall—RW18) were selected as case studies because the detailed construction-phase mapping work provided a refinement to the geological models that enabled the design of key elements to be changed or modified. This paper will present the detail of these two case studies and demonstrate the value of detailed site specific engineering geological assessment during construction in achieving better engineering outcomes.

Keywords

Hunter Expressway • Geology • Geotechnical • Cuttings • Abutment

67.1 Introduction

This paper presents two engineering geological mapping case studies for the eastern 13 km section of the Hunter Expressway, which was recently delivered as an alliance contract under the project name Hunter Expressway Alliance (HEA).

The Hunter Expressway provides a long-awaited relief of congestion and improvement of passenger and freight traffic movement in the region along the route between Sydney and Brisbane, the busiest road transport corridor in Australia.

This paper demonstrates the value of detailed construction-phase geological and geotechnical assessments that provided refinement, which enabled the design of key elements to be changed or modified during construction to deliver more robust and cost effective design solution for the project.

67.2 Project Geology

The project is underlain by the upper part of the Newcastle Coal Measures comprising a relatively thick unit of coarse-grained sandstone interbedded with beds of conglomerate, laminated fine-grained sandstone, siltstone, coal and tuff. Colluvium is present on the slopes and alluvium is deposited in the valley floor. The tuffaceous sedimentary rocks comprise tuffaceous sandstone, tuffaceous siltstone and tuffaceous claystone. The tuffaceous claystones are low-strength rocks that are particularly susceptible to weathering and are highly reactive (Aryal et al. 2013).

D.J. Och (✉) · R. Kingsland · S. Aryal · H. Zhang
Parsons Brinckerhoff, 680 George St, Sydney, 2001, Australia
e-mail: doch@pb.com.au

D.J. Och
School of Biological, Earth and Environmental Sciences,
University of New South Wales, Kensington, Australia

G. Russell
NSW Transport, Roads and Maritime Services, Newcastle,
Australia

G. Lollino et al. (eds.), *Engineering Geology for Society and Territory – Volume 6*,
DOI: 10.1007/978-3-319-09060-3_67, © Springer International Publishing Switzerland 2015

Structurally these rocks are cut by a dominant high-angled joint set trending NNE to NE with a subordinate joint set trending to the west. Also associated with the dominant joint set are some low to moderately angled joints, usually clay lined. This formation also forms prominent escarpments alongside with deep valleys which have exploited prominent tectonic joint sets. Valley cutting and associated horizontal stress relief of these thicker rigid sandstone units has resulted in irregular inter-bed fractures. Colluvial debris of variable thickness masks the frontal slope of these escarpments. Large areas of the project alignment, including the case study locations, are underlain by former underground coal mine workings predominantly in the Borehole Seam which was mined in various collieries during the early 1900s at typical depths of 70–100 m (Kingsland et al. 2012).

67.3   Geotechnical Investigation Phases

The project had four phases of geotechnical site investigation that formed the basis of the design of the expressway. Phase 1 was the geotechnical investigation for the conceptual design and planning approval. Phases 2 and 3 were completed for the concept design, project costing and detailed design. In some areas of the project alignment, factors such as time constraints, access limitations, environmental and cultural issues controlled the level of field investigations that could be completed at targeted locations for detailed design of specific earthworks or concrete structures. In these areas the detailed design was developed based on the geotechnical data available at the time. Phase 4 was completed during the construction period. It is this forth phase of investigation, predominantly geological mapping, that will be elaborated in the project case studies discussed.

67.4   Project Case Studies

The two case studies are presented herein. In both cases, a major review of the detailed design was required due to the reason that the actual site conditions differed from the design assumptions. Detailed geological assessment of the actual ground conditions exposed during construction together with further geotechnical modelling to suit the observed conditions formed the basis for the redesign of the structures. Regular site geotechnical inspections were carried out to validate the new design during construction.

67.4.1 Case Study 1: Viaduct 3 Abutment

Viaduct 3 comprises twin continuous, three span, single cell box girder bridge superstructures, approximately 199 m in length, carrying the eastbound and westbound carriageways of the expressway across a deep valley with steep side slopes within a densely forested rugged terrain—the Sugarloaf Range.

The detailed design of the substructures supporting the bridge was based on the subsurface investigation data available at the time of design and included bored piled foundations under the abutments (Fig. 67.1a). After the site was cleared for construction of the abutments, the area became accessible and surficial features including rock outcrops exposed. Detailed geological mapping of the abutment and the slope areas beneath the proposed piled foundation was undertaken, which revealed that the foundation geology below the western abutment (Abutment B) comprised sound rock with no adverse rockmass structures and was assessed to be much more competent from the surface. These actual conditions encountered were considered to be equally suitable for pad foundations (Fig. 67.1b).

Subsequent checks on bearing, sliding, eccentricity and overturning in accordance with AS5100 indicated that pad footings were adequate. Allowance for potential mine subsidence was also made from structural design perspective.

The design of Abutment B was therefore changed to pad foundation on the basis of detailed engineering geological mapping (Fig. 67.1c, d), the adequacy of the geotechnical capacity and stability and accommodation of mine subsidence.

67.4.2 Case Study 2: Cut10a Extension

The Expressway cut through a steep side hill (slope 2H:1 V or steeper locally) and formed a side cut on the north side of the alignment. A piled retaining wall RW18, 60 m long and 5 m maximum height, was selected as the design solution to retain the side slope in this section taking into consideration the alignment corridor constraints, clearing limit restrictions, cutting depth and geology. The geotechnical model for the design was based on all test data available from the area including borehole BH1056 (Fig. 67.2a–c), which was the only borehole completed at the detailed design closest to the wall. This borehole was located downstream of the proposed RW18 location as access to the upstream area for geological mapping or drilling was not possible during the design stages due to steep pre-existing topography and dense forestation. Geotechnical data available at the time of detailed design indicated that, at the highest section of the wall, the slope materials that would need to be retained would comprise about 2 m thick alluvium/residual clay over 5 m of extremely low to very low strength rockmass overlying high strength rock (Fig. 67.2a, b) and this profile was adopted as the geotechnical model for the retaining wall design.

The slope above proposed RW18 became more accessible only following corridor clearing at the start of the

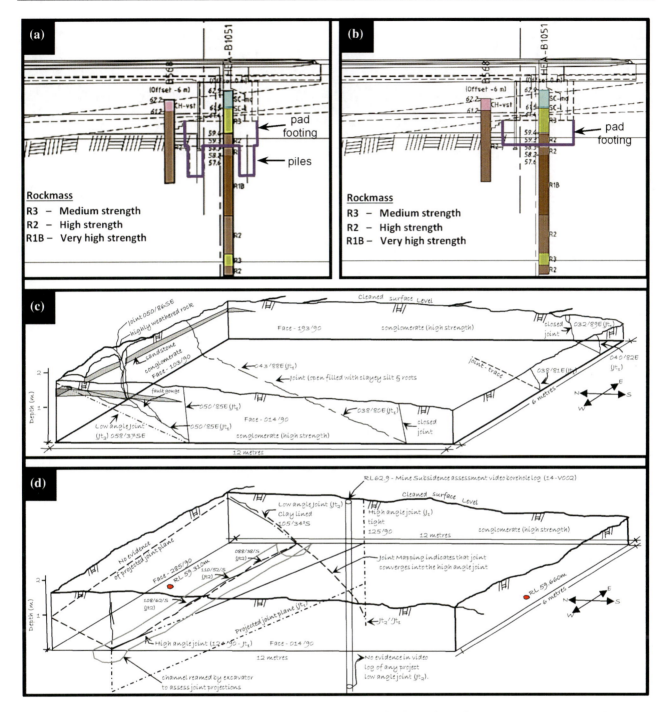

Fig. 67.1 a Original bored piled foundation design for BW11. **b** Revised pad footing design,.**c** Detailed geological mapping of excavated foundation pad—eastbound. **d** Detailed geological mapping of excavated foundation pad—westbound

construction phase. However, temporary excavation into the slope for site access road at the start of construction activities triggered localised slumping of superficial layers of in situ and colluvial materials on the upper batter section (section above the proposed wall) (Fig. 67.2d, e). The slope was stabilised by removal of slumped, dislodged and loosened

materials which inadvertently resulted in further steepening of the pre-existing steep natural slope forming the upper batter.

Subsequently, a detailed geological mapping (Fig. 67.2e) was carried out that followed further subsurface investigations including one borehole upstream of the slope and two

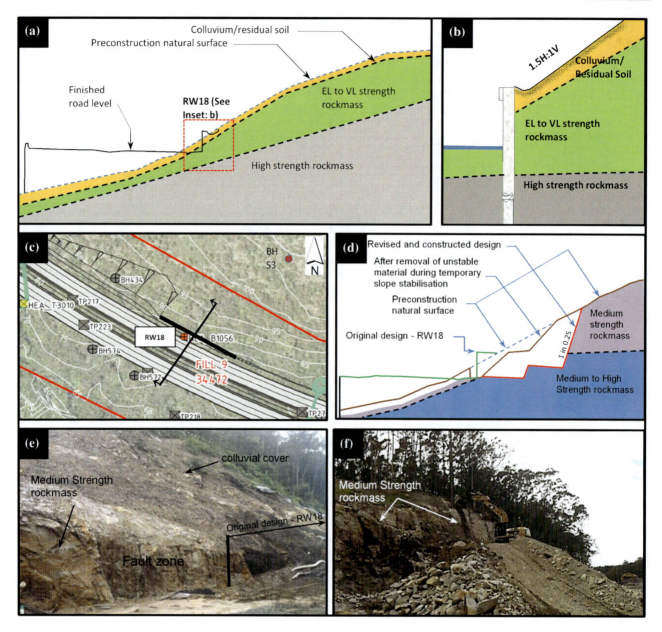

Fig. 67.2 **a** Original and revised design detail for RW18 and assumed geotechnical model. **b** Inset, with details of geotechnical model used for design at highest section of the wall **c** Location of geotechnical investigations used for original design. **d** Section detailing original design and the revised and constructed design. **e** Slope assessed during initial corridor clearing at the start of the construction phase. **f** Slope assessment which included geological mapping and test pitting

adjacent to the wall location in conjunction with progressive geotechnical inspection of the excavation and test pitting (Fig. 67.2f). The results of these investigations showed that the slope to the design road level and beyond composed of undisturbed rockmass with generally high strength sandstones and siltstones below a thin cover of residual soil and weak rock (1–2 m) at the natural surface. The outcome of investigations provided new opportunities to review the

geological models adopted to develop the retaining wall solution and revise if an alternative, more cost effective design would be possible to replace the retaining wall solution

The geological data collected during the initial investigations and during construction were reviewed to compare the actual geological conditions against design assumptions. The in situ materials forming the slope were assessed to be much more competent than the earlier design interpretation and based on these results, it was determined that the proposed retaining wall could be deleted and replaced by a steep rock batter with support/protection treatment where required (Fig. 67.2d, f).

67.5 Conclusion

The case studies presented from the Hunter Expressway (Bridge Viaduct 3 and Retaining Wall—RW18) illustrates the benefit of thorough construction phase geological mapping. The mapping executed when large areas of rock mass were exposed during construction enabled the geological models to be refined and in places considerably modified. As a result, approved geotechnical designs were able to be optimised deliver the best for the project solution with a considerable cost savings. These case studies demonstrate the critical importance of adequate engagement of engineering geologists during the construction phase to validate, challenge and modify design assumptions; ultimately achieving better engineering outcomes.

Acknowledgments The authors wish to thank Roads and Maritime Services for permission to publish and present this paper. We also want to acknowledge the support and encouragement offered by the leadership of the Hunter Expressway Alliance.

References

Aryal S, Kingsland R, Rees P, Russell G, Stahlhut O, Wheatley D (2013) Hunter Expressway, Australia: dealing with poor ground and subsidence. In: Proceedings of the institution of civil engineers, civil engineering special issue 166. ICE Publishing, London, Issue CE5, pp 22–27

Kingsland RI, Mills KW, Stahlhut O, Huang Y (2012) Mine subsidence treatment and validation strategies on Minmi to Buchanan section of the Hunter Expressway. In: Proceedings of the 11th Australia–New Zealand conference on geomechanics—ground engineering in a changing world, 15–18 July 2012, Crown Conference Centre, Melbourne, Australia

Contribution to the Behavior Study and Collapse Risk of Underground Cavities in Highly Saline Geological Formations

68

Mohamed Chikhaoui, Ammar Nechnech, Dashnor Hoxha, and Kacem Moussa

Abstract

The problem of saline soils reserved from occupation at Oran, North Algeria remained relatively unexplored or little known until recent years. Consequently, some studies were conducted, especially to characterize the real impact of an airport on these soils. The characterization of the real problems of saline soils, as well as the study of the behavior of their collapse under the coupled effect of thermal, mechanical and hydraulic, remains poorly known specially under an airport, where the instabilities and the risks of sudden collapses is an unknown problem for the authorities and citizens and the impact on the environment is not mastered. Under the action of water charged with carbon dioxide which dissolves the limestone, chalk or gypsum, many natural cavities are created. There are also pockets of dissolution filled with silt in the chalk, due to the irregularity of the contact chalk/silt. The flow of water can also enlarge the fractures at depth causing the silt that fills them and thus creating a surface subsidence due to infiltration. This phenomenon is found mainly in the dry valleys. To account for the effect of the hydro-thermomechanical coupling in predicting the collapse of saline soils, solutions were proposed for improvement of saline soil with a geosynthetic reinforcement, drainage, etc. These solutions are necessary for the proper design of airfield runways to avoid a disaster.

Keywords

Saline soils • Sebkha • Runways • Coupling • Disaster

M. Chikhaoui (✉) · A. Nechnech
Faculty of Civil Engineering, USTHB, 16111 Bab Ezzouar, Algeria
e-mail: mch_gcg16@yahoo.ca

A. Nechnech
e-mail: nechnech_a@yahoo.fr

M. Chikhaoui · A. Nechnech
Laboratory of Environment, Water, Geomechanics and Works LEEGO-USTHB, Bab Ezzouar, Algeria

D. Hoxha
Laboratory of PRISME—Faculty of Civil Engineering, Polytech'Orleans, Orleans, France
e-mail: dashnor.hoxha@univ-orleans.fr

K. Moussa
Department of Earth Sciences, FSTGAT, Oran University, Oran, Algeria
e-mail: moussakacem@yahoo.fr

68.1 Introduction

In the plain, about 6 km in the south of the town of Oran and Northern of the great Sebkha of Oran, Es-Senia airfield is located (Fig. 68.1). While the Es-Senia Oran airport is located near the large of Sebkha (western Algeria). Its extension has required the completion of a second runway. The presence of the Sebkha (salty lake) induces a rather complex geotechnical environment. This complexity is duo to the presence of dissolution cavities, of different sizes, in gypsum and the flat topography, promoting water stagnation.

G. Lollino et al. (eds.), *Engineering Geology for Society and Territory – Volume 6*,
DOI: 10.1007/978-3-319-09060-3_68, © Springer International Publishing Switzerland 2015

Fig. 68.1 Aerial view of the great Sebkha and view of the airfield of Oran, Chikhaoui et al. (2011)

68.2 Geological Factors and the Water Chemistry on Sebkha

Two important factors have governed the development of depression tectonics and climates conditioning subsequently the hydrographic network and vegetation fixation. Sedimentological analysis revealed that the climate of the sebkha, switches from sub humid to a semi-arid climate. The dynamics of Sebkha was reconstructed from the elements and geomorphological structures, Moussa (2007).

The active tectonics in Algeria is located in the northern part of the country. This region is located at the Africa-Eurasia boundary; the tectonic activity expresses the ongoing convergence between these two tectonic plates. Offshore, deformation affects the abyssal plain located close to the continent, by folding the Plio-Quaternary sediments cover. Along the slope and the continental platform, active structures with a continental extension crosscut this region. The coastal tectonics generates the coastal uplift with an average uplift of 0.50 m, Yelles-Chaouche et al. (2006). The climate of the Sebkha is Mediterranean in terms of daily and seasonal variations of precipitation, but steppe character in terms of the average temperature, annual rainfall and its seasonal distribution; it is the result of a coastal Mediterranean climate and a desert climate shelter, Moussa (2007).

Indeed, the region has a semi-arid climate characterized by an irregular rainfall, with dominance of brutal rain showers and prolonged dry season during which the heat causes intense evaporation. In winter, rain dissolves deposits of evaporites seconding materials salt solids. In summer, evaporation and capillary action causes a rise of salt resulting in the formation of efflorescence at surface.

Hydrogeology shows the superposition of two bodies of water, one deep freshwater (110 m) and located at the Messinian limestone, the other shallow salt water (82 m) located at the Pliocene formations. All around the Sebkha, several marshes are made per capillary of salt water back to the surface of the lake, Moussa (2007). The human activities have a significant influence on the physico-chemical quality of groundwater. This interaction affects the content of major elements (Ca^{+2}, Mg^{+2}, Na^+, K^+, Cl^-, SO_4^{2-}, HCO_3^-...etc.).

The Sebkha is represented by a recent alluvial deposit with the composition: sodium chloride 20 %, calcium sulfate 5 %, silica (SiO_2) 50 %, alumina and iron 20 %, carbonate of lime and magnesia 5 %, sodium chloride 1/5 % of salt Sebkha. The lake basin is supplied by sodium chloride brines down ravines of Tessala and especially by a Triassic layer that would contain per liter about 1.78 g of sodium chloride, 11 g of calcium chloride and 4 g of potassium chloride a rate of 193 g of salt, Boualla et al. (2011, 2013).

The northern side of the basin receives fresh water, located at the average depth of 4 m; the Triassic water is six times saltier than the Mediterranean water. This indeed gives 30 g/l of salt, while the one of the great Sebkha closes 180 g/l of chlorides, so it's a real salt mine; the Triassic water rises to the surface by capillary action and distributes its salts throughout the thickness of alluvium it crosses.

The decrease of the dry residue of floodwaters, cause increased levels of Ca^{++}, HCO_3^- and a subsequent decrease of the Na^+ and Cl^-. Sodium chloride (NaCl) salt is the most dominant in the water and soil of the Sebkha. The waters of the Sebkha are sodium chlorides and sulphates, Moussa (2007).

68.3 Geotechnical Problems and Soil Improvement

Under the action of water charged with carbon dioxide which dissolves the limestone, chalk or gypsum, many natural cavities are created (Fig. 68.2). One may observe dissolution pockets filled with silt in the chalk, due to the irregularity of the contact chalk/silt, Chikhaoui et al. (2011, 2013). The presence of underground cavities beneath a airfield pavement can be very dangerous, as collapse or subsidence occurs very likely.

Fig. 68.2 Photos of various cavities at Es-Senia airport, Chikhaoui et al. (2011)

The method that has been advocated for the detection of its cavities is the forced compaction along the track. This is followed by an excavation to a depth of 1.35 to 1.5 m to eliminate the surface cavities, then high energy compaction using a 50-t compactor at the bottom, to obtain good density at the base but also to identify cavities.

To avoid risk of collapse at the long term, the use of geosynthetics (geogrid 30/30 and geotextile 400/50) has become indispensable. The out of balance occurs when the subgrade can no longer exert an equal reaction to the applied load, the upper layers in turn unleash. There will be punching of the ground seat. Figure 68.3a, b show that for a given load, there is a relationship between the bearing capacity of the soil and the thickness of the pavement, Chikhaoui et al. (2011).

The principle is based on the assumption that the introduction of geotextile and geogrid may contain a cavity of 1 m in diameter with an overload of 345 kN/m^2. After treatment of the pavement foundation layer the thickness will be circa 69 cm, a reduction of 50 % of the total thickness, Chikhaoui et al. (2011, 2013). The aim is to divide all surchanges over a larger area and reduce pressure on the foundation soil to increase the bearing capacity. The incorporation of biaxial geogrids allows a lateral confinement of the foundation and reduction in the amount of aggregates in the range of 40–60 %, while offering the same capacity and the same functionality.

Geotextiles are often used in road and airport works as a tool for separation and filtration. Their performance as reinforcement depends on the geotextile-soil interaction. The study of geotextile-soil interaction under cyclical load (wheel), carried out by Bhandari and Han (2010), shows that the geotextile placed at a depth of 25.0 mm led to the registration of smaller displacement (Fig. 68.4a), than the unarmed case (Fig. 68.4b).

68.4 Dissolution Mechanism

It is known that the bicarbonate ion is the basis of the formation of calcium carbonate ($CaCO_3$) which is poorly soluble in water, and it is the main component of limestone:

$$Ca^{2+} + 2HCO_3^- \rightarrow CaCO_3 + H_2CO_3$$

where H_2CO_3 is carbonic acid.

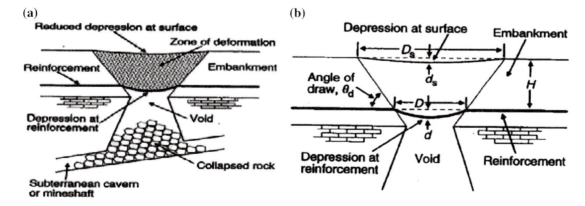

Fig. 68.3 **a** Conceptual role of reinforcement in limiting surface deformations due to subsidence, BS-2006. **b** Parameters used to determine reinforcement, BS-2006

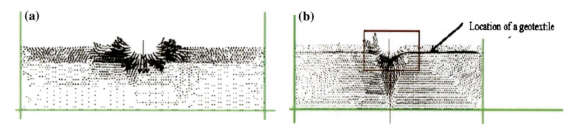

Fig. 68.4 **a** Field of displacement vectors of particles with no reinforcement geotextiles (the maximum displacement % 13.26 mm. **b** With geotextile reinforcement at a depth of 25 mm (max % 11.9 mm), Bhandari and Han (2010)

Fig. 68.5 Diagram for underground genesis of cavities in great Sebkha

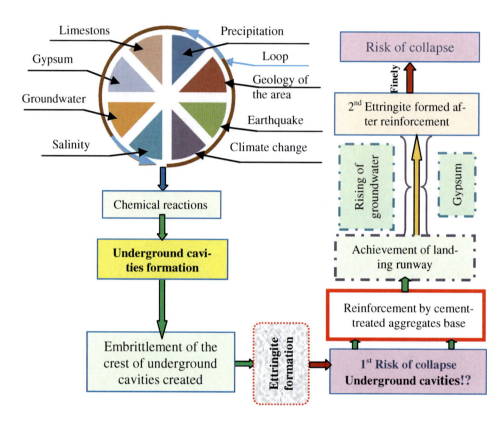

In an acidic medium, the calcium carbonate is converted to a calcium bicarbonate $(Ca^{2+}, 2HCO_3^-)$, which is very soluble in water, it is definitely a salt of a weak base $(Ca(OH)_3^-, pKa \approx 12)$ and a weak acid $(2HCO_3^-, pKa \approx 6)$, Boualla et al. (2011). The $Ca(OH)_3^-$ at atmospheric pressure and ambient temperature, gaseous and volatile is able to react with a strong acid to obtain calcium salts, water and carbon dioxide.

$$CaCO_3 + 2HCl \rightarrow CO_2 + CaCl_2$$
$$CaCO_3 + 2RCOOH \rightarrow CO_2 + CO_2 + Ca(RCOOH)_2$$

Under the simultaneous actions of groundwater leading the dissolution of gypsum and natural successive effects of earthquakes and dryness periods in Sebkha cause the creation of ettringite around weak areas of limestone. In geological times, this dislocation resulted in underground cavities which might collapse at any time under different dynamic actions (landing of airplanes) refer to Fig. 68.5.

68.5 Conclusion

The chemical nature of water depends on the path it has followed from the ground surface to the aquifer system. At first, it undergoes a surface modification due to the evaporation and then it evolves in the unsaturated zone and finally acquires the mineralization on the level of the water due to its contact more or less along the reservoir rock. All chemical analysis of samples basin Sebkha, have revealed an understanding of some parameters describing the physico-chemical water quality. The dominance of bicarbonate indicates that they constitute the bulk of the mineralization of the water, Boualla et al. (2011, 2013). In addition, the seismic activity is amplified all around the Sebkha Moussa (2007). This explains the accelerating of underground cavities formations. Subject to understand the mechanism of formation of cavities, it requires studying the exchanges matter in CO_2 at the water—atmosphere interface and those in Ca^{2+} and CO_3^{2-} ions in water interface—rock, Chikhaoui et al. (2011, 2013).

References

Bhandari A, Han J (2010) Investigation of geotextile-soil interaction under a cyclic vertical load using the discrete element method. J Geotextiles Geomembranes 28:33–43

Boualla N, Benziane A, Derrich Z (2011) Reflection of investigations physicochemical watershed Sebkha Oran. ScienceLib, Editions Mersenne: Environnement, vol 3, N°:111104

Boualla N, Moussa K, Benziane A (2013) Spatialisation des parametres physico-chimique pour la reconnaissance de l'etat des ressources en eau du bassin Sekha d'Oran. ScienceLib, Editions Mersenne: Geologie, vol 5, N°:130910

Chikhaoui M, Nechnech A, Moussa K (2013) Contribution to the study of coupling hydro-thermo mechanical for the prediction of collapses of airfield runways built on saline soils. In: 14th REAAA conference, P02A. Kuala Lumpur, Malaysia, pp 638–648, 26–28 Mar 2013

Chikhaoui M, Nechnech A, Haddadi S, Ait-Mokhtar K (2011) Contribution a la prediction des effondrements des pistes d'aerodromes construites sur des sols salins. XXIVe Congres mondial de la Route. Mexico, Ref. 0671, 26–30 Sep 2011

Moussa K (2007) Etude d'une Sebkha; la Sebkha d'Oran (Ouest algerien). Doctorat thesis Es-Sciences, Oran University, Algeria

Yelles-Chaouche A, Boudiaf A, Djellit H, Bracene R (2006) La tectonique active de la region nord-algerienne. CRAAG. Algeria. Elsevier, C.R. Geo- science, vol 338, pp 126–139

Seabed Properties for Anchoring Floating Structures in the Portuguese Offshore

69

Joaquim Pombo, Aurora Rodrigues, and A. Paula F. da Silva

Abstract

Anchoring structures in marine environments requires a good knowledge of seabed properties, based on laboratory tests and geophysical data acquisition. This work is addressed to the analysis of one anchor site located off the Portuguese coast, combining laboratory tests and high resolution seismic profiling. Seven different classes of sediments were individualised (according to United Soil Classification System), and coherent with the sedimentary layer architecture identified in the geophysical profiles; data provided awareness about the thickness of the unconsolidated sedimentary layer to the bedrock.

Keywords

Offshore • Floating structures • Seabed properties

69.1 Introduction

The recent implementation of policies for a sustainable development in the energy sector has led some countries, including Portugal, to seek new energy resources, with particular focus in renewable ones, namely those associated with the Ocean: tides, waves and offshore wind.

Due to the inaccessibility and constant site modifications, marine environments present extra difficulties for engineering geology modelling. Such complexity is expressed not only on the extremely high energy of environmental agents, both meteo and oceanographic, but also on the natural vertical and lateral variability of the marine sedimentary cover. In this context, the adequate offshore geological and geotechnical studies are due to support the implementation of those projects (Randolph and Gourvenec 2011), otherwise it can increase the cost of project or compromise its economic viability altogether.

Marine geotechnical studies are still innovative in Portugal, especially in the offshore, where the marine resources have the major potential. The first phase is underway, with the preliminary characterization of a small area of 13 km^2 and 30–60 m of depth in the Northern Portuguese shelf of S. Pedro de Muel. The purpose is to perform a multidisciplinary characterization of the upper layer of the continental shelf, considering not only the geologic parameters (bedrock topography and sedimentary layers thickness, internal stratigraphy and lateral facies variability), but also the geotechnical properties of the sediments.

This study will present the first marine geotechnical characterization of the Portuguese offshore, supported by the geologic description of the studied area.

69.2 Methods

The site investigation carried out in the study area, encompassed the following: acoustic and geophysical surveys (multibeam echo sounders, sub-bottom profiler and boomer); followed by two sediment sampling surveys with vibrocorer and grab sampler. Data and sample locations are displayed in

J. Pombo (✉) · A. Rodrigues
Instituto Hidrográfico, Lisbon, Portugal
e-mail: joaquim.pombo@hidrografico.pt

A. Rodrigues
e-mail: aurora.bizarro@hidrografico.pt

A.P.F. da Silva
CICEGe and Dept. Ciências Da Terra, Faculty of Sciences and Technology (FCT), Univ. NOVA of Lisbon, Lisbon, Portugal
e-mail: apfs@fct.unl.pt

G. Lollino et al. (eds.), *Engineering Geology for Society and Territory – Volume 6*,
DOI: 10.1007/978-3-319-09060-3_69, © Springer International Publishing Switzerland 2015

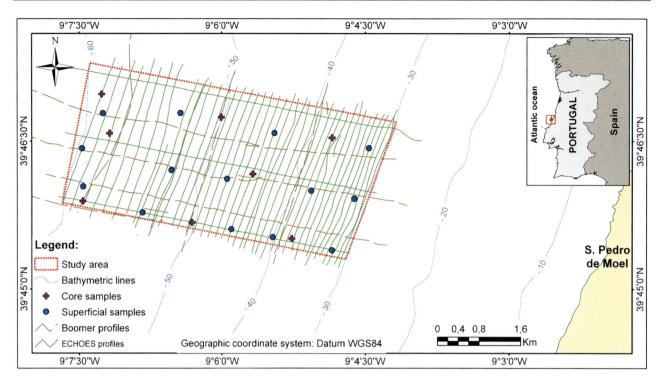

Fig. 69.1 The study area and location of seismic profiles, superficial and vertical sediments sampling

Fig. 69.1. Multibeam and sub-bottom profiling data (50 profiles) was acquired simultaneously with a Kongsberg EM710 and IXSEA Echoes 3500 chirp. Additional boomer profiles (9 profiles) were acquired with an Applied Acoustics Engineering AA200 system.

Two sampling surveys were performed using electrical vibrocorer Rossfelder P5 (8 vertical samples) and a Smith McIntyre grab (14 superficial samples).

Once in the laboratory, the corers were submitted to several tests, namely magnetic susceptibility, P-wave determination and X-ray radiography.

After extraction (2–2 cm) with a metallic ring, samples were analyzed using international normalized methods: the grain size distribution; determination of particle density; specific weight; determination of water content; carbonate content; organic matter content; Atterberg limits; and triaxial tests (CK_0D).

69.3 Preliminary Results

69.3.1 Morphology and Bottom Sediments Characterization

Multibeam data allowed the construction of a MDT surface (Fig. 69.2a), which showed that the area has a very smooth morphology with no major outcrops.

Despite the general gentle slope (about 0.3 %), a morphologic structure is recognized at 55 m of depth, consisting in a 1 m vertical displacement of the sea bottom. This feature marks the transition of a fine sedimentary deposit (between 30 and 55 m of depth) to a coarser one (between 55 and 60 m of depth), the latter also being characterized by wave ripples (NE-SE) about 3 m wavelength and 19 cm height according to Nichols (2009). In the rest of the studied area no morphological features were recognized.

The backscatter energy of the multibeam survey was analysed and processed with the GEOCODER engine (University New Hampshire), allowing the remote classification of the sea floor (Fig. 69.2b). Comparing the backscatter mosaic with the MDT surface, there is a good match between both surfaces, with the less intensity of the backscatter corresponding to the finer sediments and the higher intensity (lighter areas) corresponding to the coarser sediments.

The classification of the superficial samples and the uppermost 15 cm of the vertical samples according to Unified Soil Classification System (ASTM D2487-06), allowed the identification of three distinct sedimentary units (Fig. 69.3):

- Sand deposits poorly calibrated (SP), between 30 and 45 m of depth, characterized by fine sand in the order of 97 % and fine sediments less than 3 %; $C_u = 1.95$ and $C_c = 1.02$;

(a)

(b)

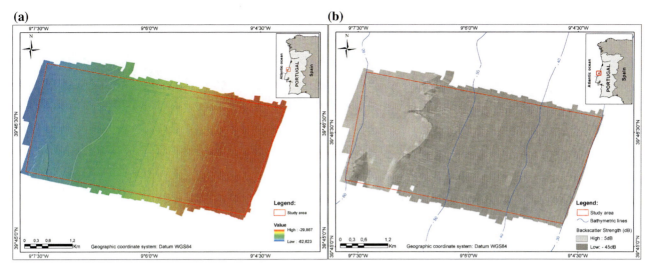

Fig. 69.2 Multibeam processed data: **a** MDT surface and **b** backscatter mosaic

Fig. 69.3 Mapping of the identified superficial sedimentary deposits (USCS): isopach of the unconsolidated sedimentary layer and location of the boomer seismic profile

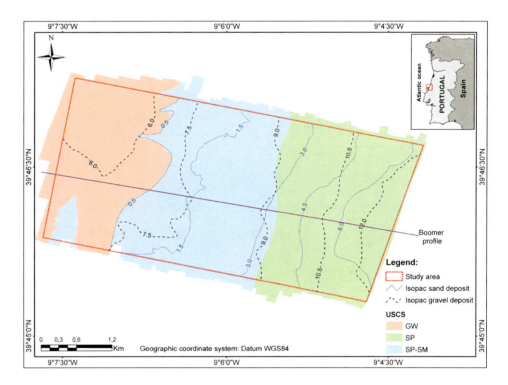

- Silty sand deposit (SP-SM) between 45 and 55 m of depth, characterized by poorly calibrated fine sands (91 %) and about 9 % of fine sediments; $C_u = 3.89$ and $C_c = 1.68$;
- Sandy gravel deposit (GW), between 55 and 60 m of depth, characterized by 78 % of well graded gravel, 19 % of sand and 3 % of fine sediments; $C_c = 2.31$ and $C_u = 5.62$.

69.3.2 Geological Structure

Geophysical profiles indicate that superficial units (identified in the backscatter) correspond to the major sequences identified in the geologic structure of the area. The basal sedimentary unit, a coarser deposit (sandy gravel sediments), covers the bedrock and is exposed in the deeper area, below

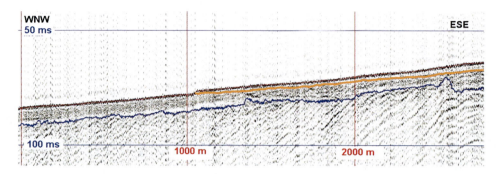

Fig. 69.4 Extract of a representative boomer seismic profile. *Blue line* corresponds to the top of bedrock and *orange line* to the top of the coarser unit and base of the finer deposit; brown reflector is the sea bed

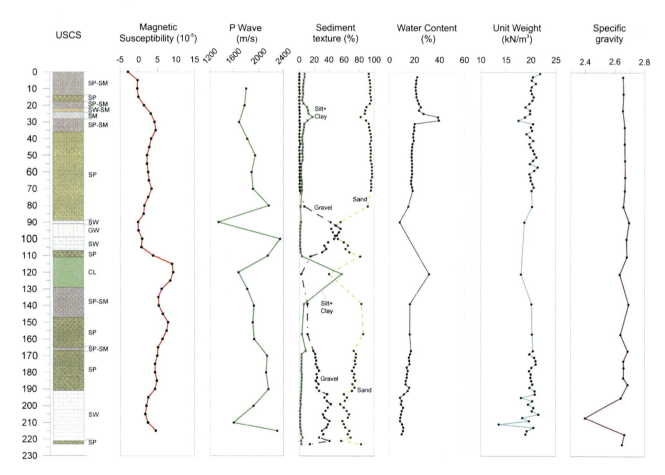

Fig. 69.5 Geotechnical proprieties of the marine deposits versus depth

55 m of depth. The more recent unit, covering the coarser deposit, is identified in the shallow area, and is composed by sandy sediments (Fig. 69.4).

According to the isopach distribution (Fig. 69.3), the sandy deposit increases its thickness towards the coast line, from 0.5 to 7 m, while the sandy-gravelly deposit, despite the difficulty in the basal delimitation, seems to be a thicker unit (5–13 m, also increasing westward).

69.3.3 Sedimentary Layer and Geotechnical Characterization

The first 15 cm of the sedimentary cover (Fig. 69.5) are characterized by negative or null susceptibilities, probably because of high pore fluid contents; this susceptibility also increase with the increasing silt-clay fraction in the sediments. The higher values (9×10^{-5} SI) are measured at

110 cm coinciding with the presence of a sandy silt level (55 % of silt-clay fraction). Deeper levels have lower susceptibility values due to the increase in the grain-size and biogenic particles.

The grain size also interferes with the compressional waves velocity (vp), as it was shown by Buckingham (2005). The lower value (1,260 m/s) was registered at 90 cm in the transition between the sandy gravel layers and finer ones. On the sandy silt level (between 110 and 128 cm), the wave velocity is 1,600 m/s, in agreement with the values observed by Hamilton (1980) for this type of sediment, while at the gravelly sand level a maximum value of 2,400 m/s was registered.

The water content along the profile is highly variable with the vertical variation of textural parameters: a content of about 20 % is typical of the sandy sediments, 17 % of the sandy gravel sediments and values between 30 and 40 % in the muddy sediments. The highest values are inversely correlated with the volume weight of the sediments as expected, due to the presence of muddy sediments.

The Atterberg Limits, determined for the sandy silt layer (between 110 and 128 cm), indicate a liquidity limit (LL) of 39 % and plasticity index (PI) of 16 %, which, according to USCS (ASTM D 2487-06), corresponds to the CL group. This type is characteristic of a sandy lean clay inorganic with medium plasticity (Burmister 1949 in Das 2006) and with active clays (A = 1.4) (Skempton 1953).

The density of the particles varies between the 2.40 and 2.69, with no apparent trend with depth. The minimum value of 2.40 is registered at the 207 cm, probably correlating with the biogenic particles.

The mechanic characterization of the sediments is currently ongoing and is performed by triaxial tests, which will allow defining the resistant parameters of the sediments.

69.4   Conclusions

The depositional architecture, recognized in the seismic profiles, was confirmed with an integrated analysis of sedimentologic and geotechnical parameters, resulting in a more accurate identification of internal interfaces and heterogeneities.

Although the data interpretation is still in a preliminary stage it was possible to identify 7 types of different sediments that reflect the source variability and equilibrium with different meteorological and oceanographic conditions, along the depositional cycle.

The use of two non-destructive methods, magnetic susceptibility and compression waves, revealed itself to be quite satisfactory for the identification of the different interfaces, although the compression waves should be made in lower intervals.

References

Buckingham MJ (2005) Compressional and shear wave properties of marine sediments: comparisons between theory and data. J Acoust Soc Am 117(1):137–152

Das MB (2006) Principles of geotechnical engineering. Toronto. ISBN:0-534-55144-0

Hamilton El (1980) Geoacoustic modeling of the sea floor. J Acoust Soc Am 68(5):1313–1340

Nichols G (2009) Sedimentology and stratigraphy, 2nd edn. Wiley-Blackwell, Hoboken. ISBN 978-1-4051-9379-5

Randolph M, Gourvenec S (2011) Offshore geotechnical engineering. Spon Press. ISBN13:978-0-415-47744-4 (hbk)

Skempton A (1953) The colloidal activity of clays. In: Proceedings of 3rd international conference on soil mechanics and foundations engineering, vol 1, pp 57–61

Model of Permafrost Thaw Halo Formation Around a Pipeline

70

Pavel Novikov, Elizaveta Makarycheva, and Valery Larionov

Abstract

Permafrost thawing around buried pipelines transporting warm hydrocarbons can result in dangerous bending of the pipeline and it's possible damage. Prediction of the permafrost-pipeline thermal interactions reflected in the thawing halo dimensions is an important problem both for design and operation of the pipeline. The aim of this work was to develop a predictive model of thawing halo formation with high ratio of the accuracy in estimating thawing halo dimensions to the quality of input data. A theoretical model that considered factors possibly influencing permafrost-pipeline thermal interaction was developed. Oil pipeline field experiments were performed in Eastern Siberia, Russia. Thaw halo extent and other local factors were measured. Then calculated thawing halo dimensions were compared with the measurements obtained in the field experiment. In addition evaluation of the individual influence of each factor considered in the model was performed in numerical studies. The developed and tested predictive model of permafrost thaw halo formation demonstrated a reasonable ratio of accuracy in estimating thawing halo dimensions to the quality of input data. The dimensions of the thawing halo were most sensitive to the temperature of the transported hydrocarbons, thermal conductivity of frozen soil and the initial temperature field of permafrost within the region of the pipeline's thermal influence.

Keywords

Permafrost • Pipeline • Thaw halo • Thermal interaction • Soil properties

70.1 Introduction

Heat released from buried pipelines transporting warm hydrocarbons progressively thaws surrounding permafrost forming a permafrost thaw halo around the pipeline. Due to the non-uniform distribution of soil properties and massive ice deposits in the permafrost, differential settling of soil under the pipeline is likely to occur. This differential settlement results in bending strain in the pipe's wall which could lead to overstress and possible damage to the pipeline.

70.2 Aim

This work was dedicated to the prediction of permafrost thaw halo formation around a pipeline. A reasonably accurate predictive model of pipeline thermal interaction with permafrost is difficult but realizable if detailed data on all of the relevant factors influencing this interaction are available (Jianfeng et al. 2009). But today obtaining full and precise information about these factors is quite difficult.

The aim of this work was to develop a predictive model that provides high ratio of the accuracy in estimating thawing halo dimensions to the quality of input data.

P. Novikov (✉) · V. Larionov
Bauman Moscow State Technical University, Moscow, Russia
e-mail: novikov-p-a@yandex.ru

E. Makarycheva · V. Larionov
Emergency Situations Research Center Ltd., Moscow, Russia

G. Lollino et al. (eds.), *Engineering Geology for Society and Territory – Volume 6*,
DOI: 10.1007/978-3-319-09060-3_70, © Springer International Publishing Switzerland 2015

Table 70.1 Input data list

Group name	Parameters
Pipeline parameters	Pipe wall temperature (t_{oil}); pipe outside diameter; pipe wall thickness; thermal insulation coating thickness (h_{ins}); thermal conductivity coefficient of thermal insulation coating; trench geometry
Ground and backfill properties	Layer thickness; thermal conductivity coefficient of thawing ground (S_t); thermal conductivity coefficient of frozen ground (S_f); heat capacity of thawing ground (C); moisture content; density; porosity; melting temperature; thawing factor; compressibility factor of thawing ground; density
Permafrost parameters	Depth; average annual temperature near the daylight surface (t_{perm})
Climate data	Average monthly temperatures; average monthly wind speed; average monthly snow depth; power density of solar radiation; air-ground heat transfer coefficient; daylight surface albedo

Fig. 70.1 Permafrost thaw halo measurement (*1* a daylight surface; *2* a pipeline; *3* a seasonally thawing layer; *4* a measured border of the thawing halo; the size of cells is 1 × 1 m)

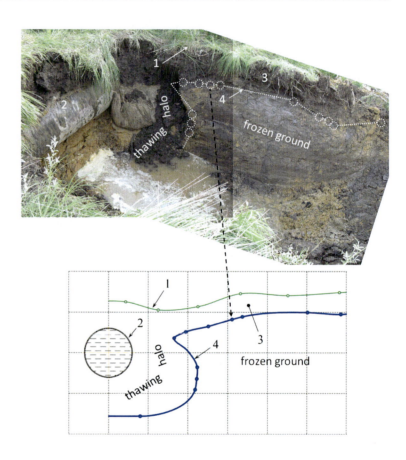

70.3 Model

A list of factors (temperature of the transported hydrocarbons, soil properties, geometry of the trench, etc.) possibly influencing permafrost-pipeline thermal interaction was constructed (Table 70.1) and from this list a model predicting this interaction was created.

70.4 Oil Pipeline Field Experiment

Permafrost thaw halo was measured around the oil pipeline located in Eastern Siberia, Russia. Measurements of thaw halo extent were taken for two sections of the pipeline after

3 years from the start of its operation (Makarycheva et al. 2013). In addition, all information about local conditions that affect permafrost-pipeline thermal interaction was collected as accurately as possible.

Measurements of thaw halo for the first section in June are shown on Fig. 70.1.

70.5 Validation

The thawing halo formation around the pipeline was calculated via the developed theoretical model using conditions recorded at the pipeline as input data. Then calculated halo dimensions were compared with the measurements of the thawing halo for each of two sections during June and

Fig. 70.2 Comparison of the measured and calculated permafrost thaw haloes (*1* a daylight surface; *2* an initial position of the pipeline; *3* a current position of the pipeline (by taking into account ground subsidence); *4* a border of the trench; *5* a measured border of the thawing halo; *6* a seasonally thawing layer; *7* a seasonally frozen layer; *8* a measured border of the seasonally frozen layer; *9* a calculated border of the thawing halo; *10* a calculated depth of the thawing halo under the pipeline; the size of cells is 1 × 1 m)

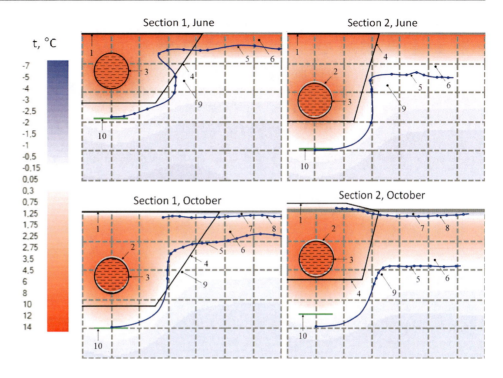

October in the third year of the pipeline's operation. Figure 70.2 demonstrates the agreement between calculated and measured results.

70.6 Demands on Initial Data Accuracy

Factors that could be managed during the pipeline design or operation were chosen among all the factors influencing the thawing halo formation. A series of numerical studies was conducted to evaluate the impact of each of these factors on permafrost thaw halo dimensions. The most influential six manageable factors are presented in the Table 70.2.

70.7 Conclusions

In conclusion, the predictive model of permafrost thaw halo formation that makes reasonable demands on initial data quality and provides sufficient accuracy of calculated thawing halo dimensions was developed and tested. According to the results of the numerical studies, the dimensions of the thawing halo are most sensitive to the temperature of the transported hydrocarbons, thermal conductivity of frozen soil and the initial temperature field of permafrost within the region of the pipeline's thermal influence. The developed model can be used as a basis for further investigations of the stress state of a pipeline in the conditions of soil subsidence.

Table 70.2 Influence of each factor on permafrost thaw halo formation

Factor[a]	−				Nominal input				+			
	Thawing halo dimensions, m				Thawing halo dimensions, m				Thawing halo dimensions, m			
	Depth under a pipe	Half-width	STL, m	SFL, m	Depth under a pipe	Half-width	STL, m	SFL, m	Depth under a pipe	Half-width	STL, m	SFL, m
Section 1, June												
C (∓ 30 %)	2.90	2.81	0.91	−	2.90	2.84	0.91	−	2.90	2.97	0.91	−
t_{oil} (∓ 30 %)	2.78	2.56	0.91	−					3.10	3.03	0.91	−
S_f (∓ 30 %)	3.10	2.69	0.91	−					2.85	2.97	0.91	−
S_t (∓ 30 %)	2.90	2.34	0.84	−					2.90	3.09	0.91	−
h_{ins} (on/off/-)	2.55	2.66	0.91	−					−	−	−	−
t_{perm}(∓ 30 %)	2.85	2.50	0.91	−					3.10	3.25	0.91	−
Section 2, June												
C (∓ 30 %)	3.93	2.13	1.41	−	3.93	2.13	1.41	−	3.93	2.13	1.41	−
t_{oil} (∓ 30 %)	3.73	1.88	1.72	−					4.10	2.38	1.84	−
S_f (∓ 30 %)	4.13	2.31	0.94	−					3.80	1.97	1.41	−
S_t (∓ 30 %)	3.88	2.00	0.97	−					3.98	2.25	1.91	−
h_{ins} (on/off/-)	3.43	1.91	1.41	−					−	−	−	−
t_{perm}(∓ 30 %)	3.78	1.94	1.09	−					4.09	2.31	1.53	−
Section 1, October												
C (∓ 30 %)	3.98	2.75	1.91	0.19	3.98	2.15	1.60	0.18	3.95	2.16	1.69	0.19
t_{oil} (∓ 30 %)	3.75	1.97	1.69	0.19					4.15	2.38	1.69	0.19
S_f (∓ 30 %)	4.20	2.25	1.63	0.13					3.80	2.06	1.69	0.19
S_t (∓ 30 %)	3.93	2.00	1.44	0.28					3.98	2.06	1.69	0.19
h_{ins} (on/off/-)	3.38	1.94	1.63	0.19					−	−	−	−
t_{perm}(∓ 30 %)	3.80	2.00	1.56	0.19					4.20	2.41	1.91	0.19
Section 2, October												
C (∓ 30 %)	3.50	2.70	1.91	0.19	3.48	2.20	1.87	0.18	3.48	2.70	1.91	0.19
t_{oil} (∓ 30 %)	3.30	2.50	1.88	0.19					3.65	3.05	1.88	0.19
S_f (∓ 30 %)	3.68	3.00	1.88	0.13					3.35	2.65	1.94	0.22
S_t (∓ 30 %)	3.43	2.28	1.66	0.28					3.50	2.90	1.91	0.09
h_{ins} (on/off/-)	3.00	2.60	1.88	0.19					−	−	−	−
t_{perm}(∓ 30 %)	3.35	2.65	1.72	0.19					3.56	2.85	2.16	0.19

STL Seasonally thawing layer, *SFL* Seasonally frozen layer

[a] see Table 70.1

References

Jianfeng X, Basel A, Ayman E, Paul J (2009) Permafrost thawing-pipeline interaction advanced finite element model. In: Proceedings of the ASME 28th international conference on ocean, Offshore and arctic engineering OMAE 2009, Honolulu, Hawaii, pp 97–102, 31 May–5 June 2009

Makarycheva EM, Larionov VI, Novikov PA (2013) Experimental studies of thawing halo for verification and calibration of forecasting mathematical models. Herald Bauman Moscow State Tech Univ 1 (48):109–116

Assessing Rock Mass Properties for Tunnelling in a Challenging Environment. The Case of Pefka Tunnel in Northern Greece

71

Vassilis Marinos, George Prountzopoulos, Petros Fortsakis, Fragkiskos Chrysochoidis, Konstantinos Seferoglou, Vassilis Perleros, and Dimitrios Sarigiannis

Abstract

The investigation of alternative solutions for the Thesaloniki Ring Road has been one of the major project design challenges in the recent years in Greece. Pefka tunnel was included in one of the alternatives that have been proposed and thoroughly examined. The total length of the tunnel was ~ 1450 m and it had two branches, with three lanes per branch. According to the geological design, the tunnel was to be excavated through a great variety of formations, such as travertine, shales and graphitic shales, alternations of meta-siltstones and graphitic shales, schistosed and intensively folded meta-siltstones, limestones, gabbros and peridotites. The large number of in situ and laboratory tests allowed (a) the reliable estimation of the intact rock properties and (b) the development of new relationships between the uniaxial compressive strength (σ_{ci}) and the point load test index (I_{s50}). These values were used for the estimation of the rock mass properties employing different methodologies. Finally, based on all available data, the anticipated rock mass behaviour in tunnel excavation is described. The key issues of this procedure are illustrated in the present paper.

Keywords

Weak rocks • Tunnel behaviour • Rock properties • Geotechnical classifications • I_{s50} correlations

71.1 Introduction

The aim of the alternative solutions for the Thessaloniki Ring Road was to decongest the traffic volume on the existing ring road and improve the access to the airport around the city. Pefka tunnel (~ 1450 m) was included in one of the alternatives that have been proposed and thoroughly examined. It had two branches, with three lanes per branch. The maximum inner width of the section is ~ 14.50 m, the maximum inner section height ~ 10.50 m and the maximum overburden height ~ 60.0 m.

Since the tunnel was to be excavated in the vicinity of the town of Pefka through a great variety of formations with different geotechnical parameters, a detailed geotechnical investigation program was carried out. The paper describes the key issues of the geological and geotechnical design focusing on the geological model, the estimation of the rock mass parameters and the anticipated ground behaviour during the excavation.

V. Marinos (✉)
Faculty of Sciences, School of Geology, Division of Geology, Aristotle University of Thessaloniki, 541 24, Thessaloniki, Greece
e-mail: marinosv@geo.auth.gr

G. Prountzopoulos · P. Fortsakis · F. Chrysochoidis · K. Seferoglou
Odotechniki Ltd, 59 3rd September Street, 10433, Athens, Greece

V. Perleros
56 Dionysou Street, 152 34, Halandri, Greece

D. Sarigiannis
Egnatia Odos S.A., 6th km Thessaloniki—Thermi, 60030570 01, Thessaloniki, Greece

71.2 Geological Setting

The outer ring road of the city of Thessaloniki is mainly situated in the foothills of Hortiatis Mountain. The Pefka Tunnel is located at the NE foothills of the Asvestochori valley along the tectonic trench of SE-NW direction.

The tunnel was planned to be driven in the Aspri Vrysi—Hortiatis unit formations of the Perirodopic zone (graphitic schists, meta-siltstones and limestones) and also in ophiolitic ones (peridotites, gabbros). The unit of Aspri Vrysi—Hortiati consists of metamorphosed old flysch series that have been transformed to clayey schists, black graphitic and light green schists, siliceous schists and metasiltstones. These formations may alternate in places. Limestones are interjected inside the schists and conclude this unit. Peridotites and gabbros are also met in separate places. Finally, in the wider area of exit portal deposits of travertine are found, originating from older karstic springs discharges of. These formations are assigned with certain codes (G1–G7) and presented in detail in the following paragraphs. The geological longitudinal section of the tunnel is illustrated in Fig. 71.1.

The graphitic schists (G2b) have clayey-silty composition and intense schistosity. Their main characteristics are the frequent shears along the schistosity and occasional clay-fillings. While the formation generally has a seamy-disturbed structure, it is sometimes found foliated with clayey zones. However, it gradually alternates with more compact and less schistosed metasiltstones (G3b). Green schists (G2a) are limited in the tunnel overburden zone. They are chlorite schists, which are disturbed with slickensided to clayey surfaces. In depth the formation becomes more compact with well-defined schistosed structure. Siliceous schists (G2c) are particularly strong and have a more compact structure. Another formation met along the tunnel is the metasiltstones (G3a). They are highly schistosed and folded, but retain compact structure. The schistosity does not separate the overall rock mass, but is contained within the blocky structure. However there are some fractured zones found along faults, where the rock mass is disturbed or even disintegrated.

Limestones (G4) are also found in the tunnel area, but their connection to other formations is not well defined. They are white, recrystallized and compact with minimum fracturing and it is believed that are found as tectonic "lenses" within the schists. The basic and ultrabasic rocks are met in the central part of the tunnel, building a compact and blocky rock mass. The formation of gabbros (G5) extends near the exit portal while the peridotites (G6) at the centre of the tunnel. In depth, they have generally compact (blocky) structure without exhibiting any schistosity or serpentinisation, while on the surface they are found more disturbed and loosened due to weathering. Finally, travertine (G1), eluvia and scree (G7) deposits are also found in the area, but they are not affecting the tunnel construction significantly.

71.3 Engineering Geological—Geotechnical Model

The geotechnical investigation for the Pefka tunnel comprised 23 sampling boreholes, one for a pressuremeter test and one for a dilatometer test, totally adding up to 1290 m. Thirteen piezometers were installed for the monitoring of the groundwater table as well as two inclinometers for the monitoring of ground movements in an area that was susceptible for a potential landslide. The geotechnical evaluation procedure included the following steps:
- Description of the geological formations in the project area (e.g. G1, G2). Different rock mass qualities could be found under the same geological formation.
- Rock mass characterization applying the GSI system on borehole cores.

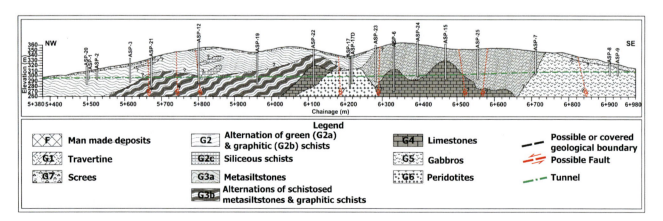

Fig. 71.1 Geological section of Pefka tunnel

- Evaluating the borehole data, five rock mass categories were defined (A, B, C, D, E) corresponding to different rock mass structure and quality of the discontinuities (Fig. 71.2).
- Statistical assessment of intact rock properties for all formations (e.g. G1, G2)
- Estimation of the rock mass parameters for every combination of geological formation and rock mass category (e.g. G2-D, G3-A).
- Description of the anticipated ground behaviour during tunnel excavation.

The rock mass quality was evaluated using the GSI rock mass characterization system (Marinos and Hoek 2000; Marinos et al. 2005) (Fig. 71.2) and the rock mass behaviour was categorized using the Tunnel Behaviour Chart (TBC) proposed by Marinos (2012). Based on this evaluation the rock mass types were:

- Rock Mass Type A: Blocky to intact rock masses with good interlocking and fair to good quality of the surfaces of discontinuities. These rock masses are expected to be stable and only local gravity failures may be observed (Geological formations: G3a, G4, G6).

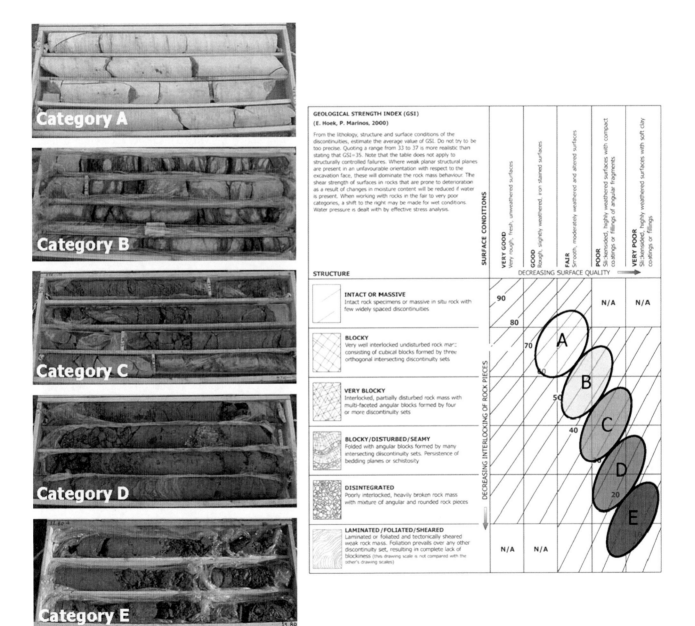

Fig. 71.2 GSI projections for every rock mass category (A–E)

Table 71.1 Engineering geological—geotechnical parameters for a number of combinations of geological formation (e.g. G1, G2) and rock mass category (e.g. A, D)

Geotechnical unit	GSI	σ_{ci} (MPa)	m_i	E_i (MPa)	γ (MN/m³)	E_m (Mpa)
G2-B	40	15	6	5,000	0.027	1,200
G2-E	15	6	6	2,500	0.025	140
G3a-A	60	18	7	9,000	0.027	4,500
G3a-E	15	12	7	6,000	0.025	250
G3b-E	15	10	7	5,000	0.025	220
G4-A	65	28	12	24,000	0.027	12,000
G5-B	45	14	27	15,000	0.027	3,000

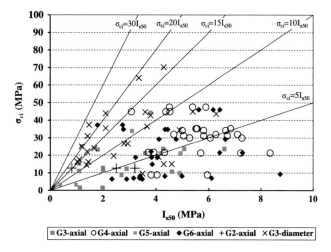

Fig. 71.3 Correlation between the results $I_{s(50)}$ and σ_{ci} values for the rocks: (i) G2: Schists, (ii) G3: Metasiltstones, (iii) G4: Limestones, (iv) G5: Gabbro, (v) G6: Peridotite

- Rock Mass Type B: Blocky to very blocky rock masses with fair to poor quality of the surfaces of discontinuities. Wedge fall or slide is the most probable failure mechanism (Geological formations: G2, G3a, G3b, G5).
- Rock Mass Type C: Blocky to seamy /disturbed rock masses with fair to poor quality of the surfaces of discontinuities. The potential failure mechanism in these rock masses would be small wedge failures or chimney type failures. (Geological formations: G2, G3b, G6).
- Rock Mass Type D: Disturbed to disintegrated rock masses with fair to poor quality of the surfaces of discontinuities. The potential failure mechanism would be chimney failure, due to the disturbance that decreases the overall cohesion, which could propagate to more extensive ravelling. In the tunnel section of high overburden, depending on the intact rock strength, small deformation could be developed. (Geological formations: G2, G3b).
- Rock Mass Type E: Sheared and laminated rock masses with poor to very poor quality of the surfaces of discontinuities. Significant deformation and face instabilities

may be developed in the area of high overburden or slope instabilities near the portals due to low rock mass strength. (Geological formations: G2, G3a, G3b).

The representative values of the engineering geological and geotechnical parameters for a number of Geotechnical Units are summarized in Table 71.1. Among a significant number of the numerous laboratory and in situ tests that were carried out, the indirect estimation of the intact rock uniaxial compressive strength (σ_{ci}) via the results of the Point Load test (I_{s50}) are presented here. Since there is a large scatter in the equations correlating Is with σ_{ci}, (e.g. Kahramman 2001; Tsiambaos and Sabatakakis 2004), a case specific correlation was established. Based on the results that are presented in the Fig. 71.3 the average value of the $\sigma_{ci}/I_{s(50)}$ ratio varied for the different formations from 5.25 to 9.90.

71.4 Conclusions

The paper is addressing a methodology developed for assessing rock mass properties in view of tunnel excavation in a challenging environment, using the case of Pefka Tunnel in northern Greece. The tunnel alignment was designed to cross eight distinct geological formations, each demonstrating different rock qualities. The methodology used consisted a four step assessment plan that evaluated the separate parameters (geological formations, intact rock properties, rock mass type, rock mass properties and rock mass behaviour) and achieved to designate the tunnel sections with similar anticipated rock mass behaviour and potential failure mechanisms, allowing the simplification and the optimization of the temporary support design.

References

Kahraman S (2001) Evaluation of simple methods for assessing the uniaxial compressive strength of rock. Int J Rock Mech Min Sci 38:981–994

Marinos P, Hoek E (2000) GSI—a geologically friendly tool for rock mass strength estimation. In: Proceedings of the GeoEng2000 at the international conference on geotechnical and geological engineering, Melbourne, Technomic publishers, Lancaster, pp 1422–1446

Marinos V (2012) Assessing rock mass behavior for tunnelling. J Environ Eng Geosci 18(4):327–341

Marinos V, Marinos P, Hoek E (2005) The geological strength index: applications and limitations. Bull Eng Geol Env 64:55–65

Tsiambaos G, Sabatakakis N (2004) Considerations on strength of intact sedimentary rocks. Eng Geol 72:261–273

Vítor Santos, A. Paula F. da Silva, and M. Graça Brito

Abstract

As known, geological conditions are a complex challenge in any civil engineering work, however this is not as relevant as in tunnelling where ground knowledge usually has many uncertainties at the beginning of the construction phase. The prediction of the ground quality of advancing tunnel face represents, itself, a key step during tunnelling, allowing to prepare and mobilize the most appropriate means to proceed with the works, with an adequate level of safety, quality and efficiency. The authors tested a quantitative methodology for prediction of rock mass quality during drill and blast excavation. The paper is based on the application of three mathematical prediction methods for the estimation of rock mass quality: linear regression, geostatistical kriging and neural networks algorithms. Additionally, a fourth empirical method was also applied, based on engineering geologist's expertise, aiming to assess the deviation of the quality prediction. The methodology was tested using RMR information of several tunnels, dug in a granitic rock mass, integrating the power generation reinforcement of Picote dam, located in the northeast of Portugal. The obtained results revealed, for all cases, a significant correlation between the estimated RMR and the observed value, thus raising good expectations for the progress of the ongoing research works.

Keywords

RMR prediction • Mathematical modelling • Advance tunnel face

72.1 Introduction

Rock mass characterization for tunnel excavation results from geological and geotechnical surface mapping, complemented by indirect and direct prospecting methods. At the beginning of the construction phase, usually there is still a high level of uncertainty in the geological (including hydrogeological and tectonic) model, leading to unpredictable hazardous situations. To mitigate such lack of information, the authors tested a quantitative methodology for prediction of rock mass quality during drill and blast (D&B) excavation, aiming to select an adequate quantitative methodology for predicting rock mass rating (RMR) values and to anticipate the rock mass quality of the advancing tunnel face, providing considerations on the geotechnical behaviour of the ground. The earlier determination of the rock mass quality of the ground to be dug allows for the timely mobilization of necessary resources to ensure the stability of the tunnel, safeguarding quality, safety and effectiveness of construction.

The selected methodology should anticipate the behaviour and natural heterogeneity of the rock mass during the tunnel construction phase and must, also, be easy to

V. Santos (✉) · A.P.F. da Silva · M.G. Brito
CICEGe, Faculdade de Ciências e Tecnologia (FCT), Univ. NOVA de Lisboa, Lisbon, Portugal
e-mail: silva.santos@gmail.com

A.P.F. da Silva
e-mail: apfs@fct.unl.pt

M.G. Brito
e-mail: mgb@fct.unl.pt

A.P.F. da Silva · M.G. Brito
Dept. Ciências da Terra, Faculdade de Ciências e Tecnologia (FCT), Univ. NOVA de Lisboa, Lisbon, Portugal

implement and to produce immediate results, in order to respond to the dynamics of the D&B excavation. Hereafter, the results of the application of three mathematical prediction methods and of one empirical method are presented and their validity is discussed.

72.2 Case Study Presentation

The Picote II Repowering Scheme is located in the northeast of Portugal, near the Spanish border, in International Douro River, around 25 km south of Miranda do Douro (Fig. 72.1). Table 72.1 presents the tunnels that integrate the scheme works and summarizes their geometric characteristics.

In the study area, there are porphyritic two-mica granites, classified during the construction with the geomechanical classification (Bieniawski 1989). The RMR was registered at each section from the main tunnels integrating the hydraulic circuit.

72.3 Methodology

Usually, D&B excavation methods use daily dynamic excavation cycles allowing the prediction of rock mass quality and reducing the need for complex methods of treatment to improve rock mass quality.

This paper presents the results of the application and test of the benefits of using a set of mathematical models (geostatistics and neural network models) to assess the quality of rock mass and predicting the geotechnical behaviour of the advancing tunnel face during D&B. Geostatistical models have been widely used in the fields of geosciences, namely in the geotechnical domain (Brito et al. 1997), as well as neural network models have in geological (Leu and Adi 2011) and geotechnical (Miranda et al. 2007) studies.

The methodology adopted is described by the two following steps:
1. Estimation of the RMR of the advancing tunnel face by using (i) linear regression correlation, (ii) ordinary kriging geostatistical method, and (iii) neural networks method;
2. Comparison of the RMR values estimated by the mathematical models and the RMR obtained by the empirical knowledge of an engineering geology expert.

The progress of the excavation is simulated by repeating the procedure successively for each of the tunnels previously excavated sections, corresponding to an average length of about 3 m; this means that the prediction of ground quality corresponds to a distance of approximately 3 m beyond the excavation face (Fig. 72.2). This process is repeated for each of the tunnels of the case study, allowing the simulation of the RMR value of the advancing tunnel face.

The linear regression method was applied by considering the data concerning the last excavated sections: 3 ($n, n - 1, n - 2$), 5 ($n, n - 1, ..., n - 4$) and 10 ($n, n - 1, ..., n - 9$),

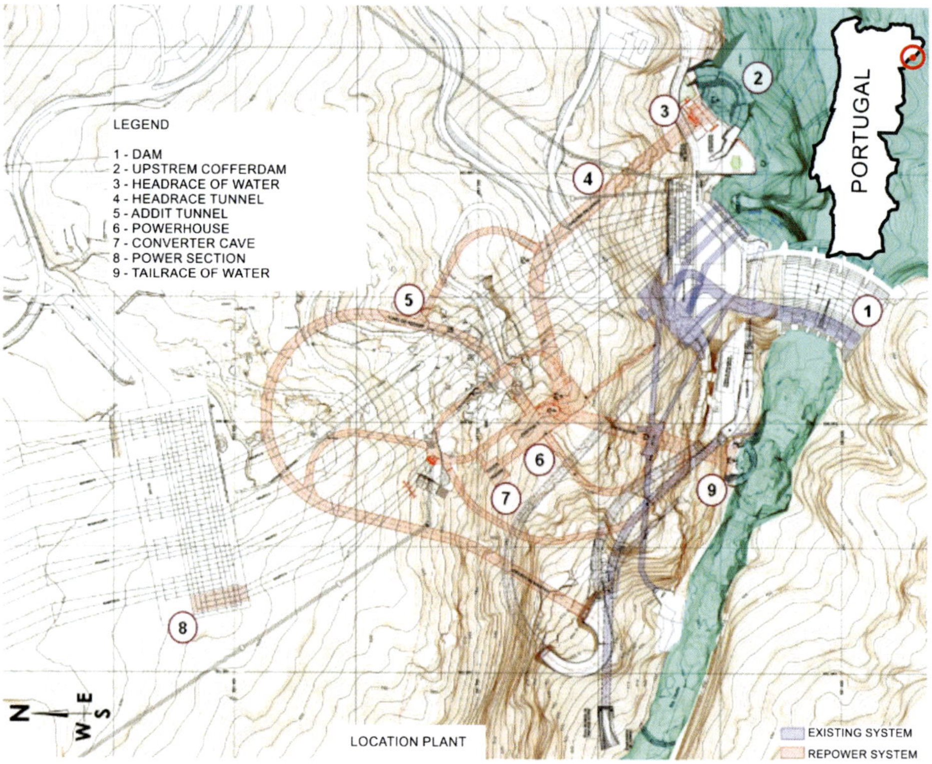

Fig. 72.1 Components of Picote II (Portugal) repowering scheme (EDP 2006, modified)

LEGEND

1 - DAM
2 - UPSTREM COFFERDAM
3 - HEADRACE OF WATER
4 - HEADRACE TUNNEL
5 - ADDIT TUNNEL
6 - POWERHOUSE
7 - CONVERTER CAVE
8 - POWER SECTION
9 - TAILRACE OF WATER

PORTUGAL

LOCATION PLANT

EXISTING SYSTEM
REPOWER SYSTEM

Table 72.1 Geometry of the underground main structures (adapted from EDP 2006)

Tunnel	Code	Length (m)	Diameter (m)	Height (m)	Section area (m^2)
Headrace	GC	300	12	12	113
Tailrace	GR	150	12	12	113
Powerhouse addit	TAC	625	8.5	8.1	62
Headrace addit	GAGC	141	5	5.5	26
Top powerhouse addit	GAAC	204	5	5.5	26
Tailrace addit	GAR	252	5	5.5	26
Gates chamber addit	GACC	121	5.5	5.8	29
Ventilation and safety	GVS	155	3.7	4	14.5

Fig. 72.2 Scheme for data collection of rock mass quality

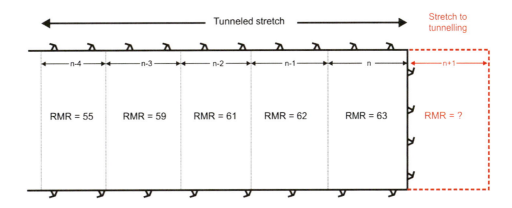

where section n corresponds to the last excavated tunnel face. Prediction of RMR values are obtained from the linear regression between observed—obs, and estimated RMR values, for a 3 m extension for the next $n + 1$ tunnel face.

Geostatistical kriging modelling started with the structural analysis of RMR behaviour along the sequential excavated tunnel sections ($n; n - 1; n - 2; ...; 1$). The obtained variogram was adjusted with a spherical model for an amplitude a, corresponding to the distance at which RMR values are no longer correlated. Ordinary kriging is used to estimate RMR values for a face at distance of $n + 1$.

For the application of neural networks methodology (neu), five independent columns containing the RMR value of the last five excavated tunnel sections ($n, n - 1, ..., n - 4$) and one dependent column that corresponds to the RMR estimated for the next ($n + 1$) section were considered. For each advancing face of the tunnel, new neural networks are generated, based in regression-based automated network search algorithms. The type of the neural networks used was the multi parallel layer (MPL) type, with $2 - 20$ hidden units. The active functions to develop the hidden neurons in MPL were the identity, logistic and the exponential, using a decay weight in a range between 0.0001 and 0.001. The number of neural networks generated was 20; of these, the one that showed less error in test is selected.

The empirical (exp) methodology is based on the analysis of RMR values along the excavated length. This methodology was applied to the case study using an experienced geological and geotechnical monitoring operator. All RMR values observed until section n were considered and arranged in a graphic presentation of RMR *versus* the extension excavated.

For evaluation of results, the correlation between the estimated RMR values *versus* in situ RMR$_{obs}$ was determined and the respective relative and absolute estimation errors were analyzed. The signal of this errors was also discriminated as, in this context, it represents an over or underestimation of rock mass quality and, consequently, of its stability/safety. To understand the influence of the tunnels geometry (length and cross sections area) in the RMR errors, the trending of errors along the tunnels sections was also analyzed.

72.4   Results and Considerations

The RMR estimated values obtained were subjected to statistical analysis as summarized in Fig. 72.3, where it is possible to refer that: (i) the values estimated by empirical method (RMR$_{exp}$) are the most similar to the RMR observed values (RMR$_{obs}$); (ii) the regression method applied to the last 3, 5 and 10 excavation sections (RMRregr3, RMRregr5

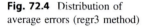

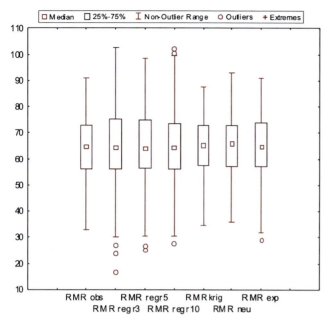

Fig. 72.3 Box-plot for estimated RMR values

and RMRregr10) presents extreme values in addition to the observed ones; and (iii) the RMR values estimated by ordinary kriging (RMR_{krig}) and neural networks (RMR_{neu}) methods do not represent the maximum and lowest values of real RMR (RMR_{obs}).

Regarding the average values, all methodologies revealed values close to those observed and, additionally, 50 % of the estimated values showed identical values to RMR_{obs}.

The methodologies that showed the highest correlation (88 %) between the estimated and the observed RMR were the empirical (*exp*) and ordinary kriging (*krig*). Nonetheless, other approaches have also presented satisfactory results (correlations above 79 % for the remaining ones).

The interpretation of the estimation errors was performed for each of the methodologies, and Figs. 72.4 and 72.5 show, as an example, the analysis applied to the estimated errors obtained by the regression method in the last three sections (*regr3*). Comparing the graphs obtained for all methods it can be stated that: (i) the average errors obtained with the different estimation methods are very similar and show a normal distribution; (ii) kriging method (*krig*) presents the smallest error amplitude; (iii) neural network method (*neu*) presents the highest RMR error values; (iv) in general, the dispersion of the error is broadly identical throughout the range of RMR with some exceptions for RMR values ranging between 40 and 60; and (v) there is not a clear correlation between the errors and the length /cross section of the excavated tunnels.

The empirical method (*exp*) reveals, itself, as the most effective, showing the relevance of an expert in the monitoring of tunnel construction. However, the use of mathematical tools represents also a good helper for less experienced operators.

Fig. 72.4 Distribution of average errors (regr3 method)

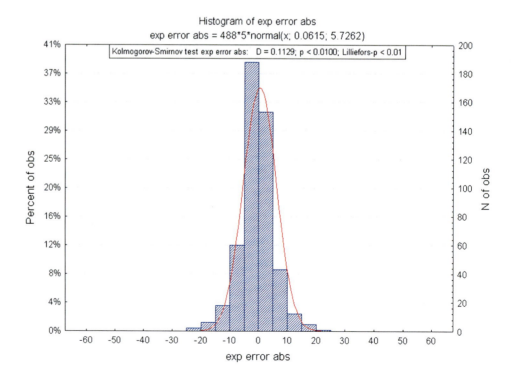

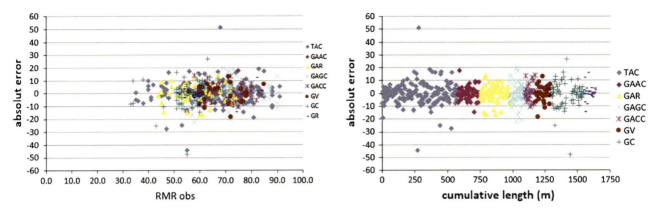

Fig. 72.5 Behaviour of absolute error of RMR: observed by expert, *left*, and by extension of different types of tunnels, *right* (regr 3 method)

72.5   Conclusions

The mathematical methodologies used for predicting the quality of a granitic rock mass by estimating the values of advancing tunnel face during D&B excavation works presented very good results, with high correlation between the estimated and the observed RMR values. Nonetheless, the skills and knowledge of an engineering geologist expert is still the most efficient and accurate method for monitoring geotechnical behaviour of rock mass excavation and avoid potential unexpected situations that may occur during tunnelling.

Acknowledgments The authors thank EDP for the authorization to use the information collected during construction of Picote II Repowering Scheme.

References

Bieniawski Z (1989) Rock Mass Classifications. Wiley, Hoboken, p 251

Brito MG, Durão F, Pereira HG, Rogado JQ (1997) Classification of heterogeneous industrial rocks: Three different approaches. In: Pawlowsky Glahn V (ed) IAMG'97–3rd annual conference international association mathematical geology. Barcelona, pp 875–879

EDP (2006) Aproveitamento hidroeléctrico do Douro Internacional - Reforço de Potência de Picote - Estudo Geológico Geotécnico, p 63

Leu S, Adi T (2011) Probabilistic prediction of tunnel geology using a Hybrid Neural-HMM. Eng Appl Artif Intell 24:658–665

Miranda T, Correia G, Sousa LR (2007) Use of AI techniques and updating in geomechanical characterisation. In 11th congress of the international society for rock mechanics: courses. ISRM, Lisboa, p 31

The Medium- to Long-Term Effects of Soil Liquefaction in the Po Plain (Italy)

73

Elio Bianchi, Lisa Borgatti, and Luca Vittuari

Abstract

The 2012 Emilia seismic sequence has shed light on some unusual geomorphological processes and related landforms observed in the Po Plain between the provinces of Modena and Bologna, namely small-scale sinkhole formation, in a non-karstic setting. In some of the areas previously affected by sinkholes, during the Emilia earthquakes, widespread coseismic effects were observed, as soil liquefaction, sand venting and ground cracks. Before 2012, these effects have been seldom observed in the Po Plain, mainly because of moderate seismicity. Known historical earthquakes, or eventually older events, could have been the original triggering factor of liquefaction of susceptible soils at shallow depth and formation of dikes and sills, as precursors of future sinkholes. To test this model, data collection on boundary conditions and a number of further field experiments is ongoing. In particular, the research is focused on three main issues: the setup of a geological model of the area, taking into account structural a tectonic features; the analysis of surface displacements horizontal (geodynamic) and vertical displacements (natural and/or artificial subsidence) and their relationships with the development of sinkholes. This is performed through the exploitation of SAR interferometric data and GPS data; geological and geotechnical characterization of soils, through a number of continuous boreholes, trenches and CPT tests for building cross sections and 3D models of areas prone to sinkhole development. These pieces of information are used for set up a numerical model and simulating the process of sinkhole triggering ad evolution in the River Po alluvial plain.

Keywords

Liquefaction • Sinkholes • GPS • InSAR • Po plain

73.1 Introduction and Methods

In Italy, natural sinkhole phenomena which are not directly connected to fluvial or karst processes are relatively uncommon Caramanna et al. (2008). On the basis of previous works, as well as available geological, geomorphologi-cal and geotechnical data (Bonori et al. 2000, 2010; Castellarin et al. 2006), the possible triggering factors and the evolution of these phenomena are described. An inventory carried out some years ago accounted for 28 areas affected by this type of sinkholes (Fig. 73.1).

To unravel the long- and short-term evolution of these landforms, a geological model of the area is built, taking into account structural a tectonic features; the analysis of surface displacements horizontal (geodynamic) and vertical displacements (natural and/or artificial subsidence) and their relationships with the development of sinkholes. This is performed through the exploitation of SAR interferometric data and GPS data; geological and geotechnical characterization of soils, through a number of

E. Bianchi (✉) · L. Borgatti · L. Vittuari
Dipartimento di Ingegneria Civile, Chimica, Ambientale e dei Materiali DICAM, Università di Bologna, Viale Risorgimento 2, 40136, Bologna, Italy
e-mail: elio.bianchi2@unibo.it

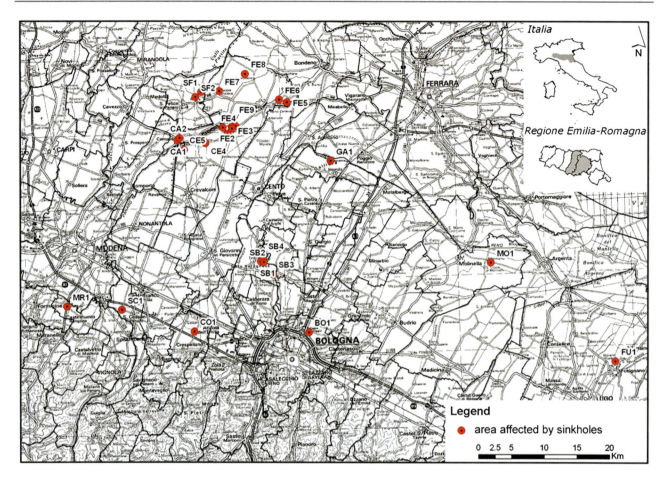

Fig. 73.1 Areas affected by sinkhole phenomena. Most of the areas are located in the Provinces of Modena and Bologna

Fig. 73.2 *Top* Cretaceous 3D structural model built in Rms (by ROXAR)

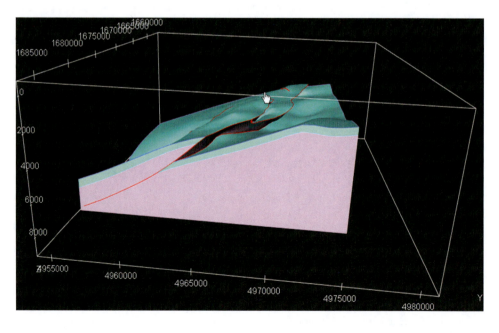

continuous boreholes, trenches and CPT tests for building cross sections and 3D models of areas prone to sinkhole development. These pieces of information are to be used to set up a numerical model and simulate the process of sinkhole triggering ad evolution in the River Po alluvial plain.

73.2 Structural Model

In this work we created a three-dimensional geometrical model of a sector of the Ferrara-Romagna fold and thrust system buried under the Plio-Pleistocene sediments of the Po Plain in order to model numerically the active stress field of the region (namely, the Mirandola fold). The input data derived from interpretation of seismic lines and geological sections reported in the literature (Boccaletti et al. 2003; Fantoni and Franciosi 2009; Pieri and Groppi 1981). A quality control was made using deep wells for oil exploration. The main tectonic elements in the model consist of a set of ramps that converge downward to the main detachment likely located within the Upper Triassic units (Fig. 73.2). A 3D model of deep geological units is needed to understand which are the structures which have generated ancient and recent earthquakes and their level of activity. This is an important issue to understand whether there is a correlation between the geomorphological phenomena observed before and after the earthquakes of May 2012.

73.3 PSInSAR Data Over the Study Area

Through monitoring 2003–2010 period, two datasets of satellite images (ascending and descending) were analyzed. The presence of the two datasets acquired with different geometry on the same area made it possible to carry out the decomposition of the motion, that is to calculate for each measuring point, the vertical component (up-down) and horizontal (E–W) of the movement starting from the original measurements carried out along the Line of Sight to the satellite (LoS). For each acquisition geometry (ascending and descending), the velocity of the movement is referred to

a point located on the vertical projection of the buried thrust above the surface, that is, where the thickness of the alluvial deposits is lower and therefore the Mesozoic substratum is more superficial. The spatial distribution of the velocity of surface displacement (horizontal and vertical) indicates a gradual increase moving away from the peak of the buried thrust in the direction of N–N—and S–SW (Fig. 73.3a, b). This technique allows us to measure and distinguish the phenomena of compaction induced by human activities (i.e., water pumping), and natural subsidence, and measure instantaneous effects, which depend on the movement of deep structures and which occur with earthquakes that may cause raising and lowering of the surface and locally soil liquefaction. The rate field has highlighted the presence of the effects of subsidence of the order of a few mm/y. The cause of such behaviour was not analyzed in detail, however, by analogy with other cases examined, it is possible to relate this increase in speed with the thickness variation of the compressible alluvial deposits. The decomposition of the motions (2003–2010 period), showed that the area of study is concerned essentially of vertical movements, with speeds generally lower than 5 mm/yr, while the horizontal component of the E–W is practically negligible.

Permanent deformation in the Po Valley from GPS data resulted in shortening with limited rates of a few mm/y Devoti et al. (2011). After the two earthquakes of May 2012, the maximum horizontal offset was observed at MO05, the site nearest to the epicenter of the main shock (30 mm horizontal and 73 mm vertical movement. Devoti (2012). This information, combined with the interferometric data, can help us to distinguish deep geodynamic effects from superficial and more local effects due to geomorphological processes and geotechnical characteristics of the soils.

(a)								**(b)**

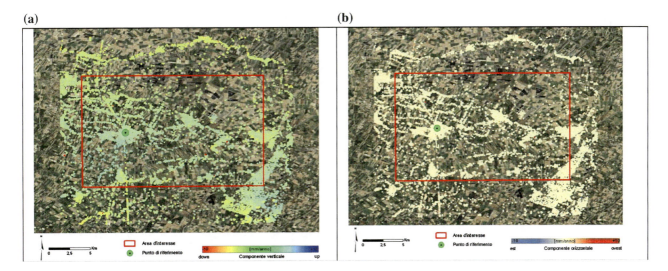

Fig. 73.3 **a** Vertical component of the velocity field. **b** Horizontal component E–W direction of the velocity field

73.4 Geological and Geotechnical Characterization of Sinkoles

These phenomena tend to develop in different geological settings. From the geomorphological point of view, some sinkhole occur near the apex of apenninic large alluvial fans, as well as in the lower alluvial plain (elevation from 64 to 6 m a.s.l.); generally, the areas fall within relatively short distances from streams, rivers and artificial channels. In most cases, the affected soils have silty-sandy texture related to alluvial ridges. The sinkholes appear both on arable land (61 %) and in orchards (25 %). Some of the sinkholes have formed in areas characterized both by low and by high rates of subsidence (from 5 to 50 mm/yr in the period 1992–2006, ARPA Ingegneria Ambientale 2007). On the basis of Cone Penetration Tests interpretation based of Schmertmann (1978) (Fig. 73.4), some of the affected areas have been characterized with reference to lithology, stratigraphy and relevant geotechnical parameters (Borgatti et al. 2010). The top unit is an overconsolidated cohesive soil, with sufficient clay fraction to form and retain shrinkage cracks in the so-called active zone (that locally can reach 5 m in depth). At a depth in the order of 6–10 m, a loose sandy unit can be typically found. The formation of seismically-induced liquefaction features is favored by: presence of liquefiable sediments (preferably a clean sand with a thickness of 1 m or more), an overlying low-permeability cap of silt and clay at least 1 or 2 m and less than 10 m in thickness, and a shallow water table (see Obermeier and Pond 1999). As these areas appeared to be prone to liquefaction, the evolution of sinkholes may be eventually related to liquefaction events.

The predisposing factors of modern sinkhole phenomena, thus, appear to depend on the stratigraphy and grain size distribution of recent loose alluvial sediments at shallow depth, and on specific hydraulic conditions of shallow semiconfined aquifers in the distal sectors of alluvial ridges. Ground shaking can cause liquefaction of susceptible soils and subsequent upsurge and eventual ejection of sands along pre-existing or newly formed ground cracks, eventually due to shrinkage in the active zone and/or coseismic fracturing in the overconsolidated cap. With time, at shallow depths, in clastic planar dikes and sub-horizontal sills, resedimentation and packing occur and small-scale proto-chambers may evolve, by successive collapses and enlargement, also accelerated by the erosion of infiltrated water in permeable materials. The ultimate triggering factor for the formation of the sinkhole, also after a relatively long time-lag from the original liquefaction event, may be related to local accidents that cause the final collapse of roofs (new seismic shaking, water table sudden drawdown, heavy vehicles transit etc.).

73.5 Conclusions

The 2012 Emilia seismic sequence has shed light on some peculiar processes and landforms. Previously, liquefaction phenomena have been seldom observed in this area, mainly because of moderate seismicity. During the Emilia earthquakes widespread coseismic effects were observed, as soil liquefaction and ground cracks. On the basis of the data gathered so far, a sinkhole evolution model is proposed: these landforms may be considered as secondary medium-term

Fig. 73.4 Geotechnical model of one of the case studies. *1* arable land, with clay, silt and sand, overconsolidated at places (qt = 0.6÷6.0 MPa); *2* low to medium consistency clay and silt (qt = 0.5÷1.5 MPa); *3* consistent clay and silt (qt > 1.5 MPa); *4* low consistency to soft clay and silt (qt ≤ 0.5 MPa); *5* sand and sandy silt with varying relative density (1.0 < qt < 10.0 MPa). In *red* the curve of the CPT; in *blue* the groundwater level, as measured in superficial wells

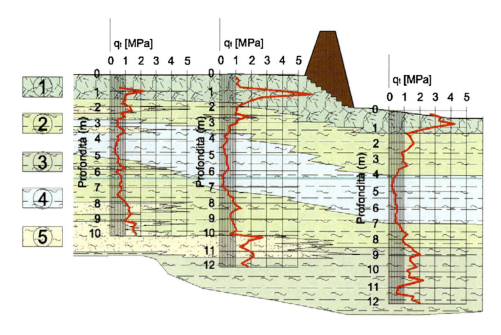

effects of earthquake-induced liquefaction. Known historical earthquakes, or eventually older events, could have been the original triggering factor of liquefaction of susceptible soils at shallow depth and formation of dikes and sills, as precursors of future sinkholes. In order to confirm this conceptual and geotechnical model, field and lab work is ongoing, also in order to set up a numerical model to simulate the process.

References

ARPA Ingegneria Ambientale (2007) La subsidenza in Emilia-Romagna. Il monitoraggio tramite interferometria satellitare. Esperienze a confronto. Arpa rivista 1/2008

Boccaletti M, Bonini M, Corti G, Gasperini P, Martelli L, Piccardi L, Tanini C, Vannucci G (2003) Carta sismotettonica della regione Emilia-Romagna. Scala 1:250.000. Regione Emilia-Romagna

Bonori O, Ciabatti M, Cremonini S, Di Giovanbattista R, Martinelli G, Maurizzi S, Quadri G, Rabbi E, Righi PV, Tinti S, Zantedeschi E (2000) Geochemical and geophysical monitoring in tectonically active areas of the Po Valley (Northern Italy). Case histories linked to gas emission structures. Geogr Fis Dinam Quat 23:3–20

Borgatti L, Bianchi E, Bonaga G, Gottardi G, Landuzzi A, Marchi G, Mastrangelo A, Rodorigo S, Vico G, Vittuari L (2010) Fenomeni di sprofondamento del piano di campagna in pianura padana: il ruolo del contesto geologico, geomorfologico e geotecnico. In: Nisio S (ed) 2° Workshop internazionale I Sinkholes, Roma, 2009, pp 181–201

Caramanna G, Ciotoli G, Nisio S (2008) A review of natural sinkhole phenomena in Italian plain areas. Nat Hazards 45(2):145–172

Castaldini D (1987) F° 75 Mirandola: un esempio di cartografia geomorfologica. Atti della Riunione dei Ricercatori di Geologia, Milano

Castellarin A, Rabbi E, Cremonini S, Martelli L, Piattoni F (2006) New insights into the underground hydrology of the eastern Po Plain (northern Italy). Boll Geof Teor ed Appl 47:271–298

Devoti R, Esposito A, Pietrantonio G, Pisani AR, Riguzzi F (2011) Evidence of large-scale deformation patterns from GPS data in the Italian subduction boundary, Earth Planet. Sci Lett 311:1–12. doi:10.1016/j.epsl.2011.09.034

Devoti R. (2012) - Combination of coseismic displacement fields: a geodetic perspective, Annals of Geophysics, 55 (4); doi:10.4401/ag-6119

Fantoni R, Franciosi R (2008) Geological framework of Po Plain and Adriatic foreland system. In: Proceedings of the 70th EAGE conference and exhibition, Rome

Fantoni R, Franciosi R (2009) Mesozoic extension and Cenozoic compression in Po Plain and Adriatic foreland. Rendiconti online Soc Geol It 9:28–31

Obermeier SF, Pond EC (1999) Issues in using liquefaction features for paleoseismic analysis. Seismolog Res Lett 70(1):34–58

Pieri M, Groppi G (1981) Subsurface geological structure of the Po Plain, Italy. Progetto finalizzato Geodinamica- Sottoprogetto 5, Modello strutturale, C.N.R., Pubbl. 414, Roma

Schmertmann JH (1978) Guidelines for CPT: performance and design. Report FHWA-TS-78-209, Federal Highway Administration, Washington DC

Part VII

Engineering Geological Problems in Deep Seated Tunnels

Convener Prof. Kurosch Thuro—*Co-convener* Heiko Käsling

Deep seated tunnels suffer both from high stress conditions and adverse rock conditions. In this session, contributions are welcome which address such problems during tunnel works using conventional as well as mechanical excavation. Focus may be on general engineering geological conditions, rock mechanical problems, stress-induced problems such as rock bursting and spalling, excavation problems or modelling of such conditions. Another focus may be on the TBM performance under high stress conditions, generating penetration problems or facing instability during TBM or conventional tunnelling, as well as connected problems including tool wear or machinery demolition. Contributions may include case studies, comprehensive views or methodical approaches.

Leaching Characteristics of Heavy Metals from Mineralized Rocks Located Along Tunnel Construction Sites

74

Nohara Yokobori, Toshifumi Igarashi, and Tetsuro Yoneda

Abstract

Soil and groundwater pollution caused by acid rock drainage (ARD) containing heavy metals leached from mineralized rocks is a serious environmental problem. Mineralized rocks are widespread throughout Hokkaido, Japan, and several tunnels for the Hokkaido Bullet Train Line are planned to be constructed through these mineralized areas. In this study, batch and column leaching experiments were conducted to investigate the leaching characteristics of heavy metals from the mineralized rock samples collected in these areas. The results showed that the mineralized samples contained substantial amounts of sulfide minerals (e.g., sphalerite (ZnS)), and that cadmium (Cd) was incorporated in some of these sulfide minerals. Moreover, the leaching concentrations of lead (Pb), Cd and arsenic (As) were higher than the Japanese environmental standards. The results of the column leaching experiment showed that these mineralized rocks could continuously release high concentrations of heavy metals for a long time. Therefore, such rocks should be disposed of properly to prevent the contamination of the surrounding environment.

Keywords

Heavy metals • Mineralized rocks • Sulfide minerals • Leaching

74.1 Introduction

Acidic leachate loaded with heavy metals and toxic metalloids is a very serious environmental problem because of the rapid deterioration of the surrounding soil and groundwater. This problem is usually limited to mining sites and mine tailings dams, but recent tunnel projects in Japan have excavated rocks that generated similarly acidic and heavy metals/metalloids loaded leachates when exposed to the environment (Tabelin and Igarashi 2009; Tatsuhara et al. 2012). The problem associated with these tunnel excavated rocks is similar to those of pyrite-rich mine wastes because the rocks are excavated along mineralized areas rich in sulfide minerals (Salinas Villafane et al. 2012a, b). When exposed to surface oxidizing conditions, these sulfide minerals, especially pyrite, are destabilized releasing acidity and high concentrations of heavy metals and metalloids (Younger et al. 2002). These phenomena are expected in the construction of the Hokkaido Bullet Train Line because tunnels will be excavated through mineralized areas of the island. Hazardous elements leached from excavated rocks have potential for soil and groundwater contamination without proper treatment. Therefore, it is important to understand the leaching characteristics of heavy metals from these rocks for the proper waste management of the tunnel excavated rocks. In this study, we evaluated the relationships between the mineral compositions and leaching characteristics of heavy metals from mineralized rocks.

N. Yokobori (✉)
Graduate School of Engineering, Hokkaido University, Sapporo, 060-8628, Japan
e-mail: yokobori@trans-er.eng.hokudai.ac.jp

T. Igarashi · T. Yoneda
Faculty of Engineering, Hokkaido University, Sapporo, 060-8628, Japan

G. Lollino et al. (eds.), *Engineering Geology for Society and Territory – Volume 6*,
DOI: 10.1007/978-3-319-09060-3_74, © Springer International Publishing Switzerland 2015

74.2   Materials and Methods

74.2.1 Site Description

Nine rock samples were collected near the planned tunnel construction sites. Figure 74.1 shows the sampling locations and the planned Hokkaido Bullet Train Line.

74.2.2 Chemical and Mineralogical Analyses

Chemical composition of the rock samples was quantified using X-ray fluorescence spectrometer (XRF) while the mineralogical composition of rock samples was determined using X-ray diffractometer (XRD). Surface observation and analysis was conducted using optical microscopes and electron probe microanalyser (EPMA).

74.2.3 Batch Leaching Experiments

Leaching characteristics of heavy metals from the samples were evaluated by conducting batch reactor-type experiments. The leaching experiments were done by mixing 15 g of crushed rock (<2 mm) and 150 ml of deionized water at 120 rpm for 6, 24 and 168 h using a reciprocal shaker. After shaking, the pH, electrical conductivity (EC), redox potential (ORP) and temperature of the suspensions were measured. The suspensions were then filtered through 0.45 μm membrane filters. The concentrations of heavy metals and coexisting ions in the filtrates were measured using an inductively coupled plasma atomic emission spectrometer (ICP-AES) and ion chromatographs, respectively.

74.3   Results and Discussion

74.3.1 Chemical and Mineralogical Properties of the Rocks

Chemical and mineralogical compositions of the rocks are summarized in Tables 74.1 and 74.2, respectively. Sulfur and Zn contents of samples K1-1, K1-3 and K1-4 were greater than 10 wt%, consistent with the detection of sphalerite (ZnS) in these samples. The Cd contents of K1-1, K1-3 and K1-4 were also substantial at 1,120, 919 and 3,510 mg/kg, respectively. The Pb contents of samples containing galena were also substantial reaching wt% levels

Fig. 74.1 Sampling locations and the planned Hokkaido Bullet Train Line

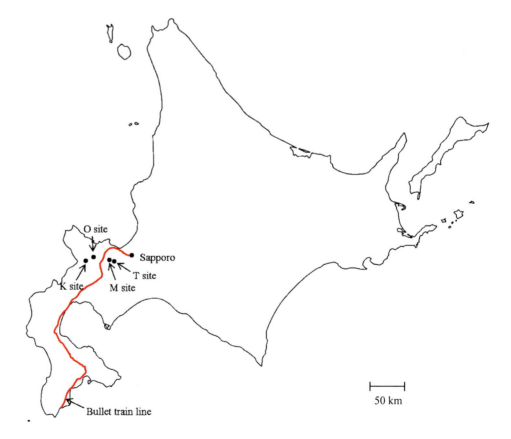

Table 74.1 Chemical compositions of mineralized rocks

Sample	SiO$_2$ (wt%)	Al$_2$O$_3$ (wt%)	Fe$_2$O$_3$ (wt%)	MnO (wt%)	S (wt%)	Zn (wt%)	Pb (mg/kg)	Cd (mg/kg)	As (mg/kg)
K1-1	–	0.4	0.52	0.107	33.5	25.4	287,000	1,120	1
K1-2	71.2	12.9	2.42	0.009	0.93	0.069	124	5.3	7.2
K1-3	37.1	3.3	2.18	0.186	13.5	26.9	456	919	438
K1-4	4	0.7	1.7	0.365	22.2	41.8	42,600	3,510	1,580
K3-1	76.2	4.5	1.34	0.094	1.15	0.46	2,890	33.8	16.8
O-1	32.2	–	8.2	33.7	1.34	1.46	4,720	44.4	32
T-1	41.2	0.037	0.69	0.493	0.109	0.052	309	5.68	5.14
M-1	46.1	0.089	5.21	0.005	2.82	0.005	22	6.2	32
M-2	46.2	2.27	0.649	0.004	0.171	0.001	113	7.43	4.82

Table 74.2 Mineralogical composition of the rocks

Sample	Identified minerals
K1-1	Sphalerite, galena, anglesite, susannite
K1-2	Quartz, pyrite, muscovite
K1-3	Quartz, sphalerite, barite
K1-4	Sphalerite, galena, cerussite, barite
K3-1	Quartz
O-1	Quartz, muscovite
T-1	Quartz
M-1	Quartz, pyrite
M-2	Quartz

(e.g., K1-1 and K1-4). Very high As contents were measured in samples K1-3 and K1-4, which were ca. two and three orders of magnitude higher than background levels, respectively. Although Cd and As contents of some of the rocks were relatively high, mineral phases containing these elements were not detected by XRD, indicating that they are present in the rocks as impurities in the other mineral phases.

Figure 74.2 shows the results of the optical microscopic observations of sphalerite. Numerous small pyrite grains are found around sphalerite. The sphalerite grain is also not homogenous as illustrated by the different hues under polarized light. Another sphalerite grain was analyzed using EPMA, and the results are shown in Fig. 74.3. The elemental maps of Zn and S were coincident with each other, indicating that this mineral is indeed sphalerite. Cadmium was also found inside this sphalerite grain, suggesting that it exists in the rock as an impurity of sphalerite. Moreover, Cd is not the only impurity found in sphalerite. As illustrated in Fig. 74.3d, As is also preferentially distributed in the core of the sphalerite grain. These results suggest that Cd and As could be released from the rock during the destabilization of sphalerite.

74.3.2 Leaching Characteristics of Arsenic, Cadmium and Lead

Table 74.3 shows the leachate pH and the leaching concentrations of heavy metals (As, Fe, Cd, Pb and Zn) after 24 h shaking. Among the rock samples, M1-1 had the lowest pH at 3.94, which could be attributed to its substantial pyrite content. In contrast, the other samples had slightly acidic to neutral pH values. The concentrations of As in the leachated were all lower than the environmental standard of Japan (10 μg/L) except for K3-1. These very low As leaching concentrations could be attributed to its immobilization via adsorption/co-precipitation with Fe-oxyhydroxides that precipate at pH > 4. The leaching concentrations of Cd exceeded the environmental standard (10 μg/L) from five of the samples evaluated. Likewise, seven samples had

Fig. 74.2 Photomicrographs of sphalerite taken with reflected (**a**) and polarized light microscopy (**b**)

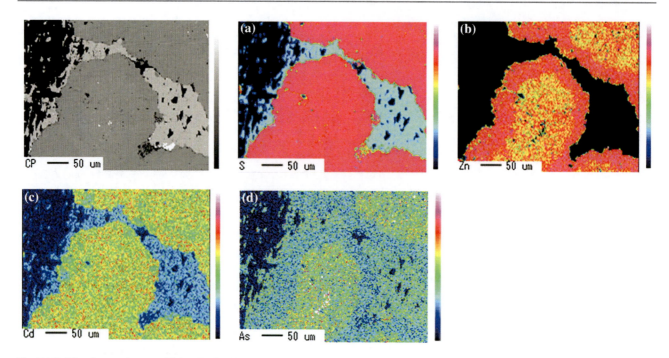

Fig. 74.3 The elemental maps of S (**a**), Zn (**b**), Cd (**c**), and As (**d**) at a magnification of 50 μm. The color intensity shows the relative abundance of the elements in the regions with black and white colors representing 0 and 100 %, respectively

Table 74.3 The pH and leaching concentrations of heavy metals after 24 h shaking

Sample	pH	As (μg/L)	Fe (mg/L)	Cd (mg/L)	Pb (mg/L)	Zn (mg/L)
K1-1	5.18	2.04	2.04	0.243	7.78	60.6
K1-2	4.15	1.98	9.75	0.028	0.255	5.18
K1-3	6	2.32	ND	0.019	0.126	27.9
K1-4	6.06	1.68	ND	0.436	7.53	51.7
K3-1	5.29	32.5	0.183	0.025	4.42	4.6
O-1	7.21	1.8	ND	ND	0.119	0.16
T-1	7.62	ND	ND	ND	ND	ND
M-1	3.94	ND	39.2	ND	1.06	1.47
M-2	4.75	ND	1.98	ND	ND	0.001

ND Not detected

Fig. 74.4 Comparison of the leaching concentrations of Pb and Cd from sample K1-4 (high sphalerite content) and K3-1 (no sphalerite content); **a** shaking time of 24 h, **b** shaking time of 168 h

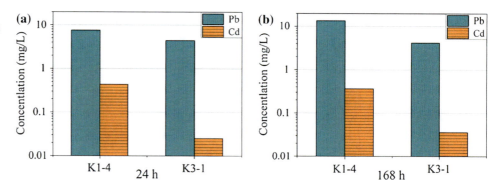

leaching concentrations of Pb higher than the environmental standard (10 μg/L). Figure 74.4 shows that the leaching concentration of Cd from K1-4 (high sphalerite content) was ca. ten-times higher than that of K3-1 (no sphalerite content), regardless of the shaking time. There is also a strong positive correlation between Cd and Zn as illustrated in Fig. 74.5. These results indicate that the leaching of Cd is caused primarily by the weathering of sphalerite. The study shows

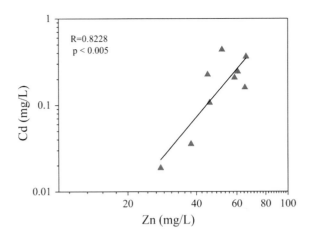

Fig. 74.5 Correlations of the leaching concentrations of the Cd and Zn

that sphalerite could be used as an index of the risks of Cd leaching for the tunnel construction of the Hokkaido Bullet Train Line.

74.4 Conclusions

Tunnel excavated rocks from mineralized areas contain substantial sulfide minerals that could release acidic leachates loaded with heavy metals and toxic metalloids when exposed to the environment. This study found that the leaching of these heavy metals and toxic metalloids was closely related to the mineral composition of the rocks.

Although pyrite was not detected in most of the samples, substantial Cd and As were still measured because they existed in the rocks as impurities of sphalerite. The leaching concentrations of As were insignificant in most of the samples because of the weakly acidic to neutral pH values of the leachates. However, Cd and Pb concentrations were substantial and exceeded the environmental standard of Japan. In particular, the release of Cd is closely related to the weathering of sphalerite. Thus, sphalerite could be used as an index of the risks of Cd leaching potential of the rocks excavated from the tunnels of the Hokkaido Bullet Train Line.

References

Tabelin CB, Igarashi T (2009) Mechanism of arsenic and lead release from hydrothermally altered rock. J Hazard Mater 169:980–990

Tatsuhara T, Arima T, Igarashi T, Tabelin CB (2012) Combined neutralization–adsorption system for the disposal of hydrothermally altered excavated rock producing acidic leachate with hazardous elements. Eng Geol 139–140:76–84

Salinas Villafane OR, Igarashi T, Kurosawa M, Takase T (2012a) Comparison of potentially toxic metals leaching from weathered rocks at a closed mine site between laboratory columns and field observation. Appl Geochem 27:2271–2279

Salinas Villafane OR, Igarashi T, Harada S, Kurosawa M, Takase T (2012b) Effect of different soil layers on porewater to remediate acidic surface environment at a closed mine site. Environ Monit Assess 184:7665–7675

Younger PL, Banwart SA, Hedin RS (2002) Mine water: hydrology, pollution, remediation. Kluwer Academic, London

Hydrogeological Controls on the Swelling of Clay-Sulfate Rocks in Tunneling

75

Christoph Butscher

Abstract

The swelling of clay-sulfate rocks often poses a severe threat to tunnels. It causes serious damage and produces high additional costs during construction and operation. The swelling of clay-sulfate rocks is triggered by water access to anhydrite-bearing layers. Therefore, we propose that groundwater flow is a key factor controlling the swelling process. A case study from the Jura Mountains in Switzerland is presented that uses numerical groundwater models to calculate flow rates at the anhydrite level in different tunnel sections. The approach assumes that an increase of groundwater flow rates into anhydrite-bearing layers after tunnel excavation corresponds to an increase in swelling. A sensitivity study analyzes the impact of hydraulic parameters on calculated flow rates. Analyzed parameters include the hydraulic conductivity of geological units, properties of the excavation-damaged zone and the hydraulic potential in aquifers near the tunnel. Implications for site investigation and potential measures to counteract the swelling problem are suggested.

Keywords

Swelling • Clay-sulfate rocks • Groundwater flow • Tunnel engineering

75.1 Introduction

Tunneling in clay-sulfate rocks often leads to engineering problems because of swelling of such rock (Einstein 1996). Swelling results in heave of the tunnel invert and damage of the lining. The clay-sulfate rocks of the Triassic Gipskeuper ("Gypsum Keuper") formation are very often impacted by swelling. Examples are known from the Swiss and French Jura Mountains and the Stuttgart metropolitan area in Germany (Steiner 1993). Swelling problems in clay-sulfate rocks are also reported from other countries, including Spain (Alonso et al. 2013), Saudi Arabia, Poland, Italy and Texas/USA (Yilmaz 2001, and references therein).

Different mechanisms are interacting in the swelling of clay-sulfate rocks, including osmotic water uptake and hydration of clay minerals and the transformation of anhydrite into gypsum (e.g., Madsen and Nüesch 1991). These swelling mechanisms require water inflow. Because water inflow into clay-sulfate rocks is a prerequisite for swelling, it is likely that groundwater flow is an important controlling factor for the swelling process, albeit groundwater flow rates in clay-sulfate rocks are often small and may not be realized during tunnel construction.

The present study is based on the work of Butscher et al. (2011). They investigated the effects of tunneling on groundwater flow at the Chienberg tunnel in the Swiss Jura Mountains. In two tunnel sections built in the Gipskeuper, heavy swelling occurred after tunnel excavation, while in two other sections with similar geological configuration observable swelling was absent (Fig. 75.1). The study used finite element groundwater models (code FEFLOW (Diersch 2009)), and showed that groundwater flow was strongly increased after tunneling in the tunnel sections with swelling, but not in the sections without swelling. High flow rates

C. Butscher (✉)
Division of Engineering Geology, Karlsruhe Institute of Technology (KIT), Institute for Applied Geosciences, Karlsruhe, Germany
e-mail: christoph.butscher@kit.edu

in anhydrite-bearing layers were therefore considered to indicate high swelling risk.

The present study adds a parametric study to the previous study. It aims at determining the sensitivity of model parameters. "Sensitive" model parameters are those strongly affecting calculated flow rates in the anhydrite-bearing layers near the tunnel. Sensitive parameters are therefore considered to have strong impact on the swelling risk. Knowledge of the sensitive parameters provides indications for exploration and planning of the tunnel and points at measures that are most promising to reduce the swelling risk.

75.2 Methods

75.2.1 Numerical Groundwater Models

Four transverse cross-sections (A to D) perpendicular to the axis of the Chienberg tunnel and one longitudinal cross-section (E) parallel to the tunnel axis were constructed (Butscher et al. 2011). Cross-sections A and C represent the geological situation of the two swelling zones in the tunnel (c.f., Fig. 75.1). Cross-sections B and D represent the geological situation of the zones with similar geological configuration as in the swelling zones, but without observable swelling. The geological cross-sections were taken as a basis for steady-state, two-dimensional, finite element groundwater models. Figure 75.2 schematically shows the setup of the groundwater model of the transverse cross-section A. Model setup, boundary conditions and hydraulic properties of the models as well as model calibration are described in detail by Butscher et al. (2011).

75.2.2 Parametric Study

The parametric study is specifically aimed at answering two questions: (1) What parameters should be determined by experiments in order to receive accurate model results? (2)

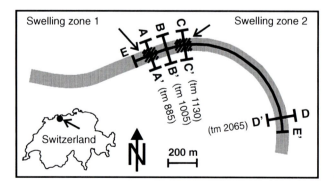

Fig. 75.1 Analyzed cross-sections and swelling zones of the Chienberg tunnel

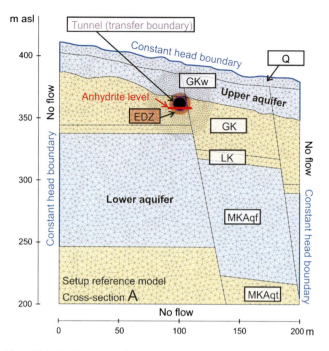

Fig. 75.2 Model setup for cross-section A. *Q* Quaternary, *GKw* weathered Gipskeuper, *GK* Gipskeuper, *LK* Lettenkeuper, *MKAqf*, Muschelkalk Aquifer, *MKaqt* Muschelkalk Aquitard

Which measures during tunnel construction or remediation have most impact on groundwater flow and are therefore expected to be most effective in reducing the swelling risk?

Typical values of hydraulic properties were assumed in a reference model. These values were varied in different scenarios. Six different parameter sets with five different parameter values (scenarios) each were used (Table 75.1). The sensitivity of the parameters, i.e., the impact of the parameters on model results, was determined by calculating flow rates at the anhydrite level in the analyzed cross-sections for the different scenarios and comparing the results of the scenarios.

75.3 Results

Table 75.2 indicates the sensitivity of investigated parameters. Most important parameters, having strong impact on flow rates into anhydrite-bearing layers, are the hydraulic conductivity of the upper aquifer and of the excavation-damaged zone (EDZ), as well as the hydraulic head in the upper aquifer.

Figure 75.3 exemplarily illustrates how the sensitivity (Table 75.2) has been estimated, using the head in the upper and in the lower aquifer as an example. Variation of the hydraulic head in the upper aquifer strongly changes flow rates at the anhydrite level, indicating a high sensitivity of this parameter (Fig. 75.3 top). For example, the flow rate in transverse cross-section A doubles if the hydraulic head in

Table 75.1 Scenarios and corresponding parameters

	Scenario 1	Scenario 2	Scenario 3	Scenario 4	Scenario 5
K upper aquifer (m/s)	1e−4	1e−5	**1e−6**	1e−7	1e−8
K Gipskeuper (m/s)	1e−10	1e−11	**1e−12**	1e−13	1e−14
K EDZ (m/s)	**1e−6**	1e−7	1e−8	1e−9	1e−10
Thickness EDZ (m)	0	1	**2**	4	6
Head upper aquifer (m)	RM −20	RM −10	**RM**	RM +10	RM +20
Head lower aquifers (m)	RM −20	RM −10	**RM**	RM +10	RM +20

Bold reference scenario. *K* hydraulic conductivity, *EDZ* excavation-damaged zone, *RM* reference model

Table 75.2 Summary of sensitivity of model parameters

Model parameter:	K upper aquifer	K Gipskeuper	K EDZ	Thickness EDZ	Head upper aquifer	Head lower aquifers
Sensitivity:	+	−	++	-	+	−

K hydraulic conductivity, *EDZ* excavation-damaged zone, ++ very high sensitivity, + high sensitivity, - moderate sensitivity, − no sensitivity

the upper aquifer is assumed 20 m higher than in the reference model. By trend, such an observation can be made in all transverse cross-sections (A to D) and in the corresponding sectors of the longitudinal cross-section. In contrast, variation of the hydraulic head in the lower aquifers (Fig. 75.3 bottom) does not influence calculated flow rates (i.e., flow rates in each section are equal in all scenarios), indicating that flow rates are not sensitive to this parameter.

75.4 Discussion and Implications for Tunneling in Clay-Sulfate Rocks

This study assumes that high flow rates in anhydrite-bearing layers indicate a high swelling risk. In order to quantify flow rates adequately using numerical groundwater models, it is important to determine sensitive model parameters at the study site by field experiments. Sensitive parameters found in this study include the hydraulic conductivity of the aquifer above the tunnel, which can be determined by pumping and/or Lugeon tests, as well as the hydraulic head in this aquifer, which can be measured by pore water pressure or water level measurements.

Some of the sensitive parameters found in this study can be influenced by engineering measures. For example, the hydraulic head in the upper aquifer strongly influences flow rates into anhydrite-bearing layers. If the hydraulic head would be lowered by adequate drainage systems or by active pumping, flow rates and accordingly the swelling potential could possibly be reduced. In this study, also the properties of the EDZ strongly influence flow rates into anhydrite-bearing layers. Excavation techniques that reduce the formation of an EDZ, for example using a tunnel boring machine or a road header instead of blasting, could therefore be advantageous with respect to the swelling problem.

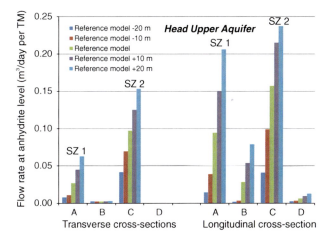

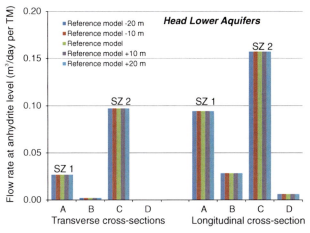

Fig. 75.3 Results of parametric study for the parameters "hydraulic head of upper aquifer" (*top*) and "hydraulic head of lower aquifers" (*bottom*). The graphs show flow rates after tunneling at the anhydrite level for different scenarios (parameter values). The parameter "hydraulic head of upper aquifer" shows high sensitivity (flow rates within a section vary strongly between scenarios). The parameter "hydraulic head of lower aquifers" is not sensitive (same flow rate within a section for all scenarios). *TM* tunnel meter, *SZ* swelling zone

References

Alonso EE, Berdugo IR, Ramon A (2013) Extreme expansive phenomena in anhydritic-gypsiferous claystone: the case of Lilla tunnel. Geotechnique 63(7):584–612

Butscher C, Huggenberger P, Zechner E, Einstein HH (2011) Relation between hydrogeological setting and swelling potential of clay-sulfate rocks in tunneling. Eng Geol 122:204–214

Diersch H-JG (2009) FEFLOW—Finite element subsurface flow and transport simulation system, Reference manual. DHI-WASY, Berlin

Einstein HH (1996) Tunnelling in difficult ground—swelling behaviour and identification of swelling rocks. Rock Mech Rock Eng 29 (3):113–124

Madsen FT, Nüesch R (1991) The swelling behaviour of clay-sulfate rocks. In: ISRM 7th international congress on rock mechanics, vol 1, Aachen, pp 285–288

Steiner W (1993) Swelling rock in tunnels: Characterization, effect of horizontal stresses and construction procedures. Int J Rock Mech Min Sci 30(4):361–380

Yilmaz I (2001) Gypsum/anhydrite: some engineering problems. Bull Eng Geol Environ 59:227–230

Marlène C. Villeneuve

Abstract

The disc cutting process for TBM excavation is dependent on the ability of the discs to initiate and propagate fractures into the tunnel face. At any depth, the geomechanical characteristics of the rock will determine how efficiently the fracture initiation and propagation processes occur. In deep tunnels the stresses induced at the tunnel boundary can lead to stress-induced failure mechanisms such as spalling and bursting. This paper examines the impact of geomechanical characteristics in combination with induced stresses at the tunnel face on the disc cutting process. TBM performance data and tunnel face maps were combined with mineralogy, grain size and fabric for deep tunnels in granitic and foliated massive rocks to determine how induced stresses enhance or hinder the fracture initiation and propagation processes. The impact of the induced stress varies with different geomechanical characteristics depending on the orientation and relative magnitudes of the stresses. In addition, stress rotation and relaxation ahead of the face can lead to stress-induced fracture creation at the face, which acts to precondition the rock prior to the cutters excavating. These results show that the sensitivity of different rock types to stress-related enhancement or hindrance of disc cutting must be taken into account for deep tunneling projects and are used to propose a geomechanical characterisation approach to identify potential for increased or reduced disc cutting efficiency in deep tunnels.

Keywords

Tunnel boring machine • Deep tunnels • Stress

76.1 Introduction

The excavation process for a hard rock TBM involves fragmentation occurring between disc cutters, which apply cyclical pressure on concentric rings, or kerfs, in the tunnel face (Roxborough and Phillips 1975). This fragmentation can comprise the creation of chips or fines, the former being the more efficient fragmentation process. The chipping process is the generation of chips when tensile fractures are induced into the rock, and then propagate parallel to the tunnel face. Grinding is the generation of fines when

M.C. Villeneuve (✉)
Department of Geological Sciences, University of Canterbury, Christchurch, New Zealand
e-mail: marlene.villeneuve@canterbury.ac.nz

fractures do not propagate through the rock and only comminution occurs at the cutter-rock interface.

Numerical modelling of the chipping process (Villeneuve et al. 2012) has shown that geomechanical characteristics, which enhance fracture initiation and propagation will favour chipping over grinding. In particular, the results showed that increased mica content and decreased quartz to feldspar ratio will promote fracture initiation and propagation, and that fractures will propagate easiest along fabric.

Observations in deep TBM tunnels show that spalling can affect the tunnel face, however, this does not necessarily correspond to spalling at the tunnel wall. The two photos in Fig. 76.1 show two examples of tunnel faces in deep Alpine tunnels. While both tunnels have smooth walls without spalling, the tunnel on the right has spalling in the face. The stress-rock interaction at the tunnel face is different than at the walls, leading to this difference in behaviour. TBM

Fig. 76.1 Photos of tunnel face in massive rock in deep Alpine tunnels excavated by TBM. (*left*) stable face, where kerfs are distinctly visible; (*right*) unstable face affected by stress-induced spalling (note the smooth, stable walls)

performance in areas of face spalling can be improved or reduced depending on the severity of spalling (Kaiser 2005), irrespective of the impact of spalling at the walls. Consideration must, therefore, be taken for spalling at the face, even if spalling at the walls is not anticipated. This paper shows how stress and geomechanical characteristics affect excavation and presents a characterisation scheme to identify conditions where advance rates can be impacted by spalling at the tunnel face.

76.2 TBM Performance in Deep, Highly Stressed Tunnels

TBM performance data (penetration rate, thrust, torque and rpm) were analysed to assess TBM performance in zones with recorded face instability. To account for the loss of time arising from the need to clear the TBM head of rock blocks (not to be confused with utilisation: maintenance, support, cutter change, etc.) the TBM data were examined for stopped thrust with continued head rotation: Thrust = 0, RPM ≠ 0. The active driving time is defined as the time during which the TBM head was turning, regardless of whether or not there was thrust. The net advance rate (NAR) is a measure of the distance travelled by the TBM in one stroke divided by the amount of time the TBM was actively driving. Ease of chipping is represented by the drillability index (DI), which is the penetration rate divided by the thrust. A higher DI value represents rock that is easier to excavate, either due to a more efficient chipping process or preconditioning of the face by induced stresses.

NAR is non-unique, where low NAR can either arise from low penetration rate or face instability. To identify face instability NAR is compared to DI to differentiate between low NAR values arising from tough excavation conditions (correspondingly low DI) and arising from face instability (correspondingly high DI). This was validated with depth of failure maps from deep TBM Alpine tunnels (Fig. 76.2).

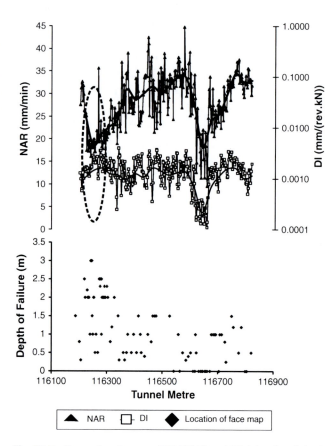

Fig. 76.2 Comparison between TBM NAR and DI (*above*) and depth of face failure (*below*) along a section of deep tunnel. High depth of failure occurs at low NAR and high DI locations (*dashed ellipse*), and no face failure occurs at low NAR and low DI

76.3 Investigation of Stress Impacts on Face Instability and Chipping

Geomechanical characteristics were recorded in a deep Alpine tunnel in massive rock over a continuous length of nearly 400 m, of which 10 % experienced face instability leading to decreased TBM performance. The characteristics were grouped into domains according to dominant

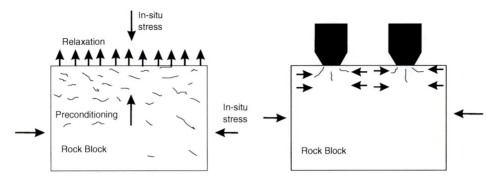

Fig. 76.3 Schematic of numerical models used to investigate the effect of stress on: (*left*) preconditioning at face; (*right*) fracture initiation and propagation at cutter-rock interface; face is at *top*

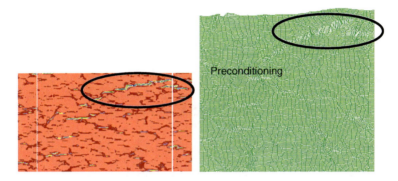

Fig. 76.4 Numerical model outputs of strain (*left*) and deformed grid (*right*, 5x vertical magnification) after 'excavation' showing potential face instability; model is 100 mm wide, face is at *top*

mineralogy, grain size, fabric, metre scale variability and stability of the tunnel walls (Villeneuve et al. 2007). The geological domains most associated with face instability have high mica content (typically 10–20 %), bimodal grain size distributions defined by clasts and matrix, fabric defined by schistosity or cleavage oriented between 0–21° to the tunnel face, greater than 10 m scale variability, and 5–30 % spalling of the tunnel walls.

Numerical modeling of the cutting process (using FLAC, as in Villeneuve et al. 2012) was used to refine the field observations. The models simulated the stress changes at the face arising from the excavation process, and leading to preconditioning (Fig. 76.3, left). The models then simulated the action of the cutters on different rock types (Fig. 76.3, right), with and without preconditioning.

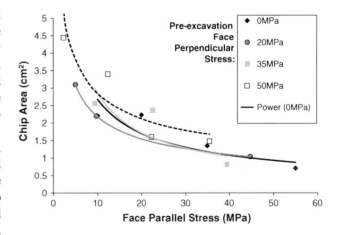

Fig. 76.5 Chip area compared to face parallel stress for samples with different magnitudes of preconditioning stress and fabric oriented and 0° to the tunnel face; UCS ≈ 100 MPa

76.4   Geomechanical and Stress State Impact on TBM Performance

The modeling shows that relaxation of sufficiently high in-situ stresses at the face leads to an outward deformation, which can initiate tunnel face-parallel fractures (Fig. 76.4). This "preconditioning" of the face by newly initiated fractures can improve the chipping process, but can also lead to spalling and face instability. Preconditioning improves chipping most where the in-situ stress perpendicular to the tunnel face is >50 % of UCS, and depends less on the in-situ stress parallel to the tunnel face (Fig. 76.5 white squares). Stress parallel to the tunnel face greater than 20 % of UCS can hinder fracture propagation from the cutter-rock interface, particularly for fabric sub parallel to the tunnel face (Fig. 76.5).

Table 76.1 Characterisation scheme for stress-induced impact on TBM performance

Characteristics	Preconditioning	Chipping performance
Face perpendicular stress (>50 % UCS)	Possible; face instability possible at high stress/strength ratio	Improved if preconditioning occurs
Face-parallel stress (>20 % UCS)	No impact	Reduced if fracture propagation is hindered
Fabric	More likely in rocks with fabric oriented sub-parallel (0–30°) from tunnel face	Improved; reduced by face-parallel stress, notably if fabric is sub parallel (0–30°) to face
Mica content	More likely if >10 %	Improved if >12 %

Rocks with fabric have greater preconditioning than isotropic rocks, especially where the fabric is sub parallel to the face, with less impact where it is oblique ($\sim 30°$), and limited impact where it is at high angle (>60°) to the face. Preconditioning is low to moderate in isotropic rock, is concentrated in mica grains, and increases slightly with mica content, but does not vary greatly with grain size (Villeneuve 2008). Rocks with fabric and higher mica content tend to improve chipping performance (Villeneuve et al. 2012). When stress is taken into consideration preconditioning leads to even greater chipping performance.

Table 76.1 can be used to identify the risk of poor TBM performance due to stress-induced face instability or poor fracture propagation. TBM projects at risk of this behaviour should consider the time required to clear blocks, additional wear on the TBM, and the disconnect from wall spalling. TBM performance prediction should also account for the potential for face-parallel stresses to reduce penetration rate.

76.5 Conclusions

Field and numerical modelling investigations show that TBM performance can be reduced by certain geomechanical characteristics and in-situ stresses. Stress relaxation at the tunnel face can lead to preconditioning the rock or, if sufficiently severe, spalling and face instability. Preconditioning will improve TBM performance by creating new fractures for chip generation, but face instability will decrease TBM performance due to the need to clear blocks. Rocks with mica content >10 % and fabric at low angle to the face are at greatest risk. The proposed characterisation scheme can be used to identify projects at risk so that investigations, design and project planning can address this behaviour.

References

Kaiser PK (2005) Tunnel stability in highly stressed, brittle ground–rock mechanics considerations for Alpine tunnelling. In: Löw S (ed) Geologie und Geotechnik der Basistunnels am Gotthard und am Lötschberg. vdf Hochschulverlag AG and der ETH Zürich, pp 183–201

Roxborough FF, Phillips HR (1975) Rock excavation by disc cutter. Int J Rock Mech Min Sci Geomech Abstr 12:361–366

Villeneuve MC (2008) Examination of geological influence on machine excavation of highly stressed tunnels in massice hard rock. PhD thesis, Queen's University, Kingston, Canada

Villeneuve MC, Diederichs MS, Kaiser PK, Frenzel C (2007) Geomechanical characterisation of massive rock for deep TBM tunnelling. In: Proceedings of rock mechanics: meeting society's challenges and demands, Vancouver, BC, Canada

Villeneuve MC, Diederichs MS, Kaiser PK (2012) Effects of grain scale heterogeneity on rock strength and the chipping process. Int J Geomech 12:632–647 (special issue)

Geotechnical Design of an Underground Mine Dam in Gyöngyösoroszi, Hungary

77

Vendel Józsa, Zoltán Czap, and Balázs Vásárhelyi

Abstract

The investigated copper mine is located in the north of Hungary at the Mátra mountain range. There are several mines in this volcanic mountain, which have been quarried since the early medieval ages. This abandoned heavy and non-ferrous metal mine is about 340-m deep. Mine closure had become necessary due to environmental reasons, because the stored water was highly acidic. The goal of this paper is to present a complex rock engineering design for infilling the drift ways. First of all, a rock mass characterization was necessary for determining the different rock-mass parameters, along with the hydrogeological conditions. Secondly, the stability of the dam was verified by different lengths, applying a geotechnical software using the finite element method. The filling material (ash) originated from a nearby power plant. The dam was designed with a corresponding drainage system and drainage capacity. The filling ash was retained with geogrid-reinforced gabion walls.

Keywords

Underground mine dam • Retaining structure • FEM analysis • Design

77.1 Introduction

Gyöngyösoroszi is located in North Hungary, at the southern slopes of the Mátra mountains, which is the highest mountain in Hungary. The copper mining started as far back as in 1767–1769. At the beginning of 1980, parallel with the mine operation, a special mine inflow operating system was developed against the acidic water. In 2012 the mine was definitively closed and it was decided to be infilled because of some environmental problems, such as acidic water among others. A drainage system was also necessary to be developed along with the dam construction. The goal of the dam design was to (a) hold the infill material and (b) to allow the mine

water to flow away (i.e. working like a filtering). The global factor of safety applied for the static calculations was $\gamma_R = 2$.

The designed cross section area with the shaft of the mine is presented in Fig. 77.1. The subsurface is around 762 m above sea level. There are 7 floors in different depths: 1st floor at 424 m down to the 7th floor at 708 m above sea level (see Fig. 77.1).

In the first case, the 1st floor was infilled and the dam design was focused on this level. The rock mass properties were determined by in situ measurements and laboratory tests. These parameters with the slurry and the dam materials are summarized in Table 77.1. The rock mass parameters were calculated by using the rock mass classification, according to Hoek et al. (2000) and the suggestion of Rowe (2001).

77.2 Design of the Underground Mine Dam

During the design of the underground mine dam, the following steps were calculated and taken into consideration:
• the phase of infilling

V. Józsa (✉) · Z. Czap
Department of Geotechnics, Budapest University of Technology and Economics, Budapest, Hungary
e-mail: jozsavendel@gmail.com

B. Vásárhelyi
Department of Structural Engineering, University of Pécs, Pécs, Hungary

Fig. 77.1 Shaft and mine system in Mátraszentimre (surface ~760 m above sea level)

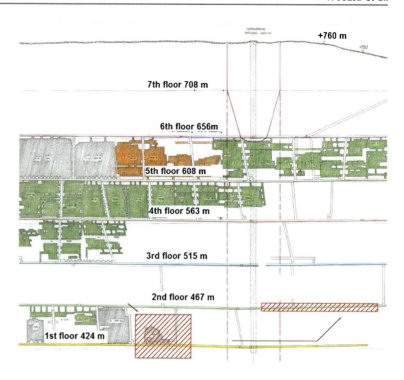

Table 77.1 Mechanical parameters

		Rock mass	Liquid slurry	Solid slurry	Aggregated rock
Inner friction angle (ϕ)	Degree	25		25	40
Cohesion (c)	kPa	1,100			
Deformation modulus (E)	MPa	1,000		20	100
Poisson ratio (v)	–	0.37	0.5	0.49	0.25
Sat. density (γ)	kN/m^3	24	14.2	14.2	20

- end of the infilling,
- permanent state.

Plaxis 3 D Tunneling finite element method (3-dimensional) was used for these calculations as the classical 2-dimensional geotechnical calculations cannot be used here because of the complexity of the geometry and stress states. It was important to model the evolving vault system and the changing stress field in the rock mass. It was also important to analyze the different stress states, such as when the dam is in a hydrostatic condition, or when the water pressure is continuously changing due to the drainage system.

It was not possible to calculate with the solidified slurry due to the maximum volume of the infill material, the capacity of the transportation, the possibility of the sectionalizing and the optimal infilling technology (Blight 2010).

Figure 77.2 presents the finite element model of the investigated mine dam (for the cases of using L = 5, 10 and 15-m long dam).

The material of the slurry consists of sand and crushed rock mixture without silt and clay. The mechanical parameters of this mixture are better than the parameters of natural aggregate systems. It was assumed that the interface ratio between the slurry and the rock mass was 2/3.

At the first step of the calculation, 8 MPa environmental pressure was used on the studied area in the primary stress state. At the second step the secondary stress state was modeled with calculating the stress field after the mine construction. This is the initial state for further calculations.

Firstly, the 5-m long dam was implemented in the finite element model (see Fig. 77.2). Due to this, the stress field had changed. The inner pressure of the dam was increased up to 284 kPa connecting to the safety level. The axial deformation with displacement is presented in Fig. 77.2. This calculation was repeated with dam lengths of $L = 10$ m (1.25 MPa) and $L = 15$ m (4.41 MPa). Figure 77.3 shows the capacity of the dam plotted after the different calculation steps.

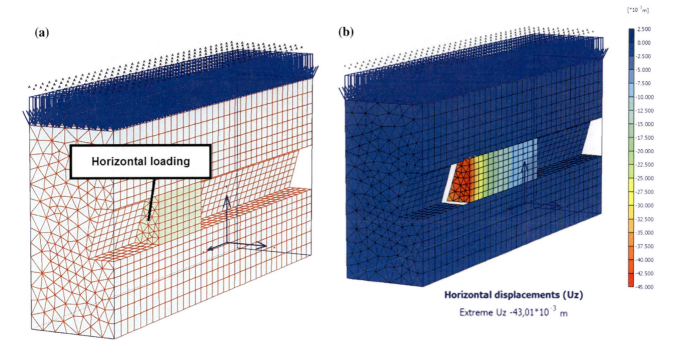

Fig. 77.2 **a** The modeled dam and **b** the deformation in axial direction

Fig. 77.3 Capacity of the different dam lengths after the calculation steps and the relationship between the capacity and the length of the dam

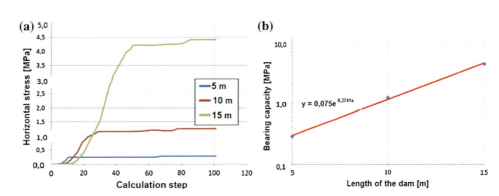

It can be seen that the capacity of the dam (CA) increased significantly with its length (*L*). Based on the calculation, the following exponential equation represents the correlation between the dam length and capacity:

$$CA = 0.075 \exp(0.2741\,L) \qquad (77.1)$$

Using this equation, the minimum length of the dam was 5.8 m. Drainage is possible during the operation of the dam, so only an effective stress can develop. This effective stress was calculated from the different depths of the first and second mine levels (H).

The next calculation process was defined to solidify the slurry. It had to be taken into account that some part of the mine had become sealed. The water pressure increases to the surface level of the model. The model was expanded with a vertical shaft for investigating the influence of vertical stresses. Due to the extremely high pressure, a soil-hardening model could be used. In Fig. 77.4 deformation, compaction and stress change were plotted due to this pressure change.

The locations of the dams were changed and their shape was also modified because the tunnel becomes smaller towards the saved area. Both the roof and the side of the

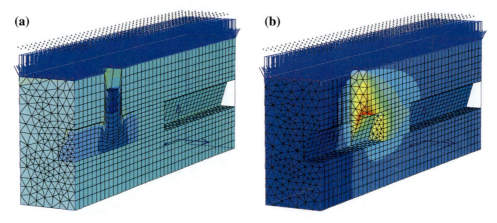

Fig. 77.4 L = 5 m long dam after infilling. **a** Compaction due to 8 MPa pressure and **b** the axial deformation and displacement (the slurry material is not presented)

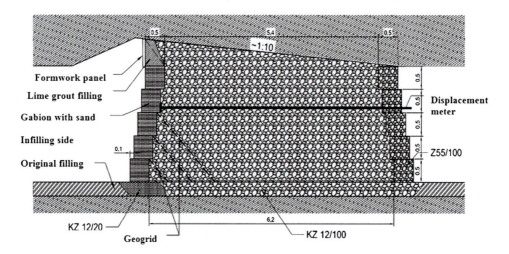

Fig. 77.5 Cross section of the dam

Fig. 77.6 Front view of the dam (June 2012)

tunnel decrease by approximately 1:10. The cross section of the design mine dam is presented in Fig. 77.5.

The construction of the dam in its almost finished stage is presented in Fig. 77.6. The stability of the retaining structure was not decreased, the drainage system of the dam is still working. The pressures of the dam, arising out of the slurry, have decreased after the solidification of the slurry (Wolkersdorfer 2008).

77.3 Summary

The solution we proposed has proven to be good both theoretically and practically.

By using the finite-element test method, it has been demonstrated that the retaining capacity of the dam depends on its length not linearly but exponentially (Fig. 77.3, Eq. 77.1). Its behaviour is analogous with the Terzaghi method used when the tunnel face stability is checked and when the loads to the lining are determined (silo pressure). We have managed to design a well-cleanable drainage system, where the risk of clogging is almost zero. The mine-water cannot bypass the dam, which is an existing threat in the event of watertight concrete/reinforced-concrete structures. The rock material used for the construction of the dam is more corrosion-proof than concrete, which renders any loss of strength impossible in the course of time.

The feasibility of building and operating a system like this has been established; it can be used for the closure of other floors too, and it can be put into action in other cases as well. As regards costs, it is very advantageous too.

References

Blight G (2010) Geotechnical engineering for mine waste storage facilities. Taylor & Francis, London

Hoek E, Kaiser PK, Bawden EF (2000) Support of underground excavations in hard rock. Balkema, Amsterdam

Rowe RK (2001) Geotechnical and geoenvironmental engineering handbook. Kluwer, Dordrecht

Wolkersdorfer Ch (2008) Water, management at abandoned flooded underground mines: fundamentals tracer tests, modelling, water treatment (mining and the environment). Springer, Berlin, p 465

An Approach on the Types and Mechanisms of Water Inrush in Traffic Tunnel Constructions in China

78

Li Tianbin, Zuo Qiankun, Meng Lubo, and Xue Demin

Abstract

Water inrush is a kind of most common geological hazards during tunnel constructions. Based on a lot of water inrush cases, this paper discusses the types and mechanisms of water inrush during traffic tunnel constructions in China. According to the groundwater source and geological structures of aquifers, tunnel water inrush can be classified into five types, namely, gushing by cutting through surface water, gushing by cutting through an aquifer, water inrush from fault zones, water inrush from Karst pipes, water inrush from other structural fracture zones. Based on case studies and mechanical analyses, the mechanisms of tunnel water inrush are revealed as cutting through and gushing free, hydraulic splitting of fissures, hydraulic fracturing of impermeable rock wall, and static-dynamic disturbance of the discontinuities of rockmasses.

Keywords

Tunnel • Water inrush • Types • Mechanisms

78.1 Introduction

Water inrush is a kind of most common geological hazards during tunnel construction, which greatly endangers tunnel construction from safety, time and economic aspects (Wang et al. 2001; Shamma et al. 2003). In recent years, based on the theoretical study and practical application, types and mechanism analysis of water inrush had been obtained some achievements and experiences. Wang et al. (2001) proposed several types of water inrush, that is, water inflow, water inrush from Karst pipes, water inrush from fault zones, water inrush for hydraulic fracturing, water inrush for bulging

fracturing. Jiang et al. (2006) discussed the mechanism of water inrush from the aquifer parallel to the tunnel. Zhang et al. (2008) presented water inrush can be caused by tension failure of impermeable layer, shear failure of fissures, hydraulic expansion of fissures, failure of key block of rocks. A lot of engineering practices show that a correct understanding of the mechanism of water inrush is much important for its mitigation. Therefore, based on collecting and analyzing a lot of water burst cases, this paper summarizes and presents systematically the main types of water inrush and its mechanisms during traffic tunnel construction in China.

L. Tianbin (✉) · Z. Qiankun · M. Lubo · X. Demin
State Key Laboratory of Geohazard Prevention and
Geoenvironment Protection, Chengdu University of Technology,
Chengdu, 610059, Sichuan, China
e-mail: ltbsklgp@qq.com

Z. Qiankun
Sichuan Provincial Transport Department Highway Planning
Survey, Design and Research Institute, 1#, Wuhou Hen Jie St,
Chengdu, 610041, China

78.2 Tunnel Water Inrush Types

78.2.1 Gushing by Cutting Through Surface Water or Groundwater

If the hydrostatic pressure exceeds the strength of rock between the working face of tunnel and surface water or groundwater, or excavation and blasting build hydraulic connections between them, water inrush will happen, as water inrush occurred at K328 + 567.2 ∼ +575.4 in Jinyunshan Tunnel and at K348 + 262 ∼ +250 in Zhongliangshan Tunnel.

78.2.2 Gushing by Cutting Through an Aquifer

Groundwater is gushing directly or along water-conducting fissures when tunnels cut through aquifers. For instance, water inrush at K42 + 786 ∼ K42 + 796 in Shizizhai Tunnel of Dashan expressway was caused by cutting through an aquifer in sandstone layers.

78.2.3 Water Inrush from Fault Zones

Extensional faults developed in thick aquifers have large cavity with loose fragments and store abundant groundwater from both hanging wall and heading wall. Once they are revealed, water inrush will happen just like the one at K12 + 200 in Dabashan Railway Tunnel.

Water inrush often occurs in water-conducting faults under the condition of highly permeable stratum interbedded with slightly permeable stratum, in which hydraulic connection is established by fault zones. For example, water inrush occurred in Bieyancao Tunnel of the Yichang-Wanzhou road was caused by the pressures of groundwater from a fault zone.

In addition, water inrush can happen in the hanging wall of impermeable compressional fault zone nearby, where groundwater is rich correspondingly due to the high crushing of hanging wall and the water stop of fault zone. Water inrush occurred in the hanging wall of a regional fault is a typical case in Dayaoshan Tunnel of the Beijing-Guangzhou railway.

78.2.4 Water Inrush from Karst Pipes

It has been indicated by a lot of cases that tunnel water inrush is highly risky in the growth area of Karst. Once excavation cuts through the water-rich Karst pipes in tunnels, water inrush will occur, e.g., the one at ZK32 + 927 of Huayingshan Highway Tunnel in 1997 and 1998.

78.2.5 Water Inrush from Other Structural Fracture Zones

The fracture zones in the axis parts of anticline and syncline are often rich in water, where once excavation cuts through, water inrush is easy to happen. The water inrushes at K297 + 100 ∼ K297 + 800 in Maoba 1[#] Tunnel (Ma 2010)

and at DK354 + 235 ∼ DK361 + 764 in Yuanliangshan Tunnel (Liu 2004) are typical cases of anticline and syncline fracture zones, respectively.

When water-rich interstratified fracture zones in sedimentary rocks are cut through by tunnel constructions, water inrush will occur. For example, the work face blasting at K42 + 375 in Shizizhai Expressway Tunnel triggered water inrush of 8950 m^3/d from interstratified fracture zone between quartz sandstone layer and silty mudstone layer.

78.3 Tunnel Water Inrush Mechanisms

78.3.1 "Cutting Through and Gushing Free" Mechanism

This kind of water inrush mechanism often occurs in the condition of cutting through the water-rich pipes or fissures in surrounding rocks. When excavation reveals water-rich Karst pipes or water-bearing fissures with good connectivity around the surrounding rocks, groundwater is gushing free. In general, initial water inrush flows along many discontinuities during non-exposed fully pipes or fissures; water flows along one or two orifices after exposed fully pipes or fissures and increase rapidly in short time.

78.3.2 "Hydraulic Splitting of Fissures" Mechanism

As the impermeable rockmass become thinner due to excavation, and water-bearing fissures are constantly expanding under high water pressures, the impermeable rockmass is split, thus leading to water inrush. Joints or fissures in stratum can open, slide and be spilt under high water pressures. Mechanical sketch of hydraulic splitting of fissures is shown in Fig. 78.1.

According to the theory of fracture mechanics, there are three fissure types during fissure evolution process. TypeIis "open type" caused by tensile stresses perpendicular to the fissure surface. TypeIIis "slide type" caused by shear stresses parallel to the fissure surface. Type III is "spilt type" caused by shear stresses perpendicular to the fissure surface. I-II compound type appears when the normal stresses on the fissure surface are tensile stresses. Fissure failure criterion can be expressed by: $K = K_\xi$. Where K and K_ξ are strength factor and limit strength of fissure, respectively. K is given by:

$$k_I = \sigma_n \sqrt{\pi a} \tag{78.1}$$

$$k_{II} = \tau \sqrt{\pi a} \tag{78.2}$$

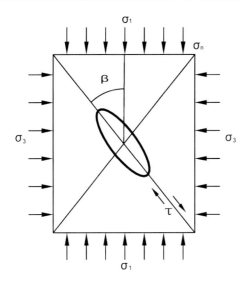

Fig. 78.1 Mechanical mechanism of hydraulic splitting of fissures

where σ_n is the normal stresses on the fissure surface, τ is the shear stresses on it, a stands for 1/2 fissure length.

78.3.3 "Hydraulic Fracturing of Impermeable Rock Wall" Mechanism

According to the different space combinations of impermeable rock wall and tunnel, hydraulic fracturing of the rock wall can be described as rockmass bending failure or shearing failure.

It is observed from Fig. 78.2 that the impermeable rock wall with high strength and adequate thicknesses can initially bear the water pressures, whereas it becomes thinner during excavation process, it cannot, thus leading to bending failure and water inrush. The failure criterion of the impermeable rock wall is given by:

$$\sigma \geq [\sigma] \tag{78.3}$$

where σ is the largest tensile stress of rock wall, $[\sigma]$ is the allowable tensile stress of rock wall.

The impermeable rock wall with lower strength or unfavorable discontinuities under water pressures fractures generally with plastic shear feature, leading to water inrush, as shown in Fig. 78.3. The shear failure criterion of the impermeable rock wall is given by:

$$\tau \geq \tau_f \tag{78.4}$$

$$\tau_f = \sigma \tan(\phi) + c \tag{78.5}$$

where τ is the shear stresses on the most unfavorable surface of rock wall, τ_f is the shear strength of rock wall, σ is the normal stress on the shear surface, c and φ are the cohesion and internal friction angle of rock wall, respectively.

78.3.4 "Static-Dynamic Disturbance of the Discontinuities of Rockmasses" Mechanism

Dynamic disturbance caused by blasting and surrounding rock stress redistribution can lead to fracture expansion, relaxation and opening of compressive fault zone and instability of surrounding rocks. Once these unfavorable geological structures connect with aquifers, water inrush will occur in tunnels.

As shown in Fig. 78.4, the hard surrounding rocks have different space geometry due to joints and fractures in them. After excavation, the key block of rockmass initially become instability due to blasting disturbance and stress releases, and then other blocks are deformed or collapse, consequently, aquifers are connected with tunnels and then water inrush appears.

With the influences of blasting and excavation, larger additional tangential stress and confining pressure changes within a certain range of compressive fault result in fault relaxation and opening. When this action makes tunnel to connect with aquifer or water-rich hanging wall, then water inrush appears (Fig. 78.5).

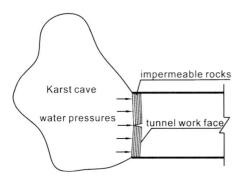

Fig. 78.2 Water inrush due to bending failure of impermeable rock wall

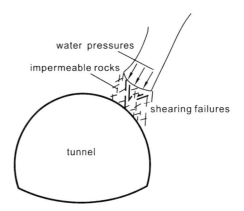

Fig. 78.3 Water inrush due to shearing failure of impermeable rock wall

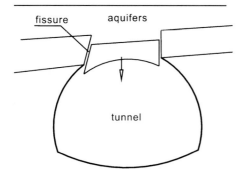

Fig. 78.4 Water inrush due to the instability of key block of rockmass

78.4 Conclusions

(1) According to the groundwater source and geological structures of aquifers, tunnel water inrush can be classified into five types, namely, gushing by cutting through surface water, gushing by cutting through an aquifer, water inrush from fault zones, water inrush from Karst pipes, water inrush from other structural fracture zones.

(2) Complex geological conditions and rockmass structures and variable interaction between tunnel and aquifer structure lead to complicated mechanism of water inrush in tunnels. Based on collecting and analyzing a lot of water burst cases, the mechanisms of tunnel water inrush are revealed as cutting through and gushing free, hydraulic splitting of fissures, hydraulic fracturing of impermeable rock wall and static-dynamic disturbance of the discontinuities of rockmasses.

(3) Analyses of the types and mechanisms of water inrush provide theoretical basis for water inrush prediction and mitigation in tunnels and play an important role in the effective control of water inrush.

Acknowledgments This study is financially supported by National Natural Science Fundation of China (41172279), Key Program for Research Group of SKLGP (SKLGP2009Z002), and Cultivating

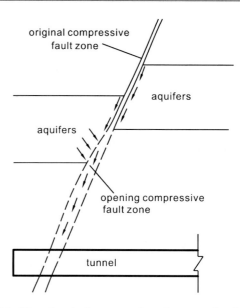

Fig. 78.5 Water inrush due to opening of compressive fault zone (Jiang et al. 2006)

Program of Excellent Innovation Team of Chengdu University of Technology (HY0084). These supports are gratefully acknowledged.

References

Jiang JP, Gao GY, Li XZ (2006) Mechanism and mitigation of water inrush in railroad tunnels. China Railway Sci 27(5):76–82

Liu ZW (2004) Water inrush mechanism and mitigation in Karst Yuanliangshan tunnel Ph.D. thesis. China University of Geosciences, Beijing

Ma XH (2010) Geological hydrology analysis of water inrush mechanism and mitigation in Maoba 1# Tunnel. Sichuan Build Mater 36(4):139–141

Shamma Duke D, Fordham S, Freeman E, Tempelis J (2003) Tom arrowhead tunnels: assessing groundwater control measures in a fractured hard rock medium. In: Proceedings-rapid excavation and tunneling conference, New Orleans, pp 296–305

Wang JX, He J, Yang LZ (2001) Hydrogeological analysis for karst water inrush in large-scale underground engineering. Hydrol Eng Geol 4:49–52

Zhang W, Li ZG, Wang QS (2008) Causes and mitigation of water inrush in Karst tunnels. Tunnel Constr 28(3):257–262

Numerical Analysis of the Influence of Tunnel Dimensions on Stress and Deformation Around Tunnels in Rocks

79

G.E. Ene, C.T. Davie, and C.O. Okogbue

Abstract

Numerical studies of a generic arched-roof profile tunnel was carried out in order to investigate the influence of geometric size on stress distribution and deformation in rockmass surrounding tunnels which can be optimized in design and construction of underground works. Results show that increasing the aspect ratio of the arched-roof tunnel will cause corresponding increase in the magnitude and size of zone of adverse compressive stress concentration at the tunnel sidewall while the extent of de-stressing zone at the invert/crown decreased. The converse is true for increasing span to height ratio. Similarly, the horizontal displacement of rock mass in the vicinity of the tunnel sidewall shows an increasing trend with increasing aspect ratio of the tunnel. In contrast, the tunnel seems not to experience significant variation in vertical displacement of the floor and roof at increasing aspect ratio. These observed trends in variations of the phenomena of stress re-distribution and deformation in rocks surrounding tunnels with tunnel dimensions demonstrate that stress induced instabilities can be effectively regulated by adopting appropriate dimensions relative to rockmass properties and engineering objective.

Keywords

Dimensions • Tunnels • Stress-distribution • Deformation

79.1 Introduction

In most civil engineering structures such as highways, power generation and distribution facilities, mainline haulage in mines (stopes and shafts), hazardous waste repositories and aqueducts that integrate tunnels or caverns, the horse-shoe tunnel cross-sectional profile are often preferred due to ease of construction and engineering objectives (Hoek and Brown 1980; Brady and Johnson 1989; Hoek 2000; Zhu and Zhao 2004; Cai et al. 2007; Lunardi 2008). However, the arched-roof cross-section which usually consists of crown with circular outline resting on a rectangular base is prone to instabilities at the crown, foot and shoulders of the ribs and failures in associated civil and mining structures often result from invert heave, even when supported (Hoek 2001; Hsu et al. 2004). The ground on support suffers from loading and bending moments from unstable slabs caused by delayed failures. In all situations, failures are linked to induced stress effects and the response of rockmass in terms of stress re-distribution and deformation (Brady and Brown 1993; Jeager et al. 2008). The key issues in design and construction of the tunnels and caverns in rocks, therefore, is the prediction of the pattern of stress re-distribution and amount of deformation that may result during and after construction. Such prediction is important for selection of appropriate excavation technique and advance rate and effective support system in order to reduce project cost, avoid inadequate or excessive support situation and prevent unstable excavation (Yeung and Leong 1997; Hoek 2001; Hsu et al. 2004; Lunardi 2008). Unfortunately, precise prediction of the two

G.E. Ene (✉) · C.O. Okogbue
Department of Geology, University of Nigeria, Nsukka, Nigeria
e-mail: ezekwesiliene@yahoo.com

C.T. Davie
School of Civil Engineering and Geosciences, University of Newcastle, Newcastle upon Tyne, UK

G. Lollino et al. (eds.), *Engineering Geology for Society and Territory – Volume 6*,
DOI: 10.1007/978-3-319-09060-3_79, © Springer International Publishing Switzerland 2015

parameters is often difficult and constitutes a complex geo-mechanic problem exacerbated by highly variable and indeterminate rockmass properties, transient in situ stress and fluctuating groundwater conditions (Hoek and Brown 1980; Whittaker and Frith 1990; Chen and Zhao 2002; Pariseau 2007; Lunardi 2008). This paper reports a numerical investigation of the effect of tunnel dimensions on stress re-distribution and deformation processes in surrounding rockmass which can enable tunnel dimensions to be optimized in design to modify and regulate the stress-induced failures.

79.2 Methodology

The tunnel shape for the generic model is an arched-roof rectangular cross-section which consists of an arch (5 m radius) profile roof with 7 m high sidewalls and 10 m span. The probable effect of tunnel dimension on deformation and stress re-distribution mechanism in rocks around the tunnel is studied using the height (H) to width (W) ratio. To implement this, the generic tunnel shape (Fig. 79.1) H/W ratio, 1.2 (12 m/10 m), is to be varied to 0.5, 0.6, 0.75, 1, 1.2, 1.4, and 2.0 in each simulation. It is important to point out that the tunnel shape effect could be duplicated in such simulation and can result in spurious results. To curtail such effect, the arch profile geometry is maintained and adjustments to simulate for size influence are effected by increasing or reducing the arch radius and/or the sidewall.

The problem domain is defined by Ubiquitous Joint constitutive model and massively jointed shale characteristics from FLAC Rock Data bank with the following rock properties: density = 2,700 Kg m^{-3}, shear modulus = 8.81E9 Pa, bulk modulus = 4.3E9 Pa, cohesion = 3.84E7 Pa, friction = 14.4°, tensile strength = 14.5 MPa, dilation = 0.0 and rockmass properties: joint angle = 30°, cohesion = 0.5E6 Pa, friction = 36°, tensile strength = 0.0 and dilation = 0.0. The tunnel is a relatively deep one, located at 500 m below the surface. The tunnel was excavated by unloading process. The stress state of rockmass and displacements at the mid-point of the roof, floor and sidewall were studied. Model solution validation (Fig. 79.2) compares numerical results of a circular profile tunnel with values of calculated tangential stresses using closed form analytical solution (Hoek and Brown 1980). Both solutions show good agreement at some locations and the deviations at the other locations are attributed to some yielding numerical solution in contrast to elastic analytical equation. Since the results of the numerical solutions and, for the purpose of possible extrapolation to other domain which may be characterized with rockmass properties that can vary considerably, a sensitivity analysis of the model input strength parameters was carried out. The possible variations

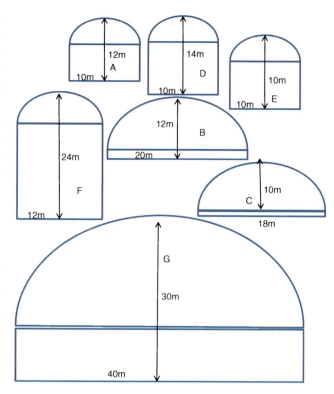

Fig. 79.1 Arched-roof profile tunnel of different height to width (aspect) ratio investigated in this study. Aspect ratio of *A* 1.2; *B* 0.6; *C* 0.5; *D* 1.4; *E* 1.0; *F* 2.0; and *G* 0.75 (*Scale* 1 cm: 3 m)

in the value of the material properties were scaled from measured and reported values based on Goodman (1989) and AASHTO 1989. The stability of the tunnel is judged by the maximum value of horizontal displacement of the right sidewall. In all the simulations, it is only the parameter under investigation that is allowed to vary within in each case, others parameters and the model set-up are caused to remain constant (Bardsley et al. 1990; Wu et al. 2001). The sensitivity of the following parameters: bulk modulus, shear modulus, cohesion, tensile strength and angle of internal friction were considered. In order to compare the sensitivity of the various parameters, the sensitivity factor, S_k, (Bedford and Cooke 2001; Zhu and Zhao 2004) was employed. The sensitivity factor expresses the ratio of the relative change in a system property P ($\eth p = |dp|/p$) to the change in certain parameter k ($\eth k = |dk|/k$) thus, $S_k = \frac{(|dp|/p)}{(|dk|/k)}$.

79.3 Results and Discussion

79.3.1 Influence of Tunnel Size and Dimensions

The influence of the ratio of tunnel height to width (aspect ratio) on the stress patterns is shown graphically in Fig. 79.3. From the figure, it is highlighted that the loading of

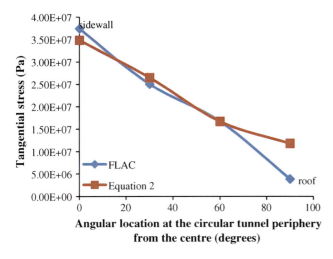

Fig. 79.2 Tangential stress at the periphery of a circular tunnel in biaxial stress field (k = 0.63) from similar locations solved using closed form equation and FLAC

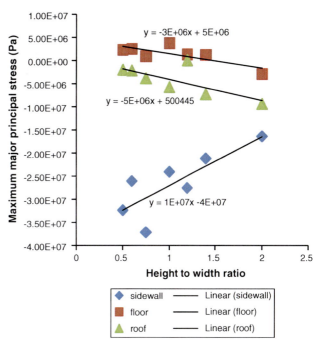

Fig. 79.3 Variation of maximum principal stress with aspect ratio of the arched-roof tunnel

compressive stress at the sidewall increases with increasing aspect ratio of the excavation. On the other hand, the tensile stress at the floor and the arched roof show decreasing trends with increasing height to width aspect ratio. In contrast, a reversed trend of decreasing compressive stress magnitude at ribs and increasing tensile region at the invert/crown is observed for width to height aspect ratio. At the same aspect ratio, similar trends in variation of stress re-distribution and type can be demonstrated by increasing the cross-sectional area of the excavation (increased excavation volume). A typical example is presented in Fig. 79.4a, b for aspect ratio of 1.2.

The aspect ratio of the underground excavation seems to have little or no effect on the vertical displacement of rock mass around the excavation perimeter under the present biaxial stress state and material model as illustrated in Fig. 79.5a. However, a remarkable sharp deviation from the general insignificant variation trend at an aspect ratio of 0.75 is worthy of note. For instance, a closer check of the model aspect ratio relative to its dimensions shows that the tunnel which exhibits the anomalous maximum displacement has dimensions of 40 m span and 30 m height. Despite the relatively low value of the aspect ratio (0.75), greater displacements were recorded due to the larger size of the excavation. The implication of this observation is that the displacement of rock mass under the given stress field is probably less dependent of the aspect ratio but rather on the volume of the excavation. This suggestion is investigated further and confirmed using the aspect ratio of 1.2 and increasing only the size of the tunnel. The results show reasonable changes in maximum vertical displacement that will result from enlarging the tunnel size at the same aspect ratio (Fig. 79.6). Indeed, at the tunnel floor, the maximum vertical displacement tends to increase with increasing size

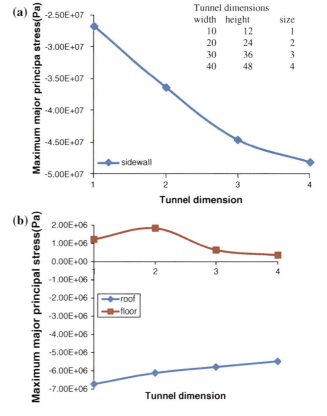

Fig. 79.4 a Variation of maximum principal stress with excavation volume of the arched-roof tunnel. **b** Variation of maximum principal stress magnitude and type with varied height to width ratios

of the excavation while the roof will experience decreasing values as the tunnel dimension is increased (see Fig. 79.6a). Contrary to these observed trends in vertical displacements, the horizontal displacement can vary considerably with variation in opening dimensions at constant aspect ratio. Figures 79.5a and 79.6b exemplify that horizontal displacement at the tunnel sidewall exhibit significant increase in maximum value with increasing aspect ratio and increasing size at the same aspect ratio respectively.

79.3.2 Parametric Analysis

The parametric analysis of input parameters for index of rockmass geomechanical properties reveals two categories of sensitivity relative to maximum horizontal deformation of the unsupported tunnel rib. The first group is made up of rockmass cohesion and friction and joint cohesion that show insignificant variation in sidewall deformation at varied values and therefore considered insensitive to the tunnel stability. The second group includes bulk and shear moduli of rockmass and orientation and friction of joints that show significant sensitivity to tunnel deformation at varied amounts. Figures 79.7 and 79.8 show the sensitivity function plots for the second group of parameters. The function plots

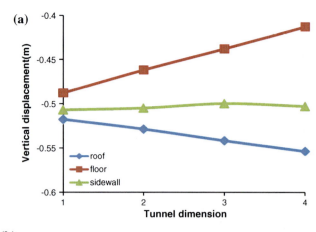

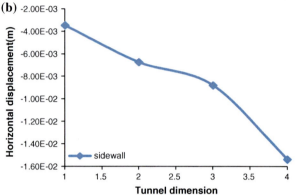

Fig. 79.6 a Variation of maximum vertical displacement with excavation volume of the arched-roof tunnel for aspect ratio of 1.2. **b** Variation of maximum vertical displacement with excavati volume of the arched-roof tunnel for aspect ratio of 1.2 (sidewall)

are adopted for the purpose of interpreting, statistically, the relative variability in the sensitivity of material properties which is of important concern in assessing the stability of the tunnel and optimization of testing schemes (Wu et al. 1991).

Figure 79.7 illustrates the sensitivity function plots for shear and bulk modulus. As presented in the figure, the sensitivity curves of both parameters show remarkable disparity. The sensitivity of the bulk modulus to maximum displacement of the tunnel sidewall increases sharply with increasing

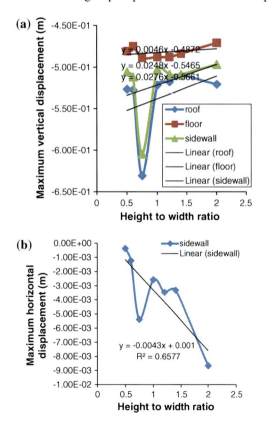

Fig. 79.5 a Variation of maximum vertical displacement with varied tunnel aspect ratio. **b** Variation of maximum horizontal displacement with varied tunnel aspect ratio

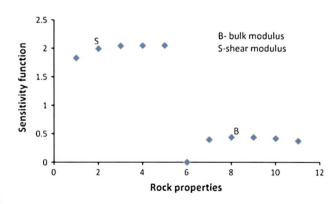

Fig. 79.7 Variation of sensitivity factor with bulk and shear modulus

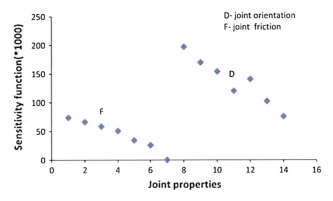

Fig. 79.8 Variation of sensitivity factor with joint dip angle and friction

bulk modulus values at lower values but appears to stabilize at higher values. The sensitivity factor of the shear modulus increases with increasing values of shear modulus up to an optimal value. Beyond this value, the sensitivity of shear modulus decreased with increasing shear modulus values. However, for the basic parameter values of 8.81 GPa for bulk modulus and 4.3 GPa for shear modulus, the corresponding calculated sensitivity factors are 2.04 and 0.43 respectively. It follows, therefore, that if the assessment of the stability of the generic tunnel should be based on the maximum horizontal displacement of the tunnel sidewall, then the most sensitive rock strength parameter will be bulk modulus because of its high sensitivity value compared to the other parameters. The implication of the results of the sensitivity evaluation of rock parameters is manifest on the effect of under or over-estimation of such parameter on the stability of the tunnel (Wu et al. 1999). For example, whereas a 30 % relative errors in measurement of cohesion and angle of internal friction values of the rock mass will result in insignificant error in the predicted maximum horizontal displacement of the right tunnel sidewall, similar relative error in the bulk modulus and shear modulus will lead to 61.2 % (2.04 × 30 %) and 12.9 % (30 % × 0.43) relative error in the predicted maximum horizontal sidewall displacement (Zhu and Zhao 2004).

The result of the sensitivity analysis of joint strength parameters is illustrated graphically in Fig. 79.8. From the figure, it is evident that both the joint dip angle and joint friction angle display increasing sensitivity with increasing parameter values. However, for the basic parameter set, 36° for joint friction and 30° for joint dip angle, the corresponding sensitivity factors are 0.14 and 0.05 respectively. This means that a 50 % error in the measurement of both parameters will result in 7 and 2.5 % error respectively in the predicted maximum deformation of the tunnel sidewall. In that case, the stability of the generic tunnel under the present model is more sensitive to the friction angle of the dip compared to the angle of dip of the joint.

79.4 Conclusion

The numerical studies of a generic arched-roof profile tunnel using finite difference code and constitutive model of Ubiquitous Joint have allowed for the evaluation of the effect of tunnel dimensions design parameters on stress distribution and deformation in surrounding rockmass as well as parametric analysis of rockmass properties. Increasing the tunnel aspect ratio (height to width) caused corresponding increase in adverse stress compressive stress magnitude and extent on the ribs and decreasing trend for tensile region at invert/crown. The converse is true for increasing width to height ratio. The horizontal displacement of rock mass in the vicinity of the tunnel sidewall increases with increasing aspect ratio of the tunnel. Also, at increasing aspect ratio, the tunnel seems not to experience significant variation in vertical displacement of the floor and roof, except for increased tunnel size which will show both increasing stress concentration and deformation.

Acknowledgment The authors are grateful to the Petroleum Technology Development Fund (PTDF) for sponsoring this research work and the University of Newcastle, Upon Tyne for providing the software.

References

Bardsley WE, Major TJ, Selby MJ (1990) Note on a Weibull property for joint spacing analysis. Int J Rock Mech Min Sci Geomech 27:133–134

Bedford T, Cooke R (2001) Probabilistic risk analysis: foundations and methods. Cambridge University Press, Cambridge

Brady BHG, Brown ET (1993) Rock mechanics for underground mining, 2nd edn. Chapman & Hall, London, pp 23–48

Brady TM, Johnson JC (1989) Comparison of finite difference code and finite element code in modelling an excavation in an underground shaft pillar. In: Pietruszczak S, Pande GN (eds) Proceedings of 3rd international symposium on numerical modelling in geomechanics (NUMOG III), Niagara Falls, pp 608–619

Cai M, Morioka H, Kaiser PK, Minami M, Maejima T, Tasaka Y, Kurose H (2007) FLAC/PFC Coupled numerical simulation of a large-scale underground excavations. Int J Rock Mech 44(4):550–564

Chen SG, Zhao J (2002) Modelling of tunnel excavation using a hybrid DEM/BEM method. Comput Aided Civi Infrastruct Eng 17:381–386

Cundall PA (2001) A discontinuous future for numerical modelling in geomechanics? Proceedings of Institution of Civil Engineers. Geotech Eng 149:41–47

Goodman RE (1989) Introduction to Rock mechanics. Wiley, New York, pp 23–67

Hoek E (2001) Big tunnels in bad rocks 2000 Terzaghi Lecture. J Geotech Geoenviron Eng ASCE 127(9):726–740

Hoek E, Brown ET (1980) Underground excavations in Rock. Institution of Mining and Metallurgy, London, pp 1–68

Hsu SC, Chang SS, Cai JR (2004) Failure mechanisms of tunnels in weak rock interbedded with structures. Int J Rock Mech Min Sci 41(3):01–06

Jeagar JC, Cook NGW, Zimmerman RW (2007) Fundamentals of Rock mechanics. Blackwell Publishing, London, pp 345–380

Lunardi P (2008) Design and construction of tunnels: analysis of controlled deformation in Rocks and soils (ADECO-RS). Sprnger-Verlag, Berlin, pp 2–78

Pariseau WG (2007) Design analysis in Rock mechanics. Taylor & Francis/Balkema, Leiden, pp 34–98

Whittaker BN, Frith RC (1990) Tunnelling design stability and construction. Institution of Mining and Metallurgy, London, pp 4–38

Wu C, Hao H, Zhao J, Zhou XX (2001) Statistical analysis of anisotropic damage of the Bukit Timah granite. Rock Mech Rock Eng 34:23–28

Yeung MR, Leong LL (1997) Effects of joint attributes on tunnel stability. Int J Rock Mech Min Sci 34:3–4

Zhu W, Zhao J (2004) Stability analysis and modelling of underground excavations in fractured rocks Elsevier, Amsterdam, pp 289

Analysis of Stress Conditions at Deep Seated Tunnels—A Case Study at Brenner Base Tunnel

Johanna Patzelt and Kurosch Thuro

Abstract

Deep seated tunnels like the Brenner Base Tunnel suffer from high in situ or primary stress conditions. This has significant effects on the deformation pattern and failure mode and therefore on the construction of the tunnel. High in situ stress conditions are a result of the rock cover which reaches approximately 1,700 m at the Brenner Base Tunnel. In collaboration with the Brenner Base Tunnel Society BBT SE, in situ stress conditions were modeled in a multitube tunnel system in different rock conditions. A rather ductile rock type, the Innsbruck quartzphyllite, and a rather brittle rock type, the Tux Central Gneiss, were used to calculate several models with the 2D-finite element code Phase2 and the 3D-finite element code RS3 (rocscience). Due to residual tectonic stresses caused by the collision of the African and Eurasian plates horizontal stresses are often increased. For the Innsbruck quartzphyllite section a 70° dip angle was used, within the section of the Tux Central Gneiss no deviation from the vertical was assumed due to geological and geophysical results. Apart from the rock cover and the direction of the main principle stress, further parameters were needed for modeling: young's modulus, uniaxial compressive strength, poisson ratio, friction angle and cohesion of the rock types have been used from site investigation reports. The performed numerical modeling showed that the secondary stresses have the same orientation as the primary stresses, hence the maximum stress appeared in the tunnel walls and minimum stress is concentrated in invert and crown.

Keywords

Brenner base tunnel • High rock cover • Rocscience

80.1 Introduction

In the course of a scientific cooperation of the TU Munich Department of Engineering Geology with the Brenner Base Tunnel BBT SE modeling were done at eight different cross sections in the region of the Innsbruck quartzphyllite and the Tux central gneiss. Modeling of the secondary stresses and the associated deformation were done by using the 2D finite element program phase2 and RS3 (rocscience).

80.2 Location and Characteristics of the Cross Sections

Five different cross sections were modeled in the region of Innsbruck quartzphyllite and three cross sections are located in the region of the Tux central gneiss. The cross sections in the Innsbruck quartzphyllite were distinguished by their different tube arrangements with connecting tunnels and various expansions in the range of the multifunction station Innsbruck Ahrental.

J. Patzelt (✉)
Pöyry Infra GmbH, Salzburg, Austria
e-mail: johanna.patzelt@poyry.com

K. Thuro
Technische Universität München, Munich, Germany
e-mail: thuro@tum.de

G. Lollino et al. (eds.), *Engineering Geology for Society and Territory – Volume 6,*
DOI: 10.1007/978-3-319-09060-3_80, © Springer International Publishing Switzerland 2015

80.3 Parameters

For the modeling parameters were given by Brenner Base Tunnel (BBT/Geoteam 2008) as well as from the Inn Valley tunnel (Thuro 1996). In a case study different rock parameters were varied (Patzelt 2010).

Following input parameters were used for modeling of rock:
- Unconfined compressive strength
- Poisson's ratio
- m_i (Hoek-Brown failure criterion, Hoek et al. 2002) Following parameters were used to characterize the rock mass:
- young's modulus
- Disturbance factor D as a function of the exploration method (blasting, TBM)
- Geological Strength Index GSI (Hoek et al. 2002).

80.4 Modeling

The specified cross sections were digitized and the geological prognosis from preliminary studies was added in form of layers with constant widths (Fig. 80.1). Because of the geometry some rock thicknesses had to be shown exaggerated, otherwise it would come to elements notional artifacts due to the network used for the calculation of 2D and 3D finite element code.

For the competent rocks the Hoek-Brown failure criterion was used. The Hoek Brown failure criterion characterized at the best the rock mass parameters for the rocks given in this range. In none of the different failure criterions any anisotropy factors could be added. A range of 10 points of GSI

value was given by BBT/Geoteam (2008). They characterized the block sizes and this controls the rock mass quality. For the "weak" material of the Inn Valley fault zone none rock parameters were given, so only characteristic values for friction angle and cohesion are specified, that it can be characterized by the Mohr-Coulomb failure criterion.

In deep seated tunnels joints play a dominant role in the stress states and the solvability. Closed fractures or joints have only a minor role, since they are usually not mechanically effective. The results show that the tangential discontinuities have an influence on the stress transfer and deformations and lead to higher amounts of deformation. Nevertheless, only deformation amounts in the range of maximum 10 cm were determined.

80.4.1 Rock and Rock Mass Parameters

In summary, the used parameters for the modeling are listed in Tables 80.1 and 80.2. They are all based on the documents of BBT/Geoteam (2008). The complete list of parameters is in Patzelt (2010: 73 ff) for further details.

80.4.2 Primary Stress Conditions

The primary stresses are composed of the superposition of stresses and the induced stresses of the tectonic plate collision. The tectonically induced stresses lead mainly to an inclination of the resulting principal stress. The angle of inclination with the most realistic results is 70° in Innsbruck quartzphyllite region. In Fig. 80.2, the primary stress conditions for cross sections B are shown. From hydrofracturing

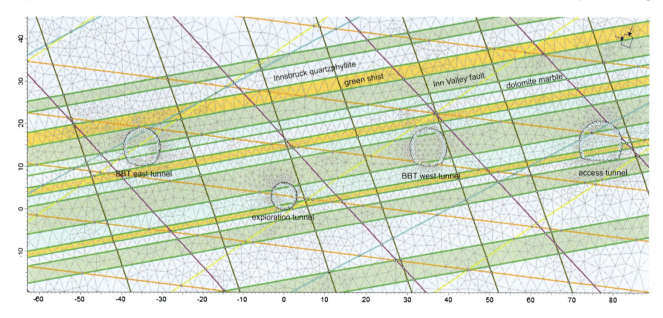

Fig. 80.1 Cross section B (part of multifunction station Innsbruck Ahrental)

Table 80.1 Summary of the used rock properties

Rock	UCS [MPa]	E-Modul [MPa]	m_i	v	GSI
Innsbruck quarzphyllite	40	5,087	12	0.19	50
Dolomite marble	30	7,662	10	0.20	65
Green shist	55	6,187	15	0.15	45

Table 80.2 Summary of the used properties in the Inn Valley fault zone

Rock	E-Modul [MPa]	c [MPa]	φ [°]	v
Inn Valley fault	1,000	0.5	28	0.19

tests it is known, that the ratio of σ1 and σ3 varies in the Wipptal between 0.3 and 1.2.

80.4.3 Stress Redistribution When Opening the Face into Sub-Areas

Further modeling should take into account the stress redistribution during successive advance of the various tunnels. In conventional propulsion of the advance in top heading, bench and invert was divided and implemented in the modeling to represent the stress redistribution (Fig. 80.3).

80.4.4 Secondary Stress Conditions

In the modeling can be seen that the secondary stresses extend in about the same direction as the primary stresses

(Fig. 80.4). The maximum secondary stresses pass through the dip angle shifted slightly in the region of eastern invert on the west elm as well as the transition to the invert. While the minimum σ1 stresses preferably spread in the layer of Inn Valley fault, the minimum σ3 stresses propagate especially in the layers of dolomite marble, which has the lowest rock mass compressive strength. The maximum σ3 stresses are arranged accordingly 90° to σ1 stresses. The minimum stresses sweep in the range between 0 and 4.5 MPa.

80.4.5 Deformation

The determined total deformation amounts—without e.g. anchor, reinforced concrete—scatter in the range of 10 cm to a maximum of 2 m, which was determined from a full-surface opening of a large face at an overburden of 890 m.

The largest deformation amounts occur in the region of Innsbruck quartzphyllite. By an overburden of 870 m is clearly seen that the rounded profile of the exploratory tunnel is the cheapest variant advance (Fig. 80.5). For the given geometry of the main tunnel tubes is seen especially at the junction of elm that increased deformations are generated.

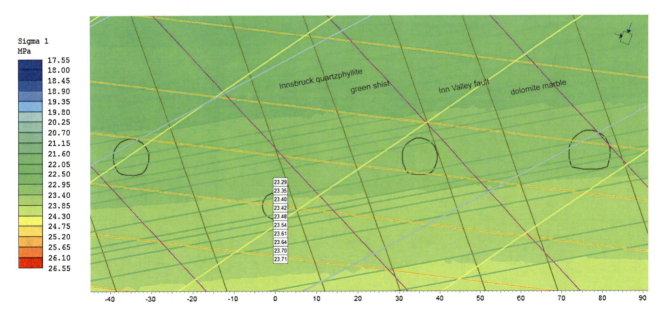

Fig. 80.2 In-situ stress conditions at cross section B

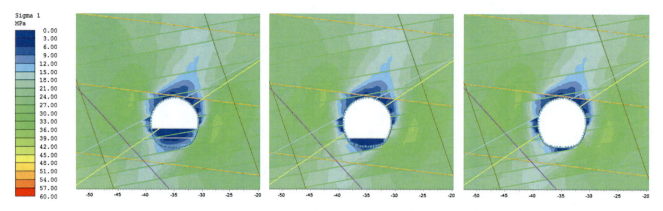

Fig. 80.3 Stress redistribution due to excavation of crown, side wall and invert applying conventional propulsion

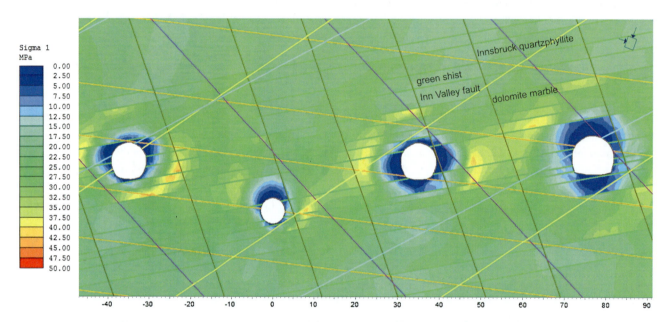

Fig. 80.4 Illustration of σ1 stresses. Minimum stresses are plotted in *dark colors*; maximum stresses are shown in *light colors*

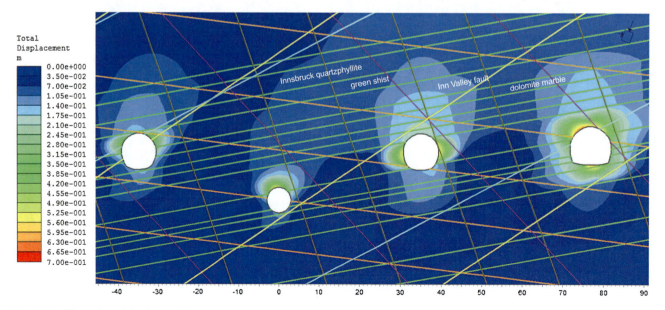

Fig. 80.5 Illustration of the deformation pattern (overburden 890 m). Maximum deformations add up to 7–10 cm at crown and tunnel sides

80.5 Conclusions

The influence of the primary stresses was examined on the stress distribution at high overburden in the Brenner base tunnel, taking account of:

- Complex geometric conditions (in particular the positions of the tunnels in a multifunction station),
- given rock and rock mass characteristics from the reports of preliminary studies,
- the orientation and training of the separation surface structure and fault zones,
- loosening of the factor D (blasting/TBM),
- and the orientation of the primary and secondary stresses.

The parameters were varied so that uncertainties in the determination or natural variations could be mapped.

References

BBT/Geoteam (2008) Ausbau Eisenbahnachse München—Verona. Brenner Basistunnel. UVE. Technische Projekt-aufbereitung. Geologie, Geotechnik, Hydrogeologie. Haupttunnel. Gebirgsarten, Gebirgsverhaltenstypen, Störzone—Haupttunnel—137 S, Technischer Bericht und Längenschnitt G1.2i-01—1 Plan 1: 25 000

Hoek E, Carranza-Torres C, Corkum B (2002): Hoek-Brown Failure Criterion—2002 Edition. In: Proceedings NARMS-TAC conference, Toronto, 1: 267—273

Patzelt J (2010) Modellierung des sekundären Spannungs-zustandes tiefliegender Tunnelbauwerke am Beispiel des Brenner-Basistunnels—Unveröffentl. Masterarbeit, Lehrstuhl für Ingenieurgeologie, Technische Universität München, 84 S, München

Rocscience (2010) Phase2 Version 7. Manual (digital)

Thuro K (1996) Bohrbarkeit beim konventionellen Sprengvortrieb. Geologisch felsmechanische Untersuchungen anhand sieben ausgewählter Tunnelprojekte—149 S, Münchner Geol. Hefte, B1, München (Technische Universität München)

Acoustic Emission Technique to Detect Micro Cracking During Uniaxial Compression of Brittle Rocks

Carola Wieser, Heiko Käsling, Manuel Raith, Ronald Richter, Dorothee Moser, Franziska Gemander, Christian Grosse, and Kurosch Thuro

Abstract

Excavation of deep underground openings causes redistribution of primary stresses and induces initiation and propagation of micro cracks. Changes in rock properties ahead of an advancing tunnel face may influence stability and penetration rates in TBM tunneling. This study is part of a PhD project where the focus is on stress-induced micro cracking and its influence on rock strength reduction. We demonstrate results from acoustic emission (AE) measurements on two homogeneous diorite samples tested under uniaxial compression. The aim was to gain information about the influence of experimental setup and settings on AE events in brittle rock. From the stress strain curve and the number of acoustic signals, main deformation stages could be determined. A three-dimensional localization of acoustic events showed a typical conjugate shear system. Future work will include tests on different rock types after inducing controlled damage by uniaxial loading. Lateral strain measurements combined with AE analysis will be applied in order to quantify rock damage and its influence on rock strength.

Keywords

Uniaxial compression • Brittle rock • Acoustic emission • Micro crack • Localization

81.1 Introduction

Any excavation of deep underground openings influences the in situ stress state in the surrounding rock mass. According to that, rock samples retrieved from highly stressed rock masses are subjected to stress changes resulting from the unloading process (e.g. Teufel 1989; Meglis 1991). As uniaxial compressive strength is usually determined several days or weeks after sampling, the relaxation process may still be in progress when the tests are performed. Deformation and strength parameters determined in the laboratory cannot be transferred to in situ stress conditions without further considerations. The detection and quantification of stress-induced damage, the time-dependence of the relaxation process as well as the influences on rock strength and deformation properties are of key interest for determining the rock strength at the time of excavation.

Acoustic emissions (AE) are elastic sound waves generated during crack formation or growth. It is well known, that loaded rock specimen emit acoustic signals which can be detected and analyzed with regard to rock damage. The waves are converted into electric signals by piezoelectric transducers which are connected to an amplification and data acquisition system. An overview of acoustic emission testing and its application is given in Grosse and Ohtsu (2010).

This paper presents the results of preliminary AE measurements on brittle rock during unconfined compression. A 3-dimensional localization of AE events was performed and deformation stages as well as stress levels in the stress–strain curve could be assessed according to the methods of Bieniawski (1967) and Martin (1997). These tests represent

C. Wieser (✉) · H. Käsling · K. Thuro
Chair for Engineering Geology, Technische Universität München,
Arcisstr. 21, 80333 Munich, Germany
e-mail: carola.wieser@tum.de

M. Raith · R. Richter · D. Moser · F. Gemander · C. Grosse
Chair of Non-Destructive Testing, Technische Universität
München, Baumbachstr. 7, 81245, Munich, Germany

the first stage of an investigation to examine and quantify the influence of fracture damage on unconfined compressive strength and deformation characteristics of brittle rocks.

81.2 Experimental Setup

We tested two specimens of homogeneous, porphyritic, fine- to medium-grained diorite from Nittenau, south-east Germany (Bavarian Forest), containing mainly feldspar, biotite and quartz (Fig. 81.1). Samples were prepared for testing with a diameter of 8 cm and a length to diameter ratio of 2:1 (Mutschler 2004). Uniaxial loading was applied under servo-control at a constant velocity rate of 0.1 mm/min. During the tests, axial strain was measured by position sensors installed between the load plates.

14 piezoelectric transducers attached to the sample surface and another two guard sensors glued to the load plates recorded background noises deriving from the hydraulic device (experimental setup shown in Fig. 81.2). The pre-amplifier included a bandpass filter with a frequency range of 10 kHz–1 MHz. Low frequencies, originating from the hydraulic pump of the loading frame were eliminated by a high-pass filter with a cutoff frequency of 10 kHz. Depending on the type of sensor, a pre-amplification of 27–33 dB was applied. AE were recorded by an automatic data acquisition system (TranAX by Elsys) using a slew rate trigger and a sampling frequency of 5 MHz.

81.3 Data Analysis

Data analysis was carried out using the software SquirrelAE which was developed at the Chair of Non-destructive Testing at Technische Universität München. Using a signal-based AE method, recorded waveforms were analyzed, although we focus on results from 3D source localization in this study. The software includes an automated picking algorithm based on the Akaike Information Criterion (AIC). The accuracy of the onset detection process is critical for a reliable localization of AE events. Localization was analyzed using the Bancroft (1985) algorithm which was developed to solve global positioning system (GPS) equations. Furthermore, the iterative algorithm of Geiger (1910) was applied which was originally established for the localization of earthquakes. Kurz (2006) gives an overview of several algorithms and their scope of application. Thus, the algorithm providing the best results depends on boundary conditions and sensor covering.

81.4 Experimental Results

In a first test, the influence of Teflon (PTFE) as a means for minimizing end effects was tested during unconfined compression. Two Teflon sheets of 2.8 mm thickness were inserted between rock specimen and load plates to reduce acoustic signals resulting from friction at the end surfaces.

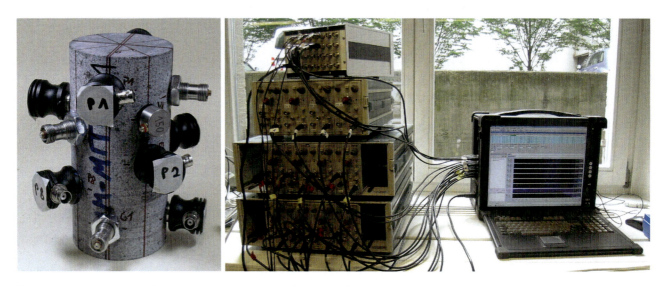

Fig. 81.1 *Left* 14 piezoelectric transducers attached to diorite sample from Nittenau (Bavarian Forest), *Right* pre-amplification and data acquisition system connected to the sample

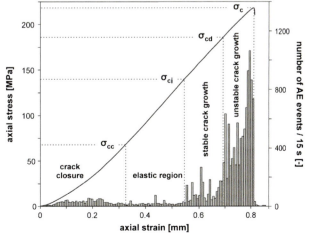

Fig. 81.2 Experimental setup of AE instrumentation and data acquisition system. Axial strain was measured by position sensors between the load plates

The rock sample was loaded uniaxially with and without Teflon. The influence of the material was visible as a decrease in the gradient of the recorded stress-strain-curve, resulting in a deformation modulus 40 % lower when PTFE was applied. As a proper determination of rock parameters was not possible with the attenuating effect, subsequent tests were carried out without Teflon. The effects and the usability of PTFE foils and plates for a reduction of end effects will be investigated more thoroughly in future tests.

A second uniaxial compression test was performed according to Mutschler (2004) with constant loading until failure without PTFE plates. From the number of AE events different stages of crack development could be determined (Fig. 81.3). According to Eberhardt et al. (1999), the first deformation stage is characterized by the closure of existing cracks aligned at an angle to the load whereas parallel cracks may open. The stress–strain curve is non-linear during this deformation stage. In our test, the crack closure threshold σ_{cc} can be set at about 31 % the peak strength (σ_c), which is 218 MPa.

The second stage is marked by the linear section of the stress–strain curve where linear elastic deformation occurs and only a few acoustic emissions were released. Event numbers increase again at a stress level of about 0.64 σ_c which represents the crack initiation threshold σ_{ci} where stable crack growth takes place (Bieniawski 1967). According to Cai et al. (2004), the crack initiation threshold

of most rocks usually ranges from 0.3 to 0.5 σ_c which is clearly lower than what our test revealed. A significant increase in events and an overall high AE activity indicates the onset of unstable crack growth and crack coalescence. This phase begins at a crack damage threshold σ_{cd} of 0.85 σ_c and thus, it is more consistent with existing studies, which state a crack damage stress level between 0.7 and 1.0 σ_c (Cai et al. 2004). With a combined analysis of the volumetric strain curve, a more precise determination of this threshold could be achieved. Thus, in the next test series, lateral strain measurements will be implemented in the experiments.

In the following, results from localization are presented. Figure 81.4 (left) shows a 3D plot of localized AE events. A typical conjugate shear pattern is representative of the typical fracture pattern observed after unconfined compression. A concentration of events occurred in the bottom and top and may be contributed to friction between end surfaces and load plates or stress concentrations on the edges of the sample. In the lower right part of the specimen a distinct accumulation of events is visible where also high energy signals are localized.

Compared to the results from localization, the tested specimen shows a different failure pattern. As shown in Fig. 81.4 (right) a rock fragment split off the sample due to a fracture sub-parallel to the stress direction. As the test was stopped shortly after failure, further major cracks were restricted which may have developed under continuing

Fig. 81.3 *Left* stress–strain curve from uniaxial compression test. Deformation stages in crack development were determined from the stress–strain curve and AE events per 15 s, *Right* final fracture pattern showing one crack aligned parallel to the loading direction

Fig. 81.4 *Left* AE event localization shows a typical conjugate shear system. Larger markers represent better correlation coefficients, red markers show higher event energies. *Right* view in z-direction with an accumulation of events on the right side where final fracture occurred

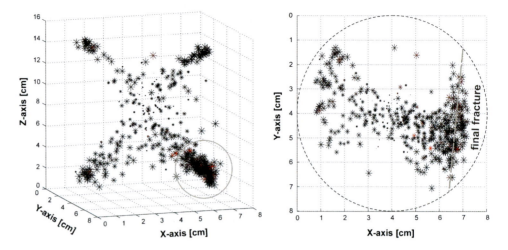

loading. Illustrated in Fig. 81.4 (left) and the sectional view in Fig. 81.4 (right), the final macroscopic fracture can also be reproduced in AE as an accumulation of events is visible in the right side of the cylinder.

81.5   Conclusion

This paper presents results of two uniaxial compression tests on homogeneous diorite (Nittenau) samples which were combined with AE measurements in order to confirm the applicability and accuracy of the experimental setup. In a preliminary test, the influence of Teflon sheets as a means for minimizing end effects was investigated. The results showed that although acoustic emission events were minimized at the end surfaces, the stress–strain curve was clearly altered due to attenuation effects. In a second uniaxial compression test, no PTFE plates were applied although friction effects were clearly visible at the end surfaces. From the stress–strain curve and AE distinct deformation stages in crack development could be defined. A 3D plot of AE events displays a typical conjugate shear system which developed during unconfined compression testing. In contrast, the final fracture pattern shows a thin rock fragment which split off the rock but did not reveal the expected conjugate shear planes. Yet, a significant accumulation of AE events on the edge of the specimen indicates the subsequent fracture pattern prior to failure.

This study demonstrates that AE analysis represents a practical non-destructive testing technique to show fracture processes inside a homogeneous rock specimen. Further research will focus on stress-induced damage, time-dependent relaxation and their influence on rock strength and deformation properties. For this purpose, AE as well as volumetric strain measurements will be carried out combined with elastic wave velocity measurements and petrographic investigations.

References

Bancroft S (1985) An algebraic solution of the GPS equations. IEEE Trans Aerosp Electron Syst AES-21 7:56–59

Bieniawski ZT (1967) Mechanism of brittle fracture of rock. Int J Rock Mech Min Sci 4:395–406

Cai M, Kaiser P, Tasaka Y, Maejima T, Morioka H, Minami M (2004) Generalized crack initiation and crack damage stress thresholds of brittle rock masses near underground excavations. Int J Rock Mech Min Sci 41(5):833–847

Eberhardt E, Stead D, Stimpson B (1999) Quantifying progressive pre-peak brittle fracture damage in rock during uniaxial compression. Int J Rock Mech Min Sci 36:361–380

Geiger L (1910) Herdbestimmung bei Erdbeben aus den Ankunfts-zeiten. Kgl Ges d Wiss Nachrichten Math-Phys Klasse 4:231–349

Grosse C, Ohtsu M (2010) Acoustic emission testing. Basics for research—applications in civil engineering. Springer, Berlin, 396 p

Kurz JH (2006) Verifikation von Bruchprozessen bei gleichzeitiger Automatisierung der Schallemissionsanalyse an Stahl- und Stahl-faserbeton. Diss., Institut für Werkstoffe im Bauwesen, Universität Stuttgart, Stuttgart, 197 p

Martin CD, Read RS, Martino JB (1997) Observation of brittle failure around a circular test tunnel. – Int J Rock Mech Min Sci 34 (7):1065–1073

Meglis IL, Engelder T, Graham EK (1991) The effect of stress-relief on ambient microcrack porosity in core samples from the Kent Cliffs (New York) and Moodus (Connecticut) scientific research bore-holes. Tectonophysics 186:163–173

Mutschler T (2004) Neufassung der Empfehlung Nr. 1 des Arbeitskre-ises "Versuchstechnik Fels" der Deutschen Gesellschaft für Geo-technik e. V.: Einaxiale Druckversuche an zylindrischen Gesteinsprüfkörpern. Bautechnik 81(10):825–834

Teufel LW (1989) Acoustic emissions during anelastic strain recovery of cores from deep boreholes. In: Khair (ed) Rock mechanics as a guide for efficient utilization of natural resources. Balkema, Rotterdam, pp 269–276

Towards a Uniform Definition of Rock Toughness for Penetration Prediction in TBM Tunneling

82

Lisa Wilfing, Heiko Käsling, and Kurosch Thuro

Abstract

In TBM tunneling, performance prediction is a major issue since calculated excavation costs and construction time of a tunnel project are mainly based on it. Prediction is dependent on the accuracy of geological and geotechnical input parameters. Besides rock strength, toughness of the excavated rock has a significant influence on penetration and cutting efficiency as increasing toughness requires greater energy to induce complete failure. Yet, existing definitions of rock toughness are not adequate or suitable for incorporation in a performance prediction model for TBM tunneling. To develop a common definition and classification system, we used standard laboratory tests (Uniaxial Compression Test, Brazilian Tensile Test). Based on this test data we analyzed several factors that can characterize rock toughness like the ratio of compressive to tensile strength (Z-coefficient), ratio of plastic to elastoplastic strain, specific failure energy and destruction work. We expect future analysis to focus on Z-coefficient but we aim to revise the classification system of Schimazek and Knatz as results showed no good correlation. Also the ratio of plastic to elastoplastic strain is a promising tool for future research. Obviously, destruction work characterized rock toughness but the determination of this parameter depends a lot on machine stiffness and settings.

Keywords

Rock toughness • TBM tunneling • Penetration prediction • Uniaxial compression

82.1 Introduction

The prediction of TBM-performance is an essential tool to estimate costs and construction time of tunnel projects. Therefore the research group ABROCK (collaboration between universities, clients and contractors) deals with the improvement of existing prediction models as well as with the development of a new, adapted model (Schneider et al. 2012).

Apart from uniaxial compressive strength (UCS), rock toughness significantly affects performance respectively penetration of a TBM (Becker and Lemmes 1984; Gehring 1995). The tougher a rock behaves, the slower cracks propagate and the more energy is needed to cause chipping and effective rock excavation. But the implementation of rock toughness in a prediction model is problematic as no suitable definition exists. Therefore, the aim of this work is the development of a common and suitable method for rock toughness characterization to gain better results in TBM performance prediction.

82.2 State of the Art

Schimazek and Knatz (1976) as well as Becker and Lemmes (1984) described in their research the first definition of toughness by the ratio Z of UCS to BTS (Brazilian tensile strength). Since then, the coefficient Z is commonly used. On the basis of their results they set the threshold of brittle ($Z > 10$) to tough ($Z < 10$) rocks at $Z = 10$. Additionally,

L. Wilfing (✉) · H. Käsling · K. Thuro
Technische Universität München, Chair of Engineering Geology,
Arcisstr. 21, 80333, Munich, Germany
e-mail: lisa.wilfing@tum.de

Thuro (1996) added the ranges of very brittle ($Z > 20$) and very tough ($Z < 5$) to this classification.

Hughes (1972) characterizes the stiffness of rocks with the *specific energy* e_s that is required to bring a rock sample to failure (Fig. 82.1). Gehring (1995) seized this idea under the term of *specific failure energy* w_f and implemented it in his prediction model for TBM performance as he observed good correlation between w_f and penetration rate at different tunnel projects. Thuro (1996) extended this method of energy-requirement to the important range beyond failure of the specimen, so called post-failure-range. Tough rocks do have a distinct post-failure-range whereas brittle rocks nearly have none. *Destruction work* W_z is calculated by the entire area (pre- & post-failure-range) beneath the stress-strain-curve and therefore shows a significant difference between brittle and tough rock. This post-failure-range has already been described by Wawersik and Fairhurst (1970). During their laboratory research program they determined two different post-failure behaviors but did not apply this observation to brittle, respectively tough rocks. The *post-failure-modulus* P_f that is determined with a regression line in the post-failure-range (Fig. 82.1), is one possibility to describe this part of the stress-strain curve (Thuro 1996).

Tough rocks show furthermore a distinct *plastic strain* ε_{pl} right before failure (Fig. 82.1). In contrast, brittle rocks demonstrate a high *elastoplastic strain* ε_{el} with minor *plastic strain* ε_{pl} before failure. The transition point where elastoplastic behavior passes over into plastic behavior is called σ_{cd} (point of irreversible crack damage, Eberhardt 1998) and is determined subjectively or by volumetric strain measurements during uniaxial compression.

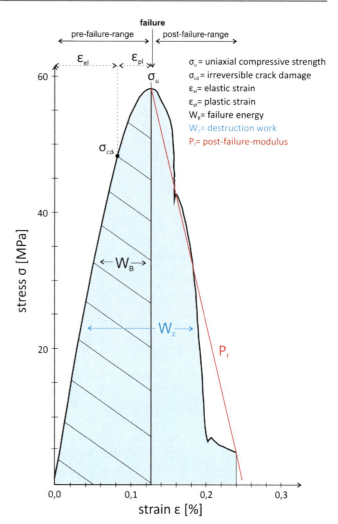

Fig. 82.1 Schematic stress–strain curve of a uniaxial compression test with the toughness defining parameters

82.3 Methodology

Basic research is done with standard laboratory tests like Uniaxial Compression Test (DGGT Testing Recommendation No.1) and Brazilian Tensile Test (DGGT Testing Recommendation No.10). To validate the toughness characterizations, following parameters are analyzed (Fig. 82.1):

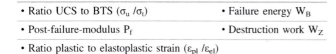

• Ratio UCS to BTS (σ_u / σ_t)	• Failure energy W_B
• Post-failure-modulus P_f	• Destruction work W_Z
• Ratio plastic to elastoplastic strain ($\varepsilon_{pl} / \varepsilon_{el}$)	

Furthermore, Acoustic Emission measurements (AE) during Uniaxial Compression Test are planned to determine σ_{cd} and to investigate whether brittle or tough rocks show different fracture propagation as well as velocities. These parameters were already analyzed by Eberhardt (1998) but only on the basis of brittle rocks.

We also expect a distinct increase of acoustic signals at the transition σ_{cd} of elastoplastic to plastic strain as from there on crack propagation is irreversible (Eberhardt 1998). Therefore AE helps to reveal the transition point with a higher accuracy and to provide a precise calculation of the ratio plastic ε_{pl} to elastoplastic ε_{el} strain for the characterization of toughness. To develop a basic definition for rock toughness, tests have been primarily made with homogenous rocks like basalt, anhydrite, diorite and granite. In the next step, also inhomogenous rocks like gneiss, amphibolite and greenschist have been analyzed. Samples have been chosen under the aspect of a wide range, so that rocks commonly known as tough/brittle have been picked out.

82.4 Results and Discussion

First results of UCS-tests verify the assumption that basalt has a brittle and anhydrite tough failure behavior. Figure 82.2 (left) shows the stress-strain curve of basalt from a quarry in Bavaria. According to ISRM (1978) the UCS of the rock is classified 'extremely high' (470 MPa) and shows by definition of Thuro (1996) an 'extremely high destruction work' (939 kJ/m^3). In contrast, the anhydrite specimen from the Haselgebirge in Austria (Fig. 82.2 right) has a UCS of 60 MPa and consequently only a moderate destruction work of 100 kJ/m^3.

Having a closer look at the stress-strain curves reveals that anhydrite with a far lesser destruction work ($\Delta = 839$ kJ/m^3) shows a much more developed post-failure-range and therefore a tougher failure behavior then basalt. Basalt as typical brittle rock has almost no post-failure-range. Hence, the characteristic of brittle rocks is that destruction work W_z is almost equivalent to failure energy W_B (Fig. 82.2 left). This comparison illustrates that destruction work as single parameter is not sufficient for toughness characterization so that more factors have to be included. Determination of destruction work also depends on stiffness and settings of the used testing machine and can at worst vary a lot between different testing laboratories.

Ratio of plastic to elastoplastic strain ($\varepsilon_{pl}/\varepsilon_{el}$) is suitable for a first, vague characterization of toughness as they can be detected visually by stress-strain curves. The tested

specimens show results for $\varepsilon_{pl}/\varepsilon_{el}$ from 0.05 (basalt) to 0.71 (anhydrite). As described in Chap. 2, these values are just an approximation because of subjective determination of the transition point σ_{cd} but show promising correlation results and have to be verified with volumetric strain measurements. Planned Acoustic Emission Tests can also define this point in a more detailed way so that an implementation in a prediction model is reasonable.

Figure 82.3 illustrates the results of 7 tested rock types with the UCS values and corresponding BTS as well as 110 test results from the rock data base of the Chair of Engineering Geology, TUM (Menschik et al. 2013). With a colored background, the classification of Schimazek and Knatz (1976) extended after Thuro (1996) is marked. Additionally the mean values of Basalt, Amphibolite and Anhydrite have been highlighted.

Tested **Basalt** samples show Z values of up to 28 and are classified as very brittle. This correlates to the failure behavior of Uniaxial Compression Test and analysis of the stress-strain curve in Fig. 82.2 left.

Amphibolite is commonly assumed to behave tough. To validate this assumption three different varieties of Amphibolite have been tested. Variety 2 and 3 show very high mean UCS values between 190 and 290 MPa. These samples derive from quarries and are used as ashlar. Amphibolite-1 is taken from a tunnel project and has lower quality (85–150 MPa). Contrary to the expectations, all amphibolite samples (except one) are not classifiable as 'tough'.

Anhydrite has Z values around 13, so according to the classification, it is called 'brittle'. Stress-strain curves (Fig. 82.2 right) in contrast show a distinct post-failure-range and a tough failure. Furthermore, all results in Fig. 82.3 (220

Fig. 82.2 Stress σ–strain ε curve of basalt (*left*) and anhydrite (*right*) as typical examples of brittle (basalt) respectively tough (anhydrite) rock failure behavior

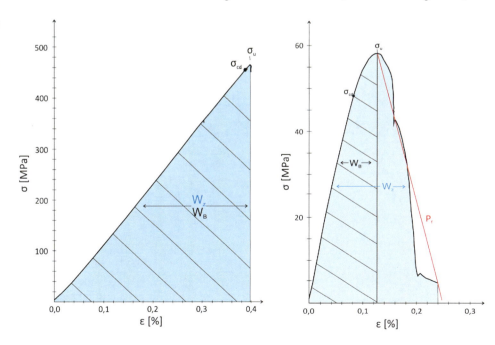

Fig. 82.3 Ratio Z (UCS/BTS) of analyzed rock types and results from TUM rock data base. Highlighted are Anhydrite/Basalt (typical example for tough/brittle) and Amphibolite. Results are plotted with the existing classification system of Schimazek and Knatz (1976), extended after Thuro (1996)

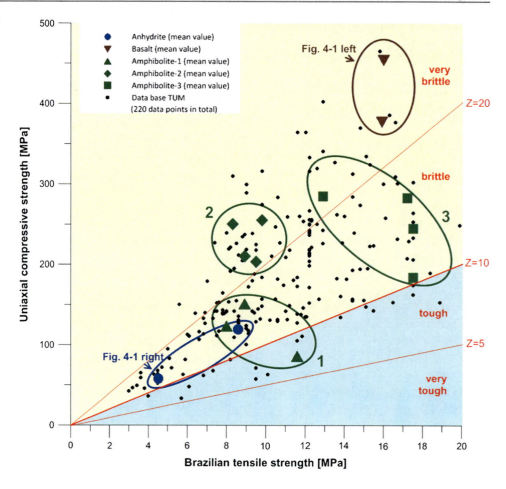

Fig. 82.4 Mean Z coefficient of analyzed rock types plotted against the ratio of destruction work to failure energy. Colored background describes post failure behavior at uniaxial compression test

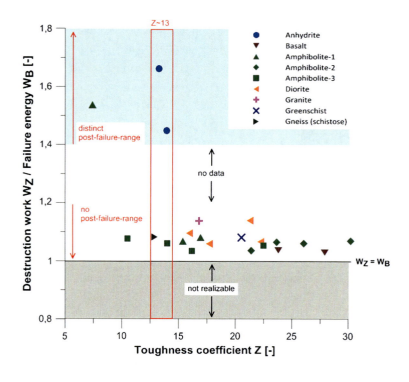

data points in total) reveal that only very few points are plotted in the classification field 'tough' or 'very tough'. These data points belong mainly to limestone with a high percentage of marl.

As the existence of a post-failure-range is one of the main characteristic of tough rocks, the ratio of destruction work W_z to failure energy W_B is illustrated in Fig. 82.4. Brittle rocks should have W_Z/W_B values of about 1 as W_Z is almost equal to W_B (Fig. 82.2 left). In contrast, values notably higher than 1 should define tough rocks as they have a significant post-failure-range ($W_Z \gg W_B$). The diagram demonstrates that 3 data points of rocks with a tough post-failure behavior do have a high W_Z/W_B value from 1.4 to 1.7. Mean Z values from all other rock samples range between 1.03 and 1.14. However, there is yet no significant trend visible that correlates low Z coefficients with high W_Z/W_B ratios. Z values around 13 show W_Z/W_B values from 1.1 to 1.7 depending on rock type (Fig. 82.4, red rectangle).

82.5 Conclusion

Most of the tested rocks had Z values higher than 10 and are classified as 'brittle' or 'very brittle' even if the stress-strain curve shows tough failure behavior. This demonstrates that the existing classification system of Schimazek and Knatz (1976) has to be revised and updated to get a better correlation with stress-strain curves. Moreover the Z values of one rock type (e.g. Amphibolite-1,-2,-3) vary a lot so that an implementation of the existing Z-coefficient into a prediction model seems unsuitable.

Additionally the investigations showed that the post-failure-range is a main characteristic of tough rocks. For a common classification, this part should be clearly described with values like the ratio of W_Z/W_B. However, these parameters have to be analyzed in detail with more rock types. Furthermore, an extended laboratory program with

acoustic emission testing is planned to gain a better understanding of fracture propagation in rocks and therefore, a distinct determination of the transition point from elasto-plastic to plastic behavior.

The topic of rock toughness has not only significance for prediction models in tunneling but also for the industry. In the future, the laboratory results should also be transferred to breakability of rocks in quarry companies. Moreover, toughness defines fracture propagation in rocks. This is an important issue for understanding the mechanism of rock falls.

References

Becker H, Lemmes F (1984) Gesteinsphysikalische Untersuchungen im Streckenvortrieb. Tunnel 2:71–76

Eberhardt E (1998) Brittle rock fracture and progressive damage in uniaxial compression. PhD thesis, Department of Geological Sciences, University of Sasketchewan, Saskatchewan, 334 S

Gehring K (1995) Leistungs- und Verschleißprognosen im maschinellen Tunnelbau. Felsbau 13–6:439–448

Hughes HM (1972) Some aspects of machining. Int J Rock Mech Min Sci 9:205–211

ISRM (1978) Suggested methods for the quantitative description of discontinuities in rock masses. Commission on standardization of laboratory and field tests, Document No. 4, Int J Rock Mech Min Sci 15:319–368

Menschik F, Thuro K, Käsling H, Bayerl M (2013) Vorstellung einer Datenbank zur Dokumentation und Analyse von Labordaten sowie im Feld gewonnen Kennwerten. 19 Tagung für Ingenieurgeologie (Munich)

Schimazek J, Knatz H (1976) Die Beurteilungen der Bearbeitbarkeit von Gesteinen durch Schneid- und Rollenbohrwerkzeuge. Erzmetall 29:113–119

Schneider E, Thuro K, Galler R (2012) Forecasting penetration and wear for TBM drives in hard rock—Results from the ABROCK research project. Geomech Tunn 5:537–546

Thuro K (1996) Bohrbarkeit beim konventionellen Sprengvortrieb. Münchner Geologische Hefte, München, 145 S

Wawersik WR, Fairhurst C (1970) A study of brittle rock fracture in laboratory compression experiments. Int J Rock Mech Min Sci 7:561–575

Stability Analysis of Accidental Blocks in the Surrounding Rockmass of Tunnels in Zipingpu Hydroelectric Project

83

Yanna Yang, Mo Xu, Shuqiang Lu, and Hong Liu

Abstract

Instable rock masses formed by structural plane boundaries can take great threaten to the underground tunnel excavation and the supporting structural stability. The boundary effect of structural planes in the rock mass should be considered during the underground tunnel excavation. The stability of the support system and rocks were determined by the preferred structural plane. Zipingpu key hydraulic project is located at Shajinba reach, upriver of Minjiang River, being built for agricultural irrigation and civil water supply. It also has integrative benefits such as preventing flood, generating electricity, environmental protection and tourism. Taking aim of stability analysis and assessment of surrounding rock mass of underground seepage tunnels in the right bank of Zipingpu hydraulic key project, and based on the survey and statistical analysis of structural planes of the diversion tunnels, structural rock mass discontinuity and the strength of fractured rock mass were studied in-depth, and three dominant directions of the structural planes included in the rock masses were divided through statistical analysis. Preferred structural planes were not long, randomly distributed distribution and extremely disadvantageous if they were connected as cutting or sliding block boundaries. The accidental rock masses were searched through the theory and method of slope rock block stability analysis system, and morphological property, stability and sensitive affected factors of them were analyzed. Rock masses which were prone to be damaged were identified and the length of the anchor bolt of the support system was designed.

Keywords

Structural planes • Accidental blocks • Stability analysis • Zipingpu hydroelectric project

Y. Yang (✉) · M. Xu (✉)
State Key Laboratory of Geohazard Prevention and Geoenvironment Protection, Chengdu University of Technology, Chengdu, Sichuan 610059, China
e-mail: yangyanna@cdut.cn

M. Xu
e-mail: xm@cdut.edu.cn

S. Lu
Hubei Engineering Research Center of Geological Hazards Prevention, China Three Gorges University, Yichang, Hubei 443002, China
e-mail: 381055335@qq.com

H. Liu
Chengdu Institute of Survey and Investigation, Chengdu, Sichuan 610081, China
e-mail: 123676858@qq.com

83.1 Introduction

One of the most important problems in tunnel excavation is to take the accidental falling of rock blocks that are formed by faults and discontinuities in the rock mass into consideration (Lee and Song 1998). Removability analysis of for falling of blocks falling should be conducted on the rock blocks based on a precise characterization of discontinuities in the rock masses around the tunnel. Analysis methods with more information about the block stability have been introduced since the block theory was suggested by Goodman and Shi (1985), which makes it possible to analyze the stability of rock blocks on slopes or around underground openings. This theory enabled us to test the combination of joint and to

Fig. 83.1 The site of Zipingpu water control project

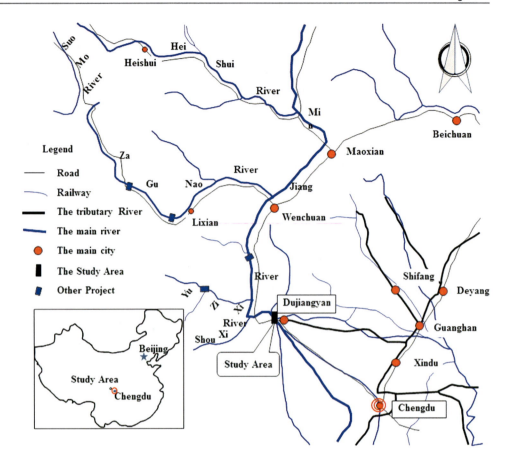

determine the safety factors of removable blocks with relatively simple and fast method because it uses only the orientation and friction angle of each joint set as the main input parameters. A three-dimensional statistical joint modeling technique was used to analyze the stability of rock blocks which were generated from the joints and the tunnels were analyzed for their volume, height, perimeter, safety factor, and probability of occurrence (Jae-Joon Song et al. 2001).

The natural stress field of the surrounding rock around the underground opening has changed during or after the excavation of the underground opening, and such a change brings instabilities or potential instable blocks to the original stable underground rock mass. The random blocks formed entirely by the mutual combination of random structural planes formed unknown relationship with position, scale and the underground project. The shape, number and scale of the random blocks in the surrounding rock were controlled by the layout of the underground project and the structural features of the rocks, while the stability of the random blocks was determined by the project characteristics of the rock, the stress state of the surrounding rock and the mechanical property of the structural plane. At present, the deterministic model and the generalized model of the surrounding rock structure and the research findings of the rock mechanical parameters, together with the openings (openings group) layout condition were mainly used to analyze the stability of

the blocks formed by mutual surrounding limitation of each level of structural planes. The evaluation for random block stability in the opening surrounding rock was based on geological investigation and exploration. For the surrounding rocks that generate special structural planes, block analysis method was used to find out the disadvantageous combination with other structural planes, and to determine the sliding direction, the sliding plane, the incision plane and the area of the incision plane, the possible volume and weight of the instable block (Ju 2005). Further, under gravity and the stress of surrounding rock, block ultimate balance theory was used to calculate the partial stability of the block assembled by structural planes, which provided proofs for the consolidating procedure. The evaluation method mainly included stereographic projection based on block theory, entity proportion projection, vector calculation method, etc.

Zipingpu water control project was located in Shajinba which was a section from the upstream of Minjiang River to Dujiangwan Weir river, 60 km from the northwest of Chengdu, Sichuan, China, and the downstream of this project was 9 km from Dujiangyan City (as Fig. 83.1 shown). This project was mainly for agricultural irrigation and civil water supply, as well as for integrative benefits such as preventing flood, generating electricity, environmental protection and tourism, and further, and it was a water adjusting project between the irrigation area of Dujiangyan and

Fig. 83.2 The direction of the tunnels is intersected with the tectonic *line* of Shajinba syncline

Chengdu City. Zipingpu water control project was one of the key projects of infrastructure construction during China's Tenth Five Year Plan Period and also the landmark project in China's Development of the West Regions. After the reservoir was completed, it could play the role of adjustment in the period of high water flow and low water flow. Therefore it greatly relieved the water supply conflict between the irrigation area of Dujiangyan City and Chengdu City. This project had a 156 m-high dam, 877 m normal impounded level, 11.12×10^8 m^3 total reservoir capacity and 760,000 KW installed capacity. The strip-like mountain ridge on the right bank of the project area took on north and east arc-strip shapes in the plane before the construction, with a length of more than 1,000 m, the width of the bottom being (with the waterside as the boundary) 400–650 m and the width of the top being (according to normal high water level) 50–250 m.

The strip-like mountain ridge on the right bank of the project area was a complete north-oriented and east-oriented pitched syncline—Shajinba syncline (as Fig. 83.2 shown). The sequences of the rock layers are normal, with no large fault going through. The bedrock was a typical flysch construction of are nacreous shale with coal of Xujiahe Group of upper Triassic Period. Still, the inside of the bed rock was mostly thick layered and hard fine packsand, with a small proportion of shale. The hydraulic structures in the project junction area were arranged within the strip-like mountain ridge on the right bank, including the ground factory buildings (four installations) behind the dam, open spillways adjacent to the right end of the dam, four underground openings that generate power by diverting water, one sand-washing underground opening and two flood-discharging and sediment-releasing underground openings transformed from two flow-guiding underground openings. Among these projects, the four underground openings that generate power by diverting water were respectively 1#, 2#, 3#, 4#, and they were sequentially arranged from the mountain to the outside of the mountain. The thickness of the rock wall between the

two underground openings that generate power by diverting water was 12–12.5 m. The maximum thickness of the two adjacent underground openings was increased to about 40 m after vertical three-dimensional intersection evacuation. The direction of the opening axis was intersected with the angle of the rock layer direction, and the geological conditions passed through by them are quite similar. The stability analysis of the random block masses was very important for the integral and partial stability researches of the surrounding rock of the opening under the highly intensive project layout and the special surrounding stress field.

83.2 Methodology

In the design period of the Zipingpu water control project, it had started from the angle that the intensity of the redistributed surrounding stress of the underground opening project being the lowest and the number of the blocks being the fewest, and it was best for the integral stability of the opening groups, and the best trending direction of the underground opening groups axis was integrally analyzed and optimally determined as NE25°. After the project layout and the opening scale were fixed, "window technique" (or network measuring method, usually 2×2 m^2) to track were used to investigate and measure different types of structural planes shown in the construction field surrounding rock, and make fisher cluster analysis to determine the level of the advantageous structural planes and the main development direction in the surrounding rock of the openings. Next, according to the development scale, intensity, shape and the project property of the openings in the practically measured structural planes, equatorial horizon projection and entity proportion projection methods were adopted to search all the possible structural planes one by one in the underground opening areas, and then all the random blocks, their shapes and scales were obtained. Further, mobility theory was used to determine the formed and potential mobile blocks.

Underground-opening sidewall block stability analysis program USASW (Ju 2005) worked out by State Key Laboratory of Geohazard Prevention and Geoenvironment Protection (Chengdu University of Technology) was used to make a fast evaluation about the stability of the surrounding rock blocks of the openings to effectively improve the safety of construction and optimize the design for system roof bolt.

As the excavation face height of the underground opening of the project was only 8–9 m, and the width is 12 m, the possibility that the top of the opening forms large-scaled mobile blocks was very little. The formed small-scaled blocks had a little effect on the stability of the openings, and such an effect could be omitted in analyzing the blocks on the top arch of the opening.

83.3 Results

83.3.1 Development Characteristics of Structural Planes of Dominant Group and the Shear Resistance Index Value

According to the project geology grading in the structure investigation, there was no type I fault-zone structural plane developed in the investigated openings. Type II and type III structural planes were mainly fault zones and interlayer compressive zones made of some shuttered zones or soft materials. Generally, the direction was clear and could be regarded as deterministic structural plane. Types IV and V fissure structural planes developed in the surrounding wall were the main boundaries to cut surrounding rock to form random blocks, and the IV and V fissure structural planes including the structural planes developed along the sandstone level, the long and big cut-layer structural planes developed in the vertical level and a number of matrix fissures developed randomly. The IV and V fissure structural planes were used as the main objects for measuring in the use of statistical method. The structural planes shown in the surrounding rock of the four underground openings that conduct water were precisely measured, and 1,400 pieces of detailed information about the fissure were gained. The results of the fisher cluster analysis displayed that there were four dominant groups of structural planes developed in the surrounding rock of the underground opening groups of the Shajinba syncline, and the properties of the structural planes were shown in Table 83.1. Referring to existed rock physical mechanics testing result, shear resistance index value of the structural plane was gained. From the perspective of safety, the cohesive strength value C of the shear resistance index in the structural plane was 0, because the structural planes in the inlet and outlet of the openings had many argillized interlayers and low cohesive force.

83.3.2 Stability Evaluation of Random Blocks

The structural planes developed in each opening gained in the investigation of practical field geology were combined with each other. Corresponding occurrences and parameters of possible sliding plane, cutting planes of the top surfaces and sliding boundary planes were put into the analysis program of the underground opening block stability. Through calculation, the geometrical shape, volume and stability coefficient of the blocks were gained. The axis orientation of the underground opening group was 132°. The side walls of the upstream and downstream were nearly vertical when the block was determined and the block stability was calculated. The occurrence of the sidewall on the right wall was $N48°W/NE\angle90°$ and on the left wall was $N48°W/SW\angle90°$. Image theory was used to determine the development and distribution of the sidewall blocks on the left and right walls.

The calculation and analysis results showed that the left wall of the underground opening group inlet at the syncline NW side did not form blocks easily, and the right wall possibly formed instable blocks, with most of the blocks being sliding planes with low stability which was only 0.283. Instable blocks were possibly formed on the left and right walls of the underground opening group inlet at the syncline SE side. The left wall of the opening was mostly intersected with the opening-oriented at small angles. The structural plane trending towards the exposure surface was the sliding surface whose stability coefficient was only 0.233. For the right wall of the opening, the dual sliding surface was formed by the orthogonal intersection between the bedding layers whose stability coefficient was 1.065. The volumes of the instable blocks were not too large to form serious hazards to the construction and support designs of the underground. The geometric shape of the block didn't vary greatly with the gap changing of the structural plane, and its stability was irrelevant to the gap of the structural plane.

The distribution and stability of blocks in different parts of the underground openings were shown in the Table 83.2. As it could be seen, the instable blocks distribution density gradually reduced from the inside of the ridge to the outside. The proportion of the number of the instable blocks of each opening in the total number was approximate. The volumes of the instable blocks were not large. The maximum thickness of the instable block was 9.03 m, and the minimal was about 1.5 m.

For the blocks formed randomly by types IV and V structural planes, the method of combining system anchor rod with reinforced anchor rod was preferable. The maximum thickness of the instable block was 9.03 m. The random blocks could be anchored by 8–12 m system anchor rod.

Table 83.1 The dominant groups of structural planes developed in the surrounding rock of the underground tunnels of Shajinba syncline and their shear resistance index value

Serial number	Occurrence	Position in the project	Property of the structural plane	Index of shear resistance for structural surface		
				Friction coefficient (f)	Internal friction angle $\varphi(°)$	Cohesive force C (MPa)
IV bedding fissure	N7°E/SE∠54°	The inlet section of underground opening at syncline NW	The bedding fissures mainly develop in the inlet and outlet of the underground opening groups and the two edges of the syncline. The maximum occurrence inclination angle in the NW side reach about	0.40 ~ 0.45	18 ~ 22	0
IV bedding fissure	N76°E/NW∠45°	The outlet section of underground opening at syncline SE	70° and the largest occurrence inclination angle in the SE side is 85°. The clearer the fissure is to the core, the even the inclination angle is. The angle is only 25° in the core of the syncline. The extending length is more than 10 m, going through the three walls of the opening, flat and smooth. The openness degree is varying from 50 to 100 mm, and parts of the fissures are filled with some secondary interlined soil and some calcites. The fissures are distributed with some argillite interlayers. The gap in the medium and fine sand of the fissures is 0.3–06 m, with the largest gap being 2 m	0.40 ~ 0.45	18 ~ 22	0
IV bedding fissure	N30°E/NW∠34°	The inlet section of underground opening at syncline NW	The variation of the occurance is large. The entension length is more than 10 m, going through the three walls of the opening, flat and smooth. There is no filling in the structural plane, with most of the planes being closed. The gap among the fissures is 0.4-0.6 m. The gap is only about 0.15 m when the development is intensive	0.50 ~ 0.55	22 ~ 24	0
	N32°E/SE∠70°	The outlet section of underground opening at syncline SE		0.50 ~ 0.55	22 ~ 24	0
V matrix (randomly) structural plane	N60° ~ 85°W/ SW∠54° ~ 70°	The inlet and outlet of the underground opening groups and the two edges of the syncline	They develop randomly, with great variation in occurance and property. The trace length on the opening wall is 3–8 m around, with a small number being more than 20 m. The plane is flat and a little rough. Some of the planes have secondary soil, sandstone detris, shed coal, etc. The cohesive degree is low, and the openness is 2–5 mm. The gap among the fissures varies largely, with 0.2–0.6 m averagedly among the medium and fine sandstone, and part of the powder sandstone being more than 1 m. As the direction of the structural planes are approximately parallel to to axis of the opening, the effect on the partial stability of the opening is large	0.55	24	0

Table 83.2 Distribution and stability analysis of the accidental rock masses of the tunnels in Zipingpu hydroelectric project

Tunnel NO.	Length of the opening investigated (m)	Number of the blocks		Volume of the instable block (m³)		The minimum value of the stability	The maximum thickness of the instable block (m)
		Total number	Instable number	Maximum value	Minimum value		
Tunnel 1#	271	8	4	7.26	0.09	0.222	9.03
Tunnel 2#	305	7	3	4.79	1.22	0.249	7.9
Tunnel 3#	236	6	3	0.19	0.4	0.188	1.44
Tunnel 4#	262	1	1	13.1	13.1	0.327	7.57

83.4 Discussion

In the design of underground opening group of the strip-shaped ridge on the right bank of Zipingpu water control project, although the instability of blocks in construction had been taken into consideration, it was still possible that there were unknown control large-scaled blocks formed by types IV and V random structural planes with types II and III deterministic structural planes due to the condition limitations in the geological investigation.

References

Goodman RE, Shi G (1985) Block theory and its application to rock engineering. Prentice-Hall, Englewood Cliffs, p 338

Ju N (2005) Systematical engineering geological study on the stability of surrounding rock mass with large span high side-wall underground caverns and their support proposals. PhD thesis of Chengdu University of Technology(in Chinese)

Lee C-I, Song J-J (1998) Stability analysis of rock blocks around a tunnel. In: Proceedings of 3rd international conference mechanical jointed faulted rock (MJFR3), pp 443–448

Mc Cullagh P, Lang P (1984) Stochastic models for rock instability in tunnels. J R Stat Soc B 46 Ž2 pp 344–352

Song JJ, Lee CI (2001) Estimation of joint length distribution using window sampling. Int J Rock Mech Min Sci 38:519–528

Engineering Geological Problems Related to Geological Disposal of High-level Nuclear Waste

Convener Prof. Weimin Ye—*Co-conveners* Yujun Cui, S. Tripathy, Xu Yongfu, Tang Chaosheng, Yonggui Chen

The ultimate disposal of high-level nuclear waste (HLW) requires their isolation from the environment for a long time. The most favoured method is burial in stable geological formations more than 500 metres deep. In order to ensure the safety of such a disposal system for a long time (generally recognized as 10 000 years), issues related to the engineering geological properties of host geological formations and buffer/backfill materials have been widely investigated in the world. This session will offer a multidisciplinary platform for researchers to exchange current achievements made in their investigations on host geological formations, groundwater, buffer/backfill materials, etc., of the geological repository.

Feasible Study of the Siting of China's High-Level Radioactive Waste Repository in an Area of Northwest China

Yuan Gexin, Zhao Zhenhua, Chen Jianjie, Jia Mingyan, Han Jimin, and Gao Weichao

Abstract

With the fast development of national nuclear industry, it is extremely urgent to disposal the high-level radioactive waste (HLW) properly. Aqishan area lies to the south of Turpan in Xinjiang Province. As one of the important candidate sites for China's HLW repository, it has many potential advantages, such as arid climate, water poverty, sparse population, large-scale granite body, and high crustal stability. The preliminary work proves that: (1) Granite batholith is widely distributed over the Aqishan area, which had formed in Late Hercynian—Early Indosinian period, with a thickness of over one thousand meters; (2) The Aqishan area is in a state of peneplain, with the latest fault activity in Middle Pleistocene and the seismic intensity of VI degree; (3) The total dissolved solids (TDS) of groundwater is up to 100 g/L because of high evaporation intensity, and the isotope data indicate that the groundwater is mainly recharged from atmosphere precipitation. Through an overall evaluation of the Aqishan area, it is found to be the feasible site for China's HLW disposal.

Keywords

HLW geological disposal • Hydrological condition • Crustal stability • Granite pluton

84.1 Introduction

Disposal of high-level radioactive waste (HLW) is generally implemented by deep geological disposal (Min 1998). Geological disposal stores long-lived radioactive materials in a stable geological repository, a geological unit which is required to remain stable for tens of thousand years. In this way, the risk of accidental waste exposure caused by human or natural disturbance can be reduced to a significantly low level. The principles for siting of an HLW repository mainly include sparse population; stable geological condition with no mineral resources; and host rocks with sufficient thickness and area, simple hydrogeological environment, low porosity, high thermal conductivity, great mechanical strength, and high thermal and radiation stability. According to the *Guideline of Site Preselection for HLW Repository* issued by the Commission of Science, Technology and Industry for National Defense and in view of the favorable factors such as non-permanent residents, arid climate, and large granite pluton in a northwest region, the granite zone in Aqishan area of northwest China is presently thought to have potential advantages for building a geological HLW repository in terms of climatic condition, geographical environment, granite pluton distribution, water resource distribution, crustal stability, and HLW safety management.

In this study, comprehensive geological studies have been focused on the large intact Xianshuigou granite pluton in this region. Surface geological mapping, seismic method, and transient electromagnetic exploration, combined with rock fracture statistics and engineering property testing, are used to study the distribution of granite pluton and the characteristics of rock minerals, as well as the lithology of surrounding strata; structural characteristics and integrity of granite pluton; and the morphology and type of deep granite

Y. Gexin · Z. Zhenhua (✉) · C. Jianjie · J. Mingyan · H. Jimin · G. Weichao
Northwest Institute of Nuclear Technology, 710024 Xi'an, China
e-mail: starzzh@gmail.com

Z. Zhenhua
Nanjing University, 210093 Nanjing, China

pluton. Associated hydrogeological characteristics are analyzed with relevant topographical and surface-water chemical data and groundwater isotope values; regional crustal stability is evaluated according to the characteristics of tectonic movement combined with those of seismic activity and geophysical field within and surrounding the northwestern region; the Quaternary geology and climate—environment change trend are evaluated by surveying the periods of regional tectonic activity and the characteristics of paleo-environmental evolution.

84.2 Traffic, Physical Geography, and Society

The northwest region is located in southeast Turpan, Xinjiang Uygur Autonomous Region. This region is approximately 160 km distant from Turpan City. The West-East Natural Gas Transmission Pipeline crosses the north part (Fig. 84.1) and connects to Dikaner County via a Class III highway. The terrain is relatively flat and most areas are reachable with vehicles.

Topographically, there are mainly denuded hills and plains in the northeastern region. The altitude of the flat terrain is 1050–1150 m above sea level. The climate is arid and water sources are lacking. The drainage system is poorly developed with no permanent runoff. Annual precipitation is 20–60 mm and the total precipitation time is less than 20 days. Annual average evaporation capacity is up to 2250–2900 mm. There are no permanent residents. To solve water shortage problems, domestic water is generally supplied from other regions.

84.3 Geological Characteristics of Xianshuigou Granite Pluton

There are mainly five granite zones surrounding the Aqishan area, which are generally composed of Middle—Late Variscan and Indosinian intrusive rocks with a total area of approximately 1200 km^2. The present survey focuses on Xianshuigou granite pluton, a regular oval-shaped pluton with an area of approximately 300 km^2. Few dykes are developed in the granite zone, which exert less impact on the integrity of pluton.

In the Late Carboniferous—Permian period, Tarim and Junggar plates collided; associated independent intrusions and regional dykes from different sources were widely distributed. In the Permian period, regional curst was uplifted due to NS extrusion; ductile crust below the uplift belt underwent selective melting due to the reduction of pressure (caused by the uplift of brittle crust) and the increase in heat energy (transformed from kinetic energy of tectonic movement); then, molten magma was uplifted and localized in the NE-SW-trending fracture zone after differentiation, leading to the formation of Xianshuigou granite pluton (Xinjiang Bureau of Geology and Mineral Resources 1997).

According to the results of remote sensing, aeromagnetic interpretation, and field geological survey, Xianshuigou granite zone mainly consists of four intrusions (Fig. 84.1). The outcropped sedimentary strata consist of hornfels and skarns of the Lower Carboniferous Gandun Formation and alluvial gravel and alluvial aeolian deposits of the Quaternary Holocene strata. Flesh red medium-grained syenogranite

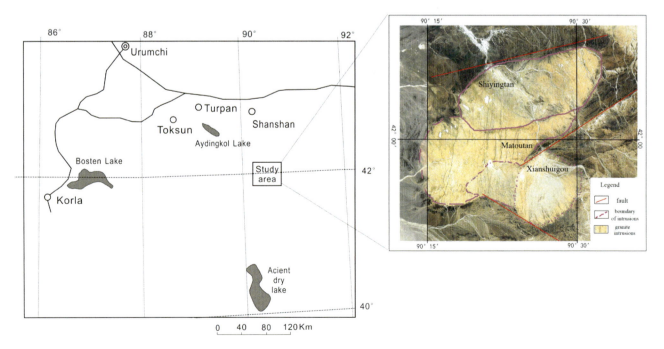

Fig. 84.1 Sketch map of the location of granite pluton in the area

Fig. 84.2 Resistivity profile of Xianshuigou granite pluton determined by transient electromagnetic method

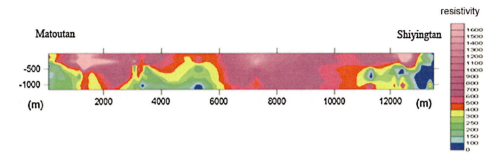

forms the pluton with strongest γ-ray spectrum field in the whole zone. In the granite rock outcrops near Xianshuigou, the latest strata intruded are Lower Permian strata with the zircon U-Pb isotopic age of 200.1 Ma (Xinjiang Bureau of Geology and Mineral Resources 1997). Together the characteristics of geophysical field reflected by the pluton, we preliminarily identify Xianshuigou granite pluton as Early Indosinian intrusive rocks (according to *prospecting report of Xianshuigou granite zone*, 2004).

The development of internal fault in Xianshuigou granite pluton is explored with an ATEM Transient Electromagnetic System. The exploration profile is 600 m long and nearly EW trending, with depth inversion at 200 m. The resistivity is high in the upper part of the profile, which decreases with increasing depth. Within the depth of 150 m, the resistivity is greater than 1000 Ωm and shows even electrical distribution. The electric property is evenly distributed with no substantial changes in the horizontal direction. There are no abnormal segments with abrupt vertical changes. No anomalies of fault structure are found in the profile.

The morphology of granite pluton on the plane is investigated by remote sensing-based geological interpretation and site exploration; the vertical morphology of deep pluton is explored mainly using transient electromagnetic technique and shallow high-resolution seismic reflection wave method.

Through the exploration with transient electromagnetic (Fig. 84.2) and seismic methods and comparative interpretation of geological profile, we propose that the pluton may belong to large-scale granite batholith; there may exist underlying strata at 400–900 m depth of the granite pluton at 4–5 km north of Matoutan; there are no large fractures within the granite pluton; and the thickness of the granite pluton is greater than 1000 m.

84.4 Hydrogeological Conditions of Xianshuigou Granite Zone

Xianshuigou granite pluton is located in an area with a typical continental climate, dry weather and less rain, windy spring and autumn, and large temperature difference between day and night. There is no perennial water; all valleys are seasonal ravines which only have temporal floods flowing after rainstorms; the evaporation is rapid.

The major type of groundwater within the pluton is bedrock fissure water. Due to insufficient supply, granite fissure water is lacking. According to the *1:500000 regional hydrogeological survey report of Shanshan—Aqishan region* (1978), groundwater has been drilled in the north at the depth less than ten and a few meters; the unit water inflow is 19 m³/d, and groundwater salinity is 33.31–125.27 g/l, i.e., saline–brine water.

Topographically, the study area is high in south and low in north, with surface runoff flowing toward a northwest lowland, Aydingkol Lake (Fig. 84.3). There is no perennial surface runoff. Groundwater recharge completely depends on the infiltration of atmospheric precipitation, i.e., noncontinuous supply. This area is arid with less rain; average annual rainfall is 20–60 mm only, mostly concentrated in summer; temporary surface floods are commonly formed after occasional storms.

In the arid and hot climate, annual evaporation is a hundred times more than precipitation. Thus, groundwater discharge is dominated by evaporation in this region.

Stable hydrogen (δD) and oxygen (δ¹⁸O) isotope data of groundwater indicate that the source of groundwater recharge is infiltration of atmospheric precipitation. The δD-δ¹⁸O relationships of groundwater samples from different areas are compared. As compared to those from the central area, water samples from the south and north margin of Kuruketage Mountains have small δD/δ¹⁸O slope, indicating that groundwater migrates from the center area to two sides.

84.5 Crustal Stability

Regional crustal stability is evaluated by analyzing the data of regional geophysical field, deep fault zone, seismic activity, and neotectonic unit division, combined with the results of field geological survey.

An average Bouguer gravity anomaly map compiled by the Geophysical Exploration Team of Xinjiang Bureau of Geology (1983) shows that there is a nearly EW-trending gravity anomaly in the south of the study area whose

Fig. 84.3 3D terrain map of study area

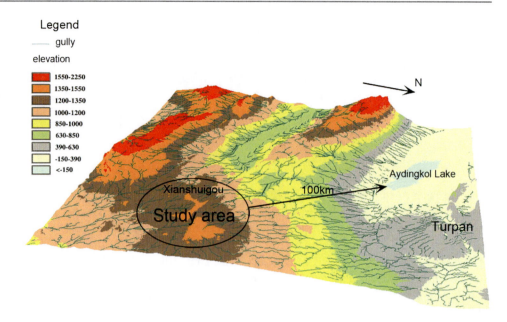

amplitude is $(-110$ to $-170) \times 10^{-5}$ m/s^2. The study area is situated in a nearly EW-trending gravity gradient wide—gentle zone with the amplitude of $(-120$ to $-130) \times 10^{-5}$ m/s^2; this indicates that there is no large faults crossing this area. According to the 1:50,000 regional aeromagnetic ΔT plane anomaly map of Xinjiang (1989), the aeromagnetic anomaly is wide and flat, indicating that the geological body is homogeneous in the granite zone. According to the Moho depth contour map (1997) in *Xinjiang Geological and Mineral Chronicles*, the study area is located on the margin of an EW-trending gradient zone; the crust thickness is 46 km which changes smoothly.

As shown in the regional epicenter distribution map (Fig. 84.4), Xianshuigou granite pluton lies in the north of Aqikekuduke fault (Fig. 84.4, No.1) where seismic events are relatively rare and the earthquakes are generally less than magnitude 5.0. As shown in the seismic intensity map of Xinjiang, the seismic intensity of the study area is magnitude VI. Overall, the study area is associated with weak seismic activity.

There are mainly two regional deep faults developed around the study area (Fig. 84.4), mostly thrusting strike-slip or reverse faults. Ophiolite, mictite, and acidic granite are distributed along the faults.

Aqikekuduke fault (Fig. 84.4, No.1) is approximately 30 km south to the study area. This fault is NW-EW-NE trending and more than 1400 km in full length. It is overall south-dipping at 50°–80°, as classified as a dextral strike-slip and reverse fault. In the south of the study area, the profile of the fault shows that a gravel layer overlies the fault zone. Thermoluminescence dating samples collected from the find sand lens of sand—gravel layer are estimated to be (143.70 ± 12.21) ka BP (according to *regional crustal stability report of northeast of Korla*, 2010). According to the profile, Bolokenu—Aqikekuduke fault has never been active since the Late Pleistocene period, thus belonging to Early—Middle Pleistocene fault.

Yamansu fault (Fig. 84.4, No.2) is approximately 40 km north of the study area, part of which is associated with pluton emplacement. Regionally, this fault is nearly EW

Fig. 84.4 Regional network distribution of earthquake epicenters (January 1970–2005, M $\geq$ 2.5)

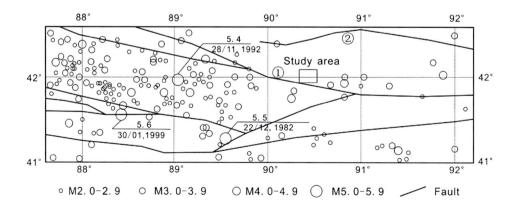

trending and extends in a soothing wavy pattern. It is generally north-dipping at 60°–70° and belongs to high-angle thrust fault, i.e., brittle—ductile fault. This fault does not dislocate the Middle—Late Pleistocene strata. Therefore, Yamansu fault is identified as an Early Pleistocene fault.

Together the above results indicate that the study area is located in a granite zone with less aeromagnetic anomalies, no abnormal abrupt changes in the EW distribution, and wide—gentle gravity gradient. This area is located in a crustal structure on the margin of an EW-trending gradient belt, which has relatively simple structure with no large faults. The study area lies in a slight uplift zone with weak seismic activity and low seismic intensity (<VI). That is, the neotectonic activity is weak and regional crust is relative stable in the study area.

84.6 Preliminary Assessment

In the northeastern region, Xianshuigou granite pluton belongs to early Indosinian intrusive rock; this pluton is affected by post-tectonic movement to a relatively low degree; there are no large faults in the pluton, with few faults and dykes developed only; the geological repository has complete surrounding rock mass, high engineering strength, and uniform stable engineering performance. Xianshuigou granite pluton is of batholith type, whose thickness is generally greater than 1000 m. The granite pluton has an area of more than 300 km^2, providing enough space for engineering disposal. Regional structure has little impacts on the pluton. The geophysical field is relatively stable and the seismic activity is weak, with seismic intensity of magnitude VI. Therefore, the preselected site is a relatively stable area conducive to long-term storage of HLW. The northeastern region is a water-poor area where the arid climate and pluton characteristics prevent the infiltration of surface precipitation. Therefore, the water system has little impact on the repository. The groundwater flow path is quite long, approximately 100 km to Aydingkol Lake in the northwest direction.

Overall, the northeastern region has good prospects as a preselected area of geological repository for HLW disposal.

Future feasibility study of the northwestern region as a preselected area of geological repository for HLW disposal needs to investigate the lithological evolution and distribution patterns of deep granite pluton; combined with geophysical data, more deep drillings should be carried out.

References

Min M (1998) Principles of the disposal of the radioactive waste [M]. Atomic Energy Press, Beijing

Xinjiang Bureau of Geology and Mineral Resources (1997) Regional geology of Xinjiang Province [M]. Geology Publishing House, Beijing

A New Apparatus for the Measurement of Swelling Pressure Under Constant Volume Condition

85

C.S. Tang, A.M. Tang, Y.J. Cui, P. Delage, and E. De Laure

Abstract

A new constant volume cell was developed, allowing the measurement of swelling pressure without any strain adjustment and any effect of the stiffness of the testing device. By employing this new cell, the swelling behavior of compacted soil specimen under different suctions (57, 38, 9 and 0 MPa) was investigated. The results show good repeatability, indicating the reliability of the new cell and validating the test procedures adopted. Moreover, the developed cell is quite convenient to study the long-term swelling behavior of soil since no load adjustment is necessary. The obtained results show that, during the progressive wetting by applying successively the suctions of 57, 39, 9 and 0 MPa, the swelling pressure increases to 0.17, 0.31, 0.46 and 0.89 MPa, respectively. The swelling pressure and the time required to reach equilibrium are function of suction. Vapour-wetting and water-wetting show different hydration mechanisms and result in different swelling behavior: the swelling pressure develops slowly and gradually to finally reach stabilization upon vapour-wetting, while it increases quickly to a peak value and followed by a small decrease upon water-wetting.

Keywords

Swelling pressure • Expansive soil • Oedometer test • Suction

85.1 Introduction

Expansive soil is a "problematic" soil, and widely distributes in many regions in the world. Generally, the parameter of swelling pressure is commonly employed to characterize the swelling properties of expansive soils (Shi et al. 2002). In the past decades, swelling pressure was defined in many ways and the measured value depends significantly on the test procedures adopted (Bilir et al. 2008).

The constant-volume swell test is one of the popular methods for measuring the soil swelling pressure. It is usually performed through strain-controlled technique in the laboratory (Tripathy et al. 2004). There are two common strain-controlled techniques: (1) applying small incremental load to compress the specimen to bring its height to the initial value as the specimen begins to swell upon wetting. Once the swelling pressure reaches equilibrium, the final applied total load is regarded as the swelling pressure (Al-Shamrani and Dhowian 2003); (2) restraining the specimen

C.S. Tang (✉)
School of Earth Sciences and Engineering, Nanjing University, Hankou Road 22, Nanjing, China
e-mail: tangchaosheng@nju.edu.cn

C.S. Tang · A.M. Tang · Y.J. Cui · P. Delage · E. De Laure
Ecole des ponts—ParisTech, U.R. Navier/CERMES, 6-8 av. Blaise Pascal, Cité Descartes, 77455, Marne-la-Vallee, France
e-mail: anhminh.tang@enpc.fr

Y.J. Cui
e-mail: yujun.cui@enpc.fr

P. Delage
e-mail: delage@cermes.enpc.fr

E. De Laure
e-mail: delaure@cermes.enpc.fr

G. Lollino et al. (eds.), *Engineering Geology for Society and Territory – Volume 6*,
DOI: 10.1007/978-3-319-09060-3_85, © Springer International Publishing Switzerland 2015

vertical movement by a rigid reaction frame. The rigid frame reacts against the sample through a high capacity load cell, which was used to measure the swelling pressure of soil sample (Aiban 2006). However, in practice, to obtain or ensure a strict constant-volume condition is not of easy task. For the former technique, each incremental load corresponds actually to a compression on the specimen, and may result in the specimen's structure changes in each cycles of swelling-compression. Moreover, in order to compress to the initial volume, the applied load should overcome the friction between specimen and oedometer/ring wall, resulting in higher measured swelling pressure. For the later one, Tang et al. (2011) found that the "constant-volume" was significantly dependent on the stiffness of the load cell and test equipment, and the measured swelling pressure error can reach 1–2 MPa.

For the purpose of improving the measurement accuracy of constant-volume swell test, a new apparatus was developed. The details of the apparatus and test procedure in controlling the constant-volume condition are presented in this paper. Using this apparatus, swelling tests over a wide range of suction were performed and the obtained results were discussed.

85.2 Development of New Constant Volume Cell

For better controlling the constant-volume condition in swelling tests, a new constant volume cell was developed and illustrated in Fig. 85.1. This cell mainly consists of three stainless steel parts: top, middle and bottom parts. The pressure sensor used to measure the swelling pressure was fixed inside the top part and directly in contact with the top surface of the soil specimen. The cylindrical soil specimen (70 mm in diameter and 10 mm in height) was placed inside the middle part. A porous stone was placed on the bottom part and in contact with the bottom surface of the soil specimen. Two inlets in the bottom part ensured the circulation of water (liquid or vapor) for the suction control. In

addition, two outlets of the middle part ensured the water/vapor evacuation from the top surface of soil specimen.

85.3 Material Used

The soils used were taken from Bure (North-eastern France). It contains 40–45 % clay minerals (illite–smectite interstratified minerals being the dominant clay minerals), 30 % carbonates and 25–30 % quartz and feldspar. The specific gravity of this material is 2.70. The obtain soil was air-dried ($w = 2.8\%$) and crushed to powder. The aggregate size distribution after crushing determined by dry sieving is presented in Fig. 85.2.

85.4 Experimental Method

The air-dried soil powder was statically compacted in the middle part of the constant-volume cell to the target dry density of 2.0 Mg/m^3. The specimen size is 70 mm in diameter and 10 mm in height. A total of three identical soil specimens, T1, T2 and T3, were prepared. The initial suction of the specimen was about 100 MPa (air-dried). After compaction, the three parts of the cell were fixed together by screws (Fig. 85.1a).

In the present work, swell tests were performed at different suctions (e.g. 57, 38 and 9 MPa). The experimental setup is shown in Fig. 85.3. During the test, suction was controlled by the vapor equilibrium technique using various saturated salt solutions. An air pump was used to ensure the vapor circulation in the system. The swelling pressure was recorded automatically using a personal computer. Considering that the relative humidity imposed by given salt solution is highly dependent on temperature, the room temperature was maintained at 20 ± 0.5 °C. For specimen T1, a suction of 57 MPa was initially applied. After the equilibrium was reached, a subsequent wetting was applied by a suction of 38 MPa. For specimen T2, the same procedure as that of T1 was applied to investigate the repeatability of the test. For specimen T3, a suction of 9 MPa was

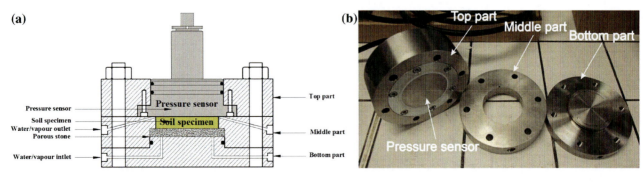

Fig. 85.1 **a** Schematic diagram of the constant-volume cell; **b** picture of the three parts of the constant volume cell

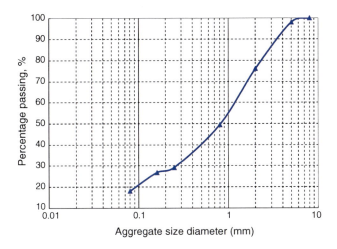

Fig. 85.2 Aggregate size distribution of the crushed soil

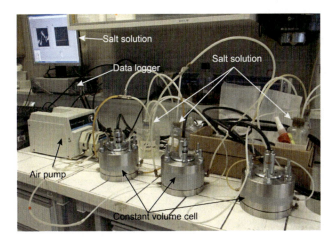

Fig. 85.3 Schematic diagram of the test setup adopted in the investigation

firstly applied. After the equilibrium was reached, distilled water was injected at a constant pressure of 15 kPa from the bottom inlet of the cell to saturate the specimen.

85.5   Results and Discussion

The results of the swelling tests are presented in terms of swelling pressure versus elapsed time (Fig. 85.4). As expected, at each suction level, the measured swelling pressure increases quickly in the beginning and then gradually reaches equilibrium upon further wetting. In Fig. 85.4a, the swelling curves of T1 and T2 are similar, indicating that the employed test procedures have good repeatability and the developed cell is reliable. For sspecimens T1 and T2, about 10 days are needed to reach equilibrium at suction of 57 MPa, and the final swelling pressure is about 0.17 MPa. The subsequent wetting path (suction controlled at 38 MPa) increases the final swelling pressure to

about 0.31 MPa and the time needed to reach the equilibrium is about 20 days. For test T3 (Fig. 85.4b), a suction of 9 MPa was applied initially. That induced a final swelling pressure of about 0.46 MPa and the time needed to reach the equilibrium is about 30 days. After that, distilled water was injected into the cell to saturate the specimen (zero suction). It can be observed that once the distilled water was injected, the swelling pressure increased suddenly to 0.92 MPa as a peak value followed by a small decrease to a stabilized value of 0.89 MPa.

The obtained results indicate that the time required to reach equilibrium depends on the suctions applied. When the specimens were gradually wetted by increasing the relative humidity (RH), water vapor was absorbed and moved towards the inside of the specimen under the effect of the suction gradient (Push 1982). But with the wetting that continues, the corresponding suction gradient gradually decreases and finally reaches equilibrium as well as the rate of vapor migration. The required equilibrium time therefore depends on two factors: the rate of vapor migration and the initial suction gradient. At a higher initial suction gradient, although the initial vapor migration rate inside the specimen is higher, more time would be required to reach suction equilibrium because more water was needed. This explains the observed behavior shown in Fig. 85.4: only about 10 days was required to obtain a stabilized swelling pressure at the applied suction value of 57 MPa but 30 days at 9 MPa.

Figure 85.4 also shows that the swelling pressure is function of suction. This phenomenon is associated with the hydration mechanisms in the level of swelling clay minerals. As presented in the previous studies, the swelling behavior of soil is the results of two combined processes: (1) the progressive absorption of successive layers of molecules in the interlayer spaces inside the clay particles results in an enlargement of interlayer distance; (2) the subdivision of particles into thinner ones that are made up of a smaller number of stacked layers, and leads to larger inter-particles pores (Delage 2007). However, the adsorption of water molecules between the layers inside the clay particles is function of suction; a decrease of suction gives rise to an increase of absorbed water layers. In the present work, the applied suction changes from 57 to 9 MPa, the corresponding RH increases from 66 to 93.7% at 20 °C. According to the results of Chipera et al. (1997), the maximum absorbed water layer in clay particles may be up to 2 layers.

In Fig. 85.4b, the development of swelling pressure curve during water-wetting stage is much sharper than that during vapor-wetting stage. Moreover, a slight decrease of swelling pressure after the peak value is observed. It may be attributed to the different hydration mechanisms between the vapor-wetting and the water-wetting. The hydration mechanism of vapor-wetting has been described above. The water

Fig. 85.4 Evolution of swelling pressure with elapsed time at different suction levels **a** specimens T1 and T2 and **b** specimen T3

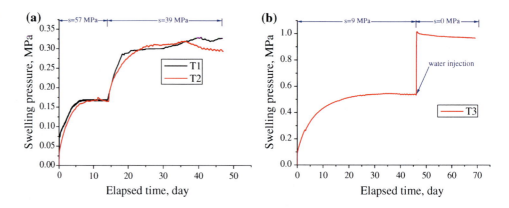

vapor is initially absorbed in small intra-aggregate pores with relative low mitigation rate. However, during water-wetting, water enters the inter-aggregate pores quickly. On one side, the aggregates begin to expand due to hydration, giving rise to a sudden increase of swelling pressure; on the other side, the continued hydration process simultaneously weakens the bonds between aggregates and soil skeleton loses its stiffness. Consequently, structure collapse occurs and causes a drop after the peak swelling pressure (Fig. 85.4b). This observation is consistent with the results presented in the previous researches (Push 1982; Komine and Ogata 1994; Cui et al. 2002).

85.6 Conclusions

1. An apparatus was developed for better performing constant swell test. The pressure sensor instead of traditional load cell was used to measure the swelling pressure of specimen during wetting, which can effectively minimize the measurement errors from the deformation of the system.

2. By applying the developed apparatus, the swelling pressure of compacted soil specimens under suction-controlled condition was measured. The results show good repeatability, indicating the reliability of the new apparatus and validating the test procedures adopted.

3. The final swelling pressure of specimen increased with decreasing suction, while the equilibrium time increased with decreasing suction.

4. Vapour-wetting and water-wetting show different hydration mechanisms and resulted in different swelling behavior.

Acknowledgments This work was gratefully supported by the National Natural Science Foundation of China (Grant No. 41072211; 41322019), Natural Science Foundation of Jiangsu Province (Grant No. BK2011339), Opening Fund of State Key Laboratory of Geohazard Prevention and Geoenvironment Protection (Chengdu University of Technology) (Grant No. SKLGP2013K010).

References

Aiban SA (2006) Compressibility and swelling characteristics of Al-Khobar Palygorskite, eastern Saudi Arabia. Eng Geol 87:205–219

Al-Shamrani MA, Dhowian AW (2003) Experimental study of lateral restraint effects on the potential heave of expansive soils. Eng Geol 69:63–81

Bilir ME, Sari D, Muftuoglu YV (2008) A computer-controlled triaxial test apparatus for measuring swelling characteristics of reconstituted clay-bearing rock. Geotech test J 31:1–6

Chipera SJ, Carey JW, Bish DL (1997) Controlled-humidity XRD analyses: application to the study of smectite expansion/contraction. In: Gilfrich JV et al (eds) Advances in X-ray analysis, vol 39, pp 713–721

Cui YJ, Loiseau C, Delage P (2002) Microstructure changes of a confined swelling soil due to suction controlled hydration. Unsaturated soils. In: Proceedings of the 3rd international conference on unsaturated soils (UNSAT 2002), Recife, Brazil (ed. Jucá, J.F.T., de Campos, T.M.P. and Marinho, F.A.M.), Lisse: Swets and Zeitlinger, vol 2, pp 593–598

Delage P (2007) Microstructure features in the behaviour of engineered barriers for nuclear waste disposal. In: Schanz T (ed) Experimental unsaturated soils mechanics, proceedings of international conference on mechanics of unsaturated soils. Springer, Weimar, Germany, pp 11–32

Komine H, Ogata N (1994) Experimental study on swelling characteristics of compacted bentonite. Can Geotech J 31:478–490

Push P (1982) Mineral-water interactions and their influence on the physical behaviour of highly compacted Na bentonite. Can Geotech J 19:381–387

Shi B, Jiang H, Liu Z, Fang HY (2002) Engineering geological characteristics of expansive soils in China. Eng Geol 78:89–94

Tang CS, Tang AM, Cui YJ, Delage P, Barnichon JD, Shi B (2011) Study of the hydro-mechanical behaviour of compacted crushed argillite. Eng Geol 118(3–4):93–103

Tripathy S, Sridharan A, Schanz T (2004) Swelling pressures of compacted bentonites from diffuse double layer theory. Can Geotech J 41:437–450

2D and 3D Thermo-Hydraulic-Mechanical Analysis of Deep Geologic Disposal in Soft Sedimentary Rock

86

Feng Zhang and Yonglin Xiong

Abstract

In deep geological disposal for high level radioactive waste (HLW), one of the most important factors is to study the thermo-hydraulic-mechanical (THM) behavior of the natural barrier, usually a host rock during heat process and hydraulic environment change. In this paper, based on a thermo-elasto-viscoplastic model for soft rock proposed by Zhang and Zhang (2009), a finite element method (FEM) has been developed to simulate the THM behavior of geological disposal. Considering the different cooling period before the disposal of HLW, two cases of 2D and 3D analyses are conducted to estimate long-term stability of the host rock. From the simulated results, the cooling period before the disposal of HLW is very important to the safety of the waste sealing construction.

Keywords

THM • FEM • High-level nuclear waste • Soil-water coupling

86.1 Introduction

In deep geological disposal for high level radioactive waste, one of the most important factors is to study the thermo-hydraulic-mechanical (THM) behavior of the natural barrier, usually a host rock during heat process and hydraulic environment change. The high level radioactive materials might permeate with underground water through the barrier systems to biosphere. The temperature effect on soft sedimentary rock due to the heat emitting from the nuclear waste canisters also needs to be investigated. The water absorption may induce a swelling of geo-materials that might lead to a damage of the nuclear waste containers. All these THM behaviors of the natural barrier need to be well understood in order to guarantee the safety and the efficiency of the waste sealing construction in long time. In this paper, Based on a thermo-elasto-viscoplastic model proposed by Zhang and Zhang (2009), a program called as 'SOFT', using FEM for spatial discretization and the finite difference method (FDM) for time domain in soil-water-heat coupling problem, has been developed to simulate the THM behavior of the host rock.

86.2 2D and 3D THM Simulations

Due to the symmetric geometry and loading conditions, only a half area is considered in the numerical simulation. 2D FEM mesh is showed in Fig. 86.1, in which the area is a rectangle of 210 m × 520 m, the right and left side are fixed in x direction, while the bottom side is restricted in vertical direction. The initial stress is calculated by gravitational analysis. For thermal condition, the initial temperature of whole considered area is 20 °C. The ground surface is always kept 20 °C all time and the heat insulation is assumed for other three sides. The initial total water head is given as 520 m. All the boundaries except the surface are undrained condition. The nuclear waste repository is a circle with 20 m in diameter and its center lies at the place 300 m below the surface.

F. Zhang (✉) · Y. Xiong (✉)
Department of Civil Engineering, Nagoya Institute of Technology, Showa-Ku, Gokiso-Cho, Nagoya, 466-8555, Japan
e-mail: cho.ho@nitech.ac.jp

Y. Xiong
e-mail: ylxiong1986@hotmail.com

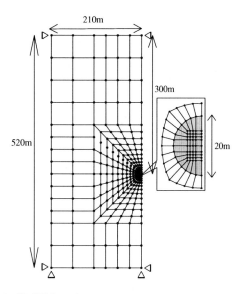

Fig. 86.1 2D FEM mesh

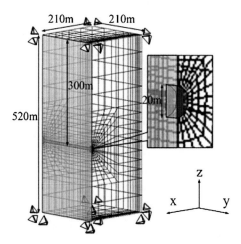

Fig. 86.2 3D FEM mesh

Figure 86.2 shows 3D FEM mesh that consists of 3,380 eight-node hexahedrons whose vertical section is the same as the 2D mesh. The initial and boundary conditions of the stress, the temperature and water head are also the same as those in 2D analysis. The thickness of the repository is the half of the diameter, that is, 10 m.

In the FE-FD analysis, a durative time of 300 years is simulated. In reality, the repository ratio of HLW is only about 1.6 % (Thunvik and Braester 1991). In present simulation, therefore, 1.6 % of heat emission is given and is showed in Fig. 86.3. In the THM analysis with FE-FD scheme, the influence of the convection of water flow is negligible because the maximum velocity of pore water is less than 10^{-8} (m/s) within the soft rock. The soft rock is

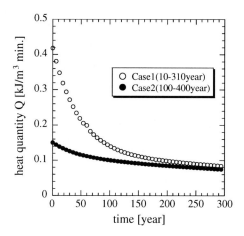

Fig. 86.3 Heat emission of HLW as the input in FEM

considered to be saturated. The technological and engineering barriers are assumed to be elastic in the THM simulation and the parameters of the materials used are listed in Table 86.1.

First of all, the 2D analysis will be discussed. Figure 86.4 shows the distribution of temperature at specified times. From the figure, it is known that the temperature is increasing initially due to the emission of the heat. In Case 1, the temperature begins to decrease during 75–150 years, and then cools down to a certain value. There is no clear further decrease of temperature within the calculated time. In addition, the generated temperature in Case 1 is much higher than that in Case 2.

Figure 86.5 shows the distribution of total water head at specified times. With the increase of temperature, the water head in two cases is also increasing with time, because the thermal expansion coefficient of water is much higher than that of the rock. And later, as the migration of the pore water, the excessive water is allowed to dissipate and consequentially turns to hydrostatic pressure.

Figure 86.6 shows the distribution of the plastic shear strain $\sqrt{2I_2^p}$ at specified times, where I_2^p is the second invariant of deviatoric plastic strain tensor. It is known from the figure that the maximum shear strain occurs in the area of 45° with horizontal direction.

By comparing the results from the two cases, it is found out that the magnitude of the increase of total water head and the plastic shear strain in Case 1 is much larger than that in Case 2 due to the difference of the generated temperature.

3D analysis is then discussed in the following section. Figures 86.7, 86.8 and 86.9 show the distribution of temperature, water head and the plastic shear strain at specified times respectively. It is found out from these figures that the same tendency observed in 2D analysis is confirmed in 3D analysis, except that the magnitude in 3D analysis is much smaller than that in 2D analysis.

Table 86.1 Physical properties and material parameters of soft sedimentary rock

Parameters			Physical properties		
Young's modulus E (MPa)	1,000.0		Pre-consolidated yield stress p_c (MPa)	0.30
Poisson's ratio v	0.120		Thermal expansion coefficient α_T (1/K)	8.0×10^{-6}
Stress ratio at critical $R_{CS}(=\sigma_1/\sigma_3)$	5.5		Permeability k (m/s)	10^{-9}
Plastic stiffness E_p	0.015		Thermal conductivity K_t (kJ m^{-1} K^{-1} Min^{-1})	0.2
Potential shape parameter β	1.1		Specific heat C (kJ Mg^{-1} K^{-1})	840
Time dependent parameter α	0.5		Heat transfer coefficient of air boundary α_c ((kJ m^{-2} K^{-1} Min^{-1})	230
Time dependent parameter C_n	0.025			
Over consolidation parameter **a**	3,000		*Specific heat of water C_w* (kJ Mg^{-1} K^{-1})	4,184
Reference void ratio N (e_0 at σ_{m0} = 98 kPa)	0.50			

Fig. 86.4 Distribution of temperature (°C) at specified times (2D)

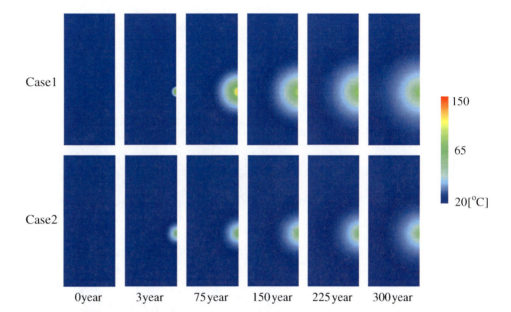

Fig. 86.5 Distribution of water head at specified times (2D)

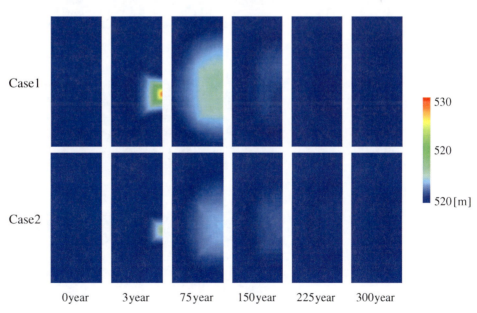

Fig. 86.6 Distribution of $\sqrt{2I_2^p}$ at specified times (2D)

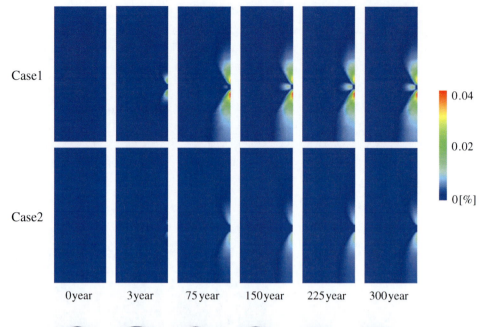

Fig. 86.7 Distribution of temperature (°C) at specified times (3D)

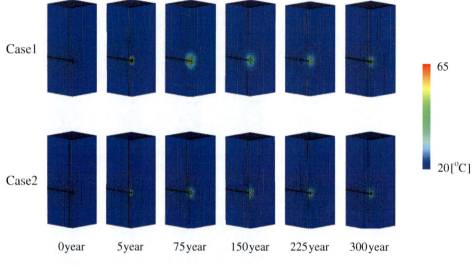

Fig. 86.8 Distribution of water head at specified times (3D)

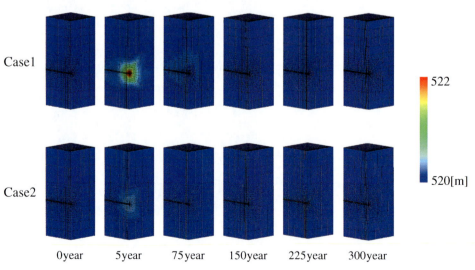

Fig. 86.9 Distribution of $\sqrt{2I_2^p}$ at specified times (3D)

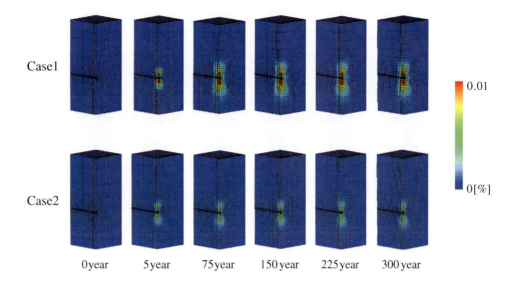

86.3 Conclusions

In this paper, 2D and 3D THM analyses on deep geological disposal of HLW with FE-FD scheme have been carried out the following conclusions can be given:

In 2D analysis, two cases are considered for different cooling period in the air before the geological disposal. It is revealed that the calculated physical quantities, such as the temperature, that total water head, the plastic shear strain in Case 1, are much larger than those in Case 2. In other words, the cooling period is very important factor for the long-term stability of the waste sealing construction.

In 3D analysis, the same tendency is observed. But the THM quantities are much smaller than those in 2D analysis. The reason is quite clear because in 3D analysis, the heat source is a point source, and the heat emission/transferring are conducted in spatial domain, resulting in a relative small concentration of heating.

References

Thunvik R, Braester C (1991) Heat propagation from a radioactive waste repository (SKB 91 reference canister). Technical report, pp 91–26

Zhang S, Zhang F (2009) A thermo-elasto-viscoplastic model for soft sedimentary rock. Soils Found 49(4):583–595

Anisotropy in Oedometer Test on Natural Boom Clay

87

Linh-Quyen Dao, Yu-Jun Cui, Anh-Minh Tang, Pierre Delage, Xiang-Ling Li, and Xavier Sillen

Abstract

The mechanical behaviour of Boom Clay has been studied for many years in the context of geological disposal of radioactive waste in Belgium. The aim of this study is to investigate the anisotropic behaviour of Boom Clay in terms of compressibility and hydraulic conductivity. Oedometer tests (with effective vertical stress (σ'_v) up to 32 MPa) were carried out on samples of various orientations: parallel, perpendicular and inclined 45° to the bedding plane. The compressibility index (C_c) and swelling index (C_s) were compared. Only a slight difference between these parameters was observed, suggesting that the anisotropic behaviour of Boom Clay cannot be revealed under the test conditions adopted. The hydraulic conductivity (k) was also determined by the Casagrande's method for different values of vertical effective stress (σ'_v). Unlike compressibility, the hydraulic conductivity, however, showed a clear anisotropic behaviour with $k_{ver} < k_{inc} < k_{hor}$.

Keywords

Boom clay • Anisotropy • Compressibility • Hydraulic conductivity • Oedometer test

87.1 Introduction

In Belgium, Boom Clay is considered as a potential host formation for the geological disposal of high level radioactive waste. The Boom Clay formation was deposited 30 million years ago (Oligocene Epoch). At the location of the Belgian Underground Research Laboratory (URL) HADES, close to the city of Mol, its thickness is about 100 m. Understanding the thermo-hydro-mechanical (THM) behaviour of the host formation is an important part of the feasibility and safety case for the geological disposal of high-level, heat-emitting radioactive waste. Indeed HM properties will condition the excavation techniques to be used for repository construction while THM properties will determine the long-term evolution of stresses, pore pressures and temperatures in the surrounding host rock after the emplacement of the waste.

The behaviour of Boom Clay is considered as cross-anisotropic due to the bedding resulting from deposition (Delage et al. 2007; Lima 2011). It is hence important to take into account the effects of inherent anisotropy in the design and development of structures in Boom clay such as the HADES URL, situated at a depth of about 225 m.

Most oedometer tests carried out till now on Boom clay have been made in a standard fashion with the loading/unloading direction perpendicular to bedding. In this paper, the compressibility and swelling properties of Boom Clay are investigated through high pressure oedometer tests carried out on samples oriented differently with respect to the bedding plane. The hydraulic conductivity was also determined based on Casagrande's method.

L.-Q. Dao · Y.-J. Cui (✉) · A.-M. Tang · P. Delage
Ecole des Ponts ParisTech, Navier/CERMES, Marne la Vallée, France
e-mail: yujun.cui@enpc.fr

X.-L. Li
EURIDICE, Mol, Belgium

X. Sillen
ONDRAF, Brussels, Belgium

G. Lollino et al. (eds.), *Engineering Geology for Society and Territory – Volume 6*,
DOI: 10.1007/978-3-319-09060-3_87, © Springer International Publishing Switzerland 2015

87.2   Materials and Methods

Boom Clay specimens of 50 mm in diameter and 20 mm in height were hand-trimmed from two "horizontal" cylindrical cores (cylinder axis is parallel to bedding) obtained from drillings executed from the HADES URL: core 1b 2006-5 (70 mm diameter and bored in 2006) and core 19C 49–75 (100 mm diameter and bored in 2007). Three samples with various bedding orientations were extracted from these cores: vertical sample ($\alpha = 0°$), inclined sample ($\alpha = 45°$) and horizontal sample ($\alpha = 90°$) (Fig. 87.1). The characteristics of these specimens are shown in Table 87.1. Note that the initial properties (including dry density, water content and void ratio) of the three samples are comparable. The vertical and inclined samples (in Test 1 and Test 3) were saturated (S_r is around 100 %) while the horizontal sample was partially desaturated ($S_r = 91$ %).

In order to study the anisotropy of compressibility, oedometer tests with loading/unloading/reloading were performed on three samples as shown in Fig. 87.2. The effective vertical stress (σ'_v) was stepwise increased up to 32 MPa. This stress is much larger than those experienced by the clay around the URL but this allowed a better determination of the compression index Cc. After being hand-trimmed from cores, specimens were installed in the oedometer cell and loaded up to the in situ vertical effective stress ($\sigma'_v = 2.4$ MPa) prior to being put in contact with water so as to avoid swelling and corresponding changes in the initial intact microstructure (Delage et al. 2007). The sample was saturated at 2.4 MPa with a synthetic water having a similar chemical composition to the field pore water. The composition of this solution is described in Le (2008). Afterwards, step unloading to 0.125 MPa, reloading up to 16 MPa, unloading again to 0.125 MPa, reloading up to 32 MPa and finally unloading to 0.125 MPa were performed. At the end of each step of loading or unloading, the axial strain of sample was verified so as to be consistent with the French standard: $\Delta H/H_0 < 5 \times 10^{-4}$ for a period of 8 h (AFNOR 1995).

The hydraulic conductivity was determined following Casagrande's method. For this purpose, the results were plotted in a semi-logarithmic scale [Nova 2010, see Eq. (87.1)]. The relationship between the hydraulic conductivity and the coefficient of consolidation is recalled in Eq. (87.2).

$$c_v = 0.197 \frac{H^2}{t_{50}} \tag{87.1}$$

$$c_v = \frac{k}{\gamma_w m_v} = \frac{k E_{oed}}{\gamma_w} \tag{87.2}$$

where H is the drainage length, equal to the half-height of the specimen, t_{50} is the time for 50 % consolidation, k is the hydraulic conductivity, γ_w is the unit weight of water ($\gamma_w = 9.81$ kN/m^3), m_v is the oedometric compressibility.

Based on Eqs. (87.1) and (87.2), the hydraulic conductivity k is determined as follows:

$$k = \frac{0.197 H^2 \gamma_w}{t_{50} E_{oed}} \tag{87.3}$$

87.3   Results and Discussions

87.3.1   Anisotropy of Compressibility and Swelling Index

The compression curves from the oedometer tests on three Boom Clay samples (vertical, inclined, horizontal) are shown in Fig. 87.2. Note that a slight swelling of about 0.5 % was observed during the saturation stage with synthetic solution under 2.4 MPa. In Fig. 87.2, only the results obtained after the saturation stage are presented, i.e. all curves start from the end of saturation under an effective vertical stress equal to 2.4 MPa that corresponds to in situ stress. A total of 90 days was needed for the oedometer test to be finished.

The unloading/reloading curves of the three samples exhibit a similar behaviour. Comparable loops were obtained along loading cycles. From the results of oedometer tests, the consolidation index (C_c) and the swelling index (C_s) were calculated for several loading/unloading steps: loading from 4 to 16 MPa (C_{c1}), unloading from 16 to 4 MPa (C_{s2}), then loading from 4 to 32 MPa (C_{c3}) and unloading from 32 to 4 MPa (C_{s4}). The results are presented in Table 87.2. It can

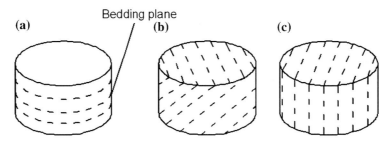

Fig. 87.1   Three types specimens used in oedometer test: **a** vertical sample ($\alpha = 0°$); **b** inclined sample ($\alpha = 45°$); **c** horizontal sample ($\alpha = 90°$). α: angle between the axis of specimen and the axis perpendicular to the bedding plane

Table 87.1 Properties of the three Boom Clay specimens: dry density ρ_d, water content w, void ratio e and degree of saturation S_r

Test	Core of origin	Distance from extrados (m)	Drilling date	Angle α (°)	ρ_d (Mg/m^3)	w (%)	e	S_r (%)
1	19C 49-75	20.1–21.4	2007	0	1.648	23.8	0.625	100
2	1b 2006-5	4.77–5.97	2006	90	1.622	22.1	0.646	91
3	1b 2006-5	4.77–5.97	2006	45	1.631	23.6	0.637	99

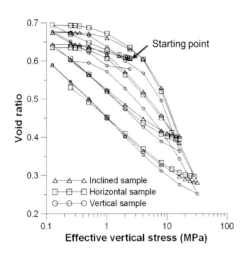

Fig. 87.2 Void ratio versus effective vertical stress for oedometer test on three Boom Clay samples (vertical, horizontal, inclined 45°)

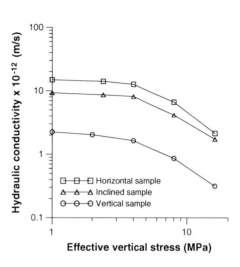

Fig. 87.3 Anisotropy of hydraulic conductivity of Boom Clay

be observed that similar values of C_c and C_s are obtained regardless of the sample orientation. This may seem to be in contradiction with the expected anisotropic behaviour of Boom Clay. This may be due to the difficulty in graphically determining Cc and Cs, the perturbation of the clay around the URL (cores are taken at different distances of the gallery), the perturbation of the samples (of which the age is different) due to preparation and the difficulty of separating primary consolidation from secondary consolidation (see Cui et al. 2009). More tests with better control such as oedometer test on less perturbed samples (fresher, taken at

greater distance from the gallery under a more realistic range of loading), and triaxial test are needed to clarify this point.

87.3.2 Anisotropy of Hydraulic Conductivity

Based on Casagrande's method, the hydraulic conductivity of the samples were determined for the loading part started from 1 MPa up to 16 MPa ($\sigma'_v = 1$ MPa $\rightarrow$ 2.4 MPa $\rightarrow$ 4 MPa $\rightarrow$ 8 MPa $\rightarrow$ 16 MPa). As shown in Fig. 87.3, the values for the three samples show the same variation trend, i.e., they decrease when the effective vertical stress (σ'_v) increases. Also, the results showed an important anisotropy in terms of hydraulic conductivity: $k_{ver} < k_{inc} < k_{hor}$.

A summary can be found in Yu et al. (2011) of values for the hydraulic conductivity of (close to) undisturbed Boom Clay at Mol, determined from a large number of laboratory and in situ measurements. Comparison shows that the horizontal conductivity obtained from oedometer tests in this study (k_{hor} around 10^{-11} m/s) is higher than that measured in situ (5×10^{-12} m/s). For the vertical hydraulic conductivity, the values obtained here ($k_{ver} = 2 \times 10^{-12}$ m/s) is however similar to that measured in situ ($k_{ver} = 2.3 \times 10^{-12}$ m/s). This suggests that the perturbation of the sample used for the determination of vertical hydraulic conductivity is less than that used for the determination of the horizontal one.

Table 87.2 Compression index (C_c) and swelling index (C_s) of three Boom Clay samples

	Vertical sample $\alpha = 0°$	Inclined sample $\alpha = 45°$	Horizontal sample $\alpha = 90°$
$\varepsilon_{v\ max}$ (%)	22.95	21.72	22.88
C_{c1}	0.339	0.337	0.346
C_{s2}	0.105	0.075	0.07
C_{c3}	0.272	0.292	0.283
C_{s4}	0.113	0.091	0.086

87.4 Conclusions

A series of oedometer tests were carried out with loading/ unloading/reloading cycles on three Boom Clay samples oriented differently. The anisotropy of compressibility and swelling properties was analysed first. It was observed that the compression and swelling index, C_c and C_s are similar for different orientations. It should be noted, however, that for better determining these indexes, a maximum vertical stress as high as 32 MPa was applied, that is much higher than the in situ stress estimated at 2.4 MPa. To assess the anisotropic behaviour of Boom Clay for conditions that are more representative of the construction, operation and post-closure evolution of a repository for radioactive waste, tests with better control and within a narrower stress range around the in situ value are recommended. The anisotropy of hydraulic conductivity of these three samples was also examined. As opposed to the compressibility properties, a significant anisotropy in terms of hydraulic properties was found. However, comparison with the in situ measurement shows that the vertical hydraulic conductivity is similar but the horizontal one is much higher, suggesting a more significant perturbation of the sample used for the determination of the horizontal conductivity. Furthermore, the age of cores used and the creep phenomenon may also be factors affecting the results obtained.

Acknowledgments ONDRAF/NIRAS (The Belgian Agency for Radioactive Waste and Enriched Fissile Materials) and ENPC (Ecole des Ponts ParisTech) are gratefully acknowledged for their financial supports.

References

AFNOR (1995) Sols: reconnaissance et essai: essai de gonflement à l'oedomètre. Détermination des déformations par chargement de plusieurs éprouvettes. XP, pp 94–091

Cui YJ, Le TT, Tang AM, Delage P, Li XL (2009) Investigating the time-dependent behaviour of Boom Clay under thermomechanical loading. Géotechnique 59:319–329. doi:10.1680/geot.2009.59.4.319

Delage P, Le TT, Tang AM, Cui YJ, Li XL (2007) Suction effects in deep Boom Clay block samples. Géotechnique 57:239–244. doi:10.1680/geot.2007.57.2.239

Le TT (2008) Comportement thermo-hydro-mécanique de l'argile de Boom. PhD thesis, Ecole Nationale des Ponts et Chaussées

Lima A (2011) Thermo-hydro-mechanical behaviour of two deep Belgian clay formations: Boom and Ypersian Clays. PhD thesis, Universitat Politèchnica de Catalunya

Nova R (2010) Soil mechanics. Wiley, New York

Yu L, Gedeon M, Wemaere I, Marivoet J, De Craen M (2011) Boom Clay hydraulic conductivity. A synthesis of 30 years of research (External Report SCK-CEN, 2011). http://publications.sckcen.be/dspace/handle/10038/7504

The OECD/NEA Report on Self-sealing of Fractures in Argillaceous Formations in the Context of Geological Disposal of Radioactive Waste

88

Helmut Bock

Abstract

Self-sealing of fractured geologic media is of prime importance in the understanding of long-term radionuclide mobility and safety of deep geological repositories for long-lived radioactive waste and spent fuel. It is often cited as one of the decisive factors favouring the choice of argillaceous formations as host rocks for deep disposals. A report of the Nuclear Energy Agency (NEA) of the Organisation for Economic Co-operation and Development (OECD) provides an overview and synthesis of the current understanding of, and conceptual approaches to, the processes that lead to sealing of natural and induced fractures in argillaceous media at typical repository depths. Systematic evidence of self-sealing is collected with reference to laboratory tests, underground research laboratory (URL) field tests and geologic and geotechnical analogues, whereby the bulk of the information stems from the URLs at Bure (Callovo-Oxfordian formation), Mol (Boom Clay) and Mont Terri (Opalinus Clay). The physical, mechanical, geochemical and hydro-mechanical processes and mechanisms are reviewed and their respective contribution to sealing of fractures assessed. It is concluded that the scientific knowledge on self-sealing has progressed to a level which, for soft and slight to moderately indurated argillaceous formations, justifies the inclusion of sealing processes in the performance assessment (PA) of deep geological repositories.

Keywords

Radioactive waste · Geological disposal · Argillaceous formation · Self-sealing · Permeability

88.1 Introduction

Deep geological repositories for high-level radioactive waste require the waste to be isolated from the bio- and hydrosphere over a time span of about 1 million years. Various programmes in the assessment of potential repository sites have shown that argillaceous formations have a very low hydraulic conductivity and that transport is dominated by diffusion. Theoretically, fractures within these formations have the potential to act as preferential flow paths however, there are indications that fractures in argillaceous formations exhibit similar hydraulic properties as intact rock. Assuming that fracturing is a dilatational process calls for an explanation of why initially hydraulic active features become tight over time. The phenomenon of fractures becoming, with the passage of time, less conductive and finally hydraulically insignificant is commonly termed "*self-sealing*".

In this context, attention is drawn to a report of the Nuclear Energy Agency (NEA) of the Organisation for Economic Co-operation and Development (OECD) in Paris (Bock et al. 2010). The Report provides a comprehensive overview and synthesis of the current understanding of, and conceptual approaches to, the processes that lead to sealing of fractures in argillaceous media at typical repository depths. The structure of the OECD/NEA Report is shown in Fig. 88.1.

The paper at hand provides an overview of some of the key aspects of the Report, in particular on:

H. Bock (✉)
Q+S Consult, Bad Bentheim, Germany
e-mail: qs-consult@t-online.de

Fig. 88.1 Flow chart of the OECD/NEA Report on self-sealing (Bock et al. 2010)

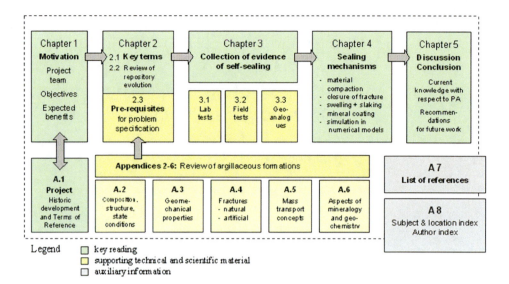

- evidence of self-sealing in argillaceous formations in laboratory tests, URL field tests and geologic and geotechnical analogues (Sect. 88.2 of this paper);
- mechanisms which contribute to sealing (Sect. 88.3 of this paper), and
- relevance of self-sealing for performance assessment (PA) in connection with the Safety Case (Sect. 88.4 of this paper).

88.2 Evidence of Self-Sealing

The Report compiles overwhelming evidence that self-sealing is a common phenomenon in a wide variety of argillaceous soils and rocks currently being considered in the context of deep geological repositories.

Self-sealing can be observed over a large spread of scales: At the millimetre to decimetre scale in laboratory test samples (see Sect. 88.2.1), at about repository scales (1–100 m range) in URL field tests (see Sect. 88.2.2) and at the kilometre scale in geologic and geotechnical analogues such as

old traffic tunnels and hydrocarbon reservoirs (see Sect. 88.2.3).

Self-sealing occurs at various rates which depend on the site-specific circumstances such as the type of clay minerals, degree of induration and chemistry of the porewater. At the tested sites there is a consistent permeability reduction trend, irrespectively of the type and dimension of the test. Typically, the reduction rate is in the order of about 10^{-1} to 10^{-2} per annum, meaning that, over a time span of 1 year, the permeability is reduced to 10^{-2} of its initial value.

88.2.1 Laboratory Tests

Figure 88.2 shows an example of fracture sealing as evidenced by inspection of the tested sample.

Beyond direct inspection of the tested samples, as in the example of Fig. 88.2, the following types of laboratory tests were considered in the Report:

1. combined isotropic loading ($\sigma_1 = \sigma_2 = \sigma_3$) and permeability tests;

Fig. 88.2 Visualisation of the sealing process of a Boom Clay sample by μCT technique— **a** before and **b** after saturation of a 38 mm ∅ sample (Bernier et al. 2004)

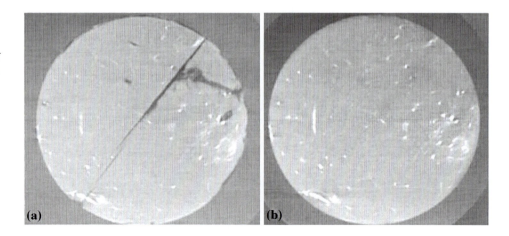

Table 88.1 Overview of URL field tests considered in the OECD/NEA Report URL /URF

URL/URF	Target formation overburden	In situ tests considered in context with self-sealing
Mont Terri, Switzerland	Opalinus Clay $\sim$ 240 m	EB Engineered barrier experiment
		EH EDZ self-sealing experiment
		EZ-A EDZ cut-off experiment
		SE SELFRAC-I and -II experiments
		HG-A gas path through host rock and along seal
		SE-H self-sealing with heat (Timodaz)
HADES Mol, Belgium	Boom Clay 230 m	Oxidation front in mine-by test (connecting gallery)
		Time-dependent hydromechanical response (SELFRAC-III)
		Collapsing borehole and instrumented core experiments (SELFRAC-IV)
Bure, France	Callovo-Oxfordian 490 m	KEY EDZ sealing barrier experiment

2. combined triaxial loading ($\sigma_1 > \sigma_2 = \sigma_3$) and permeability tests;
3. combined direct shear and permeability tests;
4. re-compaction and re-saturation in combined triaxial and permeability tests; and
5. experimental discrimination between mechanical closure of fractures by increased normal stress and swelling of the fracture wall material.

All of the above laboratory tests (sample sizes in mm- to dm-range) disclosed that self-sealing is a very common phenomenon in a wide variety of argillaceous soils and rocks which are currently considered in context with deep geological repositories.

88.2.2 Field Tests in URLs

URL field tests can be considered as a key to confidence building in the performance of argillaceous formations to host long-lived radioactive waste and spent fuel. Scientifically, many URL programmes constitute research right at the forefront of geomechanics and geo-engineering.

The structure of the Report reflects the particular importance of the Mont Terri URL in Switzerland, the HADES URF in Belgium and the Bure URL in France in considering the topic of self-sealing in argillaceous formations (see Table 88.1).

The in situ URL tests (samples in the 1–100 m range) disclosed that self-sealing is a very common phenomenon in a wide variety of argillaceous soils and rocks which are currently considered in context with deep geological repositories, from plastic clays (Boom Clay in the HADES URF) to indurated clays (Opalinus Clay at Mont Terri and Callovo-Oxfordian argillites at the Meuse-Haute Marne URL in Bure).

88.2.3 Geologic and Geotechnical Analogues

Field observations on self-sealing in argillaceous formations have been made in surface outcrops, shallow and deep boreholes and in geotechnical structures such as traffic tunnels. There is a large pool of technical and scientific methods which are employed (Table 88.2). For details and results, ref. to Bock et al. (2010).

Table 88.2 Overview of field observations on self-sealing effects in argillaceous formations

Method	Target formation	Overburden (m)	Study object
Mapping; core and borehole logging	Opalinus clay	$\leq$40	Surface outcrops; shallow boreholes
Tunnel mapping		$\leq$800	Traffic tunnels in Northern Switzerland
Core logging and permeability testing	Boda clay	$\sim$ 1,050	Structure of fault zones
Mapping and lab micro-porosity studies	Variscan claystone	Not applicable	Reconstruction of the permeability evolution of fault zone lithotypes
Mapping, core logging; modelling	Sandstone-claystone sequence	Not applicable	Specification of clay smear potential (CSP), shale gouge ratio (SGR), etc.
Borehole logging; chemical analysis	Argillaceous formations in general	100–500	Tracer concentration profiles
Mapping and joint formation theory		$< \sim$300	Surface outcrops and URLs

Table 88.3 Sealing mechanisms considered in the OECD/NEA Report and their sealing potential in argillaceous formations

Mechanism		Sealing potential	Remarks
Compaction of intact rock (pore space reduction)		Low	Limited importance as the formations considered have already been pre-compacted
Closure of fracture (subject to normal stress increase)		Moderate to high	Good theoretical knowledge
Contraction of fracture (subject to shear)		Moderate to very high	Dilatation/contraction mechanism, augmented by re-orientation of clay platelets
Creep		High to very high	Based on a bundle of micro-mechanisms
Swelling		Very high	Tends to be the most dominant factor in the sealing of fractures
Slaking	Body slaking	Low to moderate	Limited to non-existing knowledge on underlying mechanisms; poor data base, if any
	Surface slaking	Low, if any	
Mineral precipitation		Limited	Stability of precipitated material unclear

88.3 Sealing Mechanisms

It is scientific consensus that self-sealing in argillaceous media is attributable to a number of processes and mechanisms. For safety assessment and PA it is important to understand the sealing mechanisms and their possible interactions with each other. Guided by observations and evidence collected earlier, specific mechanisms were selected and considered in detail. These are listed in Column 1 of Table 88.3. The sealing potential, as disclosed in the Report, is indicated in Column 2 of that table.

88.4 Relevance of Self-Sealing for Performance Assessment (PA)

In a final step, all relevant self-sealing issues (conditions of the geological setting and mechanisms) were classified on their relevance for the performance assessment (PA) of the Safety Case. The classification scheme was in line with that developed by Mazurek et al. (2003) for the FEPCAT project of the OECD/NEA.

It turned out that there is not a single self-sealing issue left which would require "*substantially more work/thoughts, ... at least in some cases*" (Bock et al. 2010, p. 163). Obviously, the general knowledge on self-sealing has progressed to a level at which critical deficiencies are absent. It is concluded that, at least for soft and slight to moderately indurated argillaceous formations, it is justified to include the process of self-sealing in the performance assessment (PA) of deep geological repositories.

References

Bernier F et al (2004) SELFRAC—fractures and self-healing within the excavation disturbed zone in clays. Final report. 5th EURATOM Framework Progr., Brussels (EU-Commission), p 64

Bock H, Dehandschutter B, Martin CD, Mazurek M, de Haller A, Skoczylas F, Davy C (2010) Self-sealing of fractures in argillaceous formations in the context of geological disposal of radioactive waste. NEA Report 6184:1–311, Paris (OECD/NEA)

Mazurek M, Pearson FJ, Volckaert G, Bock H (2003) Features, events and processes. Evaluation catalogue for argillaceous media. OECD/NEA 4437, ISM 92-64-02148-5

Permeability and Migration of Eu(III) in Compacted GMZ Bentonite-Sand Mixtures as HLW Buffer/Backfill Material

89

Zhang Huyuan, Yan Ming, Zhou Lang, and Chen Hang

Abstract

Compacted GMZ bentonite-sand mixtures are studied as a feasible buffer/backfill material for high-level radioactive waste (HLW) disposal in China. This paper is concentrated on the hydraulic conductivity and the migration of Eu(III) in the compacted mixtures of Gaomiaozi bentonite added with quartz sand with various ratio from 0 to 50 %. Permeability tests were conducted using a flexible wall permeameter with the influent of distilled water and 2.0×10^{-5} mol/l Eu(III) solution, respectively. Test results indicated that the hydraulic conductivities measured varied about 10^{-10} cm/s, and no significant change was found with sand ratio increasing from 0 to 50 %. The migration of Eu(III) through compacted specimen was studied by effluent monitoring and Eu(III) extraction from specimen sections after permeability test. The apparent diffusion coefficients of Eu(III) estimated by Eu(III) concentration profile in the specimens varied at the level 10^{-14} m^2/s, when the compacted mixtures had a sand ratio ranging from 0 to 50 %.

Keywords

HLW • Eu(III) solution • Hydraulic conductivity • Diffusion coefficient

89.1 Introduction

Compacted bentonite-sand mixtures have been proposed as a suitable buffer/backfill material in the HLW repository in China. The function of buffer/backfill material is to retard convective water and the migration of radionuclides through buffer by adsorption to the matrix. Therefore, the hydraulic conductivity and diffusion coefficient of radionuclides on compacted bentonite-sand mixtures play an important role in assessing the performance of buffer/backfill material.

Low permeability is the key factor to fulfill the barrier function for buffer/backfill material. Zhang et al. (2011) has studied the permeability of bentonite-sand mixtures with different sand ratio. Sivapullaiah et al. (2000) has studied the effect of sand ratio, dry density and porosity on hydraulic conductivities.

The migration of radionuclides in compacted bentonite has been studied by many researchers. Most of the diffusion tests were mainly focused on the effect of sand ratio (Iida et al. 2011), and solution concentration (Wang et al. 2004), etc. on the diffusion coefficient. In this paper, permeability test and the solution migration test were conducted on the cylindrical specimens with flexible wall permeameter (US Hombolt). Hydraulic conductivities were measured and the effect of Eu(III) solution on permeability was discussed to provide a reference for the design of buffer/backfill material for HLW disposal in China.

89.2 Materials and Methods

89.2.1 Materials

The constituent of the mixtures evaluated in this test includes a processed GMZ bentonite, from Inner Mongolia, and the quartz sand from Yongdeng County in Gansu Province,

Z. Huyuan (✉) · Y. Ming · Z. Lang · C. Hang
School of Civil Engineering and Mechanics, Lanzhou University, Lanzhou 730000, China
e-mail: zhanghuyuan@lzu.edu.cn

Y. Ming
e-mail: yanm12@lzu.edu.cn

Table 89.1 Physical properties of GMZ bentonite

Physical properties						
Particle diameter	Montmorillonite content	Specific surface area	Air-dried water content	Plastic limit	Liquid limit	Specific gravity
<2 μm	74 %	570 m^2/g	10.53 %	32.43 %	228 %	2.71

China. The physical properties of the bentonite are summarized in Table 89.1. The particle density of quartz sand used in this research is 2.65 g/cm^3 and particle diameter ranges from 0.5 to 1.0 mm.

The sand ratio, R_s, is defined as the dry mass ratio of the quartz sand to bentonite-sand mixtures. Quartz sand is uniformly added into the bentonite with R_s = 0, 10, 20, 30, 40 and 50 %, respectively. And then a spray of water is performed to achieve even objective water content, followed by compaction of cylindrical specimens with the size of 100 mm in diameter for the tests.

89.2.2 Methods

Two separate tests were conducted, namely, permeability test and Eu(III) migration test, to measure the hydraulic conductivity of compacted bentonite-sand mixtures with various densities and sand ratio, and to understand the migration of radionuclide in the bentonite-sand mixtures barrier. The flexible wall permeameter was used in this test (Zhang et al. 2011). Table 89.2 lists the parameters of specimens used in permeability test and Eu(III) migration test.

89.2.2.1 Permeability Test
After the cylindrical specimen was positioned into the permeameter cell, distilled water was used to saturate the specimen with "back-pressure saturation" method: the back pressure and

confining pressure were applied step by step. After saturation, hydraulic conductivity was measured at the seepage stage.

The water head in the standpipe was monitored as a function of time, and the hydraulic conductivity of the specimen was calculated from the falling head method according to Darcy's laws.

89.2.2.2 Eu(III) Migration Test
The specimens were first saturated with distilled water for 23d as the same as that in permeability test by back-pressure method. After saturated, the distilled water was switched to solution containing 2.0×10^{-5} mol/l Eu(III) to start the Eu (III) migration test. Effluent was collected periodically, filtered through 0.45 μm pore size and chemically analyzed by ICP-MS. After 300d, experiment was terminated, followed by slicing the specimen into thin sections at about 0.50 mm in depth. The Eu(III) in each slice was extracted and measured with ICP-MS (Zhou et al. 2013).

89.3 Results and Discussions

89.3.1 Hydraulic Conductivities

Figure 89.1 shows hydraulic conductivities of bentonite-sand mixtures when distilled water was used as influent. Generally, hydraulic conductivities decrease with time and finally become stable. In the early permeation stage, hydraulic conductivities

Table 89.2 Parameters of specimens for permeability tests and Eu(III) migration tests

Test	Specimen	Sand ratio (%)	Diameter (cm)	Height (cm)	Dry density (g/cm^3)	Water content (%)
A	M0	0	10.20	3.28	1.61	16.40
	M10	10	10.16	3.26	1.72	16.30
	M20	20	10.02	3.48	1.79	19.20
	M30	30	10.31	3.93	1.64	9.80
	M40	40	10.15	3.66	1.91	13.50
	M50	50	10.22	3.91	1.97	11.00
B	N0	0	10.20	2.18	1.56	17.10
	N10	10	10.20	2.15	1.59	14.13
	N20	20	10.20	2.15	1.58	12.65
	N30	30	10.20	2.12	1.63	10.92
	N40	40	10.20	2.11	1.63	9.62
	N50	50	10.20	2.12	1.63	8.74

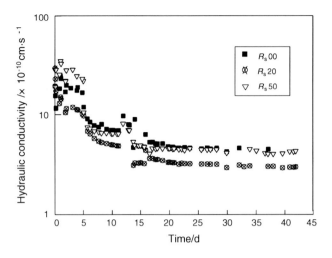

Fig. 89.1 Hydraulic conductivity with distilled water vs. time

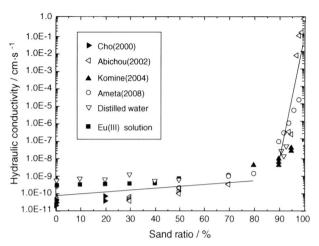

Fig. 89.2 Hydraulic conductivity vs. sand ratio

decrease obviously. This is because that montmorillonite in bentonite swell with water continuously, leading to block mixtures' internal pore. In the later stage of permeation, hydraulic conductivities remain almost the same, regarded as final hydraulic conductivities of bentonite-sand mixtures. The measured data shows that hydraulic conductivities of mixtures are at the range of $3.34–8.85 \times 10^{-10}$ cm/s, which means that hydraulic conductivities are at the same order of magnitude, even so the sand ratio varies from 0 to 50 %. As a result, the original impermeability of pure bentonite can be kept while the sand ratio is less than 50 %. Similarly, hydraulic conductivities with Eu(III) solution are at the range of $2.07–5.23 \times 10^{-10}$ cm/s, at the same order of magnitude as permeated with water.

89.3.2 Effect of Sand Ratio on Hydraulic Conductivity

Sand ratio is a major factor influencing permeability of bentonite-sand mixtures. Figure 89.2 displays that the hydraulic conductivities of bentonite-sand mixtures have little change with the sand ratio increases when sand ratio is less than 85 %. In contrast, hydraulic conductivities of compacted bentonite-sand mixtures have a rapidly growing with sand ratio increases from 85 % (Zhang et al. 2011). This research indicates that the original low permeability of pure bentonite can be kept while the sand ratio is less than 50 % in case that both distilled water and Eu(III) solution permeated through the mixtures.

89.3.3 Eu(III) Concentration Profiles

The migration test was performed to investigate the migration characteristics of Eu(III) in compacted bentonite-sand mixtures. The equation of solution is (Zhou et al. 2013):

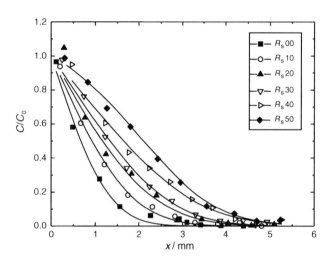

Fig. 89.3 Eu(III) concentration profiles and the fitting curves after 300d permeation

$$\frac{C}{C_0} = \frac{1}{2}\mathrm{erfc}\left[\frac{x - (v/R_\mathrm{d})t}{2\sqrt{D_\mathrm{a}t}}\right] \quad (89.1)$$

where C is the concentration (liquid phase) of Eu(III) at a distance x from the solution-bentonite interface at time t, and C_0 is the concentration of Eu(III) at the interface between the bentonite and the bulk solution, v is the seepage velocity, R_d is the retardation factor, D_a is the apparent diffusion coefficient.

Concentration profiles based on Eq. 89.1 for Eu(III) with different sand ratio were given in Fig. 89.3. Solid lines were the fitting curve from Eq. 89.1. For different sand ratio, it was observed that the fitting curves described well the experimental data.

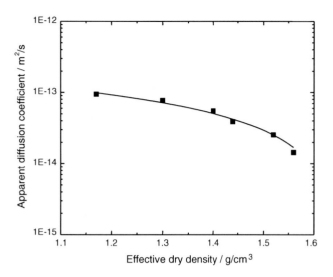

Fig. 89.4 Effect of effective dry density on the apparent diffusion

89.3.4 Effect of Effective Dry Density on Diffusion Coefficient

Figure 89.4 presents that the effect of effective dry density as described by Dixon et al. (1999) on apparent diffusion coefficient, D_a. An exponential reduction in apparent diffusion coefficient with the effective dry density was observed, which is similar to the conclusion of Sawatsky and Oscarson (1991). The reduction in D_a with increasing effective dry density was mainly attributed to an increase in the tortuosity as the effective dry density increased (Yu and Neretnieks 1997). Increasing clay content in per unit-volume of the mixtures tends to enhance the adsorption ability of bentonite-sand mixtures, and decline the diffusion coefficient. Moreover, the reduction in D_a with increasing in effective dry density was largely ascribed to an increase in the anion exclusion volume in the bentonite with increasing effective dry density, which likely decreases the effective size of the pores and so to influence the tortuosity or geometric factor (Sawatsky and Oscarson 1991).

89.4 Conclusions

1. The hydraulic conductivities of compacted bentonite-sand mixtures with sand ratio ranging from 0 to 50 % are $K = 3.34–8.85 \times 10^{-10}$ cm/s when permeated with distilled water, and $K = 2.07–5.23 \times 10^{-10}$ cm/s when permeated with Eu(III) solution. This means that pure bentonite can be mechanically improved by addition of quartz sand without increasing its impermeability.

2. The apparent diffusion coefficients of Eu(III) in compacted bentonite-sand mixtures are $D_a = 1.44–9.41 \times 10^{-14}$ m²/s. The decrease in D_a with increasing effective dry density is attributed to an increase in the tortuosity, anion exclusion volume and adsorption ability of bentonite-sand mixtures as bentonite content is increased.

Acknowledgments The present work has been carried out under the Project lzujbky-2013-k03 supported by the Fundamental Research Funds for the Central Universities.

References

Dixon DA, Graham J, Gray MN (1999) Hydraulic conductivity of clays in confined tests under low hydraulic gradients. Can Geotech J 36:815–825

Iida Y, Yamaguchi T, Tanaka T (2011) Experimental and modeling study on diffusion of selenium under variable bentonite content and porewater salinity. Nucl Sci Technol 48:1170–1183

Sawatsky NG, Oscarson DW (1991) Diffusion of technetium in dense bentonite. Water Air Soil Pollut 57–58:449–456

Sivapullaiah PV, Sridharan A, Stalin VK (2000) Hydraulic conductivity of bentonite-sand mixtures. Can Geotech J 37(2):406–413

Wang XK, Chen YX, Wu YC (2004) Diffusion of Eu(III) in compacted bentonite-effect of pH, solution concentration and humic acid. Appl Radiat Isot 60:963–969

Yu JW, Neretnieks I (1997) Diffusion and sorption properties of radionuclides in compacted bentonite. SKB, Stockholm, technical report TR97-12

Zhang HY, Zhao TY, Lu YT et al (2011) Permeability of bentonite-sand mixtures as backfill/buffer material in HLW disposal under swelling conditions. Chin J Rock Mech Eng 30:3149–3156 (in Chinese)

Zhou L, Zhang HY, Yan M et al (2013) Laboratory determination of migration of Eu(III) in compacted bentonite-sand mixtures as buffer/backfill material for high-level waste disposal. Appl Radiat Isot 82:139–144

Correlative Research on Permeability and Microstructure of Life Source Contaminated Clay

Liwen Cao, Yong Wang, Pan Huo, Zhao Sun, and Xuezhe Zhang

Abstract

Based on the test results deriving from laboratory simulation experiment device which consists of a waste layer, a drainage layer, a compacted clay layer, from the top down, as well as a contaminants recirculation system, and a few test instruments, falling head permeability test and scanning electron microscopy (SEM) are conducted to determine the permeability and microstructure of the contaminated clay at four pollution depths: 22.5, 45.0, 67.5 and 90.0 cm respectively. Consequently, quantitative relationship among such parameters as permeability coefficient k, total pore area A, area void ratio e, and mean pore size d_0 with pollution diffusion depths are explored. Finally, the evolution mechanism of the permeability and microstructure of contaminated clay related to life source is revealed. The researches show as follows: with the increase of pollution depth, permeability coefficient k, total pore area A, area void ratio e as well as mean pore size d_0 are increasing, and their relationship can be summarized as $k = 7\ln(A) - 200e^2 + 160e + 2d_0^2 - 12d_0 - 8$; Under such effects as ions exchange and adsorption, organic materials and suspended solids coagulation as well as complex chemical reactions between contaminants and clay, with pollution depth increasing, the clay permeability is improved due to the decreased influence of contaminants and their degradation products on clay, as well as the increased total pore area and void ratio.

Keywords

Life source contaminated clay • Permeability • Microstructure

The compressibility and shear strength of industry contaminated clay especially polluted by waste alkaline liquor, salt, acid and alkali, were studied by the predecessor researchers (Zhu et al. 2011), meanwhile the migration and recrystallization were touched (Negim et al. 2010).The microstructure of clay was detected by TEM and SEM combined with such computer image processing technologies as Photoshop and Mapinfo (Bogas et al. 2012; Shi and Jiang 2001; Al-Mukhtar et al. 2012). In this paper, such techniques as geotechnical test, scanning electron microscopy (SEM) and Matlab software are performed to determine correlation among the permeability coefficient and microstructure of contaminated clay by life source contaminants.

90.1 Samples of Contaminated Clay Related to Life Source

An experimental device constructed with domestic waste layer, gravel drainage layer and compacted clay from the top down is developed. The domestic waste layer principally consists of readily degradable kitchen waste, leaf, furnace cinder as well as waste paper and fruit. The gravel drainage layer filled by sand ranging from 2 to 10 mm in size is installed below the domestic waste layer. The clay is compacted in 100 cm thick for study.

L. Cao (✉) · Y. Wang · P. Huo · Z. Sun · X. Zhang
School of Resource and Earth Science, China University of Mining and Technology, Xuzhou 221116, China
e-mail: caoliwen@cumt.edu.cn

Through a series of physical, chemical and biological reactions, leachate is produced by domestic waste, and then, contaminated clay is formed by leachate seeping. After 2 years seeping, clay samples at four pollution depths: 22.5, 45.0, 67.5 and 90.0 cm are extracted from the experimental device. Thus, four groups contaminated clay samples, as well as one group of original clay sample are prepared.

90.2 Permeability and Microstructure of Contaminated Clay Related to Life Source

90.2.1 Permeability of Life Source Contaminated Clay

The permeability coefficient of contaminated clay related to life source at such four pollution depths as 22.5, 45.0, 67.5 and 90.0 cm, are indicated in Table 90.1. The experimental results show that the permeability coefficient of the contaminated clay declines by contaminants seeping.

90.2.2 Microstructure Parameters Extraction

The SEM microstructures of contaminated clay related to life source at the depths 22.5, 45.0, 67.5 and 90.0 cm see Fig. 90.1. We can conclude that the pore in the upper part of the clay is less than that of in the lower part one, and, the grain in upper part distributes more even. Then, based on optimal threshold values, the SEM images are quantified by Matlab software.

90.2.2.1 Basic Structure and Pore Size Extraction
Such basic structure parameters as total particle area, total pore area, mean particle area, mean pore area, area porosity and area void ratio are extracted by Bwlabel and Sum function. And then, such pore size parameters as maximum, minimum and mean pore size are extracted by Region props function, meanwhile, conversion coefficient is used so that the unit of binary image (pixel) can be converted to length unit. On the basis of conversion coefficient, the unit of the pore size parameters extracted is transformed to length unit. Maximum, minimum and mean pore sizes which are

transformed are increasing with the pollution depth deepening. Now, taking area void ratio as an example, it's changing in pore size distribution with the increase of pollution depth being showed in Table 90.2.

Combining with the precious experience of predecessors in respect of pore size distribution as well as Find and Ismember function, the pore size distribution in contaminated clay are classified into six categories: macro-pore, medium-pore, minor-pore and fine-pore, micro-pore, tiny-pore.

90.2.2.2 Microstructure Analysis
There exists a noteworthy growth in total pore area, area porosity and mean pore size with the pollution depth increasing in life source contaminated clay, which is identical with the intuitive analysis of SEM images. The feature of pore distribution is distinction at different pollution depth. Firstly, the contaminated clay at 22.5 and 45.0 cm depth mainly contain unimodal minor-pore, as well as bimodal fine/micro-pore without macro-pore. However, the contaminated clay at 90.0 cm depth mostly consists of unimodal macro-pore, as well as bimodal macro/medium-pore. Consequently, undergoing the life source contaminants, the clay displays transformation from macro-pore to micro-pore. Furthermore, there exhibit a structure evolution from minor-pore to macro-pore in contaminated clay with the pollution depth increasing. Although ostensibly changed slightly in quantity, the micro-pore universally exists in various pollution depths of clay, but there have actually experienced the progress from larger pore and to smaller pore since the contaminants seeping.

90.2.3 Correlation Between Microstructure and Permeability

According to the experimental date, such correlations as between permeability coefficient(k) and total pore area (A), area void ratio (e), mean pore size (d_0) are taken respectively. Then, $r = \ln(A)$, $s = (e - 0.4)^2$ and $t = (d_0 - 2.98)^2$ are ordered for the linear correlations between k and r, as well as s and t, then, formula (1) is obtained:

$$k = 7\ln(A) - 200e^2 + 160e + 2d_0^2 - 12d_0 - 8 \quad (1)$$

Available accuracy which is defined as $100\% - |\text{in situ value} - \text{predicted value}|/\text{in situ value}$, is used to compared in situ with the predicted value of k, and, the predicted value comes from the formula (1), then the available accuracy is 91.21, 96.26, 97.69 and 97.93 % at 22.5, 45.0, 67.5 and 90.0 cm pollution depth clay respectively.

Table 90.1 Permeability coefficient of contaminated clay related to life source

Pollution depth cm	22.5	45.0	67.5	90.0	0
Permeability coefficient k 10^{-7} cm/s	8.57	26.5	32.5	35.1	52.1

Fig. 90.1 Microstructure images of contaminated clay at the depths. **a** Microstructure image of clay at pollution depth 22.5 cm. **b** Microstructure image of clay at pollution depth 45.0 cm. **c** Microstructure image of clay at pollution depth 67.5 cm. **d** Microstructure image of clay at pollution depth 90.0 cm

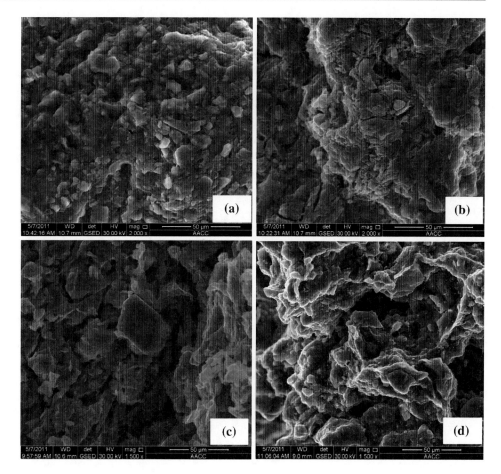

Table 90.2 Area void ratio and pore size distribution of contaminated clay at the depths

Pollution depth cm	Macro-pore ≥ 40 μm	Medium-pore $40 \sim 20$ μm	Minor-pore $20 \sim 5$ μm	Fine-pore $5 \sim 2$ μm	Micro-pore $2 \sim 1$ μm	Tiny-pore < 1 μm
22.5	0.0000	0.0000	0.0137	0.0109	0.0055	0.0053
45.0	0.0000	0.0000	0.0714	0.0306	0.0037	0.0072
67.5	0.1345	0.0578	0.1466	0.0096	0.0047	0.0038
90.0	0.2246	0.0836	0.0539	0.0080	0.0030	0.0019

90.3 Mechanism Analysis on the Permeability of Contaminated Clay Related to Life Source

The effect of chemical corrosion as well as biological degradation on clay will reveal during the clay saturated by contaminants seeping. Thus, it is necessary to interpret the modification mechanism of the contaminated clay permeability from three stages.

90.3.1 Primary Seepage Stage

Ion exchange plays a major role in the primary seepage stage. For example, such low ion with bigger crystal radius as Na^+, Ca^{2+} are exchanged by high valence ion as Fe^{3+} during the life source contaminants seeping, which will result to thin the electrical double layers on the clay surface, and lead to grow the attractive force between clay particles, then, cause clay particles flocculating, moreover, to increase the pore area, porosity and the clay permeability, accordingly.

90.3.2 Medium Seepage Stage

Adsorption reveals obviously with the development of life source contaminants seeping. The suspended solid and organic matter generated through anaerobic decomposition of microorganism can be constantly adsorbed on the surface of clay particles nevertheless. Furthermore, A great variety of metabolites can be readily adsorbed on the clay particles and suspended solids intercepted without decomposing, resulting in a notable decline in porosity and in permeability.

90.3.3 Saturated Seepage Stage

With life source contaminants seeping, the clay may interact with the contaminants by means of chemical bonds, promoting soluble salts to insoluble compounds (carbonates and phosphates) through chemical reactions under a certain oxidation-reduction potential and pH value. The precipitates filling the clay pores can prevent fluid from permeating. Additionally, large amounts of reproduction of microorganisms in clay may preclude seeping. Consequently, the permeability of clay will still decrease when undergoing the saturated seepage by contaminated fluid.

Clay undergoes from saturated seepage stage to medium seepage stage, and to primary seepage stage. First of all, the upper part of clay which is near to the waste layer will reach to saturated stage, where suspended solids, organic matters and insoluble carbonates from chemical reactions between clay and the life source contaminants are created, then, clay pore will be clogged, and the distribution of macro-pore and medium-pore will be lessened. In summary, specifically, such role as ion exchange, adsorption, the deposition of suspended solid and organic matter, the dissolution of soluble salts and insoluble salt precipitation by complex chemical reactions between the clay and life source contaminants take diverse influence on the clay permeability during the different seeping stage.

90.4 Conclusions

The clay suffered from the life source contaminants decrease in permeability because of a significant reduction in porosity and pore size. The clay presents a gradual enhancement in permeability for the transformation of well-to-poorly saturated effects with the pollution depth increasing. The correlation between microstructure and permeability coefficient is developed.

In the three seeping stage, such actions as ion exchange, adsorption, the deposition of suspended solid and organic matter, the dissolution of soluble salt and insoluble salt precipitation play a different role on the clay permeability.

Acknowledgements This research is jointly supported by the National Natural Science Foundation of China No. 41372326 and No. 41072236, the Fundamental Research Funds for the Central Universities NO.2014ZDPY27, as well as the Priority Academic Program Development of Jiangsu Higher Education Institutions.

References

Al-Mukhtar M, Khattab S, Alcover JF (2012) Microstructure and geotechnical properties of lime-treated expansive clayey soil. Eng Geol 139–140:17–27

Bogas JA, Gomes A, Gomes MG (2012) Estimation of water absorbed by expanding clay aggregates during structural lightweight concrete production. Mater Struct/Materiaux et Constructions 45(10):1565–1576

Negim O, Eloifi B, Mench M et al (2010) Effect of basic slag addition on soil properties, growth and leaf mineral composition of beans in a cu-contaminated soil. Soil Sediment Contam 19(2):174–187

Shi B, Jiang HT (2001) Research on the analysis techniques for clayey soil microstructure. Chin J Rock Mech Eng 20(6):864–870 (in Chinese)

Zhu CP, Liu HL, Shen Y (2011) Laboratory tests on shear strength properties of soil polluted by acid and alkali. Chin J Geotech Eng 33 (7):1146–1152 (in Chinese)

Diffusion of La^{3+} in Compacted GMZ Bentonite Used as Buffer Material in HLW Disposal

91

Yonggui Chen, Lihui Niu, Yong He, Weimin Ye, and Chunming Zhu

Abstract

It has been studied that the GMZ bentonite which has a chemical barrier property can play an important role in retarding the migration of radioactive nuclide. In order to understand the diffusion properties of nuclide in GMZ bentonite, a series of experiments were conducted to obtain the apparent diffusion coefficient of La^{3+} in GMZ bentonite under different dry density and pH value conditions using a autonomous ion diffusion instrument. The results show that the apparent diffusion coefficient of La^{3+} decreases significantly with the increasing dry density or pH value. The influence of dry density on the diffusion is very apparent for a low dry density, and the influence gradually slow down with the increase of dry density. At last the influence mechanism of the dry density and pH value on diffusion coefficient was also analyzed.

Keywords

GMZ bentonite • Diffusion property • Lanthanum ion

91.1 Introduction

The deep geological disposal which based on a multi-barrier system is the most appropriate means for safe disposal of high-level radioactive waste (HLW). Bentonite has been chosen as the buffer/backfill material for this disposal because of good mechanical buffer resistance, low permeability, better swelling property and adsorption property (Wen 2006).

The chemical barrier properties of bentonite are mainly manifested in retarding the migration of radioactive nuclide. Considering most of the nuclide uranium, thorium, plutonium in disposal repository are actinide elements, and lanthanide and actinide elements have similar electronic configuration inside and outside layer which determine their similar physical and chemical properties. Therefore the lanthanide ions are always selected instead of radionuclide in the study. In this work, effects of the dry density and pH value on the diffusion property of La^{3+} in GMZ bentonite were studied.

91.2 Experimental

91.2.1 Materials

The GMZ bentonite derived from inner mongolia China was used in this experiment. Its major mineral is (Wang and Su 2006): montmorillonite 75.4 %, quartz 11.7 %, crystaballite 7 %, feldspar 4.3 %, calcite 0.5 %, kaolinite 0.8 %. It has been studied that GMZ bentonite has strong cation exchange capacity, hydration ability and strong adsorption ability.

Y. Chen (✉) · L. Niu · Y. He · W. Ye · C. Zhu
Tongji University, 1239 Siping Road, Shanghai 200092, People's Republic of China
e-mail: cyg@tongji.edu.cn

Y. Chen · Y. He
Changsha University of Science and Technology, Changsha 410114, People's Republic of China

G. Lollino et al. (eds.), *Engineering Geology for Society and Territory – Volume 6*,
DOI: 10.1007/978-3-319-09060-3_91, © Springer International Publishing Switzerland 2015

91.2.2 Experiment

The instrument used in this experiment is a self-designed ion diffusion instrument consists of an organic glass shell and two stainless steel rings. Before the diffusion, the soil was saturated and then the bottom of the instrument was controlled being impervious. During diffusing, the ion migrated along a single direction. Therefore the flow velocity of water in the soil is zero, which can be regarded as pure diffusion of ions.

The GMZ bentonite samples with the diameter of 20 mm and the thickness of 30 mm were prepared with different dry density. Each test would last about 90 days at a given dry density and pH value. After diffusion, the bentonite was cut into 2 mm thick slices and then dissolved in 30 ml 0.5 M HCl solution for 24 h. The amount of La^{3+} in the slice was measured by ICP. One-dimensional non-steady-state diffusion of La^{3+} in compacted GMZ bentonite can be described by Fick's second law:

$$\frac{\partial c}{\partial t} = D \frac{\partial^2 c}{\partial x^2} \qquad (1)$$

where D is the diffusion coefficient in compacted bentonite (m^2/s), C is the La^{3+} concentration in diffusion solution (mol/l), t is the diffusion time (s) and x is the distance from the diffusion source (m).

91.3 Results and Discussion

The results are illustrated in Table 91.1. Table 91.1 shows that the apparent diffusion coefficient of La^{3+} for the soils with the dry density of 1.3, 1.5 and 1.7 g/cm^3 under different conditions.

91.3.1 Effect of Dry Density on the Diffusion

It can be seen from the Table 91.1 that the apparent diffusion coefficient of La^{3+} decreases notably with increasing dry density. It is obvious that the dry density paly an important role on the diffusion coefficient for the low dry density soils.

With the increase of dry density, the influence gradually slows down. The mechanism of the influence of dry density on the ion diffusion basically can be explained as following: (1) Changes of dry density affects ion distribution coefficient which affect the ion diffusion coefficient. The soil with a high dry density has a large surface area, and it could adsorb more cationic relatively than that with a low dry density. This resulted in the ion distribution coefficient increased thus the diffusion coefficient decreased (Muurinen and Lehikoinen 1995). (2) Elevated dry density decreases the effective porosity of the bentonite which results in a decrease in diffusion coefficient (Sawatsky and Oscarson 1991). (3) Changes of dry density affects the layer spacing. With the increase of dry density, layer spacing of bentonite decreases, and lead to the ion diffusion coefficient decreases (Kozaki et al. 2005). (4) Changes of dry density cause the main diffusion process changed which lead to the changes of diffusion coefficient (Wang 2003).

91.3.2 Effect of PH on the Diffusion

Table 91.1 also showed that the apparent diffusion coefficient of La^{3+} decreases slightly with increasing pH value.

The influence mechanism of pH value on the ion diffusion coefficient can be explained that the change of pH value affects the absorption of ions in bentonite. Quite a few scholars have done some works on the effect of pH value on ions diffusion. Wang (2003), Wang and Chen (2004) and Wang and Liu (2004) studied the effects of pH value on the diffusion of Cs, Sr and Eu ions in compacted MX-80 bentonite and found the adsorption of ions increased with the increasing of pH, which lead to the diffusion coefficient of ions decrease. Giannakopoulou and Haidouti (2007) studied the diffusion of Cs ion under different pH and found the similar results. The adsorption of Cs is less under low pH value and, the adsorption is maximum when the pH increase to 8. Because of the competitive adsorption of H^+ and Cs^+ leading to the adsorption of Cs^+ decreases under low pH. With the increase of pH, the negative charge on the surface of the bentonite increased and then resulted in the Cs^+ adsorption increased sharply.

91.4 Conclusions

1. A self-designed apparatus was conducted to study the diffusion characteristics of rare earth ions in the compacted GMZ benotnite used as Chinese buffer/backfilling material in the HLW disposal.
2. The diffusion coefficient of La^{3+} in the compacted GMZ bentonite decreases notably with increasing dry density.

Table 91.1 The results of La^{3+} diffused in compacted GMZ bentonite

Test	Dry density (g/cm^3)	Solution concentration (mol/L)	pH	Diffusion time (s)	D_a ($\times 10^{-12}$ m^2/s)
(1)	1.3	5.2×10^{-3}	7.5	6.840×10^6	15.82
(2)	1.5	5.2×10^{-3}	7.5	5.849×10^6	3.018
(3)	1.7	5.2×10^{-3}	8.9	8.533×10^6	2.513
(4)	1.7	5.2×10^{-3}	7.5	7.236×10^6	2.708

3. The diffusion coefficient of La^{3+} in the compacted GMZ bentonite decreases slightly with increasing pH value.

Acknowledgements The present work has been carried out under the financial supports of "National Natural Science Foundation of China (41272287 and 41030748)", "the Fundamental Research Funds for the Central Universities", "Pujiang Program of Shanghai (13PJD029)" and the Key Project supported by Scientific & Technical Fund of Hunan Province (2014FJ2005).

References

Giannakopoulou F, Haidouti C (2007) Sorption behavior of cesium on various soils under different pH levels. J Hazard Mater 149:553–556

Kozaki T, Fujishima A, Saito N et al (2005) Effects of dry density and exchangeable cations on the diffusion process of sodium ions in compacted montmorillonite. Eng Geol 81:246–254

Muurinen A, Lehikoinen J (1995) Evaluation of phenomena affecting diffusion of cations in compacted bentonite. Nuclear Waste Commission of Finnish Power Companies, YJT-95-05, p 70

Sawatsky NG, Oscarson DW (1991) Diffusion of technetium in dense bentonite. Water Air Soil Pollut 57–58:449–456

Wang J, Su R (2006) Deep geological disposal of high-level radioactive waste in China. Chin J Rock Mech Eng 25(4):649–658

Wang XK, Chen YX (2004) Diffusion of Eu(III) in compacted bentonite—effect of pH, solution concentration and humic acid. Appl Radiat Isot 60:963–969

Wang XK, Liu XP (2004) Effect of pH and concentration on the diffusion of radiostrontium in compacted bentonite—a capillary experimental study. Appl Radiat Isot 61:1413–1418

Wang XK (2003) Diffusion of ^{137}Cs in compacted bentonite: Effect of pH and concentration. J Radioanal Nucl Chem 258(2):315–319

Wen ZJ (2006) Physical property of China's buffer material for high-level radioactive waste repositories. Chin J Rock Mech Eng 25 (4):794–800

Soil Mechanics of Unsaturated Soils with Fractal-Texture

92

Yongfu Xu and Ling Cao

Abstract

The mechanical properties of unsaturated soils are a function of matric suction, and can be obtained based on currently available procedures. However, each procedure has its limitations and consequently cares should be taken in the selection of a proper procedure. Fractal approach seems to be a potentially useful tool to describe hierarchical systems and is suitable to model the pore structure of unsaturated soils. In this paper, the soil–water characteristics (SWC), unsaturated hydraulic conductivity function and unsaturated shear strength were derived from the fractal model for the pore-size distribution (PSD), and were expressed by only two independently physical parameters, the fractal dimension and air-entry value. The predictions of the proposed soil–water characteristics, unsaturated hydraulic conductivity and unsaturated shear strength were in good accord with the published experimental data.

Keywords

Unsaturated soil • Fractal dimension • Pore-size distribution • Soil–water characteristics • Hydraulic conductivity • Shear strength

92.1 Introduction

The matric suction is an important stress state variable to express the mechanical properties of unsaturated soils. The relationship between the matric suction (ψ) and the radius (r) of the incurvated surface between pore-air and pore-water is expressed by the Young-Laplace equation, here is written as $\psi = 2\sigma \cos\alpha/r$, here ψ is the matric suction, σ is the surface tension, α is the contact angle. The mechanical properties of unsaturated soils are correlated to the microstructures of soils through the Young-Laplace equation.

The pore-size distribution (PSD) can be described by a fractal model because pores are invariable at any characterized scale used to examine them (Xu and Sun 2002). Fractal dimensions can be defined in connection with real-world data, and they can be measured approximately by means of experiments (Mandelbrot 1982). In this paper, the pore-size distribution was studied using fractal theory, and soil–water characteristics (SWC), hydraulic conductivity and shear strength of unsaturated soils were deduced from the fractal model for the soil pore-size distribution.

92.2 Soil–Water Characteristics

For fractal pore-size distribution (PSD), a power-law existing between the scale of measurement (r) and the pore phase of Eq. (1).

$$N(A, r) = Cr^{-D} \tag{1}$$

for some positive constant C. The total volume of the pores with the radius less than r is given by

$$V_p = \int_0^r N 4\pi r\, dr = Ar^{3-D} \tag{2}$$

Y. Xu (✉) · L. Cao
Shanghai Jiao Tong University, 800 Min Hang Road,
Shanghai 200240, China
e-mail: yongfuxu@sjtu.edu.cn

where $A = 4\pi CD/(3 - D)$. Fractal dimension spans a large range from 1.0 to 3.0.

The pore water within unsaturated soils can be divided into three forms (Wheeler and Karbue 1996) (Fig. 92.1): bulk water within those void spaces that are completely flooded, meniscus water surrounding all inter-particle contact points that are not covered by bulk water and absorbed water (which is tightly bounded to the soil particles and acts as parts of the soil skeleton). When the volumetric water content is less than its residual value, the pore-water is tightly absorbed by soil particle and cannot move freely and seen as a part of soil particles. The contribution of the filled pores with radius $r \rightarrow r+dr$ to the water content is given by

$$dA = N4\pi r^2 dr / V_T \qquad (3)$$

where A is the relative volumetric water content, and $A = \theta - \theta_r$, θ and θ_r are the actual and residual volumetric water content, respectively. The residual volumetric water content is the volumetric water content at which the effectiveness of matric suction to cause further removal of water requires vapour migration. Combining Eqs. (1) and (3) with the Young-Laplace equation, the soil–water characteristic curve is obtained as follows

$$S_e = \left(\frac{\psi}{\psi_e}\right)^{D-3} \qquad (4)$$

where S_e is the effective degree of saturation, and $S_e = A/A_s$, ψ_e is the air-entry value, $\psi_e = 2\sigma\cos\alpha/R$, R is the maximum radius of soil pores. The residual volumetric water content θ_r is not always made routinely, in which case it has to be estimated by extrapolating available soil–water characteristics data towards lower water content, such as shown in Fig. 92.1.

Figure 92.2 shows the soil pore-size distribution obtained from mercury intrusion porosimetry (Watabe et al. 2000). It is obtained that the fractal dimensions of the pore-size distribution of S-02, S-03 and S-04 are 2.66, 2.63 and 2.51, respectively. The maximum radius is the radius at which the pore volume reaches the maximum value, and is defined as the intersection between the fitting line and the line of $V_p/V_T = 100$ %, here V_T is the total volume of soil pores. The maximum radium of the soil pores of S-02, S-03 and S-04 are 0.125 mm, 0.06 mm and 0.006 mm, respectively. According to the Young-Laplace equation, the air-entry values of S-02, S-03 and S-04 are 12.5, 2.5 and 25 kPa, respectively.

The soil–water characteristic curves are predicted using Eq. (4) for the soils of S-02, S-03 and S-04. The air-entry value and fractal dimension of S-02, S-03 and S-04 are obtained from Fig. 92.2. Predictions of Eq. (4) show in good accord with the experimental data of the soil–water characteristic curve in Fig. 92.3.

92.3 Unsaturated Hydraulic Conductivity

The relatively hydraulic conductivity (k_r) is the ratio of the hydraulic conductivity at any degree of saturation (k_w) to the hydraulic conductivity at saturation (k_s). The relative hydraulic conductivity is related to the radius of the soil pores and the pore-size distribution (Mualem 1976). Using the fractal pore-size distribution and the Young-Laplace equation, Xu (2004a) gave the relative hydraulic conductivity

$$k_r = \left(\frac{\psi}{\psi_e}\right)^{3D-11} \qquad (5a)$$

$$k_r = S_e^{(3D-11)/(D-3)} \qquad (5b)$$

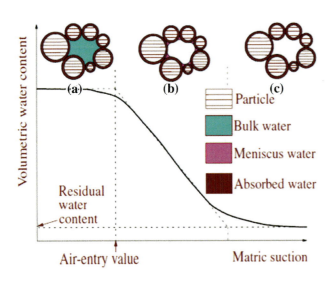

Fig. 92.1 Pore-water formation in soils

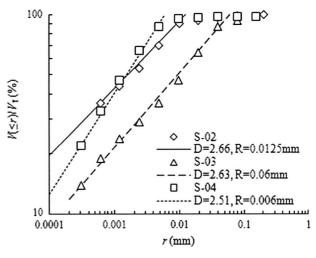

Fig. 92.2 Fractal model for PSD

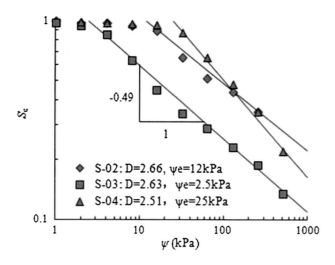

Fig. 92.3 Experiments versus prediction of SWC

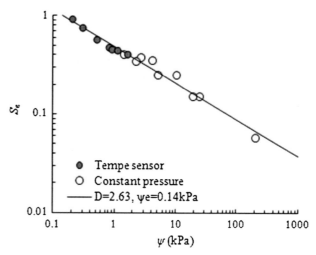

Fig. 92.4 D and ψ_e obtained from SWC

The soil–water diffusivity (d) can be obtained from the unsaturated hydraulic conductivity and the soil–water characteristics, and is written as

$$d = k_s k_r |d\psi/d\theta| \qquad (6)$$

Substituting Eqs. (5a) and (5b) into Eq. (6), it is obtained

$$d = k_s \psi_e \left(\frac{\psi}{\psi_e}\right)^{2D-7} \qquad (7a)$$

$$d = k_s \psi_e S_e^{(2D-7)/(D-3)} \qquad (7a)$$

Smettem and Kirkby (1990) gave the experimental data of the soil–water characteristics for haploxeroll loam, shown in Fig. 92.4. The fractal dimension is 2.63, and the air-entry value is 0.14 kPa obtained from Fig. 92.4. Using the fractal dimension and the air-entry value, the unsaturated hydraulic conductivity and the soil–water diffusivity can be predicted. The comparisons between the experimental data and the predictions of unsaturated hydraulic conductivity and the soil–water diffusivity are shown in Fig. 92.5.

92.4   Effective Stress and Unsaturated Shear Strength

The effective stress of unsaturated soils was given by Bishop and Blight (1963)

$$\sigma' = (\sigma_n - u_a) + \chi\psi \qquad (8)$$

where ψ is matric suction, χ is a parameter related to the water content, and is written as (Xu 2004b)

$$\chi = \left(\frac{\psi}{\psi_e}\right)^{D-3} \qquad (9)$$

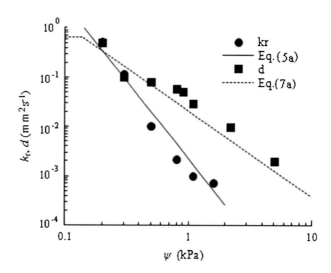

Fig. 92.5 Prediction of k_r and d

Substituting Eq. (9) into Eq. (8), effective stress of unsaturated soils is written as

$$\sigma' = (\sigma_n - u_a) + \psi_e^{3-D}\psi^{D-2} \qquad (10)$$

According to the Mohr-Coulomb criteria and effective stress formula, the shear strength of unsaturated soils is given by

$$\tau_f = c' + (\sigma_n - u_a)\tan\phi' + \psi_e^{3-D}\psi^{D-2}\tan\phi' \qquad (11)$$

Experimental data of the soil–water characteristics and shear strength with different matric suction and net stress of weathered granite soil were collected from Lee et al. (2005). The residual water content was nearly 10 % and increased with the net confining stress. The air-entry values were 2.3, 3.7, 5.1 and 6.5 kPa under net confining stress of 0, 100, 200

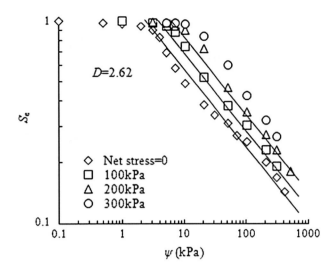

Fig. 92.6 D and ψ_e obtained from SWC

and 300 kPa, respectively given by Lee et al. (2005). Using the residual water content and the air-entry value, the fractal dimension of weathered granite soil can be evaluated from the soil–water characteristics. The fractal dimension of weathered granite soil was 2.60 obtained from Fig. 92.6. The effective shear strength parameters of weathered granite soil were also given by Lee et al. (2005). The effective cohesion (c') and internal friction angle (φ') were 20 kPa and 41.4°, respectively for weathered granite soil.

Using fractal dimension and the air-entry value, effective stress can be calculated from Eq. (10). The relationship between shear strength and effective stress was predicted and shown in Fig. 92.7. Experimental data of shear strength with

different matric suction and net normal stress were nearly located at the same line in the plane of τ–σ'. It is implied that effective stress expressed in Eq. (10) can well express the relationship between shear strength and effective stress. The prediction of shear strength using Eq. (11) was conducted using the effective shear strength parameters, the air-entry values and fractal dimensions. The effective shear strength parameters and air-entry values were given by Lee et al. (2005), and fractal dimensions were obtained from Fig. 92.6.

92.5 Conclusion

A new theoretical method is given to predict the soil–water characteristics, unsaturated hydraulic conductivity, unsaturated soil–water diffusivity, effective stress and shear strength in this paper. According to the results of mercury intrusion porosimetry, the soil pore-size distribution can be described by fractal model. Under the assumption of absorbed water being a part of the soil particle, the fractal model for the pore-size distribution is equivalent to that for the soil–water characteristics. That is, the fractal dimension and the air-entry value have the same values obtained from the pore-size distribution and the soil–water characteristics. The fractal dimension and the air-entry value obtained from the soil–water characteristics were successfully used to prediction unsaturated hydraulic conductivity and the soil–water diffusivity.

A new theoretical formula for the effective stress of unsaturated soils is presented in this paper. Two parameters, the fractal dimension and the air-entry value, in the effective stress formula can be obtained from the pore-size distribution and the soil–water characteristics. The proposed effective stress was verified by the experiments of unsaturated shear strength.

Acknowledgments The National Nature Science Foundation of China (Grant No.41272318) is acknowledged for its financial support. State Key Laboratory of Ocean Engineering is also acknowledged.

References

Bishop AW, Blight GE (1963) Some aspects of effective stress in saturated and unsaturated soils. Geotechnique 13:177–197

Lee IM, Sung SG, Cho GC (2005) Effect of stress state on the unsaturated shear strength of a weathered granite. Can Geotech J 42:624–631

Mandelbrot BB (1982) The fractal geometry of nature. W.H. Freeman, New York

Mualem Y (1976) A new model for predicting the hydraulic conductivity of unsaturated porous media. Water Res Res 12:513–522

Smettem KRJ, Kirkby C (1990) Measuring the hydraulic properties of a stable aggregate soil. J Hydrol 117:1–13

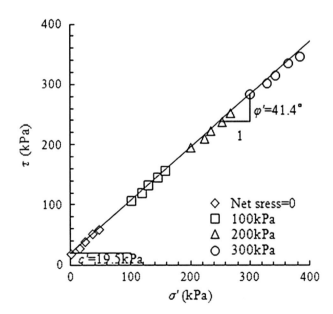

Fig. 92.7 Verification of shear strength

Watabe Y, Leroueil S, Le Bihan J-P (2000) Influence of compacted conditions on pore-size distribution and saturated hydraulic conductivity of a glacial till. Can Geotech J 37:1184–1192

Wheeler SJ, Karbue D (1996) Constitutive modeling. In: Alonso EE, Delage P, Proceedings of the 1st International Conference Unsat Soils AA. Balkema, Rotterdam

Xu YF, Sun DA (2002) A fractal model for soil pores and its application to determination of water permeability. Physica A 316:56–64

Xu YF (2004a) Calculation of unsaturated hydraulic conductivity using a fractal model for the pore-size distribution. Comput Geotech 31:549–557

Xu YF (2004b) Fractal approach to unsaturated shear strength. J Geotech Geoenviron Eng, ASCE 3:264–274

Thermal Effects on Chemical Diffusion in Multicomponent Ionic Systems

Hywel R. Thomas and Majid Sedighi

Abstract

A theoretical formulation for diffusion of multiple ions is presented in this paper which considers the effects of electrochemical and thermal diffusion potentials. The work presented is of relevance to applications such as the geological disposal of high level radioactive waste, where limited experimental information is available on ionic transfer in compacted clay buffer under non-isothermal conditions. The proposed approach incorporates the overall charge conservation in the formulation of multicomponent chemical diffusion. Thermal diffusion, i.e. Soret effect is studied in more detail by considering an explicit approach to include this process in the formulation. A detailed description of the theoretical developments is provided. A series of simulations using the proposed formulation is presented which involves pure diffusion of multiple ions under thermal gradients. The results are compared with experimental data reported in literature.

Keywords

Multicomponent chemicals • Thermal diffusion • Clay barrier • Theoretical modeling

93.1 Introduction

Understanding of chemical processes in clay buffer under thermal or thermo-hydraulic conditions is of key importance for safety assessment and design of engineered barrier systems proposed for geological disposal of high level radioactive waste. The transport of chemical ions in low permeability clays and more specifically in expansive clays is dominated by the diffusion mechanism. This in fact, is highly affected by the specific microstructure evolution and surface electrostatic forces of the clay minerals (Pusch and Yong 2006). As the result, the diffusion rate of an ionic species varies with the type of moving ion in the system. This requires the modelling of the process under different effective diffusion coefficients for

different ions and under electrochemical potential effects (e.g. Appelo and Wersin 2007).

Under non-isothermal conditions, further developments and investigations of the diffusive behaviour of multicomponent chemicals, due to the effects of combined electrochemical and thermal diffusion potentials is desirable, in order to obtain a better understanding of the chemical processes under coupled THCM behaviour. This paper presents advances to the theoretical formulation of multicomponent chemical diffusion. In particular, a framework for coupling thermal diffusion with diffusion due to electrochemical potential is presented. This latter potential arises from the transport of ions with different rates of diffusion. A general equation for diffusion of multicomponent chemicals has been developed which considers coupled electrochemical and thermal diffusion potentials in multi-ionic systems. An approach has been incorporated for estimating the thermal diffusion coefficient or the Soret coefficient in the absence of sufficient experimental data in clays.

A comprehensive description of the new formulation developed and model application has been provided by Thomas et al. (2012). This paper aims to revisit the latter

H.R. Thomas (✉) · M. Sedighi
Geoenvironmental Research Centre, Cardiff University, Queen's Buildings, Newport Road, CF24 3AA, Cardiff, UK
e-mail: ThomasHR@cf.ac.uk

M. Sedighi
e-mail: SedighiM@cf.ac.uk

development with further focus on theoretical approach, governing equations and testing the model developed. A series of validation tests on the developments against experimental results reported in the literature are presented and a comparison between developed theoretical formulation and experimental data is provided.

93.2 Theoretical Model Development

It is well established that under non-isothermal conditions, a temperature gradient can also induce mass flow due to thermal diffusion, i.e. Soret-Ludwig effect or Soret effect. Therefore, electrochemical diffusion has to be coupled with thermal diffusion in order to obtain an overall understanding of the combined effects. Modelling the diffusion of multi-component chemicals considering solely the electrochemical effects (i.e. under isothermal conditions) has been well developed and applied for aqueous solutions, porous media and clays (e.g. Lasaga 1979; Appelo and Wersin 2007). Further development of the modelling of the combined diffusion process due to electrochemical and thermal diffusion potentials (i.e. under non-isothermal conditions) in multiple ionic systems is proposed here.

Thomas et al. (2012) showed that the diffusion flux presented in Eq. (1) can be presented in an expanded form, following Lasaga (1979) and Balluffi et al. (2005) as:

$$J_i^{\text{Diff}} = -\frac{D_i^0 c_i}{RT}\frac{\partial \mu_i}{\partial c_i}\nabla c_i - \frac{D_i^0 c_i}{RT}Fz_i\nabla\Phi - \frac{D_i^0 c_i}{RT}\frac{Q_i^*}{T}\nabla T \quad (1)$$

where, the first term of the equation represents the diffusion flux due to the chemical potential, the second term denotes diffusion due to electrical potentials and the third term represents the thermal diffusion. D_i^0 and c_i is the tracer diffusion coefficient and concentration of the subscripted component. R denotes the gas constant. μ_i represents the chemical potential and z_i is the charge of the ion of the ith chemical component. F denotes the Faraday's constant, Φ refers to the electrical potential of the solution and T is the absolute temperature. Q_i^* represents the "heat of transport" of the ith chemical component which corresponds to the energy state of the diffusing ions in the regions with different temperatures which directly corresponds to the entropy of transport (Agar et al. 1989; Balluffi et al. 2005).

Substituting the electrical potential calculated and the derivative of chemical potential with respect to chemical concentration, the expanded form of diffusion flux equation is obtained, given as:

$$J_i^{\text{Diff}} = -D_i^0\left(1 + \frac{\partial \ln \gamma_i}{\partial \ln c_i}\right)\nabla c_i + \frac{z_i D_i^0 c_i}{\sum_{k=1}^{nc} z_k^2 D_k^0 c_k}\sum_{j=1}^{nc} z_j D_j^0\left(1 + \frac{\partial \ln \gamma_j}{\partial \ln c_j}\right)$$
$$- \frac{D_i^0 c_i}{RT}\frac{Q_i^*}{T}\nabla T + \frac{z_i D_i^0 c_i}{\sum_{k=1}^{nc} z_k^2 D_k^0 c_k}\sum_{j=1}^{nc} \frac{z_j D_j^0 c_j}{RT}\frac{Q_j^*}{T}\nabla T$$

$$(2)$$

where, γ_i is the activity coefficient of the ith chemical component. nc represents the number of chemical components in the solution.

The above equation represents the diffusion process comprising two main compartments of chemical (molecular) diffusion (the first two terms on the right hand side of the equation) and thermally induced mass diffusion (the second two terms on the right hand side of the equation) incorporated the effects related to the electrical potential sourcing from different diffusion coefficients.

Limited experimental information on thermal diffusion effects and the coefficients is available for mixed electrolyte systems and clays. In the absence of sufficient experimental data, a theoretical approach is adopted here to study the overall effects of thermal diffusion. In aqueous solutions, the heat of transport for single chemical in a dilute solution can be related to the thermal diffusion coefficient or the Soret coefficient (Agar et al. 1989). Accordingly, the heat of transport of the ith component in solution can be obtained as a function of ion tracer diffusion and its valence, given as (Agar et al. 1989):

$$Q_j^* = A z_i^2 D_i^0\left(1 + \frac{\partial \ln \gamma_i}{\partial \ln c_i}\right) \quad (3)$$

where, A is a constant value, i.e. 2.20×10^{12} and 2.20×10^{12} depending on the hydrodynamic boundary conditions (Agar et al. 1989).

In a general form, the diffusive transport of the ith chemical component in pore water of the clay, considering the diffusion flux comprising chemical and thermal components can be presented:

$$\frac{\partial \theta_l c_i}{\partial t} + \frac{\partial \theta_l s_i}{\partial t} = -\nabla \cdot \sum_{j=1}^{nc} \tau_i \theta_l D_{ij}^0 \nabla c_j - \nabla \cdot \sum_{j=1}^{nc} \tau_i \theta_l D_{ij}^T \nabla T$$

$$(4)$$

where θ_l stands for the volumetric liquid content and s_i stands for the sink/source term related to the geochemical reactions at which the ith component is produced or depleted by the reactions. τ_i is the tortuosity factor of the porous medium which can be different for anions and cations in clays. The molecular diffusion coefficient, i.e. D_{ij}^0 and thermal diffusion coefficient, i.e. D_{ij}^T, can be presented according to the theoretical developments presented above as:

$$D_{ij}^0 = -\delta_{ij} D_i^0 \left(1 + \frac{\partial \ln \gamma_i}{\partial \ln c_i}\right) + \frac{z_i D_i^0 c_i}{\sum_{k=1}^{nc} z_k^2 D_k^0 c_k} z_j D_j^0 \left(1 + \frac{\partial \ln \gamma_j}{\partial \ln c_j}\right) \quad (5)$$

$$D_{ij}^T = -\delta_{ij} D_i^0 c_i \left(\frac{A z_i^2 D_i^0}{RT^2}\right) \left(1 + \frac{\partial \ln \gamma_i}{\partial \ln c_i}\right) + \frac{z_i D_i^0 c_i}{\sum_{k=1}^{nc} z_k^2 D_k^0 c_k} z_j D_j^0 c_i \left(\frac{A z_i^2 D_i^0}{RT^2}\right) \left(1 + \frac{\partial \ln \gamma_j}{\partial \ln c_j}\right) \quad (6)$$

where, δ_{ij} is the Kronecker's delta function which is equal to 1 when $i = j$ and equal to 0 when $i \neq j$. It is noted that the term $\left(\frac{A z_i^2 D_i^0}{RT^2}\right)$ represents the "Soret" coefficient, i.e. S_i^T, which is theoretically described here. This value can also be replaced by experimentally determined values.

The described formulation described was implemented in the chemical module of a coupled thermal, hydraulic, chemical and mechanical model (THCM), developed at the Geoenvironmental Research Centre (e.g. Thomas and He 1995; Thomas et al. 2012). Further details of formulation development and implementation have been provided by Sedighi (2011) and Thomas et al. (2012).

93.3 Theory Application and Validation

Theoretical formulation developed is examined via a series of thermal diffusion simulations developed based on similar experimental tests carried out by Leaist and Hui (1990). Two series of simulations are presented including (i) thermal diffusion of NaOH in water (binary solution) at four different

concentrations and (ii) thermal diffusion of a mixed NaCl–NaOH solution in water (ternary solution). The simulations were performed once using the theoretical equation for the Soret coefficient using Eq. (3) and once using the Soret coefficients reported from results of experiments carried out by Leaist and Hui (1990). The results are compared and discussed.

The simulation exercises consisted of a cell in which the solution was subjected to a constant temperature gradient across the cell by fixing the temperature at 30 and 20 °C at the boundaries similar to the condition described in experimental research by Leaist and Hui (1990). Four initial concentrations of NaOH solutions were considered, i.e. 0.001, 0.005, 0.015 and 0.02 mol/l. The composition considered for simulations of the ternary solution of NaCl–NaOH was 0.005–0.015 (mol/l), i.e. [NaCl]/[NaOH] = 0.005/0.01g (mol/l).

The studied domain is 1 cm long which was discretised into 100 equally sized 4-noded quadrilateral elements for numerical analysis. The temperature was considered to be fixed at 20 and 30 °C at the boundaries. The initial temperature of the solution is 25 °C. Different diffusion coefficients have been considered for the involved chemical components. The tracer diffusion coefficients of Na^+, Cl^- and OH^- ions in the simulations are 13.3×10^{-10}, 20.3×10^{-10} and 52.7×10^{-10} m²/s, respectively (Lasaga 1979).

Figure 93.1 presents the results of simulations in terms of concentration variation with distance from cold boundary at steady state for NaOH solution (Case I) and NaCl–NaOH solution (Case II) analysed, respectively. The results presented by "lines" are related to simulations in which the theoretical approach for the Soret coefficient in Eq. (3) whilst the results presented by "symbols are those achieved based on the experimentally determined Soret values reported by Leaist and Hui (1990). The values of the Soret coefficient associated with each concentration in the graphs represent the average value of the Soret coefficients reported by Leaist and Hui (1990).

Fig. 93.1 Results of simulations depicting the concentration profiles at steady state of the NaOH-water binary solutions at different concentrations (Case I) and NaOH–NaCl-water ternary solution (Case II)

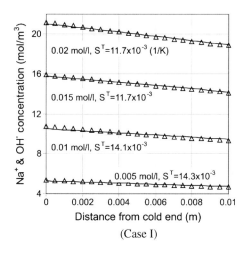

(Case I)

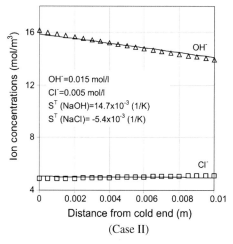

(Case II)

The simulation results show a close agreement between the theoretical approach adopted for thermal diffusion and the experimental values reported by Leaist and Hui (1990). The results show a good correlation between simulations based on the adopted approach and on the experimentally determined values for the Soret coefficient in the case of OH⁻ ions.

In the case of the ternary solution, Leaist and Hui (1990) reported negative values for the Soret coefficient for NaCl. This implies that in the ternary solutions, chloride ions tend to move toward hot side in contrast with the general trend observed in binary solutions. The trend observed in this case using the theoretical approach does not represent a reverse process; however the mode generated very close correlation with the results using the experimentally reported values for the Soret coefficient.

Comparison between the results of the simulations on ternary and binary solutions indicates that: (i) coupling the chemical and electrical effects has increased the rate of chloride diffusion away from heat source and (ii) the rate of diffusion towards the colder side for OH⁻ ions has shown an increase in compassion with the binary solution. The behaviour observed is related to the application of the electro-neutrality conditions in the formulation developed.

Since different diffusion coefficients are used in the simulation, the overall charge must be conserved during the analysis according to the formulation of diffusion developed. The model results for all simulations conserve the overall charge during the analysis and at steady state, verifying the transport equation for multicomponent chemicals under coupled electrochemical and thermal diffusion potentials.

93.4 Conclusions

Application of the theoretical approach proposed is of interested in problems such as chemical behaviour of compacted bentonite under high level radioactive waste repository conditions. Further insight into the diffusion of ions in multicomponent system of compacted clays can be achieved using presented model.

The approach proposed considers the effects of electrochemical and thermal potentials on the diffusion of multicomponent chemicals under coupled conditions. The model incorporates the overall charge conservation in the formulation of multicomponent chemical transport both on molecular and thermal diffusion. The theoretically determined the Soret coefficients provide values close to the experimentally reported values for ionic species based on the validation example provided in this paper. The approach proposed can be used and extended to study the behaviour of multiple ions present in clay-water system under thermal effects.

References

Agar JN, Mou CY, Lin J (1989) Single-ion heat of transport in electrolyte solutions: a hydrodyamic theory. J Phys Chem 93:2079–2082

Appelo CAJ, Wersin P (2007) Multicomponent diffusion modeling in clay systems with application to the diffusion of tritium, iodide and sodium in Opalinus clay. Environ Sci Technol 41:5002–5007

Balluffi RW, Allen SM, Carter WC (2005) Kinetics of materials. Wiley Interscience, Wiley

Lasaga AC (1979) The treatment of multicomponent diffusion and ion pairs in diagenetic fluxes. Am J Sci 279:324–346

Leaist DG, Hui L (1990) Conductometric determination of the Soret coefficients of a ternary mixed electrolyte, Reversed thermal diffusion of sodium chloride in aqueous sodium hydroxide solutions. J Phys Chem 94:45–447

Pusch R, Yong RN (2006) Microstructure of smectite clays and engineering performance. Taylor and Francis, London

Sedighi, M (2011) An investigation of hydro-geochemical processes in coupled thermal, hydraulic, chemical and mechanical behaviour of unsaturated soils. Ph.D. Thesis, Cardiff University, UK

Thomas HR, He Y (1995) Analysis of coupled heat, moisture, and air transfer in a deformable unsaturated soil. Géotechnique 45:677–689

Thomas HR, Sedighi M, Vardon PJ (2012) Diffusive reactive transport of multicomponent chemicals under coupled thermal, hydraulic, chemical and mechanical conditions. Geotech Geol Eng 30 (4):841–857

Unsaturated Hydraulic Conductivity of Highly Compacted Sand-GMZ01 Bentonite Mixtures Under Confined Conditions

94

W.M. Ye, Wei Su, Miao Shen, Y.G. Chen, and Y.J. Cui

Abstract

Highly compacted sand-bentonite mixtures are commonly recognized as potential buffer/backfill materials using in deep geological repository for high-level radioactive waste disposals. After water retention curve was determined, instantaneous profile method was employed in this paper for measuring unsaturated hydraulic conductivity of highly compacted GMZ01 bentonite and quartz sand mixture (7:3), with a dry density of 1.90 g/cm^3, under confined conditions. Results show that, as suction decreased from 60 MPa to zero, the measured unsaturated hydraulic conductivity firstly decreased and then turned to increase. This phenomenon can be explained in terms of microstructure changes during hydration under constant-volume conditions.

Keywords

GMZ01 bentonite • Sand-bentonite mixture • Unsaturated hydraulic conductivity • Suction • Constant volume

94.1 Introduction

As one of the key properties of buffer/backfill materials, the hydraulic property of the compacted bentonite has drawn much attention. Dixon et al. (1987) tested the hydraulic conductivity of saturated compacted bentonites and analyzed the related influencing factors; Komine (2004) predicted the saturated permeability of bentonite based on porosity

changes. For the unsaturated bentonite, Cui et al. (2008) found that with suction decreases, the unsaturated permeability decreases to a certain value and then turns to increase under both confined and unconfined conditions. Ye and Qian (2009), Ye et al. (2010) obtained the similar conclusion by testing on GMZ01 bentonite.

GMZ bentonite is a preferable buffer/backfill material for Chinese deep geological disposal program. Under confined conditions, the unsaturated hydraulic conductivity of the highly compacted GMZ bentonite (1.70 g/cm^3) changes from 1.13×10^{-13} to 8.41×10^{-15} m/s (Ye and Qian 2009). While under unconfined conditions, it is about 1.0×10^{-12}–1.0×10^{-15} m/s (Ye et al. 2010). Based on Kozeny-Carmen semi-empirical function, Niu et al. (2009) proposed a semi-empirical equation for the calculation of unsaturated hydraulic conductivity of GMZ01 bentonite with consideration of micro-structural changes.

In this paper, water retention curves of compacted quartz sand-GMZ01 bentonite (3:7) mixture are determined under confined conditions. Humidity controlled hydration test was conducted to measure the evolutions of relative humidity at different locations along axial direction of the cylindrical specimen. The instantaneous profile method was employed

W.M. Ye · W. Su (✉) · M. Shen · Y.G. Chen · Y.J. Cui
Key Laboratory of Geotechnical and Underground Engineering of Ministry of Education, Tongji University, Shanghai, 200092, China
e-mail: suweiown@163.com

W.M. Ye
e-mail: ye_tju@tongji.edu.cn

W.M. Ye
United Research Center for Urban Environment and Sustainable Development, The Ministry of Education, Beijing, China

Y.J. Cui
Laboratoire Navier, Ecole des Ponts ParisTech, Paris, France

for obtaining the unsaturated hydraulic conductivity. Mechanism was analyzed from the microstructure aspect.

94.2 Materials

A mixture of quartz sand and GMZ01 bentonite (3:7) was used. A dry density of 1.90 g/cm³ for compacted specimen was chosen in order to obtain an equivalent dry density of 1.70 g/cm³ for bentonite in the mixture.

GMZ01 bentonite contains chemical compositions of SiO_2, Al_2O_3 and H_2O. It has specific gravity of 2.66, Liquid limit and plastic limit are 276 and 37 %, respectively. Total specific surface area is 570 m²/g. Cation exchange capacity is 77.30 mmol/100 g. Dominant exchange cations are Na^+ (43.36 mmol/100 g), Ca^{2+} (29.14 mmol/100 g), Mg^{2+} (12.33 mmol/100 g) and K^+ (2.51 mmol/100 g). Main mineral compositions: 75.4 % montmorillonite, 11.7 % quartz, 4.3 % feldspar and 7.3 % cristobalite.

Quartz sand used originates from Fengyang, Anhui province, China. Chemical compositions of the sand are SiO_2 (99.20 %), Al_2O_3, Fe_2O_3 and CaO. particle size is 0.075–0.1 mm, specific gravity is 2.65.

94.3 Experiments

94.3.1 Determination of Water Retention Curves

For determination of water retention curves of specimens under constant volume conditions, the osmotic technique and vapor equilibrium technique were employed. Suctions of 309, 110, 82, 21, 9 and 4 MPa (vapor equilibrium technique), as well as 0.1 and 0.01 MPa (osmotic technique) were conducted.

94.3.2 Unsaturated Infiltration Test

Specimen preparation. Quartz sand and GMZ01 bentonite were carefully mixed (3:7) and fully stirred to a uniform state. Initial water content of the mixture was equilibrated to 9 %. Then, the mixture was compacted to 150 mm in height, 50 mm in diameter and dry density of 1.90 g/cm³ using a custom-designed compaction mold. The cylindrical specimen was compacted in three layers in order to ensure the whole homogeneity.

Apparatus and unsaturated hydraulic conductivity test. The apparatus used by Ye and Qian (2009) was employed in this study. After removing from compaction model, compacted specimen was pushed into the cell immediately and sealed in the cell, kept for days until the homogeneity was reached. Then infiltration test was started with water being injected into the specimen through the inlet at the bottom of the cell under 1 m waterhead, i.e., 10 kPa. The volume of infiltrated water was recorded using the pressure-volume controller. Variations of humidity were recorded by five sensors, among which four were fixed (equal-spaced) on the sidewall and one in the upper plate of the cell. When the humidity near water inlet approached to saturation, the test was finished.

94.4 Results and Analyses

Water retention characteristics. Figure 94.1 shows the measured water retention curve of the specimen. It is obvious that the water content of the mixture increases at a declining rate as suction decreasing, which is consistent with reports of Chen et al. (2006) about compacted GMZ bentonite.

Unsaturated hydraulic conductivity. Evolutions of humidity with time of the specimen (Fig. 94.2) shows that the relative humidity at 30 mm from the inlet reaches 90 % after 300 h hydration, 1,200 h for the same value at 60 mm. While much more time needs for other locations.

With results in Figs. 94.1 and 94.2, suction profile and hydraulic head profile can be determined. Then hydraulic gradient and water flux can be calculated. Based on Darcy's law, the variation of hydraulic conductivity with suction (Fig. 94.3) was calculated by the following equation:

$$k_w = -\frac{1}{A} \cdot \frac{q}{\frac{1}{2}[i_t + i_{t+\Delta t}]} \tag{94.1}$$

where A is cross-section area of specimen; q is water flux; i_t is hydraulic gradient at time t.

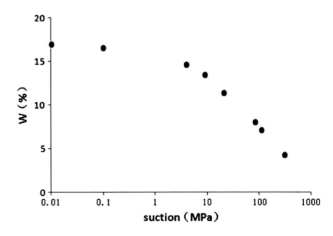

Fig. 94.1 SWRC of confined compacted sand-GMZ01 bentonite mixture

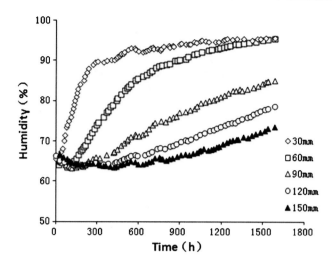

Fig. 94.2 Evolutions of humidity with time of the sand-GMZ01 bentonite mixture

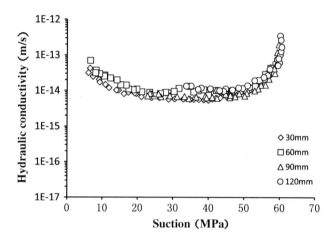

Fig. 94.3 Unsaturated hydraulic conductivity evolves with suction

94.5 Discussion

Figure 94.3 indicates that the unsaturated hydraulic conductivity of the confined specimen firstly decreases (30–60 MPa) and then turns to increase (6–30 MPa) with decreasing suction. Literatures point out that two contrasting processes and their interactions are responsible for this phenomenon. During hydration, bentonite particles expand into macro-pores during wetting under confined conditions, leading to a decrease of the macro-pore volume, thus the unsaturated hydraulic conductivity decreases (Cui et al. 2002). Meanwhile, the flowing area in the compacted mixture increases with decreasing suction, resulting in an increase of the unsaturated hydraulic conductivity (Benson and Gribb 1997).

At the beginning of hydration, the effective flow is largely dependent on the cross-sectional area of existing porous channels. As the hydration continues, on the one hand, clay expands, occupying the effective flow channels; on the other hand, gels of clay particles exfoliated from larger aggregates or grains fill the channels, especially the wider ones, and reorganize there, resulting in the clogging of water path. This phenomenon can lead to hydraulic conductivity decrease (Pusch et al. 2012). With suction sequentially reducing, the effect of bentonite expansion is getting weaker, the increasing of effective flow cross-sectional area due to the diffusional character of water makes hydraulic conductivity increase (Pusch and Yong 2003).

94.6 Conclusion

The hydraulic conductivity of sand-GMZ01 bentonite mixture under confined conditions firstly decreases and then turns to increase with the suction decreasing. This variation law is similar to the hydraulic conductivity of pure bentonite under confined conditions.

Acknowledgments The authors are grateful to the National Natural Science Foundation of China (Projects No. 41030748), China Atomic Energy Authority (Project [2011]1051) for the financial supports. Program for Changjiang Scholars and Innovative Research Team in University (PCSIRT, IRT1029) is also greatly acknowledged.

References

Benson CH, Gribb MM (1997) Measuring unsaturated hydraulic conductivity in the laboratory and field. In: Unsaturated soil engineering practice, New York, ASCE, 0-7844-0259-0. Geotechnical Special Publication No. 68, pp 113–168, 344

Chen B, Qian LX, Ye WM, Cui YJ, Wang J (2006) Soil-water characteristic curves of Gaomiaozi bentonite. Chin J Rock Mech Eng 25:788–793

Cui YJ, Loiseau C, Delage P (2002) Microstructure changes of a confined swelling soil due to suction controlled hydration. In: Jucá JFT, de Campos TMP, Marinho FAM (eds) Unsaturated soils. Proceedings of the 3rd international conference on unsaturated soils (UNSAT 2002), Recife, Brazil, vol 2. Swets & Zeitlinger, Lisse, pp 593–598

Cui YJ, Tang AM, Loiseay C, Delage P (2008) Determining the unsaturated hydraulic conductivity of a compacted sand—bentonite mixture under constant-volume and free-swell conditions. Phys Chem Earth 33:s462–s471

Dixon DA, Cheung SCH, Gray MN, Davidson BC (1987) The hydraulic conductivity of dense clay soils. In: Proceedings of 40th Canadian geotechnical conference, Regina, Saskatchewan—Canada, pp 389–396

Komine H (2004) Simplified evaluation on hydraulic conductivities of sand–bentonite mixture backfill. Appl Clay Sci 26(1–4):13–19

Niu WJ, Ye WM, Chen B, Qian LX (2009) SWCC and permeability of unsaturated gaomiaozi bentonite in free swelling condition. Chin J Undergr Space Eng 5:952–955

Pusch R, Yong R (2003) Water saturation and retention of hydrophilic clay buffer—microstructural aspects. Appl Clay Sci 23:61–68

Pusch R, Richard P, Zuzana W, Liu X, Sven K (2012) Role of clay microstructure in expandable buffer clay. J Purity Util React Environ 1(6):297–322

Ye W, Qian L (2009) Laboratory test on unsaturated hydraulic conductivity of densely compacted Gaomiaozi Bentonite under confined conditions. Chin J Geotech Eng 31:105–108

Ye W, Niu W, Chen B, Chen Y (2010) Unsaturated hydraulic conductivity of densely compacted Gaomiaozi bentonite under unconfined conditions. J Tongji Univ (Nat Sci) 38(10):1439–1443

Adsorption, Desorption and Competitive Adsorption of Heavy Metal Ions from Aqueous Solution onto GMZ01 Bentonite

W.M. Ye, Yong He, Y.G. Chen, Bao Chen, and Y.J. Cui

Abstract

The geochemical processes of adsorption, desorption and competitive adsorption on bentonite are important for the long-term safety assessment of high-level nuclear waste (HLW) repositories. In this paper, batch adsorption, desorption experiments of Cr(III) and the competitive adsorption experiment with Cu(II) were performed in aqueous solutions on Gao-miao-zi (GMZ) bentonite. Results show that the pH value significantly affect the Cr(III) adsorption/desorption on GMZ bentonite. Both adsorption and desorption isotherms are consistent with the Freundlich equation. The distribution coefficients (K_d) were calculated from the competitive adsorption test; Higher K_d values of Cr(III) were obtained than that of Cu(II), indicating the stronger retention capacity of Cr(III) on GMZ bentonite in a binary metal system.

Keywords

GMZ01 bentonite • Adsorption • Desorption • Competitive adsorption • Heavy metal ions

95.1 Introduction

Bentonite has excellent adsorption properties and possesses adsorption sites available within its interlayer space as well as on the outer surface and edges (Tabak et al. 2007). Because of the retardation for the transport of radionuclides from the repository to the environment, bentonite has attracted great interest in nuclear waste management.

W.M. Ye (✉) · Y. He · Y.G. Chen · B. Chen · Y.J. Cui
Key Laboratory of Geotechnical and Underground Engineering of Ministry of Education, Tongji University, Shanghai, 200092, China
e-mail: ye_tju@tongji.edu.cn

Y.G. Chen
e-mail: 12heyong@tongji.edu.cn

W.M. Ye
United Research Center for Urban Environment and Sustainable Development, The Ministry of Education, Beijing, China

Y.J. Cui
Laboratoire Navier, Ecole des Ponts ParisTech, Paris, France

In recent years, many studies have focused on the adsorption of heavy metals on bentonite and under different experimental conditions (Bhattacharyya and Gupta 2008; Chen et al. 2012). These works showed that soil pH, temperature, time, and ionic strength were reported to be important factors that influence the adsorption of heavy metals on bentonite. Adsorption/desorption is a major process responsible for the fate of heavy metals in soils, since the mobility of heavy metals is directly related to their partitioning between soil and soil solution (Sparks 2001). Thus, the study of desorption is conducive to elucidate adsorption process on soil, to recover heavy metal ions from adsorbent, and to multiple regenerate adsorbent (Chen et al. 2013). Most soil-metal bonding information has been derived from studies conducted using single-metal system (Gomes et al. 2001). However, the presence of only one heavy metal ion is a rare situation in reality. Thus, it is important to study the affinity of bentonite toward the adsorption of specific metal ions to investigate their removal from multi-component solutions. Though application of bentonite in the field of waste water treatments has proliferated in recent years, efforts in analyzing desorption and competitive adsorption properties on bentonite, especially on GMZ bentonite in China, is still lacking. Where, the GMZ bentonite has been considered as a possible

material for construction of engineering barrier in deep geological repository for high-level radioactive waste disposal in
China (Ye et al. 2010).

The aim of this work is to review the adsorption/
desorption properties of Cr(III) on GMZ bentonite, to analyze the affinity of bentonite toward the adsorption of specific metal ions.

95.2 Experimental

95.2.1 Materials

The raw GMZ01 bentonite used in this work was extracted
from the northern Chinese Inner Mongolia autonomous
region, 300 km northwest from Beijing. The mineralogical
composition was quantitatively analyzed by using the X-ray
diffraction method. The bulk composition (mass fraction)
was determined as follows: 75.4 % montmorillonite, 11.7 %
quartz, 7.3 % cristobalite, 4.3 % feldspar, 0.8 % kaolinite,
0.5 % calcite. This shows that the proportion of montmorillonite is dominant in the GMZ01 bentonite. The cation
exchange capacity (CEC) of GMZ01 bentonite is
0.773 mmol/g and the specific surface area (SSA) is 570 m^2/
g (Ye et al. 2010). Stock solutions of Cr(III) and Cu(II) with
the concentration of 1.923×10^{-3} mol/L were prepared from
its nitrate and sulfate, respectively.

95.2.2 Batch Adsorption Studies

The adsorption capacity of Cr(III) on GMZ bentonite was
investigated by using batch technique in polyethylene centrifuge tubes sealed with a screw cap under ambient conditions. When adsorption equilibrium reached, the suspension
was centrifuged and most of the supernatant was exchanged
with the same volume of the background electrolyte solution
of NaCl. The competitive adsorption was carried out under a
varying temperature from 293 to 323 K. The concentrations
of Cr(III) solution before and after adsorption were measured
by using an ultraviolet–visible (UV–VIS) spectrophotometer, WFJ2100, at a wave length of 540 nm. The concentrations of Cu(II) solution before and after adsorption were
measured by using an atomic absorption spectrophotometer,
WFX-1F2B2.

The adsorbed amount of heavy metal ions on GMZ
bentonite was calculated from the difference between the
initial concentration and the equilibrium one:

$$q_e = \frac{(C_o - C_e)V}{m} \quad \text{or} \quad \%removal = \left(\frac{C_o - C_e}{C_o}\right) \times 100$$

$$(95.1)$$

where q_e (mg/g) is the amount of heavy metal ions adsorbed
on bentonite, C_0 (mg/L) is the initial concentration of heavy
metal ions in suspension, C_e (mg/L) is the aqueous concentration of heavy metal ions in equilibrium solution, V (L)
is the volume of heavy metal ions solution, and m (g) is the
mass of the adsorbent.

Equilibrium distribution coefficient (K_d, L/mg) used
extensively for charactering various heavy metal ions
adsorption and desorption, was calculated as:

$$K_d = \frac{q_e}{C_e} \quad (95.2)$$

95.3 Results and Discussion

95.3.1 Effect of pH on Adsorption and Desorption

Figure 95.1 shows that the adsorption capacity increases as
pH increases from 1.0 to 7.0. Above pH 7.0, its effect on the
adsorption become significant and reaches a plateau (Chen
et al. 2012). However, the desorption rate decreases gradually at pH 1–9 and maintains a high level around pH = 1
(Chen et al. 2013).

The adsorption and desorption mechanism of Cr(III) on
GMZ bentonite can be explained as follows: (i) in acidic
region both the adsorbent and adsorbate are positively
charged and the net interaction is that of electrostatic
repulsion and (ii) the positively charged metal ions face a
good competition with the higher concentration of H$^+$ ions
present in the acidic reaction mixture. It has been found that
at low pH solution most heavy metals become mobile and
adsorption onto clay particles becomes less effective.

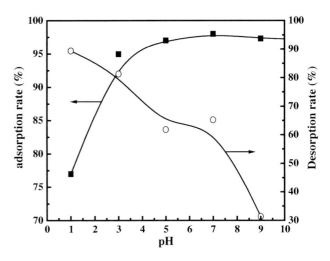

Fig. 95.1 Effect of pH on Cr(III) adsorption and desorption (Chen
et al. 2012, 2013)

95.3.2 Adsorption and Desorption Isotherms

Figure 95.2 shows the adsorption and desorption isotherms of Cr(III) on GMZ bentonite (Chen et al. 2013). The equilibrium data were correlated with the Freundlich isotherm. The Freundlich equation [Eq.(95.3)] was used:

$$q_e = k_F C_e^{1/n} \text{ or, in linear form}: \lg q_e = \lg k_F + 1/n \lg C_e$$

$$(95.3)$$

where k_F is the constant indicative of the relative adsorption or desorption capacity of the adsorbent ($mg^{1-1/n} \cdot L^{1/n} \cdot g^{-1}$) and $1/n$ is the constant indicative of the intensity of the adsorption or desorption. A plot of $\lg q_e$ against $\lg C_e$ gives a straight line, the slope and the intercept of which correspond to $1/n$ and $\lg k_F$, respectively.

The results show that the data obtained for Cr(III) adsorption and desorption fit well to the Freundlich model. Similar results have been reported by Turin and Bowman (1997), Shirvani et al. (2006) for herbicide and cadmium desorption from Casa Grande and Palygorskite soil, respectively.

95.3.3 Competitive Adsorption of Cu(II)/Cr(III) in Binary Metal Ions Solution System

Figure 95.3 shows that the K_d value for each metal in the studied samples was used to compare the adsorption capacities of bentonite for the metals. A high K_d medium value indicates a high metal retention by the solid phase through adsorption and chemical reactions.

The sequence of adsorption affinity of metals according to the distribution coefficient (K_d) toward bentonite was found

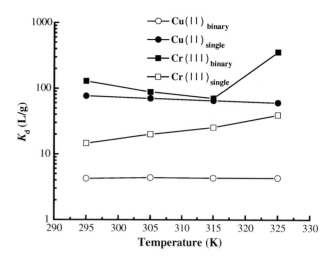

Fig. 95.3 K_d values in single or binary system

to be the following: $K_{d\ binary}$ (Cr)>$K_{d\ single}$ (Cu)>$K_{d\ single}$ (Cr)>$K_{d\ binary}$ (Cu), indicating that Cr(III) was stronger adsorbed by bentonite than Cu(II) in binary metal system. Similar results that the Cr(III) had the higher affinity to kaolinite than Cu(II) have been reported by Covelo et al. (2007).

95.4 Conclusions

In this study, the adsorption/desorption properties of Cr(III) on GMZ bentonite were reviewed and the competitive adsorption was also analyzed. The following conclusions can be drawn:

The pH value is an important factor in the Cr(III) adsorption/desorption on GMZ bentonite.

Both adsorption and desorption isotherms are consistent with the Freundlich model.

The GMZ bentonite has stronger retention capacity of Cr (III) than that of Cu(II) in a binary metal system.

Acknowledgements The authors are grateful to the National Natural Science Foundation of China (Projects No. 41030748 and No. 41272287), China Atomic Energy Authority (Project [2011]1051) for the financial supports. Program for Changjiang Scholars and Innovative Research Team in University (PCSIRT, IRT1029) is also greatly acknowledged.

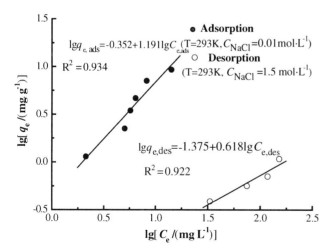

Fig. 95.2 Freundlich adsorption and desorption isotherms on GMZ bentonite (Chen et al. 2013)

References

Bhattacharyya KG, Gupta SS (2008) Adsorption of a few heavy metals on natural and modified kaolinite and montmorillonite: a review. Adv Colloid Interface Sci 140:114–131

Chen YG, He Y, Ye WM, Lin CH, Zhang XF, Ye B (2012) Removal of chromium (III) from aqueous solutions by adsorption on bentonite from Gaomiaozi, China. Environ Earth Sci 67(5):1261–1268

Chen YG, He Y, Ye WM, Sui WH, Xiao MM (2013) Desorption of Cr (III) adsorbed onto GMZ01 bentonite: effect of shaking time, ionic strength, temperature and pH. Trans Nonferrous Met Soc China

Covelo EF, Vega FA, Andrade ML (2007) Competitive sorption and desorption of heavy metals by individual soil components. J Hazard Mater 140:308–315

Gomes PC, Fontes MPF, da Silva AG, Mendonca ED, Netto AR (2001) Selectivity sequence and competitive adsorption of heavy metals by Brazilian soils. Soil Sci Soc Am J 65:1115–1121

Shirvani M, Kalbasi M, Shariatmadari H, Nourbakhsh F, Najafi B (2006) Sorption–desorption of cadmium in aqueous palygorskite, sepiolite, and calcite suspensions: isotherm hysteresis. Chemosphere 65:2178–2184

Sparks DL (2001) Elucidating the fundamental chemistry of soils: past and recent achievements and future frontiers. Geoderma 100:303–319

Tabak A, Afsin B, Aygun SF, Koksal E (2007) Structural characteristics of organo-modified bentonites of different origin. J Therm Anal Calorim 87:375–381

Turin HJ, Bowman RS (1997) Sorption behavior and competition of bromacil, napropamide, and prometryn. J Environ Qual 26 (5):1282–1287

Ye WM, Chen YG, Chen B, Wang Q, Wang J (2010) Advances on the knowledge of the buffer/backfill properties of heavily-compacted GMZ01 bentonite. Eng Geol 116:12–20

Enhanced Isothermal Effect on Swelling Pressure of Compacted MX80 Bentonite

96

Snehasis Tripathy, Ramakrishna Bag, and Hywel R. Thomas

Abstract

Laboratory isothermal swelling pressure tests were carried out on compacted MX80 bentonite at 25 and 70 °C with distilled water as the hydrating fluid. Assessment of the swelling pressure was carried out using the electrical triple-layer theory for interacting clay platelets. An increase in the temperature reduced the swelling pressure of the bentonite on account of decrease in the Gouy-layer charge, mid-plane potential and distance functions. Agreements between the calculated swelling pressures from the triple-layer theory and the experimental results were found to be reasonable for compaction dry densities of less than 1.45 Mg/m^3.

Keywords

Laboratory test • Bentonite • Swelling pressure • Temperature • Triple-layer theory

96.1 Introduction

Compacted bentonites have been proposed to be used as buffer around the waste canisters in the deep underground storage of high-level waste. The insitu boundary conditions in this case dictate exposure of compacted bentonites to elevated temperature and hydration upon fluid uptake from host rock. Therefore, research works in this context have been carried out by several researchers in the past (Pusch 1980; Pusch et al. 1990; Villar and Lloret 2004; Lloret et al. 2004; Romero et al. 2003, 2005; Jacinto et al. 2009; Ye et al. 2009, to name few).

An increase in the temperature has been shown to decrease the swelling pressure of compacted bentonites. Similarly, water retention capacities of bentonites and other swelling clays have been shown to decrease with an increase in the temperature. Pusch (1980) stated that the reduction of swelling pressure is due to the presence of less stable interlayer and inter-particle water at higher temperature. An increase in the temperature increases the kinetic energy of

the water molecules and tends to reduce the hydration of the clay particles. The thermal properties of water have been considered to explain the reductions in the water retention capacities of bentonites. For example, the surface tension of water decreases with an increase in the temperature causing a reduction in the matric suction. Romero et al. (2003) stated that the changes in water retention induced by temperature are associated mainly with temperature dependence of the interfacial tension and wetting coefficient, and with thermal expansion of entrapped air.

Mitchell and Soga (2005) stated that, an increase in the temperature has two effects, such as that the dielectric constant of the pore fluid decreases and the Debye length (i.e., the maximum possible diffuse double layer thickness) increases. Electrical double-layer calculations (Tripathy et al. 2004; Schanz and Tripathy 2009) would clearly indicate that, independent effects of various parameters should not be considered while drawing conclusions on the salient features of the electrical double layer associated with interacting clay platelets systems. An increase in the temperature and a decrease in dielectric constant of the pore fluid tend to reduce the Debye length for a given clay-water system.

Laboratory tests were carried out in order to study the influence of temperature on the swelling pressure of compacted bentonite. The electrical triple-layer (i.e., the clay surface, the Stern-layer, and the diffuse double layer)

S. Tripathy (✉) · H.R. Thomas
Cardiff University, Cardiff, UK
e-mail: Tripathys@cf.ac.uk

R. Bag
National Institute of Technology, Rourkela, India

theory or the Stern theory as applicable to interacting clay platelets and for constant surface potential case (Verwey and Overbeek 1948; Tripathy et al. 2013) was used to assess swelling pressures for a range of swollen dry densities.

96.2 Material and Method

Commercially available MX80 bentonite containing 84 % montmorillonite was used in this study. The specific gravity of the bentonite was found to be 2.76. The liquid limit, plastic limit and shrinkage limit of the bentonite were found to be 385, 43 and 15.8 %, respectively (Bag 2011).

The swelling pressure tests were carried out in an oedometer cell designed and developed at Cardiff University (see Fig. 96.1). Specimens were prepared by compacting MX80 bentonite at a water content of about 10 % for a range of dry densities between 1.25 and 1.8 Mg/m³. In an attempt to eliminate the influence of post-compaction residual lateral stresses on swelling pressures, after the compaction process, the compacted specimens were extruded and inserted back into the specimen rings prior to testing them for swelling pressures. Distilled water was used as the hydrating fluid. The tests were terminated once the measured swelling pressures were equilibrated. In total, seventeen tests were carried out (twelve tests at an ambient temperature of 25 °C and five tests at 70 °C). Based on the final water contents and the mass of the specimens, the swollen dry densities of the

specimens were calculated. The calculated swollen dry densities were found to be smaller than the compacted dry densities of the specimens prior to testing. Reductions of dry densities of the specimens are attributed due primarily to the expansion of the measuring system.

96.3 Results and Discussion

The experimental dry density versus swelling pressure results are shown in Fig. 96.2. The swelling pressure decreased with an increase in the temperature. A reduction in the swelling pressure due to an increase in the temperature was found to be about 15–20 % for the range of dry density considered in this study.

From the triple-layer theory dealt with in this study (Tripathy et al. 2013), it was noted that an increase in the temperature increases the electric potentials at the Stern plane and in the mid-plane, and decreases the Stern-layer charge, the Gouy-layer charge, the Debye length, and non-dimensional mid-plane and distance functions. Therefore, a reduction in the swelling pressure of bentonites due to an increase in the temperature is well supported from theoretical considerations (Fig. 96.2).

Reasonably good agreements can be noted between the calculated results from the theory and experimental results for dry densities less than or equal to about 1.45 Mg/m³. At a dry density of 1.45 Mg/m³, overlapping of the Stern-layers

Fig. 96.1 Schematic of the swelling pressure device used in this study

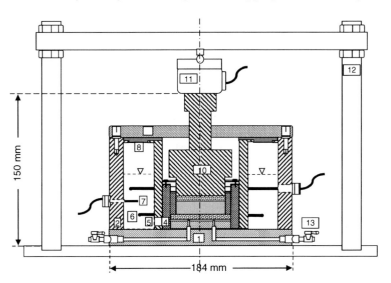

1. Device base
2. Outer casing
3. Inner separator ring
4. Inner housing unit
   Specimen ring
   Top porous disc
   Bottom porous disc
   Compacted specimen
   Locking collar
   Locking key
5. Oil chamber for elevated iso-thermal test
6. Electric coil
7. Thermocouple
8. Oil chamber cover
9. Top lid
10. Pressure pad
11. Load cell
12. Loading frame
13. Fluid supply

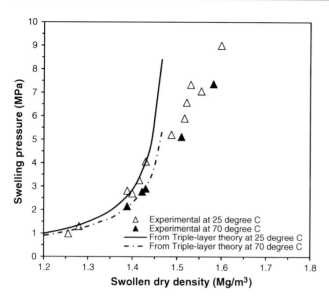

Fig. 96.2 Dry density versus experimental and theoretical swelling pressure plot

among the clay platelets occurs (i.e., the distance between two parallel clay platelets is equal to 1.0 nm). Swelling pressures at swollen dry densities greater than about 1.45 Mg/m^3 stem from forces associated with hydration of surfaces of clay platelets and the exchangeable cations.

96.4   Conclusions

Laboratory swelling pressure tests were conducted on compacted MX80 bentonite in order to study the influence of elevated temperature on the swelling pressure. The electrical triple-layer theory was used to explore various parameters responsible for a variation of the swelling pressure due to an increase in the temperature.

The study showed that an increase in the temperature reduced the swelling pressure of compacted specimens as has been noted by several researchers in the past. Electrical triple-layer calculations showed that a decrease in the swelling pressure due to an increase in the temperature is the combined influence of increase in the electric potentials at the Stern plane and in the mid-plane, and decrease in the Stern-layer

charge, the Gouy-layer charge, the Debye length, and non-dimensional mid-plane and distance functions. Agreements between the calculated and the experiment swelling pressures of compacted MX80 bentonite were very good for swollen dry density less than or equal to 1.45 Mg/m^3.

References

Bag R (2011) Coupled thermo-hydro-mechanical-chemical behavior of MX80 bentonite in geotechnical applications. PhD thesis, Cardiff University

Jacinto AC, Villar MV, Gomez-Espina R, Ledesma A (2009) Adaptation of the van Genuchten expression to the effects of temperature and density for compacted bentonites. Appl Clay Sci 42:575–582

Lloret A, Romero E, Villar MV (2004) FEBEX II project final report on thermo-hydro-mechanical laboratory tests. Publicación Técnica ENRESA 10/04, Madrid, 180 pp

Mitchell JK, Soga K (2005) Fundamentals of soil behaviour. Wiley, New York

Pusch R (1980) Swelling pressure of highly compacted bentonite. SKB Technical report, TR-80-13

Pusch R, Karlnland O, Hokmark H (1990) GMM-a general microstructural model for qualitative and quantitative studies of smectite clays. SKB technical report 90-43, Stockholm, Sweden

Romero E, Gens A, Lloret A (2003) Suction effects on a compacted clay under nonisothermal conditions. Géotechnique 53(1):65–81

Romero E, Villar MV, Lloret A (2005) Thermo-hydro-mechanical behaviour of two heavily overconsolidated clays. Eng Geol 81:255–268

Schanz T, Tripathy S (2009) Swelling pressure of a divalent-rich bentonite: Diffuse double-layer theory revisited. Water Resour Res 45, W00C12. doi:10.1029/2007WR006495

Tripathy S, Bag R, Thomas HR (2013) Effect of stern-layer on the compressibility behaviour of bentonites. Acta Geotech. doi:10.1007/s11440-013-0222-y

Tripathy S, Sridharan A, Schanz T (2004) Swelling pressures of compacted bentonites from diffuse double layer theory. Can Geotech J 41:437–450

Verwey EJW, Overbeek JThG (1948) Theory of the stability of lyophobic colloids. Elsevier, Amsterdam

Villar MV, Lloret A (2004) Influence of temperature on the hydro-mechanical behaviour of a compacted bentonite. Appl Clay Sci 26:337–350

Ye WM, Cui YJ, Qian LX, Chen B (2009) An experimental study of the water transfer through confined compacted GMZ bentonite. Eng Geol 108(3–4):169–176

Effects of Stress and Suction on the Volume Change Behaviour of GMZ Bentonite During Heating

97

Wei-Min Ye, Qiong Wang, Ya-Wei Zhang, Bao Chen, and Yong-Gui Chen

Abstract

Unsaturated compacted bentonite is considered by several countries as sealing/backfill material in the deep geological repository for high-level radioactive waste (HLW). In the field conditions, where bentonite is subjected to coupled thermo-hydro-mechanical actions, its thermo-mechanical behavior may change with suction and stress variations. This work focuses on the compacted Gaomiaozi (GMZ) bentonite, which has been chosen as potential sealing/backfill material in the Chinese repository concept. The effects of vertical stress and suction on the volume change behaviour during thermal loading (heating) were experimentally investigated. A high pressure oedometer frame permitting simultaneous control of temperature, suction and pressure was used for testing. Compacted samples were heated from the original temperature of 20 to 80 °C at total suctions ranging from 4.2 to 110 MPa, and vertical stress of 0.1 or 5 MPa. Results show that heating at constant suction and vertical stress induces either swelling or contraction. At a constant vertical stress, samples with higher suctions swell during heating, and the lower the suction the lower the swelling strain; on the contrary heating resulted in a thermal contraction with lower suctions. At a constant higher suction (110 MPa), sample swells during heating; however, for a relative lower suction, a contraction was induced by heating for a higher vertical stress of 5 MPa. In short, at high pressure and low suction, heating tends to induce contraction, while at low pressure and high suction, heating induced expansion.

Keywords

GMZ bentonite • Suction • Vertical stress • Heating • Volume change

W.-M. Ye · Y.-W. Zhang · B. Chen · Y.-G. Chen
Key Laboratory of Geotechnical and Underground Engineering of Ministry of Education, Tongji University, Shanghai 200092, China

W.-M. Ye
United Research Center for Urban Environment and Sustainable Development, The Ministry of Education, Shanghai 200092, China

Q. Wang (✉)
Centre of Excellence for Geotechnical Science and Engineering, The University of Newcastle, Callaghan NSW 2308, Australia
e-mail: wangqiong139@hotmail.com

97.1 Introduction

Thermo-mechanical behavior of expansive clays used as sealing/backfill materials for high-level radioactive wastes (HLW) disposal has attracted large attentions. As it is subjected to coupled thermo-hydro-mechanical (T-H-M) actions, induced by the water infiltration from the geological barrier, stresses generated by the swelling, and heat dissipation from the nuclear waste packages, changes of water content (suction) and stress change may have great impacts.

This work focuses on the Gaomiaozi (GMZ) bentonite, which has been chosen as potential sealing/backfill material in the Chinese repository concept. A series of heating test were performed on compacted samples with vertical stress and suction control. The results obtained allowed the analyzing of the vertical stress and suction effect on the volume change behavior during heating.

97.2 Materials and Method

97.2.1 Materials

The Gaomiazi GMZ bentonite from Inner Mongolia Autonomous Region, China, was used. With a high content of montmorillonite (74.5 %), it has an average specific gravity of 2.66, a liquid limit of 313 %, and a plastic limit of 38 %. The cation exchange capacity (CEC) is 62.59–82.06/100 g, and the total specific surface is 570 m^2/g.

All tests were performed on compacted samples. The GMZ01 bentonite powder with an initial water content of 8.6 % was used for the samples preparation. Samples were statically compacted in a thick-walled compaction cell (50 mm in internal diameter) at a controlled rate of 0.375 kN/min to the desired dry density of 1.70 Mg/m^3.

97.2.2 Experimental Methods

An oedometer frame with simultaneous control of temperature and suction was used. It consists four main parts: (1) oedometer cell (50 mm in inner diameter); (2) loading system; (3) suction controlled system and (4) temperature control system. The soil specimen, sandwiched by two porous disks, was put in the oedometer cell with two valves both at the top and bottom of the cell for water or air circulation. A suction control system by vapour equilibrium method was connected to these valves for suction imposition (4.2–110 MPa). Suctions of Saturated saline solutions at different temperatures calibrated by Tang and Cui (2005) were employed. The vertical stress was applied through a loading piston on the top, by a double arms lever system (Marcial et al. 2002); the axial displacement was monitored by a micrometer dial gauge (with a precision of 0.001 mm) fixed on the piston. A water bath with a thermostat was used for temperature control (with a precision of ±0.1 °C).

In order to study the effect of vertical stress and suction on the volume change behaviour during heating, five samples listed in Table 97.1 were tested. From the as-compacted state (with an initial water content of 8.6 %), different suctions were first imposed by vapour phase technique (4.2, 38 and 110 MPa by saturated solutions of K_2SO_4, NaCl, K_2CO_3, respectively). Once the suction equilibrium was reached, a vertical stress of 0.1 or 5 MPa was applied up to the stabilisation, i.e. a volumetric strain rate of less than 0.01 % within 8 h. Specimens were then heated at a speed of

Table 97.1 Test programme

Test No.	Suction (MPa)	Vertical stress (MPa)
T-01	4.2	0.1
T-02	38	0.1
T-03	110	0.1
T-04	38	5
T-05	110	5

10 °C/2 h to 80 °C, during which constant suction and vertical stress were maintained, and the vertical strain of specimens were measured.

97.3 Results and Discussion

Figure 97.1 presents the volumetric change of compacted GMZ01 bentonite during heating at different suctions under a constant vertical stress. It presents a significant effect of suction on the heating induced volume changes. At a constant vertical stress of 0.1 MPa (Fig. 97.1a), samples with higher suctions (110 and 39 MPa) swell during heating from 20 to 80 °C. On the contrary heating resulted in a thermal contraction with a relative lower suction (4.2 MPa). For the vertical stress of 5 MPa (Fig. 97.1b), similar phenomena were obtained: at high suction of 110 MPa, swelling was observed during heating; while, at a relative lower suction of 39 MPa, heated induced a few expansion, and then a tendency toward contraction. All results evidence the significant effect of suction on the heating induced volumetric stain: at a constant vertical stress, heating induces expansion under higher suction and a contraction under lower suction. This is consistent with the conclusion of Tang et al. (2008) from MX80 bentonite, where the suction effect was explained by the softening of the swelling clay aggregates due to suction decrease (Tang et al. 2008; Tang and Cui 2009).

From the results shown in Fig. 97.1, it can be also observed that the shift suction between dilation and contraction are different for different vertical stress: under a lower vertical stress of 0.1 MPa, heating led to swelling at 38 MPa suction; however, contraction was induced under a higher vertical stress of 5 MPa at the same suction level (i.e. 38 MPa). To further analyze the effect of vertical stress on heating induced volumetric changes, experimental results (shown in Fig. 97.1) at the same suction was re-plotted versus vertical stress in Fig. 97.2. It presents that under a higher suction of 110 MPa, heating resulted in expansion under both 0.1 and 5 MPa vertical stresses. However, it is not the case at 38 MPa suction, where a contraction was obtained for a higher vertical stress of 5 MPa. This implies that heating tends to induce a contraction at high pressure and low suction. This phenomena is in agreement with the

Fig. 97.1 Volume changes during heating at different suction **a** $p = 0.1$ MPa, **b** $p = 5$ MPa

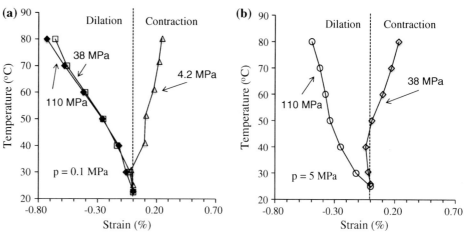

Fig. 97.2 Volume changes during heating under different vertical stress **a** $s = 110$ MPa, **b** $s = 38$ MPa

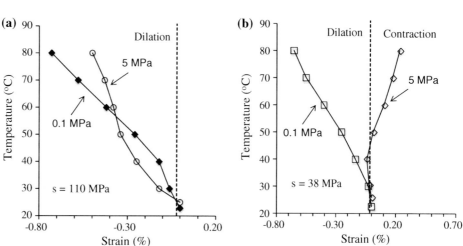

observations of Romero et al. (2005) and Tang et al. (2008) from FEBEX bentonite and MX80 bentonite, respectively.

97.4 Conclusion

The effects of vertical stress and suction on the volume change behavior of GMZ01 bentonite during thermal loading (heating) were experimentally investigated in this study. It was observed that heating at constant suction and vertical stress induces either swelling or contraction. At a constant vertical stress, samples swell at higher suction and contract at lower suction during heating. Regarding the constant suction, samples at both higher and lower vertical stress swell upon heating at a high suction; however, for a relative lower suction, contraction was induced by heating at a higher vertical stress compared to the expansive at a lower vertical stress.

In a word, heating induces expansion under low pressure and high suction; while at high pressure and low suction, heating tends to induce a contraction.

Acknowledgments The authors are grateful to the National Natural Science Foundation of China (Projects No. 41030748, 41272287), China Atomic Energy Authority (Project [2011]1051) for the financial supports. This work was also conducted within a Program for Chang-jiang Scholars and Innovative Research Team in University (PCSIRT, IRT1029).

References

Marcial D, Delage P, Cui YJ (2002) On the high stress compression of bentonites. Can Geotech J 39:812–820

Romero E, Villar MV, Lloret A (2005) Thermo-hydro-mechanical behaviour of heavily overconsolidated clays. Eng Geol 81:255–268

Tang AM, Cui YJ (2005) Controlling suction by the vapour equilibrium technique at different temperatures and its application in determining the water retention properties of MX80 clay. Can Geotech J 42:287–296

Tang AM, Cui YJ (2009) Modelling the thermomechanical volume change behaviour of compacted expansive clays. Géotechnique 59:185–195

Tang AM, Cui YJ, Barnel N (2008) Thermo-mechanical behaviour of compacted swelling clay. Geotechnique 58(1):45–54

Preliminary Assessment of Tunnel Stability for a Radioactive Waste Repository in Boom Clay

98

P. Arnold, P.J. Vardon, and M.A. Hicks

Abstract

This paper investigates the stability of tunnel galleries of a radioactive waste repository in Boom Clay, with specific reference to the current Dutch disposal concept. In this preliminary study an isotropic analytical solution has been implemented within a reliability based framework to assess the mechanical response of the Boom Clay, accounting for the aleatory uncertainty in soil property values. The performance, that is, the extent of the plastic zone with respect to a set limit state, is then quantified in terms of exceedance probability rather than using a single deterministic value. This allows for input into a quantifiable and more transparent decision making process. The effect of defining material parameter correlation within the statistical description of soil property values is illustrated via correlated and uncorrelated analyses. The results of this preliminary analytical investigation present a vital input for later two- and three-dimensional numerical modelling of the current disposal concept for a generic Dutch radioactive waste repository in Boom Clay, for both the pre- and post-closure analyses.

Keywords

Analytical methods • Boom clay • HLW • Radioactive waste • Tunnelling

98.1 Introduction

In many countries around the world, it is proposed to manage radioactive waste in the long-term via geological disposal in hard rock (e.g. granite), salt rock or clay rock. The Dutch national programme is currently investigating the use of clay rock, i.e. Boom Clay, with a proposed repository situated at over 500 m depth and with a stratum thickness of about 100 m (Verhoef et al. 2011). The Boom Clay formation exists under most of the Netherlands and the location of the repository is at present unknown, although the inherent properties of Boom Clay are known to vary with both depth and location (e.g. Barnichon et al. 2000). The geomechanical

and consequential financial feasibility of such a repository need to be investigated. Moreover, with approximately 10 km of shafts, galleries and tunnels (Verhoef et al. 2011), the cost increase associated with an increase in concrete lining thickness or increase in gallery spacing is significant. For example, the cost of the lining has been estimated to be approximately 80 % of the cost of the tunnel construction (Barnichon et al. 2000).

This paper investigates tunnel stability in Boom Clay at approximately 500 m depth. An analytical model incorporating elasto-plastic material behaviour has been developed and utilised to account for, in a simplified manner, the strain-softening behaviour which typically is encountered for Boom Clay under low confining stresses (e.g. Horseman et al. 1987; Barnichon et al. 2000). The uncertainty of material parameters has been considered by using the Monte Carlo Method (MCM), with the objective to identify the most important parameters. Analyses with both uncorrelated

P. Arnold · P.J. Vardon (✉) · M.A. Hicks
Geo-Engineering Section, Delft University of Technology,
PO Box 50482600 GA, Delft, The Netherlands
e-mail: p.j.vardon@tudelft.nl

G. Lollino et al. (eds.), *Engineering Geology for Society and Territory – Volume 6*,
DOI: 10.1007/978-3-319-09060-3_98, © Springer International Publishing Switzerland 2015

Fig. 98.2 Stresses around
excavated cavity with
impermeable concrete liner. The
stresses $\bar{\sigma}$ correspond to the EP
interface (r_{pl}/r_c) and $\hat{\sigma}$ to the RP
interface (r_{rpl}/r_c)

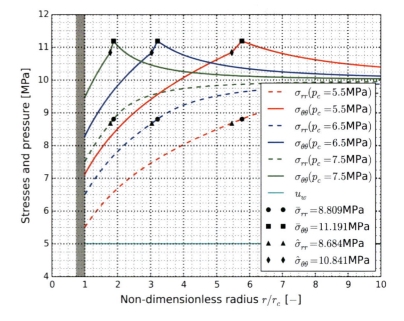

Fig. 98.3 Probability of failure,
i.e. $P_f = P[r_{pl} > r_{pl,lim}]$. The *solid
line* corresponds to the
uncorrelated case and the *dashed
line* to the correlated case

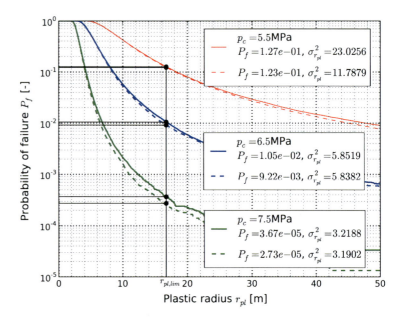

soil property values $\boldsymbol{\mu}_X$. The three zones highlighted in
Fig. 98.1 are evident, with the EP and RP interfaces
extending non-linearly with the linear decrease in cavity
pressure.

The exceedance probability for a MC analysis can be
computed by $P[G(\boldsymbol{X}) > 0] = N_f/N_r$, for which the number
of failed realisations is $N_f(G(\boldsymbol{X}) > 0)$ with respect to the
performance function $G(\boldsymbol{X})$. Given the design exceedance
criterion for the extent of the plastic zone,
$G(\boldsymbol{X}) = r_{pl} - r_{pl,lim}$, with $r_{pl,lim} = b_c/3 = 16.7$ m corresponding to a hypothetical limiting plastic radius. Fig-
ure 98.3 summarises the associated exceedance (failure)
probabilities, which, as expected, increase with decreasing

liner support. It is evident that for the correlated case the
variance of the response decreases, which thus leads
potentially to lower failure probabilities, e.g. 0.0274 % over
0.0367 % for $p_c = 7.5$ MPa. Hence, a knowledge of the
material parameter correlations reduces uncertainty for this
performance assessment.

98.4 Conclusions

A preliminary study of the mechanical response of Boom
Clay, around disposal galleries of a radioactive waste
repository, has been carried out. An elasto-plastic, strain

softening analytical solution has been implemented, within a reliability based framework and initial results have been presented. Three zones can be clearly identified; the elastic zone, the strain softening zone and the residual zone. The effect of the aleatory uncertainty on the probability of failure is illustrated, along with the impact of correlating material parameters, which, in this investigation, leads to a reduction of the exceedance probability.

Acknowledgments The research leading to these results has received funding from the Dutch research programme on geological disposal, OPERA. OPERA is financed by the Dutch Ministry of Economic Affairs, Agriculture and Innovation, and the public limited liability company Elektriciteits-Produktiemaatschappij Zuid-Nederland (EPZ), and is coordinated by COVRA.

References

Barnichon JD, Neerdael B, Grupa J, Vervoort A (2000) CORA 18: Project TRUCKII. Waste and Disposal Department, SCK·CEN, Mol, Belgium

Bernier F, Li X-L, Bastiaens W (2007) Twenty-five years geotechnical observation and testing in the tertiary Boom Clay formation. Géotechnique 57(2):229–237

Bésuelle P, Viggiani G, Desrues J, Coll C, Charrier P (2013) A laboratory experimental study of the hydromechanical behavior of Boom Clay, rock mechanics and rock engineering (online). doi:10.1007/s00603-013-0421-8

Chen SL, Abousleiman YN, Muraleetharan KK (2012) Closed-form elastoplastic solution for the wellbore problem in strain hardening/softening rock formations. Int J Geomech 12(4):494–507

Deng YF, Tang AM, Cui YJ, Nguyen XP, Li X-L, Wouters L (2011) Laboratory hydro-mechanical characterisation of Boom Clay at essen and mol. Phys Chem Earth, Parts A/B/C 36(17–18):1878–1890 (Clays in natural & engineered barriers for radioactive waste confinement)

Graziani A, Ribacchi R (1993) Critical conditions for a tunnel in a strain softening rock. In: Proceedings of the international symposium on assessment and prevention of failure phenomena in rock engineering, Istanbul, Turkey, pp 199–204

Horseman ST, Winter MG, Entwistle DC (1987) Geotechnical characterization of Boom Clay in relation to the disposal of radioactive waste. Nuclear Science and Technology, EUR 10987, Commission of the European Communities

Verhoef E, Neeft E, Grupa J, Poley A (2011) Outline of a disposal concept in clay. OPERA-PG-COV008, COVRA, NL

Wildenborg A, Orlic B, de Lange G, de Leeuw CS, van Weert F, Veling EJM, de Cock S, Thimus JF, Lehnen-de Rooij C, den Haan EJ (2000) CORA 19: transport of radionuclides disposed of in clay of tertiary origin (TRACTOR)—Final report, Utrecht, The Netherlands

Jürgen Hesser, Diethelm Kaiser, Heinz Schmitz, and Thomas Spies

Abstract

Measurements of acoustic emission have been performed since 1994 in a repository of nuclear waste (a former salt mine) in combination with deformation measurements in order to monitor the mine structure regarding stability and the integrity of the rock mass. Measuring acoustic emissions, the generation of micro- and macrocracks in the rock mass can be detected. Therefore this method provides the identification and observation of high levels of load and of load changes in large-scale. Generally, an analysis in detail requires the inclusion of additional measurements and numerical calculations. The generation of microcracks and macrocracks leads to a volume increase, which can be quantified by deformation measurements. The results show that areas of high acoustic emission activity are in good accordance with zones of larger deformations.

Keywords

Acoustic emission measurement • Deformation measurement • Micro-crack • Salt mine

99.1 Introduction

The anthropogenic invasion by the construction of underground openings (e.g. boreholes, drifts, chambers, caverns) and the utilization of these openings (e.g. heat production, exploitation of raw minerals, energy storage, waste disposal) lead to changes of the state of stress combined with rock mass deformations. If the load of the rock mass exceeds a rock mass specific limit, microcracks and macrocracks will be generated, which can cause a loosening up with the result of changes in bearing behaviour and permeability. Regarding the stability of underground bearing systems and the integrity of geological barriers, such areas with ongoing generation of microcracks and macrocracks have to be detected at an early stage and the damage development has to be predicted (Heusermann 2001).

On the one hand, microcrack generation involves the emission of seismic energy, which is transmitted through the rock mass in form of seismic waves. This acoustic emission (AE) can be detected using piezoelectric sensors. Networks of sensors enable the localization of AE. On the other hand, microcracks induce a small but irreversible volume increase (dilatancy), which can be observed by deformation measurements. The combination of both monitoring methods, AE and deformation measurements, together with other investigation methods, like numerical calculations (e.g. Fahland et al. 2005) and laboratory tests enable a detailed analysis of the geomechanical behaviour of the rock mass with regard to the evaluation of stability and integrity.

J. Hesser (✉) · D. Kaiser · H. Schmitz · T. Spies
Federal Institute for Geosciences and Natural Resources (BGR), Hanover, Germany
e-mail: juergen.hesser@bgr.de

D. Kaiser
e-mail: diethelm.kaiser@bgr.de

99.2 Measuring Methods

99.2.1 Acoustic Emission Measurements

With acoustic emmission (AE) measurements, seismic energy of high frequencies is recorded, due to generation or expansion of cracks with small dimensions in the scale of millimetres to centimetres. Elastic seismic waves propagate from these cracks into the solid rock and are recorded at frequencies of 1–100 kHz. Three networks of up to 32 sensors were installed in boreholes in the mine, in total 90 sensors. The principle of AE measurements is sketched in Fig. 99.1. The signals are digitized, analysed and stored automatically. Automatic processing includes localization and determination of the strength (magnitude) of the microcrack events (Spies and Eisenblätter 2001). Currently approximately 2,000 events are localized per hour allowing detailed monitoring of temporal and spatial variations of microcrack processes (Köhler et al. 2009; Becker et al. 2010).

AE measurements are a suitable monitoring method for highly excavated rock mass areas, because microcrack generation and therefore areas of dilatancy can be detected reliably.

99.2.2 Deformation Measurements

From the mechanical point of view, the volume increase induced by microcrack generation as well as displacements caused by macrocracks or fractures can be observed by deformation measurements. Deformations close to the openings can be determined by convergence measurements. Deformations at a greater distance to the openings can be monitored by borehole measurements, using extensometers

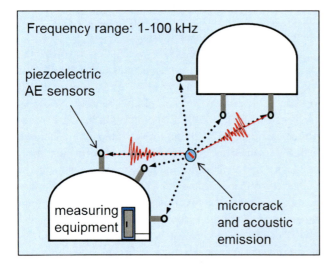

Fig. 99.1 Principle of acoustic emission measurements in mines

or inclinometers. The observation of discrete joint faces can be performed with fissurometers.

Convergence measurements are carried out to determine the deformation of openings by measuring the distance of points on the cavity perimeter. The rock mass deformation along the axis of a borehole is gathered by extensometers, repeatedly measuring the distance of anchor points in the borehole to a reference point near the cavity contour. Displacements perpendicular to the axis of a borehole in the rock mass are measured with inclinometers, determining the change of inclination of borehole segments. With fissurometers, the relative displacements of joint faces on the cavity perimeter are observed. Details on these measuring methods are described in Dunnicliff (1993).

99.3 Measuring Results

The measurements described above were performed in the central mine segment whose chambers were backfilled using salt concrete. The results of the measurements show the effects of backfilling on the rock mass behaviour. The monitored mine area is characterized by a high density of excavation and therefore by high loads of the pillars between the large chambers, which are arranged one above the others. As the preservation of the underground stability and the rock mass integrity in this mine area were the main objective, the selected chambers were backfilled in succession with salt concrete from the second half of 2003 up to the beginning of 2011. Apart from the AE sensors many geotechnical sensors were installed in the central mine region to monitor the temporal development of temperature, stress and deformation.

With the backfilling, the humidity in the salt mine increased and the hydration of the salt concrete led to a temperature increase in the rock mass. Also a strong increase of the AE activity with characteristic concentrations was observed in the vicinity of the backfilled chambers. Mostly the microcrack generation was concentrated in the roof of the chambers as well as in the pillars. Generally, high microcrack activity continued for several years. Figure 99.2 shows locations of AE events in the central mine area in a vertical cross section.

In Fahland et al. (2005) the temperature increase was identified as the cause of the microcrack generation in this area, induced by the exothermic hydration. Hesser and Spies (2007) showed with a laboratory experiment, that the increase of humidity also leads to increased microcrack activity, at least in the vicinity of the backfilled openings.

Two extensometers were installed in the roof of chamber 2n at the second level in the southern part of the chamber. The determined section deformations of the extensometers are shown in Fig. 99.3. During the whole measuring time, a vertical extension of the pillar was observed. First, in the

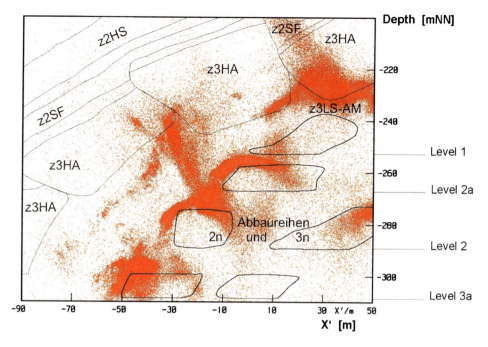

Fig. 99.2 Locations of AE events in the year 2010 in a vertical cross section in the central area of the salt mine with a sketch of the excavation chambers and the geological units

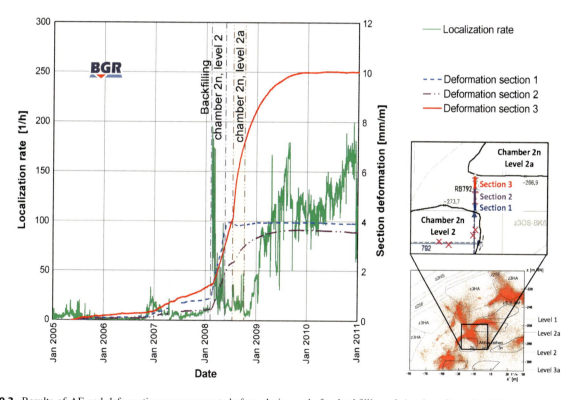

Fig. 99.3 Results of AE and deformation measurements before, during and after backfilling of chambers 2a on level 2 and level 2a

years 2005 and 2006, this extension was very small. The backfilling of chambers below the chamber 2n on the second level led to a deformation increase in 2007.

In 2008 the chambers 2n on the levels 2 and 2a were backfilled (see Fig. 99.3). This backfilling activity led to a significant increase of the vertical pillar deformation right at its beginning. In the roof of chamber 2n at the second level and in the centre of the pillar the deformation reached a value about 4 mm/m in 2009. In the floor of chamber 2n on level 2a (section 3) the vertical pillar extension of 10 mm/m at the end of 2009 was much higher than other ones.

Figure 99.3 also shows a strong rise of the AE activity in the roof of the chamber with the beginning of the backfill activity in 2008. Later on, the activity decreased up to the second half of 2008. Backfilling the chamber on level 2a led only to small changes of the AE activity but to an obvious increase of deformation in the floor of chamber 2n on level 2a. After both chambers were backfilled, the AE activity increased again to a relatively high level since the end of 2008.

If there had been only dilatant processes in the pillar, a quite symmetric deformation distribution would have been observed by the extensometer measurements, with decreasing deformation from the pillar centre to the openings. Indeed, maximum pillar extension was found in the floor of chamber 2n at level 2a. Thus the opening of a joint face or concordant relative displacements in this area can be assumed.

A detailed evaluation shows that the AE activity increase started earlier than the increase of deformation. In this case, the microcrack generation could be detected by AE measurements before measurable deformations occurred.

Nevertheless, the comparison of the AE localization rate with the deformation development in Fig. 99.3 shows that an increase of AE activity is not strongly linked to an increase of vertical deformation in the roof of the chamber. In the years 2009 and 2010 the AE activity in the pillar increased while the section deformations reached constant values.

99.4   Conclusions

For more than 18 years AE activity has been monitored in a former salt mine resulting in a large dataset of more than 90 million localized microcrack events. These measurements are an important contribution to the monitoring of the mine to assess the stability of the mine and the integrity of the rock mass. Spatial and temporal changes of the rock mass load can be identified and observed in large-scale using this method. But for a detailed analysis of the observed events

and their processes the results of other geotechnical measurement methods have to be included.

With backfilling of the excavation chambers in the central area of the salt mine, the temperature in the rock mass increased. The backfilling with salt concrete also resulted in a strong and long lasting increase of AE activity, especially in the roof of the chambers. The increased microcrack activity was associated with larger deformations. This is often the case as also in other parts of the mine the AE activity and the deformations are correlated. In one pillar the temperature increase caused the activation of a joint face. With extensometer measurements the opening of the joint face respectively relative displacements in the joint face was verified.

Acknowledgments The measurements are performed on behalf of "Bundesamt für Strahlenschutz" (German Federal Office for Radiation Protection). Part of the data was provided by „Deutsche Gesellschaft zum Bau und Betrieb von Endlagern für Abfallstoffe". Since 18 years the "Gesellschaft für Materialprüfung und Geophysik" is a reliable partner in the development of measurement systems and analysis software for AE measurements and also in scientific discussions.

References

Becker D, Cailleau B, Dahm T, Shapiro S, Kaiser D (2010) Stress triggering and stress memory observed from acoustic emission records in a salt mine. Geophys J Int 182:933–948. doi:10.1111/j.1365-246X.2010.04642.x

Dunnicliff J (1993) Geotechnical instrumentation for monitoring field performance. Wiley, New York, 577 p. ISBN 0-471-00546-0

Fahland S, Eickemeier R, Spies T (2005) Bewertung von Gebirgsbeanspruchungen bei Verfüllmaßnahmen im ERAM. 5. Altbergbaukolloquium vom 3. bis 5. November 2005. Verlag Glückauf GmbH; TU Clausthal

Hesser J, Spies T (2007) The influence of humidity on microcrack processes in rock salt. In: Wallner M, Lux K-H, Minkley W, Hardy Jr R (eds) Proceedings of SaltMech6—6th conference on the mechanical behaviour of rock salt, pp 45–52. Taylor and Francis, London

Heusermann S (2001) Beurteilung der geomechanischen Stabilität und Integrität von Endlagerbergwerken im Salzgebirge auf der Grundlage geologischer und ingenieurgeologischer Untersuchungen. Geologische Beiträge Hannover 2:159–174

Köhler N, Spies T, Dahm T (2009) Seismicity patterns and variation of the frequency-magnitude distribution of microcracks in salt. Geophys J Int 179:489–499. doi:10.1111/j.1365-246X.2010.04642.x

Spies T, Eisenblätter J (2001) Acoustic emission monitoring of closely spaced excavations in an underground repository. J Acoust Emission 19:153–161

Printed by Books on Demand, Germany